TRANSACTIONS OF THE INTERNATIONAL
ASTRONOMICAL UNION VOLUME XXIXB

PROCEEDINGS OF THE TWENTY-NINTH GENERAL ASSEMBLY HONOLULU 2015

COVER ILLUSTRATION:

Nighttime panorama of Halekalā, Maui Island, Hawai'i

Credit: Jason K. Chu

IAU SYMPOSIUM PROCEEDINGS SERIES

Chief Editor
PIERO BENVENUTI, IAU General Secretary
IAU-UAI Secretariat
98-bis Blvd Arago
F-75014 Paris
France
iau-general.secretary@iap.fr

Editor
MARIA TERESA LAGO, IAU Assistant General Secretary
Universidade do Porto
Centro de Astrofísica
Rua das Estrelas
4150-762 Porto
Portugal
mtlago@astro.up.pt

INTERNATIONAL ASTRONOMICAL UNION

UNION ASTRONOMIQUE INTERNATIONALE

TRANSACTIONS OF THE INTERNATIONAL ASTRONOMICAL UNION VOLUME XXIXB

PROCEEDINGS OF THE TWENTY-NINTH GENERAL ASSEMBLY HONOLULU 2015

Edited by

PIERO BENVENUTI
General Secretary

CAMBRIDGE UNIVERSITY PRESS
University Printing House, Cambridge CB2 8BS, United Kingdom
1 Liberty Plaza, Floor 20, New York, NY 10006, USA
10 Stamford Road, Oakleigh, Melbourne 3166, Australia

© International Astronomical Union 2019

This book is in copyright. Subject to statutory exception
and to the provisions of relevant collective licensing agreements,
no reproduction of any part may take place without
the written permission of the International Astronomical Union.

First published 2019

Printed in the UK by Bell & Bain, Glasgow, UK

Typeset in System LaTeX 2ε

A catalogue record for this book is available from the British Library Library of Congress Cataloguing in Publication data

This journal issue has been printed on FSC™-certified paper and cover board. FSC is an independent, non-governmental, not-for-profit organization established to promote the responsible management of the world's forests. Please see www.fsc.org for information.

ISBN 9781108481694 hardback
ISSN 1743-9213

Table of Contents

Chapter X

Chapter XI

Preface

The IAU *Transactions B* series report the non-scientific, administrative activities that take place during the General Assembly (GA). These reports are very important from the historical point of view, because they record all the essential decisions taken by the main governing body of the IAU: the approval of the financial accounts and of the proposed budget for the next triennium, the admission of new National Members and Individual members, the evaluation of Division and Commission reports, the approval of Resolutions, etc.

The XXIX GA has been characterised by the completion of the internal restructuring that began in 2012 at the XXVIII GA in Beijng with the approval of the new 9 Divisions. At the time, aiming at a smooth transition, the existing Commissions remained unchanged, they were only re-allocated to the new Divisions. As announced in Beijing, all the old Commissions had to terminate at the XXIX GA and new ones (which could also be a continuation of the existing) had to be approved in Honolulu.

Following a Call for Proposal for the new Commissions, they were analysed and discussed by the Executive Committee in its 95th Meeting (Padova, April 15-17, 2015) and 35 Commissions were finally approved, together with the elected Presidents and Organising Committees, in the 96th EC Meeting (Honolulu, Aug 2nd, 2015). This action completed the restructuring of the Union.

In order to mark this important milestone in the history of the IAU, it was decided to dedicate the Volume XXIXA of the *Transactions A* of the IAU to the Reports by the old Commissions. The Volume, edited by Thierry Montmerle, is entitled "Commissions legacy Reports": because of this, the Chapter VI of the current *Transactions B* Volume, which is traditionally dedicated to the Commissions Reports, has been left empty. In addition, since most of the affiliation to the new Commissions took place after the XXIX General Assembly, the Chapter XI of this *Transactions B* Volume, only lists the members of the old Commissions, updated to August 2015.

An additional novelty in the scientific programme of the XXIX GA, was the substitution of the meetings called "Joint Discussions" (JD) and "Special Sessions" (SpS) with a single type of meeting called "Focus Meetings". As a matter of fact, the distinction between the previous two type of meetings had become more and more diffuse and their replacement in a single category was in order. The proceedings of the JD and SpS were traditionally published in the Volumes called "Highlights in Astronomy": the last issue of the Series, Vol. 16, was published after the XVIII GA in Beijing while the new Series "Astronomy in Focus" was inaugurated at the XXIX GA in Honolulu and collects in two Volumes the proceedings of the new "Focus Meetings". The Proceedings of the 6 Scientific Symposia that took place during the GA were published separately by Cambridge University Press in the Series IAU Symposia.

The General Assembly welcomed the re-admission of Columbia as a National Member and of 1199 new Individual Members from 55 Countries. The XXIX GA also offered the occasion to sign the Agreement for the constitution of 5 new Regional Nodes of the Office of Astronomy for Development: the South-West Asian Regional Office at Yerevan, Armenia, the Andean Regional Office at Bogotá, Colombia and Santiago, Chile, the Arab Regional Office at Amman, Jordan, the West African Regional Office at Nsukka, Nigeria and the Portuguese Language Expertise Centre in Porto, Portugal.

Strictly connected with the growth of the world network of OAD Regional Nodes was the Resolution B1, approved by the GA, which extends the validity of the "Strategic Plan 2010-2020: Astronomy for Development" until the XXXI GA to be held in August

2021 in Busan, Republic of Korea (see Chapter II). The same Resolution also assigned to the Executive Committee to prepare an extended "Strategic Plan" to be presented for approval at the XXX GA in August 2018 in Vienna.

Finally, it is worth mentioning that the XXIX General Assembly in Honolulu was animated by some local group protesting against the construction of the Thirty Meter Telescope (TMT) on top of the Mauna Kea mountain. The protest was always pacific and it did not interfere with the scientific activities of the general Assembly, although the foreseen visits to the Mauna Kea Observatory were cancelled for preventive security measures.

A sincere thank to the American Astronomical Society, to the Hawai'ian Astronomical Institutes and to the local Government for organizing an exceptionally fruitful and pleasant General Assembly, in one of the most beautiful place in the world and most renown astronomical site!

Piero Benvenuti
General Secretary

Transactions IAU, Volume XXIXB
Proceedings IAU Symposium No. Volume XXIXB, 2018
P. Benvenuti, ed.
© International Astronomical Union 2019
doi:10.1017/S1743921318003940

CHAPTER I

TWENTY NINTH GENERAL ASSEMBLY

INAUGURAL CEREMONY

Monday, 3 August 2015, 16:00-18:00

Hawai'i Convention Center, Honolulu, HI

1. Opening Event

The XXIXth General Assembly was opened by a Oli Chant and Blessing by D. Akaka Jr.

2. Opening Address

Prof. Norio Kaifu, IAU President

Respective guests, ladies and gentlemen, and all friends, good evening, and Aloha! Welcome to the 29th IAU General Assembly here in Honolulu! And thanks for all hospitalities of Hawai'ian community. Hawai'i had been the place where the IAU General Assembly must visit, and now we are here. Hawai'i is well-known as a cultural, economic and tourism center in the Pacific region, but for we astronomers Hawai'i had been particularly important, because it is one of the most active frontiers of astronomical observations in the world for more than half a century.

This 29th General Assembly in Honolulu has some special features through the history of IAU.

Firstly, the total reconstruction of scientific organizational structure of IAU will be completed in this GA, by finishing the reform of its Commissions. All new Commissions with new Steering Committees will start activities for the coming three years after Honolulu under a fresh and inclusive mandate. Through this reform the role of the nine Divisions becomes more central to IAU activity, and the Commissions will be more flexible in responding to the rapid evolution of astronomy in the 21th Century. I express deep thanks to Division and Commission Presidents and Officers who devotedly worked for this long-term process of reforming the IAU.

Secondly, the Honolulu GA was organized with six Symposia as usual, but also with 22 Focus Meetings and a variety of Division Meetings. This new meeting structure gives us more flexibility than before. Of course we need to observe the effects by these changes carefully, watch further activities under the new structure, and find better ways for the IAU. We understand, however, these changes are an inevitable evolution for the IAU, a rapidly-growing international union of scientists.

Thirdly, we will see in this GA the IAU firmly broadening its connection with other communities in the world, including education, promotion of scientific knowledge, and contact with the general public. At this time, fifth years since the start of the IAU 10-years Strategic Plan "Astronomy for Developments", Executive Committee received an extremely positive report from the OAD Review Panel, demonstrating fantastic success of the Astronomy for Developments project. I am happy to report that the Executive Committee held on Sunday in Honolulu adopted the new OAD Agreement between IAU and National Research Foundation of South Africa, to ensure further activity of OAD.

The OAO, Office for Astronomical Outreach was established in the NAOJ, Japan, just after the previous GA in Beijing. It has been working on the IYL2015, "International Year of Light 2015" cooperating with UNESCO. The OAO also actively developing cooperation with school teachers, students and children, amateur astronomers, and general public in the world particularly through the project "NameExoWorld", which will announce during this GA the start of public voting for the names of selected Exoplanets.

Furthermore, the IAU signed the new agreement with the NASL, Norwegian Academy of Science and Letters, to open "OYA", the Office for Young Astronomers, particularly for the promotion of International School of Young Astronomers. Together with OAD and OAO, the OYA will be IAU′s third Global office to support activities to develop astronomy and science in future. The IAU is now really one of the leading international scientific unions on education, outreach and developments in the field of basic sciences, and I am very much proud of it.

During this GA there will be many discussion about the future observing facilities which will provide exciting perspectives for astronomy. The TMT, Thirty Meter Telescope planned to construct on Maunakea.

However, with the increasing scope of astronomy we acknowledge that occasional conflict between our growing scientific activities and the interests of the general public may occur. In Hawai'i, as you know, we hear voices that criticize the development of telescopes at Mauna Kea. The history of astronomy and its huge positive impact on civilization makes us confident that astronomical research will continue to open a vast new world for humankind, give us a better perspective to understand our world, and provide excitement and dreams to children. In parallel, we truly wish to respect all cultures, to remain entirely open in our activity, and to live with all nations and cultures together. As all astronomers in Hawai'i know, and as I know from my many-years experiences here, Hawai'i is a Land of Aloha. I expect you found my letter sent to all GA participants in the Registration Bag, together with some background information. Please note this letter and material, and if you have chance to contact people who wish to appeal or protest, we welcome you to talk these Hawai'ian people respectfully and frankly.

As we anticipate, this 29th IAU General Assembly in Honolulu will be one of the largest, most active, and memorable General Assemblies in the history of IAU. I wish all of us have an exciting and fruitful time during the General Assembly.

And please enjoy Hawai'i. Thank you for your kind attention.

3. Welcome address by the Governor of Hawai'i

David Y. Ige, Governor of Hawai'i

Alhoa! On behalf of the people of the State of Hawai'i, I send my warmest greetings to all members of the International Astronomical Union (IAU) attending the 2015 IAU General Assembly.

Since 1919 the IAU has been the international authority for assigning designations to celestial bodies and the surface features on them. representing the vast majority of professional astronomers, its mission is to promote and safeguard the science of astronomy through international cooperation.

I am pleased that the IAU has chosen to hold this year's assembly in Hawai'i, where interest in astronomy is deeply entrenched in our island heritage. The heavens played a central role in ancient Hawai'ian culture, and early Polynesians relied on their knowledge of the stars to successfully navigate thousadnds of miles acreoss the open ocean. Today, Hawai'i continues to be at the forefront of astronomical studies through the observatories at Maunakea on the island of Hawai'i and Haleakalā on Maui.

This triennial assembly promises to be an exciting and informative event, featuring more than 3,000 professional astronomers, and offering Symposia and Focus Meetings that cover a broad range of astronomical topics.

Mahalo [thank you] to the IAU, American Astronomical Society, University of Hawai'i at Mānoa Institute for Astronomy, and many event organizers for contributing to this worthy event. Please accept my best wishes for an enjoyable and successful meeting.

4. Welcome address by the NSF Director

France Cordova

Let me thank President Kaifu, President Urry, and Director Hasinger for their invitation to deliver the inaugural address of this XXIX General Assembly of the International Astronomical Union.

It is a great honor to represent the Obama Administration at this historic event. As an astrophysicist myself, I am delighted to say aloha to my fellow IAU members and welcome all of you to this meeting.

I am also honored to represent the National Science Foundation the premier U.S. basic scientific research agency and one of the world's leading institutions in astronomy.

For more than six decades, NSF-funded researchers and facilities have been exploring the most intriguing mysteries of the heavens. Today, among those mysteries are the origin and evolution of stars and galaxies, the formation of solar systems, the existence of habitable planets, and the nature of dark matter and dark energy.

And what do we seek to discover from our observations of the heavens?

"To know the unknowable" this quote will be familiar to native Hawaiians.

In my own case, as a girl I had an early yearning to understand the mysteries of the universe. I didn't know what astrophysics was, but I had always loved looking at the night sky and asking, "Why are there stars? How are they formed? Why are there so many and no more? Why are some bright, while others are barely visible?"

Thinking about those questions resonated with me, as I am sure it does with you.

I started my career as an X-ray astronomer. Some years later, I was honored to become NASA's first female chief scientist.

And let me take this opportunity to congratulate our NASA colleagues for the spectacular success of their New Horizons Pluto Mission talk about new mysteries to solve!

As New Horizons has so vividly reminded us, people everywhere have a deepseated yearning to understand the universe in which we live.

In October 2009, President Barack Obama invited a group of astronomers to the White House to mark the International Year of Astronomy, the 400th anniversary of Galileo's first use of a telescope to observe the night sky.

The President has long emphasized the study of science and astronomy in order to increase understanding of the natural world and to encourage greater scientific cooperation across national borders.

Nobel Prize-winning chemist Ei-ichi Negishi, in his 2010 speech accepting the award, said "The final reward for any researcher is to see his or her lifetime of work extend beyond academia and laboratories, into the mainstream of global society where it can breathe hope into the world."

Isn't that the goal that all of us in the global research community strive to achieve? And let me add that the U.S. is proud to partner with many countries around the world in exploring the heavens. Basic research is the primary focus of the National Science Foundation, including astronomical breakthroughs that have changed our understanding of the universe.

One significant NSF-funded effort resulted last year in University of Hawai'i at Mānoa astronomer Brent Tully being awarded the 2014 Gruber Cosmology Prize and the 2014 Victor Ambartsumian International Prize.

Dr. Tully led an international team of astronomers in defining the contours of the supercluster of galaxies containing our own Milky Way.

Those astronomers named the supercluster "Laniakea", meaning "immense heaven" in Hawaiian, to honor Polynesian navigators who used knowledge of the heavens to voyage across the immense Pacific Ocean.

The name was suggested by Nawa'a Napoleon, an associate professor of Hawaiian language at Kap'iolani Community College.

One new significant challenge for the National Science Foundation is the enormous increase in raw research data resulting from vastly increased computational capabilities also known as "Big Data."

The growing field of machine learning in which computers learn from large data sets and find patterns that humans don't easily recognize has great long-term implications for astronomy. For example, the image at right is from "Solar Superstorms," an ultrahigh-resolution demonstration that takes viewers into the magnetic fields and superhot plasma surrounding the Sun as it produces dramatic flares, violent solar tornadoes, and coronal mass ejections.

This groundbreaking scientific visualization is based on computations from the NSF-supported supercomputing initiative, Blue Waters, at the National Center for Supercomputing Applications at the University of Illinois.

As dramatic as the visualization is, it is only a hint of the advances Big Data may produce in years ahead.

While the National Science Foundation is widely recognized as our nation's premier basic scientific research agency, we find there are more international partnerships emerging that enable NSF to extend our ability to produce significant scientific research. We have found that global collaborations accelerate the progress of science and improve health, security, and prosperity throughout the world.

For example, the Atacama Large Millimeter/submillimeter Array or ALMA telescope has received more than $1 billion in investments from a broad international coalition including Europe, East Asia led by Japan and Chile, with North American funding led by NSF.

ALMA is providing a testing ground for theories of star birth and stellar evolution, and solar system and galaxy formation.

A remarkable ALMA image of the young star HL Tau and its protoplanetary disk reveals multiple rings and gaps that herald the presence of emerging planets as they sweep their orbits clear of dust and gas.

Another significant NSF partnership involves the Gemini team of twin 8.1-meter optical/infrared telescopes on Cerro Pachón in Chile and on Maunakea here in Hawai'i.

The International Gemini Observatory is a partnership of the U.S., Canada, Australia, Brazil, Argentina, and Chile, as well as the University of Hawai'i as the host of the northern site. The Republic of Korea joined the partnership in 2015 as a limited-term collaborator and is expected to become a full partner in 2017.

Gemini's capabilities full-sky coverage, rapid response to transients, agile scheduling, and specialized optics enabled it to capture an image of the Kronberger 61 nebula, showing an ionized shell of expelled gas resembling a soccer ball. Incidentally, the nebula was named for an amateur astronomer in Austria.

The image was made by the Gemini Multi-Object Spectrograph GMOS on the Gemini North telescope on Maunakea.

Another cutting-edge, NSF-supported observatory is the Daniel K. Inouye Solar Telescope, now under construction on Haleakalā. This next-generation solar telescope represents a collaboration of 22 institutions, reflecting a broad segment of the solar-physics community.

Once completed, it will be the premier ground-based solar observatory. Thanks to the people of Hawai'i, It will enable astronomers everywhere to glean new insights into solar phenomena, including what are the mechanisms responsible for solar storms that ultimately affect the Earth.

Furthermore, we expect that this increased understanding of the Sun will help protect vital space-based assets such as communication and weather satellites and the power grids here on Earth.

The top recommendation of the 2010 National Academy of Sciences decadal survey of astronomy was the Large Synoptic Survey Telescope LSST which is now under construction on Cerro Pachón in Chile. Just a few months ago, I participated in the exciting "first stone ceremony" to launch LSST construction.

LSST will be a wide-field "survey" telescope that photographs the entire available sky every few nights. Advanced computers will gather and analyze the millions of gigabytes of data LSST will generate each year.

A pilot project called the Deep Lens Survey uses imaging from NSF's four-meter telescopes to suggest what half a degree of sky will look like when LSST is in operation, projected to begin in 2022.

An innovative citizen-science program will involve people of all ages in LSST discoveries, making discovery opportunities available to K-12 students as easily as to the professional astronomer. This is just one example of NSF's commitment to engaging the public in the thrill of discovery and increasing public understanding of scientific research.

Far from the 2,700-meter high Cerro Pachón in Chile lies NSF's IceCube Neutrino Observatory at the Amundsen-Scott South Pole Station in Antarctica.

IceCube is the world's largest neutrino detector and is among the most ambitious scientific construction projects ever attempted. It searches for neutrinos from the most violent astrophysical sources: exploding stars, gamma-ray bursts, and cataclysmic phenomena involving black holes and neutron stars.

The highest-energy neutrino ever observed by IceCube, with an estimated energy of 1.14 peta-electron-volts (PeV), was nicknamed "Ernie" by IceCube physicists.

I recently attended the inauguration of the High Altitude Water Cherenkov or HAWC gamma-ray observatory near Puebla, Mexico.

HAWC represents a unique partnership between the National Science Foundation, the U.S. Department of Energy, and CONACYT Mexico's National Council of Science and Technology. HAWC will give scientists a new window for detecting and recording gamma rays and cosmic rays emitted by black holes, merging neutron stars, streams of hot gas moving at close to the speed of light, and other exotic phenomena in the universe.

HAWC will monitor approximately two-thirds of the sky every 24 hours with unprecedented sensitivity to the highest-energy gamma rays. HAWC will complement the operations of NASA's Fermi Gamma-ray Space Telescope and the VERITAS gamma-ray observatory.

It will also be part of the growing field of "multi-messenger astrophysics" that includes cosmic ray observatories, IceCube, and the Advanced Laser Interferometer Gravitational-Wave Observatory.

Finally, I would like to say a few words about the beautiful setting that the IAU chose for its first General Assembly in the U.S. in nearly three decades.

No doubt the IAU was attracted by Hawai'i's breathtaking beauty, unique cultural heritage, and aloha spirit of its friendly people all great reasons for holding this assembly here. The National Science Foundation and many other scientific institutions worldwide come to partner in scientific research at an extraordinary site.

With its biodiversity on land and in the surrounding oceans, its unique geological history and formations, and its high volcanic peaks, Hawai'i is one of the Earth's great scientific treasures. It is a treasure that all of us want to see honored, preserved, and protected.

The National Science Foundation has partnered with the people of Hawai'i and Hawaiian institutions for many years and takes seriously its responsibilities to be a good steward of Hawai'i's unique natural resources and cultural heritage and to be respectful of Hawai'i's people and customs.

We hope to continue our partnerships in order to create opportunity for the next generations of seekers of knowledge for many years to come.

Let me again thank the International Astronomical Union for the opportunity to be with you for this historic General Assembly. Just as the universe knows no borders, the science community's exploration of its mysteries has always been an international endeavor. We look forward to the promise of even greater cooperation among nations and institutions as we expand scientific understanding of this endlessly fascinating challenge.

Again, I wish you all a productive meeting. Mahalo!

5. Presentation of Partners, Sponsors and Exhibitors

The IAU and Organising Committees acknowledge the invaluable support of the following institutions and organisations:

Partners

Air China
Beijing Science Video Network
H3C
Microsoft Research
Star Alliance

Sponsors

Astronomy & Astrophysics
Associated Universities, Inc.
USRA - Universities Space Research Association
U.S. National Committee for IAU & The National Academies
Publications of the Astronomical Society of Japan

Monthly Notices of the Royal Astronomical Society
International Astronomical Union
American Astronomical Society
ANDOR - an Oxford Instrument Company
GMT - Giant Magellan Telescope
Spectral Instruments Inc.
The Korean Astronomical Society
National Astronomical Observatory of Japan
Korea Astronomy and Space Science Institute
ACE - Astronomical Consultants & Equipment Inc.
The Norwegian Academy of Science and Letters
AURA - Association of Universities for Research in Astronomy
SKA - Squae Kilometre Array
Ball Aerospace & Technologies Corp.

Exhibitors

American Astronomical Society
Andor Technology
Associated Universities, Inc.
Astro Haven Enterprises
Astronomical Consultants & Equipment Inc.
Australian Astronomy Group
Big Kid Science & Story Time From Space
Cambridge University Press
Center for Astronomy & Physics Education Research (CAPER)
Chile - SOCHIAS
Chinese Astronomical Society
EDP Sciences - Astronomy and Astrophysics
Elsevier
European Southern Observatory (ESO)
European Space Agency
Finger Lakes Instrumentation
Gaia Research for European Astronomy Training - GREAT
Giant Magellan Telescope
Gravitational Wave Astronomy: Opening a New Window on the Universe
Hubble 25
IAU Office of Development, Office of Outreach, International Year of Light
IAU General Assembly 2018 - Vienna
IOP Publishing
The Korean Astronomical Society
Korea Astronomy & Space Science Institute
Maunakea Observatories
MESATech
Millennium Institute of Astrophysics
NASA
National Astronomical Observatory of Japan
National Radio Astronomy Observatory (NRAO)
National Science Foundation

Officina Stellare SRL- Italy
Oxford University Press
Planck Legacy Archive
PlaneWave Instruments
QHYCCD
RADIONET3
SCAR Astronomy and Astrophysics from Antarctica
SOFIA - Stratospheric Observatory for Infrared Astronomy
Spectral Instruments
Springer
Square Kilometer Array Telescope
Sunpower Cryotel Cryocoolers
The Astronomical Society of the Republic of China (Taiwan) (ASROC)
Universities Space Research Association
University of Arizona Press - Princeton University Press - University of Hawaii Press

6. The Gruber Foundation: Presentation of the Cosmology Prize 2015 and TGF Fellowship 2015

The President of the IAU, Norio Kaifu, opened the Ceremony, in the presence of

- Patricia Murphy Gruber, President Emeritus, The Gruber Foundation
- John Carlstrom, 2015 Cosmology Prize Recipient
- Jeremiah Ostriker, 2015 Cosmology Prize Recipient
- Lyman Page, 2015 Cosmology Prize Recipient
- Wendy Freedman, Chair, Cosmology Selection Advisory Board
- Norio Kaifu, President, IAU
- Thierry Montmerle, General Secretary, IAU
- Cristbal Petrovich, Cosmology Fellow

Norio Kaifu:

In year 2000 the Gruber Foundation created its International Prize Program with its inaugural prize in Cosmology. The intent of the Prizes is to call attention to the importance of accomplishments that benefit mankind in special, important areas of endeavor. The IAU is pleased to have been chosen by the Gruber Foundation to collaborate with them on the Cosmology Prize since its inception. The IAU has an advisory role in the constitution of the selection committee and as part of our collaboration we are very fortunate to receive an annual grant of $50K from the Gruber Foundation to be awarded to a postdoctoral fellow.

It goes without saying that the prestige and reputation of any award is determined by the quality of its awardees. By this criterion the Gruber Cosmology Prize has done superbly. During the past decade the Cosmology Prize recipients alone could write as authoritative a history of the universe as any group. The careers of the Gruber Fellows have also benefitted greatly from the resources made available to them by the Gruber Foundation stipend.

In the years of the IAU General Assembly the tradition has been for the Gruber Cosmology Prize to be presented in this Inaugural Ceremony. It is my great pleasure to present to you the President of The Peter & Patricia Gruber Foundation and co-Founder & President Emeritus of The Gruber Foundation, which will continue the Gruber Prize

Program as part of Yale University, Patricia Gruber, who will introduce this year's Cosmology Prize awards ceremony.

Patricia Gruber:

Welcome to the presentation of the 16th annual Cosmology Prize, honoring a leading cosmologist, astronomer, astrophysicist or scientific philosopher for theoretical, analytical, or conceptual discoveries leading to fundamental advances in our understanding of the universe. On behalf of all of us at the Foundation, we are pleased to be here in Honolulu to present this Prize at the 29th General Assembly of the International Astronomical Union. Thank you, Norio Kaifu, for your warm welcome.

The Cosmology Prize was established in 2000 as the first Gruber international prize, and I'd like to acknowledge the founding vision and leadership of my husband Peter Gruber in establishing these prizes, as well as his passing. Peter died last fall after leading a long and productive life. Many know his story: he was born in Budapest and, after fleeing Hungary during WWII, his family settled in India where he was educated by Jesuits in the Himalayas. Later, he came to the US and went to work on Wall Street. His success as an asset manager allowed us to establish the Gruber Foundation and create the Prize Program a few years later. To further continue the legacy, an Agreement in 2011 with Yale University established the Foundation there, where the science prizes and human rights programs will continue to flourish.

The Cosmology Prize is presented in conjunction with the International Astronomical Union. It is my pleasure to introduce Thierry Montmerle, Secretary General of the IAU, who will say a few words about this fruitful collaboration.

Thierry Montmerle:

The primary goal of the IAU is the development of astronomy world-wide. To this end, the IAU is pleased to collaborate with the Gruber Foundation on the Cosmology Prize.

The collaboration between the Gruber Foundation and the IAU consists not only of the Cosmology Prize, but also an annual $50,000 Fellowship. The fellowship is administered by the IAU and awarded competitively to a postdoctoral researcher—the stipend is to be used to further his or her research.

Awards are presented to promising young scientists of any nationality to pursue education and research at a center of excellence in their field; the IAU selects recipients from applications received from around the world. The fellowship has been awarded to scientists from Poland, India, Spain, Greece, the Russian Federation, Mexico, the UK, Colombia, and the United States.

The 2015 Gruber Fellow is Cristobal Petrovich, from Princeton University, from Chile. His work focuses on the dynamics of globular clusters, with the aim of providing a more realistic dynamical paradigm for this class of stellar systems. I am happy to introduce her on this occasion and invite her to say a few words.

Cristobal Petrovich: expresses thanks and makes brief remarks.

Patricia Gruber:

Thank you Cristobal Petrovich.

We are here to honor the achievements of John Carlstrom, Jeremiah Ostriker, and Lyman Page. But first let me tell you a little about the company they are keeping.

The Foundation's prize program, established in 2000, now presents three annual $500,000 prizes in the fields of:

Cosmology; Genetics; and Neuroscience;

Each prize recognizes achievements and discoveries that produce fundamental shifts in human knowledge and culture. Until 2011, the Foundation also awarded prizes in Justice and Women's Rights. Under our succession plan with Yale University, these two prizes are now part of an exciting new program at Yale Law School.

The Genetics Prize will be presented at the annual meeting of the American Society of Human Genetics on October 9th, to Emmanuelle Charpentier and Jennifer Doudna.

On October 18th, at the annual meeting of the Society for Neuroscience, the Neuroscience Prize will be presented to Carla Shatz and Michael Greenberg.

Returning to Cosmology, the 2015 Prize recipients were was selected by a distinguished Cosmology Prize advisory board:

- Wendy Freedman (Chair)
- Sadanori Okamura
- Rashid Sunyaev
- Frans Pretorius
- Helge Kragh
- Subir Sarkar
- Andrew Fabian

Owen Gingerich and Martin Rees also serve as special cosmology advisors to the Foundation. I deeply appreciate the knowledge, commitment, and enthusiasm that the advisors bring to the judging process. Let me now invite the advisory board Chair, Wendy Freedman, to present the official citation and introduce the scientific accomplishments of our Cosmology Prize recipients

Wendy Freedman:

The Recipients of the 2015 Prize are John Carlstrom, Jeremiah Ostriker, and Lyman Page.

The official citation reads:

The Gruber Foundation proudly presents the 2015 Cosmology Prize to Jeremiah P. Ostriker for wide-ranging theoretical work over 50 years, which has clarified our understanding of galactic structure and evolution, dark matter, the intergalactic medium and high energy astrophysics; and to John Carlstrom and Lyman Page for their leadership in ground-based observational CMB cosmology, including instrumentation: from TOCO and DASI to the South Pole Telescope and the Atacama Cosmology Telescope.

John Carlstrom: Expresses thanks, very brief remarks.
Jeremiah Ostriker: Expresses thanks, very brief remarks.
Lyman Page: Expresses thanks, very brief remarks.

Patricia Gruber:

Please note that the 2015 Gruber Cosmology Prize recipients will give public lectures entitled "What do we know about dark matter? An historical approach" and "The Cosmic Microwave Background: Where We Are and Where Are We Headed" at 12:45 pm

Wednesday, in this room. Thank you for attending the 2015 Cosmology Prize ceremony. This concludes our presentation.

7. IAU-NASL Office for Young Astronomers

Oddbjørn Engvold:

The International School for Young Astronomers is an essential and highly valued educational program of the IAU. Normally three weeks long, ISYAs are international postgraduate schools for regions where students have fewer opportunities to be directly exposed to up-to-date astrophysics. The schools are intended for young astronomers who are mainly — but not exclusively — from astronomically developing countries and who have already finished their first-degree studies. The main objective of ISYAs is to provide participants with exposure to modern astrophysics through lectures from an international faculty on selected topics of astronomy, seminars, practical exercises and observations, and exchange of experiences. Since the first ISYA in 1967, 36 have been organized in 24 countries. The number of students per school has varied between 30 and 50, representing between 5 and 25 different nationalities in the host regions. The current success of the ISYA program is a result of dedicated organization and efforts by a number of individuals under the leadership of the ISYA Director, Jean-Pierre de Greve, and Deputy Director, Kam-Ching Leung. The IAU wishes to establish a robust financial and organizational basis for this highly valued educational program. To this end, the Norwegian Academy of Sciences and Letters' financial contributions to the ISYAs since 2009 allowed the IAU and NASL to establish an Office for Young Astronomers (OYA). The OYA is a virtual office, housed at the NASL in Oslo, Norway, consisting of a Steering Committee with overall responsibility for program operations. An IAU Vice-President chairs the Steering Committee; other members are the ISYA Director and Deputy Director, a representative from NASL, and the President of IAU Division C, Education, Outreach, and Heritage. The objective of this new OYA is to strengthen the overall organization of the ISYAs by functioning as a supporting structure for the Director and working with the local organizers of each individual school. The schools are organized each year by invitation of a host country, and the individual ISYAs are prepared in close collaboration with local organizations. The ISYA Director and the local organizer determine the curriculum for each school, in agreement with the OYA Steering Committee. Strong involvement of the local organizers and community of the individual ISYAs will be essential to ensure positive follow-up and growth afterwards. The establishment of the OYA is one of the latest projects in a history of collaborations between the IAU and the NASL. The IAU Executive Committee has long advised the NASL on appointments of new committee members for the international Kavli Prize in Astrophysics, and the NASL has provided financial support for the ISYAs and for the Young Astronomers Luncheons at recent IAU General Assemblies — including the one at this GA.

Transactions IAU, Volume XXIXB
Proceedings IAU Symposium No. Volume XXIXB, 2018
P. Benvenuti, ed.
© International Astronomical Union 2019
doi:10.1017/S1743921318003952

CHAPTER II

TWENTY NINETH GENERAL ASSEMBLY BUSINESS SESSIONS

FIRST SESSION: Tuesday, 4th August 2015, 16:00-18:00

Chaired by Norio Kaifu, President of the IAU
Hawai'i Convention Center, Honolulu, HI

1. Opening and Welcome

The President of the IAU, Prof. Norio Kaifu, welcomed the delegates and members to this first business session of the General Assembly. The President invited the General Secretary, Dr. Thierry Montmerle, to start the business session.

2. Representatives of IAU National Members

The General Secretary listed the representatives of the National Members (IAU member countries). They are given in the table at the end of this Chapter.

The columns list separately:
- the National Members
- the dues category
- the number of votes attached to each National Representative:
    (a) for ordinary votes, membership, etc.
    (b) for financial matters

Note: "Interim" National Members, or National Members in arrears of their dues, do not vote, even if they have appointed a Representative (countries indicated by a asterisk).
- the National Representatives, in their capacity to vote the budget and other issues on behalf of the National Members;
- the National Representative for Finance, in their capacity to discuss and vote on financial matters, as well as to propose nominations and elect the Finance Committee (formerly known as Finance Sub-Committee) for the triennium;
- the National Representatives for Membership Nominations, in they capacity to discuss and vote on individual membership issues, as well as to propose nominations and elect the Special Nominating Committee (SNC), and the Membership Committee.

For each National Member, the first line gives the names of the Representatives present during the first week of the GA, the second line, if present, gives the names of the Representatives present during the second week of the GA.

3. Adoption of Agenda

The delegates approved the agenda of the meeting as published in the Program Book.

4. Reminder of voting rules

The President reminded the delegates and the members about the voting rules, depending on the nature of the vote.

5. Appointment of Official Tellers

The Official Tellers were duly approved. They were:

- Françoise Combes *(Div. B President)*
- Lidia van Driel-Gresztelyi *(Div. E President)*
- Marry Kay Hemenway *(chair, Div. C President)*
- David Silva *(Div. B President)*
- Giovanni Valsecchi *(Div. F President)*

6. Admission of New National Members to the Union

The following new National Members were unanimously admitted to the Union:

Country: Colombia
Category: I
Adhering Organization: Observatorio Astronomico Nacional, Bogota

7. Revisions of Statutes and Bye-Laws

There were no revisions of the Statutes and Bye-Laws.

The current version of the Statutes and Bye-Laws are reported in Chapter VII of these *Transactions*.

8. Report of the Executive Committee (2009-2012)

The report of the Executive Committee was presented by the General Secretary, Dr. Thierry Montmerle, and is reported in Chapter V of these *Transactions*.

9. Report of the Special Nominating Committee

The Special Nominating Committee 2012-2015 was formed by:

President:	Norio Kaifu (chair, Japan)
Former President:	Robert Williams (USA)
EC-Selected Member:	Françoise Combes (France)
Selected Members:	Edward van den Heuvel (Netherland)
	Malcolm S. Longair (UK)
	Matthias Steinmetz (Germany)
	Gang Zhao (China Nanjing)
Advisors (GS & AGS):	Thierry Montmerle, Piero Benvenuti
Observer (PE):	Silvia Torres-Peimbert

The Special Nominating Committee proposes the following slate for IAU members for the Officers and Members of the IAU Executive Committee for the triennium 2015-2018:

President	Silvia Torres Peimbert (Mexico)
President-Elect	Ewine van Dishoeck (Netherland)
General Secretary	Piero Benvenuti (Italy)
Assistant General Secretary	Teresa Lago (Portugal)
Vice-President	Debra Elmegreen (USA)
Vice-President	Ajit Kembhavi (India)
Vice-President	Renée Kraan-Korteweg (South Africa)
Vice-President	Xiao-Wei Liu (China Nanjing)
Vice-President	Dina Prialnik (Israel)
Vice-President	Boris Shustov (Russia)
Adviser	Norio Kaifu (Japan)
Adviser	Thierry Montmerle (France)

10. Proposal to host the XXXI General Assembly in 2021

Four proposals were received to host the XXXI General Assembly in 2021. They were presented by:

- Canada (Montréal)
- Chile (Santiago)
- Korea (Rep. of, Busan)
- South Africa (Cape Town)

Representatives of the four proposals were invited to present their bid during the II Session of the 96th Executive Committee Meeting on Friday August 7th, 2015. The decision by the Executive Committee on the location of the 2021 General Assembly will be announced during the II Business Session of the GA.

Closure of Session

There being no other business items to discuss, the President declared the session closed.

Representatives of National Members (see Sect. 2)

National Member	Dues Category	Votes (a, b)	National Representatives	Finance Committee Nominations	Membership & SNC Nominations
Argentina	II	1, 3	-	-	-
Armenia	I	1, 2	Areg Mickaelian	Areg Mickaelian	Areg Mickaelian
Australia	IV	1, 5	Andrew Hopkins	Stuart Wyithe	Stuart Wyithe
Austria	II	1, 3	Manuel Güdel	Manuel Güdel	Manuel Güdel
Belgium	IV	1, 5	-	-	-
Bulgaria	I	1, 2	-	-	-
Brazil	III	1, 4	Othon Winter	Othon Winter	Othon Winter
Canada	V	1, 6	Christine Wilson	Christine Wilson	Christine Wilson
Chile	II	1, 3	Ezequiel Treister	Patricio Rojo	Ezequiel Treister
China Nanjing	VI	1, 7	Gang ZHAO	Shuangnan ZHANG	Gang ZHAO
China Taipei	II	1, 3	You-Hua Chu	You-Hua Chu	You-Hua Chu
Costa Rica	interim	0, 0			

National Member	Dues Category	Votes (a, b)	National Representatives	Finance Committee Nominations	Membership & SNC Nominations
Croatia	I	1, 2	-	-	-
Czech Rep.	III	1, 4	Petr Heinzel	Richard Wunsch	Jiri Borovicka
Denmark	III	1, 4	J. Christensen-Dalsgaard	J. Christensen-Dalsgaard	J. Christensen-Dalsgaard
Estonia	I	1, 2	Laurits Leedjärv	Laurits Leedjärv	Laurits Leedjärv
Finland	II	1, 3	Heidi Korhonen	Heidi Korhonen	Heidi Korhonen
France	VII	1, 8	Denis Burgarella	Daniel Hestroffer	Corinne Charbonnel
Germany	VII	1, 8	Matthias Steinmetz	Matthias Steinmetz	Matthias Steinmetz
Greece	III	1, 4	V. Charmandaris	V. Charmandaris	V. Charmandaris
Honduras	interim	0, 0			
Hungary	II	1, 3	Csaba Kiss	Laszlo Viktor Tóth	Csaba Kiss
Iceland	I	1, 2	-	-	-
India	V	1, 6	Prajval Shashtri	Prajval Shashtri	Prajval Shashtri
Indonesia	I	0, 0	Dhani Herdiwijaya	Dhani Herdiwijaya	Dhani Herdiwijaya
Iran	I	0, 0			
Ireland	I	1, 2	Matt Redman	Matt Redman	Matt Redman
Israel	III	1, 4	Shay Zucker	Shay Zucker	Shay Zucker
Italy	VII	1, 8	Ginevra Trinchieri	Ginevra Trinchieri	Ginevra Trinchieri
Japan	VII	1, 8	Sadanori Okamura	Sadanori Okamura	Hitoshi Yamaoka
Korea, Ref. of	II	1, 3	Hyesung Kang	Hyung Mok Lee	Hyung Mok Lee
Latvia	I	0, 0			
Lebanon	interim	0, 0			
Lithuania*	I	0, 0	-	-	-
Malaysia	I	1, 2	-	-	-
Mexico	III	1, 4	Hector Bravo-Alfaro	Miriam Pena	Miriam Pena
Mongolia	interim	0, 0			
Morocco*	I	0, 0	-	-	-
Netherlands	V	1, 6	R.A.M.J Wijers	Lex Kaper	Lex Kaper
New Zealand	II	1, 3	Karen Pollard	Karen Pollard	Karen Pollard
Nigeria	I	1, 2	-	-	-
Norway	II	1, 3	Hakon Dahle	Hakon Dahle	Hakon Dahle
Panama	interim	0, 0			
Peru*	I	0, 0	-	-	-
Philippines	I	1, 2	Cynthia P. Celebre	Cynthia P. Celebre	Cynthia P. Celebre
Poland	IV	1, 5	Andrzej Udalski	Andrzej Udalski	Andrzej Udalski
Portugal	II	1, 3	-	-	-
Romania	I	1, 2	-	-	-
Russian Fed.	V	1, 6	Dmitrij Bisikalo	Lyudmila I. Mashonkina	Andrei K. Dambis
Saudi Arabia	I	0, 0	-	-	-
Serbia	I	1, 2	-	-	-
Slovakia	I	1, 2	Ales Kucera	Ales Kucera	Ales Kucera
South Africa	III	1, 4	Patrick Woudt	Patrick Woudt	Patrick Woudt
Spain	IV	1, 5	Rafael Bachiller	Rafael Bachiller	Rafael Bachiller
Sweden	III	1, 4	Dainis Dravins	Dainis Dravins	Dainis Dravins
Switzerland	III	1, 4	Georges Meylan	Georges Meylan	Georges Meylan
Tajikistan	I	1, 2	-	-	-
Thailand	I	1, 2	Busaba Kramer	Busaba Kramer	Busaba Kramer
Turkey	I	1, 2	-	-	-
Ukraine	III	1, 4	Peter P. Berczik	Peter P. Berczik	Peter P. Berczik
UK	VII	1, 8	Martin Barstow	Martin Barstow	Adrian Michael Cruise
United States	X	1, 11	Lynne Hillenbrand	Sara Heap	Ken Kellermann
Vatican	I	1, 2	Christopher Corbally	Pavel Gábor	Christopher Corbally
Venezuela	I	1, 2	-	-	-

Transactions IAU, Volume XXIXB
Proceedings IAU Symposium No. Volume XXIXB, 2018
P. Benvenuti, ed.
© International Astronomical Union 2019
doi:10.1017/S1743921318003964

CHAPTER II

TWENTY NINETH GENERAL ASSEMBLY BUSINESS SESSIONS

SECOND SESSION: Thursday, 12 august 2015, 16:00-18:00

Chaired by Norio Kaifu, President of the IAU
Hawai'i Convention Center, Honolulu, HI

1. Opening and Welcome

The President of the IAU, Prof. Norio Kaifu, welcomed the delegates and members to the second business session of the General Assembly. The President invited the General Secretary, Dr. Thierry Montmerle, to start the business session.

2. Individual Membership

The General Secretary reported that, after screening by the Membership Committee, IAU National Members nominated 1186 new Individual Members, and Division Presidents nominated 13 new Individual Members. The list of new Members was projected on the screen.

3. Deceased Members

The Executive Committee regretfully reported the decease of 348 Individual Members of the Union, since August 2012. The General Secretary displayed their names on slides, and the General Assembly observed a one-minute silence in their respectful memory.

4. Appointment of the Official Tellers

The Official Tellers were duly appointed, and confirmed that the quorum requirements for the National Representatives were duly satisfied.

5. Resolutions

The Resolutions submitted to the General Assembly were examined by the Resolutions Committee, Chaired by Ian Corbett. They were all "B" type, i.e., with scientific content. They were submitted in English and published in the GA Newspaper ("Kai'aleleiaka - The Milky Way", issue 1, pag. 13) before the vote. The French version was made available on the IAU web site after their adoption.

These Resolutions were:

B1 on the IAU Strategic Plan 2010-2020: Astronomy for the Developing World.

B2 on recommended zero points for the absolute and apparent bolometric magnitude scales.

B3 on recommended nominal conversion constants for selected solar and planetary properties.

B4 on the Protection of Radio Astronomy Observations in the Frequency Range 76 - 81 GHz from Interference Caused by Automobile Radars.

After discussion and clarifications the four Resolution were unanimously approved.

6. Elected Division Presidents and Vice-Presidents

The General Secretary presented the names of the Division Presidents and Vice-Presidents that were elected by the Members of the respective Division.

The list is as follows:

Division A: Fundamental Astronomy
President: Anne Lemaître (Belgium)
Vice-President: Daniel Hestroffer (France)

Division B: Facilities, Technologes, and Data Science
President: Pietro Ubertini (Italy)
Vice-President: Michael Burton (Australia)

Division C: Education, Outreach, and Heritage
President: John Hearnshaw (New Zealand)
Vice-President: Susana Deustua (USA)

Division D: High Energy Phenomena and Fundamental Physics
President: Chryssa Kouveliotou (USA)
Vice-President: Elena Pian (Italy)

Division E: Sun and Heliosphere
President: Yihua Yan (China, Nanjing)
Vice-President: Sarah Gibson (USA)

Division F: Planetary Systems and Bioastronomy
President: Nader Haghighipour (USA)
Vice-President: Gonzalo Tancredi (Uruguay)

Division G: Stars and Stellar Physics
President: Corinne Charbonnel (France)
Vice-President: David Soderblom (USA)

Division H: Interstellar Matter and Local Universe
President: Bruce Elmegreen (USA)
Vice-President: Leonardo Testi (Italy)

Division	*Commission*	*Title*
A	A1	Astrometry
	A2	Rotation of the Earth
	A3	Fundamental Standards
	A4	Celestial Mechanics & Dynamical Astronomy
B	B1	Computational Astrophysics
	B2	Data & Documentation
	B3	Astroinformatics & Astrostatistics
	B4	Radio Astronomy
	B5	Laboratory Astrophysics
	B6	Astronomical Photometry & Polarimetry
	B7	Protection of Existing & Potential Observatory Sites
C	C1	Astronomy Education & Development
	C2	Communicating Astronomy with the Public
	C3	History of Astronomy
	C4	World Heritage and Astronomy
D	D1	Gravitational Wave Astrophysics
E	E1	Solar Radiation and Structure
	E2	Solar Activity
	E3	Solar Impact throughout the Heliosphere
F	F1	Meteors, Meteorites, and Interplanetary Dust
	F2	Exoplanets and the Solar system
	F3	Astrobiology
G	G1	Binary and Multiple Star Systems
	G2	Massive Stars
	G3	Stellar Evolution
	G4	Pulsating Stars
	G5	Stellar & Planetary Atmospheres
H	H1	Local Universe
	H2	Astrochemistry
	H3	Planetary Nebulae
	H4	Stellar clusters throughout Cosmic Space and Time
J	J1	Galaxy Spectral Energy Distributions
	J2	Intergalactic medium

Divisions	*Commission*	*Title*
D-J	X1	Supermassive Black Holes, Feedback & Galaxy Evolution
A-F	X2	Solar System Ephemerides

Division J: Galaxies and Cosmology
President: : Claus Leitherer (USA)
Vice-President: Matthew Malkan (USA)

The Assembly took note.

7. New Commissions

The General Secretary recalled that, following the restructuring of the Divisions at the XXVIII General Assembly in Beijing, also the Commissions had to be restructured. Following a call for confirming, cancelling or renaming the old Commissions and proposing new ones, the Executive Committee, in its 95th Meeting, approved the following Commissions:

7.1. *Division Commissions*

7.2. *Cross-Divisions Commissions*

8. Elected Commissions Presidents and Vice-Presidents

The General Secretary presented the names of the Division Presidents and Vice-Presidents that were elected by the Members of the respective Division.

Division	*Commission*	*President*	*Vice-President*
A	A1	Anthony G.A. Brown (Netherlands)	Jean Souchay (France)
	A2	Richard Gross (USA)	Florian Seitz (Germany)
	A3	Catherine Y. Hohenkerk (UK)	Brian Luzum (USA)
	A4	Cristian Beaug (Argentina)	Alessandra Celletti (Italy)
B	B1	Simon F. Portegies-Zwart (Netherlands)	Dmitrij Bisikalo (Russia)
	B2	Michael Wise (Netherlands)	Anja C. Schrder (South Africa)
	B3	Eric D. Feigelson (USA)	Prajval Shastri (India)
	B4	Gabriele Giovannini (Italy)	Anthony Beasley (USA)
	B5	Farid Salama (USA)	Helen J. Fraser (UK)
	B6	Saul J. Adelman (USA)	Antonio Mario Magalhaes (Brazil)
	B7	Richard F. Green (USA)	Constance Elaine Walker (USA)
C	C1	Beatriz Elena Garcia (Argentina)	Paulo Sergio Bretone (Brazil)
	C2	Pedro Russo (Netherlands)	Rick Fienberg (USA)
	C3	Xiaochun Sun (China Nanjing)	Wayne Orchiston (Thailand)
	C4	Clive L.N. Ruggles (UK)	Gudrun Wolfschmidt (Germany)
D	D1	Neil Gehrels (USA)	Marica Branchesi (Italy)
E	E1	Natalie A. Krivova (Germany)	Alexander Kosovichev (USA)
	E2	Lyndsay Fletcher (UK)	Paul S. Cally (Australia)
	E3	Ingrid Mann (Sweden)	Carine Briand (France)
F	F1	Jiri Borovicka (Czech Republic)	Diego Janches (USA)
	F2	Alain Lecavelier des Etangs (France)	Jack J. Lissauer (USA)
	F3	Sun Kwok (China Nanjing)	Masatoshi Ohishi (Japan)
G	G1	Andrej Prsa (USA)	Virginia Trimble (USA)
	G2	Artemio Herrero Davo (Spain)	orick S. Vink (UK)
	G3	John C. Lattanzio (Australia)	Marc Howard Pinsonneault (USA)
	G4	Christopher Simon Jeffery (UK)	Jaymie Matthews (Canada)
	G5	Ivan Hubeny (USA)	Carlos Allende Prieto (Spain)
H	H1	Eva K. Grebel (Germany)	Dante Minniti (Chile)
	H2	Thomas J. Millar (UK)	Edwin A. Bergin (USA)
	H3	Letizia Stanghellini (USA)	Albert Zijlstra (UK)
	H4	Richard De Grijs (China Nanjing)	Amanda I. Karakas (Australia)
J	J1	Denis Burgarella (France)	Cristina Carmen Popescu (UK)
	J2	Avery Abraham Meiksin (UK)	Hsiao-Wen Chen (USA)
D-J	X1	William Richard Forman (USA)	Thaisa Storchi-Bergmann (Brazil)
A-F	X2	Andrea Milani Comparetti (Italy)	William M. Folkner (USA)

9. Financial Matters

Report of the Finance Sub-Committee

Presented by Beatriz Barbuy (Chair)

Accounts 2013-2015

The Finance Committee received the Balance sheets reporting income and expenses, for the period 2013-2015, together with a report from the General Secretary Thierry Montmerle. The accounts show overall the usual pattern with however a few differences, listed below:

1. The income from countries is below expectation, due to many countries not having paid so far, with a difference of 130 KE in 2015. This involves small as well as large countries in terms of contributions. This is the same deficit amount indicated in the total budget 2013-2015. We expect that many of the default-payments will be fulfilled along 2015 (according to T. Montmerle there is a historic peak after the General Assemblies).

2. The extra expenses not foreseen in the budget approved in Beijing, concern the following items:

(*a*) Participation of Division Presidents in the Executive Committee meetings each year, amounting to an extra expense of around 90 kE. We understand that this was exceptional, due to the restructuration of the Divisions and Commissions. For the next triennium, it is expected to have one meeting in the third year, including the Division Presidents, in preparation of the following GA, as previously occurred since 2000s.
(*b*) Contribution to the International Year of Light, amount of 30 kE.
(*c*) Office for Astronomy Outreach (OAO), amount of 25 kE. T. Montmerle informed that Japan has been paying most of it.
(*d*) Office of Astronomy for Development (OAD), amounting to an average extra of 100 KE with respect to previous years when no OAD existed.

The Finance Committee recommends approval of the extra expenses 2a,b,c,d. We also recommend approval of the new budget lines on funding for other smaller items (e.g. Planet naming). A compensation of the extra expenses was obtained from the fact that the General Secretary lives in Paris, which is generally not the case, and will be needed again for the next triennium.

The FC members showed some concern as not having an external audit so far. We were informed by Thierry Montmerle that the auditor cannot close his accounts before all amounts are deposited, and it is now waiting for the final amount from CUP. We should receive the external audit report in due time in a few weeks, and certainly before preparing the final report in August.

Finally we recommend approval of the preliminary budget for 2016-2018 which includes a 2% adjustment in the due's unit to take into account inflation. The proposed unit for the years 2016, 2017 and 2018 are, respectively 2800 €, 2860 € and 3035 €.

The present report should be read together with the report received by the FC from the General Secretary Thierry Montmerle, where more details are given (see Chapter V in this Volume).

The report was submitted to the vote of National Representatives, and the budget for 2016-2018 approved with one abstention.

10. Election of the Finance Committee

The General Assembly unanimously appointed the following Finance Committee (2016-2018):

Joao Alves (Austria)	2nd term
Kate J. Brooks (Australia)	2nd term
Zhanwen Han (China Nanjing)	elected, 1st term, Chair
J.J. Kavelaars (Canada)	elected, 1st term
Laszlo Kiss (Hungary)	elected, 1st term
Nikolaos D. Kylafis (Greece)	2nd term
Tushar P. Prabhu (India)	2nd term
Lee Anne M. Willson (USA)	2nd term

11. Election of the Membership Committee

The General Assembly unanimously appointed the following Membership Committee (2016-2018):

Sonia Ehlerova (Czech Republic)	elected, 1st term
William Harris (Canada)	elected, 1st term
Myungshin Im (Korea, Rep.)	2nd term
Lex Kaper (Netherlands)	2nd term
Daniela Lazzaro (Brazil)	elected, 1st term
Helmut O. Rucker (Austria)	2nd term, Chair
Ramotholo R. Sefako (South Africa)	2nd term
David Soderblom (USA)	elected, 1st term

12. Appointment of the Resolutions Committee

The General Assembly unanimously appointed the following Resolutions Committee (2016-2018):

Katia Cunha (Brazil)	elected, 1st term
Bruce G. Elmegreen (USA)	2nd term, Chair
Toshio Fukushima (Japan)	elected, 1st term
Sergei A. Klioner (Germany)	elected, 1st term
Rene Kraan-Korteweg (South Africa)	2nd term

13. Appointment of the Special Nominating Committee

The General Assembly unanimously appointed the following Resolutions Committee (2016-2018):

• *IAU membres ex-officio:*
Silvia Torres-Peimbert (President, Mexico; Chair)
Norio Kaifu (Past President, Japan)

• *IAU Advisors:*
Piero Benvenuti (General Secretary; Italy)
Teresa Lago (Assistant General Secretary; Portugal)

•*Appointed members:*
Xavier Barcons (Spain)
Matthew Colless (Australia)
Thomas Henning (Germany)
Dimitri Pourbaix (Belgium)
Monica Rubio (Chile)

14. Election of the Executive Committee 2015-2018

The General Assembly unanimously appointed the following Executive Committee (2016-2018):

President	Silvia Torres Peimbert (Mexico)
President-Elect	Ewine van Dishoeck (Netherlands)
General Secretary	Piero Benvenuti (Italy)
Assistant General Secretary	Teresa Lago (Portugal)
Vice-President	Renée Kraan-Korteweg (South Africa)
Vice-President	Xiao-Wei Liu (China, Nanjing)
Vice-President	Dina Prialnik (Israel)
Vice-President	Debra Elmegreen (USA)
Vice-President	Ajit Kembhavi (India)
Vice-President	Boris Shustov (Russia)
Advisor	Norio Kaifu (Japan)
Advisor	Thierry Montemerle (France)

15. Date and location of the XXXI General Assembly in 2021

The XXXI General Assembly will take place a Busan, Republic of Korea, on August 16-27, 2021.

Transactions IAU, Volume XXIXB
Proceedings IAU Symposium No. Volume XXIXB, 2018
P. Benvenuti, ed.
© International Astronomical Union 2019
doi:10.1017/S1743921318003976

CHAPTER III

TWENTY NINETH GENERAL ASSEMBLY

CLOSING CEREMONY

Friday, 14 August 2015, 16:30-18:00

Hawai'i Convention Center, Honolulu, HI

1. Welcome by Silvia Torres-Peimbert, IAU President

The incoming IAU President, Silvia Torres-Peimbert, welcomed the participants to the Closing Ceremony of the XXIXth General Assembly.

2. Invitation to the XXXth General Assembly, Vienna, Austria, August 2018

A video movie was shown on the screens, presenting the venue and attractions of the next General Assembly, and welcomed the participants in advance.

3. Address by the Retiring President

Prof. Norio Kaifu

The 29th IAU General Assembly has now finished, and the newly re-organized IAU started quite successfully here in Honolulu. It was regrettable that one of the important activities planned during this General Assembly, visits to the Maunakea telescopes by worldwide astronomers, could not be realized because of the local conflicts on Maunakea. However, I believe that all of you will agree, the 29th General Assembly was very active and very successful.

I express millions of thanks to Meg Urry, Kevin Marvel, and all those associated with the American Astronomical Society, who hosted and executed this General Assembly in Honolulu so nicely. Thanks also to Guenther Hasinger, Roy Gal, Doug Simon, and the Local Organizing Committee, including all of the Maunakea Observatories, for wonderful social events and tremendous efforts behind the scenes that resulted so well. I thank all Divisions, Commissions, and organizers for such rich and fruitful scientific meetings. My thanks are also extended to the Hawaiian Convention Center and all Local Communities of Hawaii for their work toward the success of the IAU General Assembly.

As we saw in the Business Sessions, the IAU welcomed one new National Member, Colombia, and nearly 1,200 new Individual Members. The rapid increase in Individual Members is a prominent Characteristic of the IAU, demonstrating that the IAU is a growing International Scientific Union, with more and more members of the younger generation. However, I would also draw your attention to the National Membership. The number of IAU National Members is currently 74, whereas it was 61 in 1994, twenty years ago. Therefore, the National Membership has increased 21% within the past 20 years. But if we compare it with the increase in Individual Members of 44% during the same period, the increase in National Membership is considerably slower. This means that the rapid growth of astronomy is happening mostly in the economically developed countries, and less in the developing world. This fact clearly shows how important the recent IAU activities, OAD, OAO, and newly started OYA, are to develop astronomy worldwide toward its future. Also, the promotion of regional coordination, like the newly established East Asian Observatory, organization of South-East Asian coordination, and worldwide efforts to develop the SKA, will give the IAU and world astronomy a new perspective for the global growth of astronomy.

Through the past 6 years as President-Elect and President, I have learned a great deal about the truly rich activities of the IAU. I feel strongly that an important base that enabled the IAU to promote such diverse activities was the introduction of Individual Membership, formally started in 1961, and this system made the IAU an extremely active International Scientific Union. We deeply thank those previous great leaders of the IAU, and wish that the new scientific organization started here in Honolulu will provide a solid platform for a new leap toward the future of the IAU.

Finally, I express many thanks to Thierry Montmerle for his passion and tremendous work as General Secretary, a well-known position of extremely hard work. I thank Silvia Torres-Peimbert and Piero Benvenuti for working together as a wonderful team of Officers of the IAU. My deep thanks are also extended to advisors Robert Williams and Ian Corbett for their continuous support and helpful advice on many difficult occasions during the past three years. I feel very happy to remain on the next Executive Committee as an advisor, together with Thierry, to support the new officers for the coming three years.

Aloha, and Mahalo, thank you.

4. Address by the Retiring General Secretary

Dr. Thierry Montmerle

Aloha !

Dear colleagues, Dear Guests, Dear friends,

As this triennium draws to a close, my term as General Secretary will end very soon. When I took over from Ian Corbett in Beijing, I had no idea of how much the IAU would accomplish in three years ...

There were many adventures. Most of them started in the US, and are ending (in the present phase) in the US again.

Let me briefly explain. At my first EC meeting after Beijing as freshly elected Assistant General Secretary, that took place at STScI in Baltimore at the invitation of Bob Williams, I presented a few slides explaining why I, as an IAU member, was not happy with the Division structure, and boldly started inventing new Divisions and assigning Commissions to them. Bob was taken aback, and after some discussion with the EC, offered me a challenge: "Well, Thierry, OK, come back to us next time with a project".

This early trust from the EC to undertake a restructuring of the Divisions and the Commissions, followed by other initiatives taken by various people, played a key role in my term as General Secretary.

This is a story that deserves to be told, even briefly, because it illustrates what I now, here, like to call "the Aloha spirit of the IAU", and that can be summarized in three buzzwords: "We dare, We care, We share".

1. WE DARE

You have all lived the various episodes of the Division Restructuring (approved in Beijing) and the Commission Reform (approved here by the EC and the Division Presidents), so I will not repeat them here. I just want to say that it is the result of a long (6 years) and trusty collaboration between two successive ECs and two "generations" of Division Presidents (and one generation of Commission Presidents), and, ultimately, of the will of the IAU membership to adapt to a constantly evolving astronomical world. I'm proud to be able to say today: "Mission Accomplished".

The new Commissions, with their new or updated topics, with a renewed, more focussed demography, and the exceptionally large participation rate for the election of their new Organizing Committees (from 60% up to nearly 90%), demonstrate that the IAU is ready for the future, and that it is really composed of what I like to call "IAU Citizens", not just "IAU members".

The new Commissions will be in place in a couple of weeks on the IAU web site, but you can already start thinking about establishing new Working Groups. The procedure is easy, and the fact that the Division Steering Committees now incorporate ex-officio Commission Presidents should facilitate selecting the best scientific structure between Commission Working Groups and Division Working Groups.

2. WE CARE

The mission of the IAU is to be not only a forum for professional astronomers, but is also to go out to society and participate in education and development worldwide, using astronomy as a tool for fostering interest in science and technology, even in less developed countries.

The IAU has now three offices, developed in partnership with institutions from several countries.

- The Office of Astronomy for Development in Cape Town, in partnership with the South African National Research Foundation, has seen a spectacular development of its activities initiated during the previous triennium. It now has several "Regional Nodes" and "Language Centers" across the five continents, and with its revised structure will be even more efficient at selecting and helping educational projects from all over the world to thrive.
- The Office of Astronomy Outreach in Tokyo, in partnership with the National Observatory of Japan and other Asian countries, was developed and plays a key role in the projects specific to 2015 that I will briefly describe in a minute.
- The Office for Young Astronomers in Oslo, in partnership with the Norwegian Academy of Science and Letters, to continue and broaden the decade-long "International School for Young Astronomers" worldwide.

3. WE SHARE

The IAU must also go towards the public. The year 2015, and in particular this General Assembly, is seeing the culmination of two projects initiated by the Executive Committee and implemented by the Office of Astronomy Outreach: the so-called "Cosmic Light" cornerstone of the UNESCO "International Year of Light and Light Technologies", also know as "IYL2015", and the "NameExoWorlds" contest to call for the public naming of selected exoplanets and their host stars.

The "Cosmic Light" project draws heavily on the legacy of the International Year of Astronomy "IYA2009", which gave an enormous advantage in starting early IYL2015, far earlier than our physicists colleagues ! Actually, the IAU, although not in the drivers' seat in IYL2015, which was run essentially by physicists and the lighting industry, was setting the standards for many IYL2015 Cornerstones. Here, you may have seen "Cosmic Light" in action, at Stargazing parties in Ala Moana Parks, or in conferences and displays at the Bishop Museum?

The "NameExoWorlds" contest is about engaging the public worldwide to attribute names for 20 exoplanetary systems and their host stars, selected by astronomy clubs and similar non-profit organizations interested in astronomy. The launch of the vote took place in this very room last Tuesday, and I'm pleased to report that, at the time of writing this address, the number of votes received by the IAU is close to 125,000 in only three days !

Last but not least, the IAU continues to develop a very fruitful collaboration with UNESCO on "Astronomy and World Heritage" issues, more precisely in the process of nomination of astronomical sites (past or present) to the list of the World Heritage. One such project, for example, concerns "High-Altitude Observatories" across the world.

In summary, this has been a fantastically rich and exciting experience, which many, or perhaps all, of us in the Executive Committee or among Division Presidents, will remember, and is formally (but perhaps not in reality) ending here at this extra-ordinary General Assembly. I have been helped and supported by wonderful people, and I am most grateful to them.

I'm proud to be an IAU member.

I'm proud to be an "IAU citizen".

Mahalo !

5. Address by the Incoming President

Prof. Silvia Torres-Peimbert

It is a great honor to be President of the International Astronomical Union, traditionally one of the most active international scientific organizations. From 2000 to 2006 I participated as vicepresident of this Union. During that period I met IAU presidents Bob Kraft, Franco Pacini, Ronald Eckers and Catherine Cesarsky as well as with the General Secretaries Johannes Andersen, Hans Rickman, Oddbjorn Engvold and Karel van der Hucht; from them I learned some of the multiple tasks and complexities of this society. For last three years, in my role as President-Elect, I have had the privilege of working directly with Bob Williams, Norio Kaifu, Thierry Montmerle, Piero Benvenuti and Ian Corbett, as well as with 6 vicepresidents that were part of the Executive Committee. All of us have worked with enthusiasm for the same goal, the well being of the society.

I am specially grateful to Norio Kaifu for his strength and serenity and to Thierry Montmerle for his energy and enthusiasm. Both of them have decidedly lead the Union through its restructuring process which has been implemented.

The Division Presidents during this period worked outstandingly in establishing the new structure. It is now firmly in place and we are confident that it will indeed fulfill the IAU mission ?to promote and safeguard the science of astronomy in all its aspects through international cooperation?. There are now a set of very active Divisions, Commissions and Working Groups that will continue to promote the unifying tasks that our discipline requires.

My main concern as President is to involve each of the members in understanding that this organization does have an impact in our professional lives. It helps us to maintain contact with our international colleagues, and to keep us informed of the different activities and developments carried out by the astronomical community. The General Assemblies, Simposia and Regional Meetings are the backbone of the union. They play a central role in fostering astronomy around the world, as well as in attaining consensus on important decisions that affect our science. In other words, I invite all of you to participate more actively in this society.

It is our intention to strengthen the ties between the governing bodies and the membership of the union. To this end we plan to use more intensively the webpage as a means of communication, and to start a new epoch of the Information Bulletin series, probably mainly in an electronic version to all the community. In parallel we request that you keep in contact with us.

For several decades IAU has been committed to education projects, mainly through the International School for Young Astronomers. In 2009 in Rio de Janeiro, the General Assembly approved the "Astronomy for Development. Strategic Plan" which led to the creation of the Office for Astronomy Development, OAD. This office has become an important driver of astronomy education and outreach in several parts of the world. As you have heard already, through the OAD there have been 68 funded projects and 9 regional nodes or language centers across the globe. We expect for this activity to continue with increasing success.

Similarly the Office for Astronomy Outreach, OAO, based in Tokyo has extended the action of the Union to the general public and in general to amateur astronomers around the world. This initiative recognizes the importance that the union gives to these groups of friends and their activities, that until now have been disconnected from the community of professional astronomers.

At this General Assembly, more than 3000 astronomers and exhibitors from 63 countries have come to share with us their work. It has been a very rich program. There were 6 simposia, 22 focus meetings and many splinter meetings. We had the opportunity to listen to very inspiring invited and plenary talks. Moreover there were teaching and public activities. And the exhibits were very informative and attractive.

Keep in mind that in three years the next General Assembly will take place in Vienna, where we will meet again old and new friends. We will have the opportunity to communicate to each other the new results and the exciting discoveries that continue to amaze and enlighten us. We will celebrate the first centennial of this great organization!

Finally let me thank our hosts: the Hawaii Astronomers and the American Astronomical Society for their hospitality, their generosity and all the support given to us. Many thanks also to the many organizations that sponsored this meeting. We will remember this general assembly for many years to come.

Thank you all!

6. Address by the Incoming General Secretary

Prof. Piero Benvenuti

Aloha!

Dear colleagues and friends, distinguished guests:

It is a great honour for me to start holding office as General Secretary of the International Astronomical Union and I feel especially privileged to take up duty at this particular moment in time, at the XXIX General Assembly in Honolulu. There are several reasons for that. First of all, thanks to the fine and dedicated work of the previous Presidents and General Secretaries, I am inheriting a completely renewed Union: three years ago, in Beijing, the new Divisions, today here the new Commissions and the Working Groups that soon will follow suit. New energy has been injected in the activities of the Union, and we had a tangible and vivid example of them during these two weeks. At this point I cannot but commit myself to continue supporting the renewal process, offering to the Divisions and Commissions those tools - e.g. dedicated interactive web services - that will facilitate their collaborative work. Tomorrow, during the first Executive Meeting of the new triennium, I will ask the Division Presidents to indicate to me their priority requirements that we will then try to implement to the best possible extent. Talking about Divisions, in the past triennium I enjoyed their very fruitful interaction with the Executive Committee and I am convinced that it is essential for an efficient operation of the Union. Financially, we cannot afford to continue supporting directly their participation in every EC meeting, but, with the help of the ever evolving technology, we will find the way to keep them virtually present as interactive participants.

The second reason that makes me feel a privileged General Secretary, is the renewed Agreement with NRF on the Office of Astronomy for Development. I do hope each of you had a chance recently to have a look at the OAD web pages and check the great job that a small group of dedicated people, together with hundredths of volunteers have achieved: if not, do it soon and feel proud of it. Because - you see - if the IAU were just to be concerned with the development of Astronomy and with fostering international scientific collaboration, it would just do what any Scientific Union is supposed to do. But our Strategic Plan of Astronomy for Development makes the difference: it recognizes the deep root of our Science with humanity, a unique characteristics that singles it out of any other discipline. And using the natural appeal of Astronomy for promoting the human progress, brings it back to its classical function, so beautifully expressed by Plato in his dialogue "The Republic". What has been achieved so far is indeed impressive, as it was also highlighted by the recent external review Report. Just consider the 5 new OAD Regional Nodes that have been started during this General Assembly: all of them are strategically located and have great potential. But the growth alone is not enough: as Khotso Mokhele was warning us yesterday, the real work starts now. The OAD needs to consolidate its achievements and maintain the control and coordination in its expanding international network. All this requires additional resources that are well beyond the modest investment that IAU can provide annually. I am committed to pursue an aggressive fundraising campaign that would allow us to prepare, in time for the next GA in Vienna, an affordable and credible plan that goes beyond 2021, as requested by the Resolution B1 that was approved yesterday. Let me add that the fundraising campaign starts now, by asking each National Member to assure a timely payment of the dues! I am fully aware of the intervening difficulties that various Countries may encounter, but it is obvious that without that basic income, all activities of the Union will suffer and start starving. On a related subject, I repeat here my full commitment to follow strictly the Union financial Working Rules, in particular by restarting the annual audit of the accounts and increasing transparency, regularly forwarding the annual reports of the Finance Committee to the National Members.

Let me now come to a more personal reason for feeling privileged. Before coming here, I was scanning through the long list of past General Secretaries - a spine-chilling exercise - and I discovered that I am the first Italian astronomer to hold that position: an honour that I wish to dedicate to my glorious University of Padova, that, with its motto "Universa Universis Patavina Libertas" reminds all of us that Science is tantamount to Freedom. And if the motto weren't enough, you just have to glance at the original Galileo's chair, when entering its Aula Magna. That is not however the only "first" of the next triennium: for the first time in the history of the Union, the majority of the Officers is composed of women. One Italian and three women: you may start wondering if it is just a statistical fluctuation or if there are some hidden astrological connections... well, start pondering: you have three years to solve the conundrum! Now, seriously: I believe our Union is giving a great example and I wish to consider it not just a simple attempt to reach gender equality. As I believe Astronomy - so deeply rooted as it is in the grain of humanity and culture - is not just a Science as many others, similarly I believe a woman scientist is not just another scientist. Women have a special natural gift for caring and educating - and I underline the etymo of the word: e-duco, I pull out the best of someone - therefore a woman who is also an astronomer can have a greater impact on the society than a simple scientist. It is not a case that our Working Group Women in Astronomy, in the report that they have prepared at this Assembly, has indicated the desire to closely collaborate with the OAD and in particular with the Regional Nodes for contributing to a better and more just society. I am looking forward to working pleasantly and efficiently with my colleagues women Officers!

Coming to the conclusion of my address, I like to invite you all to Vienna in 2018, for the XXX General Assembly. We will be celebrating there the first 100 years of the IAU and we should mark that important anniversary with some special and significant event. We are already planning to prepare a book about the history of the Union and, given the location of the Assembly, we may consider some additional project on the relation between Astronomy and Philosophy, in particular the Philosophy of Science. Finally, I would like to express my gratitude to Norio and Thierry who, friendly and patiently, accompanied me during the past triennium.

Mahalo... und Aufwiedersehen in Wien in 2018!

7. Handing out the IAU flag

The AAS Executive Officer Kevin Marvel thanked everyone for coming to theXXIX General Assembly and introduced Gerhard Hensler of the University ofVienna, who invited all there for the XXX GA in 2018. Then Roy Gal of theUniversity of Hawai'i officiated as the IAU flag was ceremonially transferredfrom the USA to Austria.

8. Closing

The festivities were brought to a close with an oli (chant) by Kumu PattyeWright, who then played her ipu (a type of drum) while six dancers performeda mesmerizing hula. Afterward, filing out of the ballroom and ontothe 4th-floor patio for a group photo, attendees were greeted by a spectacularrainbow, as if the sky itself wanted to participate in the celebration!

9. Image Archive

Images taken during the General Assembly can be viewed and dowloaded from the IAU web site:
`https://www.iau.org/public/images/archive/category/general_assembly_2015/`

Transactions IAU, Volume XXIXB
Proceedings IAU Symposium No. Volume XXIXB, 2018
P. Benvenuti, ed.
© International Astronomical Union 2019
doi:10.1017/S1743921318003988

CHAPTER IV

TWENTY NINGTH GENERAL ASSEMBLY

RESOLUTIONS OF THE XXIXth GENERAL ASSEMBLY

1. Resolutions Committee (2012-2015)

The members of the Resolutions Committee for the 2012-2015 triennium were:

Ian Corbett (UK; Chair)
Bruce Elmegreen (USA)
Renée Kraan-Korteweg (South Africa)
Yanchun Liang (China Nanjing)
Karel van der Hucht (Netherlands)

The report of the Resolutions Committee is given in Chapter II of these *Transactions*.

2. Approved Resolutions

RESOLUTION B1

on the IAU Strategic Plan 2010-2020: Astronomy for the Developing World

Proposed by the IAU Executive Committee

The XXIX General Assembly of the International Astronomical Union,

Recognising

1. That the XXVII General Assembly, meeting in Rio de Janiero, Brazil, on 13 August 2009 unanimously passed a Resolution resolving that the IAU should approve the goals specified in the Strategic Plan: Astronomy for the Developing World as objectives for the IAU in the coming decade,

2. That to further these objectives the IAU established the Office of Astronomy for Development (OAD) in Cape Town, South Africa, as an equal partnership between the IAU and the National Research Foundation of South Africa,

3. That the OAD has successfully promoted an ambitious international programme of activities in pursuit of the objectives of the IAU Strategic Plan,

4. That a recent independent review of the OAD concluded that "its performance has been outstanding, particularly given the very limited resources that have been made available to an organisation with such ambitious terms of reference,"

Resolves

1. That the pursuit of the goals of the Strategic Plan: Astronomy for the Developing World should continue until the XXXI General Assembly to be held August 2021,

2. That the Executive Committee should present for approval at the XXX General Assembly to be held in Vienna, Austria in August 2018 an extended Strategic Plan which addresses the future of the OAD and its activities beyond 2021,

3. That the Executive Committee should consult existing and potential stakeholders in the preparation of this Strategic Plan.

RESOLUTION B2

on recommended zero points for the absolute and apparent bolometric magnitude scales.

Proposed by the IAU Inter-Division A-G Working Group on Nominal Units for Stellar & Planetary Astronomy

The XXIX General Assembly of International Astronomical Union,

Noting

1. the absence of an exact definition of the zero point for the absolute and apparent bolometric magnitude scales, which has resulted in the proliferation of different zero points for bolometric magnitudes and bolometric corrections in the literature (ranging at approximately the tenth of a magnitude level; see, e.g., Bessell, Castelli, & Plez 1998; Torres 2010),

2. that IAU Commissions 25 and 36 approved identical draft resolutions for defining the zero point for the bolometric magnitude scale (Andersen 1999), but that the resolution never subsequently reached the stage of approval by the IAU General Assembly, and was only sporadically adopted within the astronomical community,

3. that recent total solar irradiance measurements have led to a revised solar luminosity that differs slightly from the value used to set the zero point of the absolute bolometric magnitude scale in the Commission 25 and 36 draft resolutions,

Considering

1. the need for a standardized absolute and apparent bolometric magnitude scale for accurately and repeatably transforming photometric measurements into radiative luminosities and irradiances, independently of the variable Sun,

2. that multiple zero points for bolometric corrections pervade the literature due to a lack of a standard zero point for the bolometric magnitude scale,

Recommends

1. to define the zero point of the absolute bolometric magnitude scale by specifying that a radiation source with absolute bolometric magnitude[1] ($M_{bol} = 0$) mag has a radiative luminosity of exactly

$$L_o = 3.0128 \times 10^{28} \ W \tag{2.1}$$

and the absolute bolometric magnitude M_{bol} for a source of luminosity L (in W) is

$$M_{bol} = -2.5 \log(L/L_{\odot}) = -2.5 \log(L) + 71.197\,425 \tag{2.2}$$

The zero point was selected so that the nominal solar luminosity[2] ($L^N = 3.828 \times 10^{26} W$) corresponds closely to absolute bolometric magnitude $M_{bol} = 4.74$ mag, the value most commonly adopted in the recent literature (e.g., Bessell, Castelli, & Plez 1998; Cox 2000; Torres 2010). Using the proposed zero point L_o, the nominal solar luminosity ($L^N = 3.828 \times 10^{26} W$) corresponds to bolometric magnitude $M_{bol} \simeq 4.739996$ ... mag - i.e., sufficiently close to 4.74 mag for any foreseeable practical purpose.

2. to define the zero point of the *apparent bolometric magnitude* scale by specifying that $m_{bol} = 0$ mag corresponds to an *irradiance* or *heat flux density*[3] of

$$f_o = 2.518\,021\,002 ... \times 10^{-8} \ Wm^{-2} \tag{2.3}$$

and hence the apparent bolometric magnitude m_{bol} for an irradiance f (in Wm^{-2}) is

$$m_{bol} = -2.5 \log(f/f_o) = -2.5 \log f - 18.997\,351 ... \tag{2.4}$$

The irradiance f_o corresponds to that measured from an isotropically emitting radiation source with absolute bolometric magnitude $M_{bol} = 0$ mag (luminosity L_o) at the standard distance of 10 parsecs[4] (based on the IAU 2012 definition of the astronomical unit).

As the absolute bolometric magnitude zero point and astronomical unit are defined precisely, further digits for the apparent bolometric magnitude zero point irradiance f_o may be calculated beyond the digits shown, if needed. The adopted value of f_o agrees with that in common use (e.g., Lang 1974, Cox 2000) at the level of <0.1 %. Using this zero point, the *nominal solar irradiance* $S^N_{\odot}$ ($1361 Wm^{-2}$) corresponds to a solar apparent bolometric magnitude of $m_{bol\odot} \simeq -26.832$ mag.

References

Andersen, J. 1999, Transactions of the International Astronomical Union, Series B, 23, pgs. 141 & 182

Bessell, M. S., Castelli, F., & Plez, B. 1998, Astronomy & Astrophysics, 333, 231

Bureau International des Poids et Mesures, 2006, The International System of Units (SI), 8th edition, Organisation Intergouvernementale de la Convention du Métre

Cox, A. N. 2000, Allen?s Astrophysical Quantities, 4th Edition

Dyson, F. W. 1913, MNRAS, 73, 334

Kopp, G. 2014, Journal of Space Weather and Space Climate, 4, A14

Kopp, G., Lawrence, G., Rottman, G., 2005, Solar Physics, 230, 129

Kopp, G., & Lean, J. L. 2011, Geophysical Research Letters, 38, L01706

Lang, K. R. 1974, Astrophysical Formulae, A Compendium for the Physicist and Astrophysicist, Springer-Verlag

Meftah, M., Irbah, A., Hauchecorne, A., *et al.* 2015, Solar Physics, 290, 673

Schmutz W., Fehlmann A., Finsterle W., *et al.* 2013, AIP Conf. Proc. 1531, p. 624627, doi:10.1063/1.4804847

Notes

[1]The notation of M_{bol} referring to *absolute bolometric magnitude* and m_{bol} referring to *apparent bolometric magnitude* was adopted by Commission 3 (Notations) at the VIth IAU General Assembly in Stockholm in 1938

[2]Modern spaceborne total solar irradiance (TSI) instruments are absolutely calibrated at the 0.03% level (Kopp 2014). The TIM/SORCE experiment established a lower TSI value than previously reported based on the fully characterized TIM instrument (Kopp *et al.* 2005, Kopp & Lean 2011). This revised TSI scale was later confirmed by PREMOS/PICARD, the first spaceborne TSI radiometer that was irradiance-calibrated in vacuum at the TSI Radiometer Facility (TRF) with SI-traceability prior to launch (Schmutz *et al.* 2013). The DIARAD/PREMOS (Meftah *et al.* 2015), ACRIM3/ACRIMSat (Willson 2014), VIRGO/SOHO, and TCTE/STP-Sat3 (`http://lasp.colorado.edu/home/tcte/`) flight instruments are now consistent with this new TSI scale within instrument uncertainties, with the DIARAD, ACRIM3, and VIRGO having made post-launch corrections and the TCTE having been validated on the TRF prior to its 2013 launch. The cycle 23 observations with these experiments are consistent with a TSI value (rounded to an appropriate number of significant digits) and uncertainty of: $S_{\odot} = 1361(\pm 1) W m^{-2}$ (2 σ uncertainty). The uncertainty range includes contributions from the absolute accuracies of the latest TSI instruments as well as uncertainties in assessing a secular trend in TSI over solar cycle 23 using older measurements. Combining this total solar irradiance value with the IAU 2012 definition of the astronomical unit leads to a current best estimate of the mean solar luminosity of $L_{\odot} = 4\pi(1au)^2 S_{\odot} = 3.8275(\pm 0.0014) \times 10^{26} W$. Based on this, we adopt a nominal solar luminosity of $\mathcal{L}_{\odot}^{N} = 3.828 \times 10^{26} W$

[3]The terms *irradiance* and *heat flux density* are used interchangeably, both with SI units of Wm^{-2}(Wilkins 1989, Bureau International des Poids et Mesures 2006). See also https://www.iau.org/publications/proceedings rules/units/.

[4]The parsec is *the unit of distance... corresponding to a parallax of 1"* (Dyson 1913). One parsec is equivalent to 1 AU tan (1") = $3.085\,677\,581 \times 10^{16}$ m.

RESOLUTION B3

on recommended nominal conversion constants for selected solar and planetary properties.

Proposed by IAU Inter-Division A-G Working Group on Nominal Units for Stellar & Planetary Astronomy

The XXIX General Assembly of the International Astronomical Union,

Recognizing

that notably different values of the solar mass, radius, luminosity, effective temperature, total solar irradiance, of the masses and radii of the Earth and Jupiter, and of the Newtonian constant of gravitation G have been used by researchers to express and derive fundamental stellar and planetary properties,

Noting

1. that neither the solar nor the planetary masses and radii are secularly constant and that their instantaneous values are gradually being determined more precisely through improved observational techniques and methods of data analysis, and
2. that the common practice of expressing the stellar and planetary properties in units of the properties of the Sun, the Earth, or Jupiter inevitably leads to unnecessary systematic differences that are becoming apparent with the rapidly increasing accuracy of spectroscopic, photometric, and interferometric observations of stars and extrasolar planets[5], and

3. that the universal constant of gravitation G is currently one of the least precisely determined constants, whereas the error in the product $GM_{\odot}$ is five orders of magnitude smaller (Petit & Luzum 2010, and references therein),

Recommends

In all scientific publications in which **accurate** values of basic stellar or planetary properties are derived or quoted:

1. that whenever expressing stellar properties in units of the solar radius, total solar irradiance, solar luminosity, solar effective temperature, or solar mass parameter, that the nominal values $\mathcal{R}^{\mathrm{N}}_{\odot}$, $\mathcal{S}^{\mathrm{N}}_{\odot}$, $\mathcal{L}^{\mathrm{N}}_{\odot}$, $\mathcal{T}^{\mathrm{N}}_{\mathrm{eff}\odot}$, and $(\mathcal{GM})^{\mathrm{N}}_{\odot}$, be used, respectively, which are by definition *exact* and are expressed in SI units. These *nominal* values should be understood as conversion factors only — chosen to be close to the current commonly accepted estimates (see table below) — not as the true solar properties. Their consistent use in all relevant formulas and/or model calculations will guarantee a uniform conversion to SI units. Symbols such as $L_{\odot}$ and $R_{\odot}$, for example, should only be used to refer to actual estimates of the solar luminosity and solar radius (with uncertainties),

2. that the same be done for expressing planetary properties in units of the equatorial and polar radii of the Earth and Jupiter (i.e., adopting nominal values $\mathcal{R}^{\mathrm{N}}_{e\mathrm{E}}$, $\mathcal{R}^{\mathrm{N}}_{p\mathrm{E}}$,

$\mathcal{R}^{\rm N}_{e\rm J}$, and $\mathcal{R}^{\rm N}_{p\rm J}$, expressed in meters), and the nominal terrestrial and jovian mass parameters $(\mathcal{GM})^{\rm N}_{\rm E}$ and $(\mathcal{GM})^{\rm N}_{\rm J}$, respectively (expressed in units of $\rm m^3\,s^{-2}$). Symbols such as $\rm GM_E$, listed in the IAU 2009 system of astronomical constants (Luzum *et al.* 2011), should be used only to refer to actual estimates (with uncertainties),

3. that the IAU (2015) System of Nominal Solar and Planetary Conversion Constants be adopted as listed below:

SOLAR CONVERSION CONSTANTS		
$1\mathcal{R}^{\rm N}_{\odot}$	=	6.957×10^8 m
$1\mathcal{S}^{\rm N}_{\odot}$	=	$1361\,\rm W\,m^{-2}$
$1\mathcal{L}^{\rm N}_{\odot}$	=	3.828×10^{26} W
$1\mathcal{T}^{\rm N}_{\rm eff\odot}$	=	5772 K
$1(\mathcal{GM})^{\rm N}_{\odot}$	=	$1.327\,124\,4 \times 10^{20}\ \rm m^3 s^{-2}$

PLANETARY CONVERSION CONSTANTS		
$1\,\mathcal{R}^{\rm N}_{e\rm E}$	=	6.3781×10^6 m
$1\,\mathcal{R}^{\rm N}_{p\rm E}$	=	6.3568×10^6 m
$1\,\mathcal{R}^{\rm N}_{e\rm J}$	=	7.1492×10^7 m
$1\,\mathcal{R}^{\rm N}_{p\rm J}$	=	6.6854×10^7 m
$1\,(\mathcal{GM})^{\rm N}_{\rm E}$	=	$3.986\,004 \times 10^{14}\,\rm m^3\,s^{-2}$
$1\,(\mathcal{GM})^{\rm N}_{\rm J}$	=	$1.266\,865\,3 \times 10^{17}\,\rm m^3\,s^{-2}$

4. that an object's mass can be quoted in nominal solar masses $\mathcal{M}^{\rm N}_{\odot}$ by taking the ratio $(GM)_{\rm object}/(\mathcal{GM})^{\rm N}_{\odot}$, or in corresponding nominal jovian and terrestrial masses, $\mathcal{M}^{\rm N}_{\rm J}$ and $\mathcal{M}^{\rm N}_{\rm E}$, respectively, dividing by $(\mathcal{GM})^{\rm N}_{\rm J}$ and $(\mathcal{GM})^{\rm N}_{\rm E}$,

5. that if SI masses are explicitly needed, they should be expressed in terms of $(GM)_{\rm object}/G$, where the estimate of the Newtonian constant G should be specified in the publication (for example, the 2014 CODATA value is $G = 6.67408\,(\pm 0.00031) \times 10^{-11}\,\rm m^3\,kg^{-1}\,s^{-2}$),

6. that if nominal volumes are needed, that a nominal terrestrial volume be derived as $4\,\pi\,\mathcal{R}^{\rm N\ 2}_{e\rm E}\ \mathcal{R}^{\rm N}_{p\rm E}/3$, and nominal jovian volume as $4\,\pi\,\mathcal{R}^{\rm N\ 2}_{e\rm J}\ \mathcal{R}^{\rm N}_{p\rm J}/3$.

Explanation

1. The need for increased accuracy has led to a requirement to distinguish between Barycentric Coordinate Time (TCB) and Barycentric Dynamical Time (TDB). For this reason the *nominal solar mass parameter* $(\mathcal{GM})^{\rm N}_{\odot}$ value is adopted as an exact number, given with a precision within which its TCB and TDB values agree (Luzum *et al.* 2011). This precision is considered to be sufficient for most applications in stellar and exoplanetary research for the forseeable future.

2. The *nominal solar radius* $\mathcal{R}^{\rm N}_{\odot}$ corresponds to the solar photospheric radius suggested by Haberreiter *et al.* (2008)[6], who resolved the long-standing discrepancy between

the seismic and photospheric solar radii. This $\mathcal{R}^{\rm N}_{\odot}$ value is consistent with that adopted by Torres *et al.* (2010) in their recent compilation of updated radii of well observed eclipsing binary systems.

3. The *nominal total solar irradiance* $\mathcal{S}^{\rm N}_{\odot}$ corresponds to the mean total electromagnetic energy from the Sun, integrated over all wavelengths, incident per unit area per unit time at distance 1 au — also measured contemporarily as the *total solar irradiance* (TSI; e.g., Willson 1978) and known historically as the *solar constant* (Pouillet 1838). $\mathcal{S}^{\rm N}_{\odot}$ corresponds to the solar cycle 23-averaged TSI ($S_{\odot} = 1361\,(\pm\,1)$ W m^{-2}; 2σ uncertainty; Kopp *et al.*, in prep.)[7].

4. The *nominal solar luminosity* $\mathcal{L}^{\rm N}_{\odot}$ corresponds to the mean solar radiative luminosity rounded to an appropriate number of significant figures. The current (2015) best estimate of the mean solar luminosity $L_{\odot}$ was calculated using the solar cycle-averaged TSI[3] and the IAU 2012 definition of the astronomical unit[8].

5. The *nominal solar effective temperature* $\mathcal{T}^{\rm N}_{\rm eff\odot}$ corresponds to the effective temperature calculated using the current (2015) best estimates of the solar radiative luminosity and photospheric radius, and the CODATA 2014 value for the Stefan-Boltzmann constant[9], rounded to an appropriate number of significant figures.

6. The parameters $\mathcal{R}^{\rm N}_{e\rm E}$ and $\mathcal{R}^{\rm N}_{p\rm E}$ correspond respectively to the Earth's "zero tide" equatorial and polar radii as adopted following 2003 and 2010 IERS Conventions (McCarthy & Petit 2004; Petit & Luzum 2010), the IAU 2009 system of astronomical constants (Luzum *et al.* 2011), and the IAU Working Group on Cartographic Coordinates and Rotational Elements (Archinal *et al.* 2011). If equatorial vs. polar radius is not explicitly specified, it should be understood that *nominal terrestrial radius* refers specifically to $\mathcal{R}^{\rm N}_{e\rm E}$, following common usage.

7. The parameters $\mathcal{R}^{\rm N}_{e\rm J}$ and $\mathcal{R}^{\rm N}_{p\rm J}$ correspond respectively to the one-bar equatorial and polar radii of Jupiter adopted by the IAU Working Group on Cartographic Coordinates and Rotational Elements 2009 (Archinal *et al.* 2011). If equatorial vs. polar radius is not explicitly specified, it should be understood that *nominal jovian radius* refers specifically to $\mathcal{R}^{\rm N}_{e\rm J}$, following common usage.

8. The *nominal terrestrial mass parameter* $(\mathcal{GM})^{\rm N}_{\rm E}$ is adopted from the IAU 2009 system of astronomical constants (Luzum *et al.* 2011), but rounded to the precision within which its TCB and TDB values agree. The *nominal jovian mass parameter* $(\mathcal{GM})^{\rm N}_{\rm J}$ is calculated based on the mass parameter for the Jupiter system from the IAU 2009 system of astronomical constants (Luzum *et al.* 2011), subtracting off the contribution from the Galilean satellites (Jacobson *et al.* 2000). The quoted value is rounded to the precision within which the TCB and TDB values agree, and the uncertainties in the masses of the satellites are negligible.

9. The nominal value of a quantity Q can be transcribed in LaTeX with the help of the definitions listed below for use in the text and in equations:

```
\newcommand{\Qnom}{\hbox{$\mathcal{Q}^{\rm N}_{\odot}$}}
\newcommand{\Qn}{\mathcal Q^{\rm N}_{\odot}}
```

References

Archinal, B. A., A'Hearn, M. F., Bowell, E., *et al.* 2011, Celestial Mechanics and Dynamical Astronomy 109, 101

Haberreiter, M., Schmutz, W., Kosovichev, A. G. 2008, ApJ, 675, L53

Harmanec, P., Prša, A. 2011, PASP, 123, 976

Jacobson, R. A., Haw, R. J., McElrath, T. P., & Antreasian, P. G. 2000, J. Astronaut. Sci. 48(4), 495

Kopp, G. 2014, Journal of Space Weather and Space Climate, 4, A14

Kopp, G., Lawrence, G., Rottman, G., 2005, Solar Physics, 230, 129

Kopp, G., & Lean, J. L. 2011, Geophys. Res. Letters, 38, L01706

Luzum, B., Capitaine, N., Fienga, A., *et al.* 2011, Celestial Mechanics and Dynamical Astronomy, 110, 293

McCarthy, D. D. & Petit, G. 2004 IERS Technical Note No. 32, 1

Meftah, M., Irbah, A., Hauchecorne, A., *et al.* 2015, Solar Physics, 290, 673

Petit, G., Luzum, B. (Eds.) 2010 IERS Technical Note No. 36

Pouillet, C. S. M. 1838, *Memoire sur le chaleur solaire*, Paris, Bachelier

Prša, A. & Harmanec, P. 2012, Proc. IAU Symp. 282, Cambridge Univ., Press, 339

Schmutz W., Fehlmann A., Finsterle W., *et al.* 2013, AIP Conf. Proc. 1531, p. 624???627, doi:10.1063/1.4804847

Torres, G., Andersen, J., Giménez, A. 2010, A&A Rev., 18, 67

Willson, R. C. 2014, Astrophysics & Space Science, 352, 341

Willson, R. C. 1978, Journal of Geophysical Research, 83, 4003

Notes

[5] Note, e.g., that since projected rotational velocities of stars ($v \sin i$) are measured in SI units, the use of different values for the solar radius can lead to measurable differences in the rotational periods of giant stars (see Harmanec and Prša 2011).

[6] Haberreiter *et al.* (2008) determined the solar photospheric radius, defined to be where $\tau_{\rm Ross} = 2/3$, to be 695 658 ($\pm$ 140) km. The adopted $\mathcal{R}^{\rm N}_{\odot}$ is based on this value, quoting an appropriate number of significant figures given the uncertainty, and differs slightly from the nominal solar radius tentatively proposed by Harmanec & Prša (2011) and Prša & Harmanec (2012).

[7] The TSI is variable at the ~0.08% ($\sim$1 W m^{-2}) level and may be variable at slightly larger amplitudes over timescales of centuries. Modern spaceborne TSI instruments are absolutely calibrated at the 0.03% level (Kopp 2014). The TIM/SORCE experiment established a lower TSI value than previously reported based on the fully characterized TIM instrument (Kopp *et al.* 2005, Kopp & Lean 2011). This revised TSI scale was later confirmed by PREMOS/PICARD, the first spaceborne TSI radiometer that was irradiance-calibrated in vacuum at the TSI Radiometer Facility (TRF) with SI-traceability prior to launch (Schmutz *et al.* 2013). The DIARAD/PREMOS (Meftah *et al.* 2015), ACRIM3/ACRIMSat (Willson 2014), VIRGO/SoHO, and TCTE/STP-Sat3 (http://lasp.colorado.edu/home/tcte/) flight instruments are now consistent with this new TSI scale within instrument uncertainties, with the DIARAD, ACRIM3, and VIRGO having made post-launch corrections and the TCTE having been validated on the TRF prior to its 2013 launch. The Cycle 23 observations with these experiments are consistent with a mean TSI value of $S_{\odot} = 1361$ W m^{-2} (± 1 W m^{-2}; 2σ). The uncertainty range includes contributions from the absolute accuracies of the latest TSI instruments as well as uncertainties in assessing a secular trend in TSI over solar cycle 23 using older measurements.

[8] Resolution B2 of the XXVIII General Assembly of the IAU in 2012 defined the astronomical unit *to be a conventional unit of length equal to 149 597 870 700 m exactly.* Using the current best estimate of the TSI (discussed in endnote 3), this is consistent with a current best estimate of the Sun's mean radiative luminosity of $L_{\odot} = 4\,\pi\,(1\,\mathrm{au})^2\,S_{\odot} = 3.8275\,(\pm\,0.0014) \times 10^{26}\,\mathrm{W}$.

[9] The CODATA 2014 value for the Stefan-Boltzmann constant is $\sigma = 5.670\,367\,(\pm\,0.000\,013) \times 10^{-8}\,\mathrm{W\,m^{-2}\,K^{-4}}$. The current best estimate for the solar effective temperature is calculated to be $T_{\mathrm{eff},\odot} = 5772.0\,(\pm\,0.8)\,\mathrm{K}$.

RESOLUTION B4

on the Protection of Radio Astronomy Observations in the Frequency Range 76 - 81 GHz from Interference Caused by Automobile Radars.

Proposed by IAU Commission 40 (Radio Astronomy)

The XXIX General Assembly of the International Astronomical Union,

Recognizing

1. that the International Astronomical Union is a Sector Member of the Radiocommunication Sector of the International Telecommunication Union (ITU-R),

2. that radio astronomy observations are protected in their allocated bands from interference caused by active radio services by national regulations based on the Radio Regulations (RR) adopted by the International Telecommunication Union (ITU),

3. that the frequency ranges 76 - 77.5 GHz and 79 - 81 GHz are allocated to the radio astronomy service on a primary basis (Article 5 of the RR),

4. that Article 29.9 of the RR states that "In providing protection from interference to the radio astronomy service on a permanent or temporary basis, administrations shall use appropriate means such as geographical separation, site shielding, antenna directivity and the use of time-sharing and the minimum practicable transmitter power";

Considering

1. that radio astronomy observations consist of the reception of extremely weak signals from cosmic sources,

2. that radio astronomy receivers have exceptionally high sensitivity, which results in high susceptibility to interference caused by man-made radio signals,

3. that radio frequencies are a limited resource that should be shared,

4. that automobile manufacturers intend to utilize millimeter-wave radars operating in the frequency range 76 - 81 GHz for a number of purposes, that include the increasing of safety in driving,

5. that agenda item 1.18 of World Radiocommunication Conference 2015 (WRC-15) of the ITU calls for consideration of allocating the frequency range 77.5 - 78 GHz to radar applications worldwide, and that this allocation is expected to be applied worldwide in conjunction with existing allocations to radar applications in the frequency range 76 - 81 GHz,

6. that the ITU has not identified measures to protect radio astronomy observations in the frequency range 76- 81 GHz from interference caused by automobile radars.

Resolves

1. to request that WRC-15 take all possible steps to protect radio astronomy observations in the range 76 - 81 GHz from interference caused by automobile radars,

2. to express the view that the most effective protection of radio astronomy observations would be through geographical separation,

3. to send a copy of this resolution to administrations that operate or host radio astronomy observations in the frequency range 76 - 81 GHz, and where automobile radars are operating or plan to operate in the same frequency range,

4. to encourage astronomers, particularly those in countries that fall under Resolves 3, to work proactively in protecting radio astronomy observations in the frequency range 76 - 81 GHz.

3. Résolutions Approuvées

RESOLUTION B1

Sur le Plan Stratégique de l'UAI 2010-2020: Astronomie pour le Développement

Proposée par la Commission 25 de l'UAI

La XXIXe Assemblée Générale de l'Union Astronomique Internationale,

Reconnaissant

1. Que la XXVIIe Assemblée Générale, réunie à Rio de Janeiro (Brésil) le 13 août 2009, a adopté à l'unanimité une Résolution stipulant que l'UAI "approuve les buts mentionnés dans le Plan Stratégique "Astronomie pour le Développement" comme faisant partie de ses objectifs pour la décennie à venir",

2. Que pour atteindre ces objectifs l'UAI a créé un "Centre d'Astronomie pour le Développement" (Office of Astronomy for Development, OAD) au Cap (Afrique du Sud), en partenariat bilatéral entre l'UAI et le Fonds National de Recherche d'Afrique du Sud,

3. Que l'OAD a entrepris avec succès un programme international ambitieux pour répondre aux objectifs du Plan Stratégique de l'UAI,

4. Qu'un comité indépendant d'évaluation de l'OAD est arrivé à la conclusion que "son bilan est remarquable, en particulier compte tenu des ressources très limitées qui ont été mis à la disposition d'une organisation au cahier des charges si ambitieux",

Recommande

1. Que les objectifs du Plan Stratégique "Astronomie pour le Développement" soient poursuivis jusqu'à la XXXIe Assemblée Genérale de 2021,

2. Que le Comité Executif soumette à l'approbation de la XXXe Assemblée Générale, devant se tenir à Vienne (Autriche) en août 2018, une extension du Plan Stratégique précisant le futur de l'OAD et ses activités au-delà de 2021,

3. Que le Comité Exécutif consulte les parties prenantes existantes ou potentielles au cours de la préparation de cette extension du Plan Stratégique.

RESOLUTION B2

Sur la recommandation du "point zéro" des échelles de magnitude bolométrique absolue et apparente

Proposée par le Groupe de Travail Inter-Division A/G de l'UAI
"Unités Nominales pour l'Astronomie Stellaire et Planétaire"

La XXIXe Assemblée Générale de l'Union Astronomique Internationale,

Notant

1. L'absence d'une définition précise du "point zéro" des échelles de magnitude bolométrique absolue et apparente, qui s'est traduite dans la littérature par une prolifération de différents points zéro pour les magnitudes bolométriques et les corrections bolométriques (à un niveau atteignant le dixième de magnitude, voir par exemple Bessell, Castelli, & Plez 1998; Torres 2010);

2. Que les Commissions 25 et 36 ont approuvé des textes de résolution identiques pour la définition du point zéro de l'échelle de magnitude bolométrique (Andersen 1999), mais que cette résolution n'a pu se concrétiser au point de pouvoir être soumise à l'approbation de l'Assemblée Générale de l'UAI, et n'a été adoptée que ponctuellement par la communauté astronomique;

3. Que les mesures récentes de la constante solaire ont conduit à une révision de la luminosité solaire qui diffère légèrement de la valeur utilisée pour établir le point zéro de l'échelle de magnitude bolométrique, telle que préconisée par ces textes de résolution;

Considérant

1. La nécessité de définir une échelle standard pour les magnitudes bolométriques absolues et apparentes;

2. Qu'une multitude de points zéro pour les corrections bolométriques ont envahi la littérature en raison de l'absence d'accord sur un point zéro communément accepté pour l'échelle de magnitude bolométrique;

Recommnande

1. De définir the point zéro de l'échelle des *magnitudes bolométriques absolues*[10] en spécifiant qu'une source de rayonnement de magnitude bolométrique absolue $M_{Bol} =$ 0 mag émet une luminosité radiative exactement égale à:

$$L_o = 3.0128 \times 10^{28} \ W \quad (3.1)$$

et que la magnitude absolue bolométrique M_{bol} pour une source de luminosité L (in W) est donnéè par:

$$M_{bol} = -2.5\log(L/L_{\odot}) = -2.5\log(L) + 71.197\,425 \quad (3.2)$$

Le point zéro a été choisi de telle façon que la *luminosité solaire nominale*[11] ($L^N = 3.828 \times 10^{26} W$) corresponde presque exactement à la magnitude bolométrique absolue $M_{bol} = 4.74$ mag, valeur la plus couramment adoptée dans la littérature récente (voir par exemple, Bessell, Castelli, & Plez 1998; Cox 2000; Torres 2010);

2. De définir le point zéro de l'échelle des *magnitudes bolométriques apparentes* en spécifiant que $m_{Bol} = 0$ mag correspond à une *irradiance*, ou *densité de flux de chaleur*[12], égale à:

$$f_o = 2.518\,021\,002... \times 10^{-8}\ Wm^{-2} \quad (3.3)$$

et qu'en conséquence la magnitude bolométrique apparente $m_{Bol} = 0$ correspondant à une irradiance f est:

$$m_{bol} = -2.5\log(f/f_o) = -2.5\log f - 18.997\,351... \quad (3.4)$$

L'irradiance f_o correspond à celle mesurée en provenance d'une source de rayonnement isotrope et de magnitude absolue $M_{bol} = 0$ mag (luminosité L_o) à une distance standard[13] de 10 parsecs (sur la base de la définition de l'unité astronomique adoptée par l'UAI en 2012).
La valeur adoptée pour f_o est en accord avec certaines valeurs d'utilisation courante (par exemple, Lang 1974, Cox 2000), avec une différence inférieure à 0,1%. Sur la base de ce point zéro, la valeur de la constante solaire $S^N_{\odot}$ (irradiance égale à $1361 Wm^{-2}$) correspond à une magnitude bolométrique solaire apparente égale à $m_{bol\odot} \simeq -26.832$ mag.

Références

Andersen, J. 1999, Transactions of the International Astronomical Union, Series B, 23, pgs. 141 & 182

Bessell, M. S., Castelli, F., & Plez, B. 1998, Astronomy & Astrophysics, 333, 231

Bureau International des Poids et Mesures, 2006, The International System of Units (SI), 8th edition, Organisation Intergouvernementale de la Convention du Métre

Cox, A. N. 2000, Allen?s Astrophysical Quantities, 4th Edition

Dyson, F. W. 1913, MNRAS, 73, 334

Kopp, G. 2014, Journal of Space Weather and Space Climate, 4, A14

Kopp, G., Lawrence, G., Rottman, G., 2005, Solar Physics, 230, 129

Kopp, G., & Lean, J. L. 2011, Geophysical Research Letters, 38, L01706

Lang, K. R. 1974, Astrophysical Formulae, A Compendium for the Physicist and Astrophysicist, Springer-Verlag

Meftah, M., Irbah, A., Hauchecorne, A., *et al.* 2015, Solar Physics, 290, 673

Schmutz W., Fehlmann A., Finsterle W., *et al.* 2013, AIP Conf. Proc. 1531, p. 624627, doi:10.1063/1.4804847

Torres, G. 2010, Astronomical Journal, 140, 1158

Wilkins, G. A. 1989, "The IAU Style Manual (1989): The Preparation of Astronomical Papers and Reports"

Willson, R. C. 2014, Astrophysics & Space Science, 352, 341

Notes

[10]Les notations M_{bol} pour la *magnitude bolométrique absolue* et m_{bol} ont été adoptées par la Commission 3 (Notations) lors de la Ve Assemblée Générale de l'UAI à Stockholm en 1938, voir `https://www.UAI.org/static/resolutions/UAI1938 French.pdf`. Ici, M_{Bol} et m_{bol} se rapportent spécifiquement aux magnitudes bolométriques définies sur la base des points zéro de la présente Résolution.

[11]Les instruments spatiaux de mesure de l'irradiance solaire totale (ou constante solaire; Total Solar Irradiance, "TSI") font l'objet d'une calibration absolue au niveau de 0,03 % (Kopp 2014). L'expérience TIM/SORCE a fourni une valeur de la TSI inférieure à celle précédemment publiée sur la base de la caractérisation définitive de l'instrument TIM (Kopp *et al.* 2005, Kopp & Lean 2011). Cette nouvelle valeur de l'échelle TSI a été confirmée ultérieurement par l'expérience PREMOS/PICARD, le premier radiomètre TSI spatial calibré dans le vide au Laboratoire de Radiométrie TSI (TSI Radiometer Facility, TRF), incluant une traçabilité SI préalable au lancement (Schmutz *et al.* 2013). Les expériences DIARAD/PREMOS (Meftah *et al.* 2015), ACRIM3/ACRIMSat (Willson 2014), VIRGO/SOHO, and TCTE/STP-Sat3 (`http://lasp.colorado.edu/home/tcte/`) ont donné des résultats compatibles avec cette nouvelle échelle compte tenu des incertitudes instrumentales, DIARAD, ACRM3 et VIRGO ayant été l'objet de corrections après le lancement, et TCTE ayant été validé au TRF avant son lancement en 2013. Les observations effectuées au cours du Cycle 23 avec ces expériences donnent des valeurs pour la constante solaire (arrondies au nombre de chiffres significatifs appropriés) et une incertitude (2 σ) $S_{\odot} = 1361(\pm 1 W m^{-2}$. L'incertitude inclut les contributions sur la précision absolue des instruments TSI les plus récents, ainsi que les incertitudes sur l'extrapolation au Cycle 23 des valeurs trouvées par les instruments plus anciens. En combinant cette valeur de la constante solaire avec la définition de l'unité astronomique donnée par l'UAI en 2012, on trouve la meilleure valeur actuelle de la luminosité solaire moyenne, $L_{\odot} = 4\pi(1au)^2 S_{\odot} = 3.8275(\pm 0.0014) \times 10^{26} W$. Sur cette base, on adoptera la *luminosité solaire nominale* $\mathcal{L}_{\odot}^{N} = 3.828 \times 10^{26} W$

[12]The terms *irradiance* and *heat flux density* are used interchangeably, both with SI units of $W m^{-2}$(Wilkins 1989, Bureau International des Poids et Mesures 2006). See also https://www.iau.org/publications/proceedings rules/units/.

[13]The parsec is *the unit of distance... corresponding to a parallax of 1"* (Dyson 1913). One parsec is equivalent to 1 AU tan (1") = 3.085 677 581 × 10^{16} m.

RESOLUTION B3

Sur les valeurs recommandées de constantes de conversion pour une sélection de propriétés solaires et planétaires

Proposée par le Groupe de Travail Inter-Division A/G de l'UAI "Unités Nominales pour l'Astronomie Stellaire et Planétaire"

La XXIXe Assemblée Générale de l'Union Astronomique Internationale,

Reconnaissant

Que des valeurs notablement différentes de la masse, du rayon, de la luminosité, de la température effective, et de l'irradiance totale du Soleil (constante solaire), ainsi que des masses et des rayons de la Terre et de Jupiter, et de la constante de gravitation newtonienne G, sont utilisées pour exprimer et déduire des propriétés stellaires et planétaires fondamentales,

Notant

1. Que ni la masse du Soleil ni les masses des planètes ne sont constantes dans le temps, et que leur valeurs instantanées sont progressivement déterminées de plus en plus précisément grâce aux progrès des techniques observationnelles et des méthodes d'analyse des données, et

2. Que la pratique répandue d'exprimer les propriétés stellaires et planétaires en unités dérivées des propriétés du Soleil, de la Terre et de Jupiter, conduit inévitablement à d'inutiles différences systématiques qui se font jour au regard des progrès rapides en précision des observations spectroscopiques, photométriques, et interférométriques d'étoiles et de planètes extrasolaires,[14] et

3. Que la constante universelle de la gravitation G est une des constantes les moins précisément déterminées, alors que l'erreur sur le produit GM est plus petite par cinq ordres de grandeur (Petit & Luzum 2010, et références incluses),

Recommande

Dans toutes les publications scientifiques dans lesquelles des valeurs **précises** de propriétés stellaires ou planétaires de base sont déterminées ou citées:

1. Que, lorsqu'il s'agit d'exprimer des propriétés stellaires en unités de rayon, d'irradiance, de luminosité, de température effective, ou de masse solaire, on doit utiliser leurs valeurs *nominales*, notées respectivement $\mathcal{R}^{\mathrm{N}}_{\odot}$, $\mathcal{S}^{\mathrm{N}}_{\odot}$, $\mathcal{L}^{\mathrm{N}}_{\odot}$, $\mathcal{T}^{\mathrm{N}}_{\mathrm{eff}\odot}$, et $(\mathcal{GM})^{\mathrm{N}}_{\odot}$, qui sont par définition exactes et exprimées en unités SI. Il faut entendre ces valeurs nominales uniquement comme des facteurs de conversion, choisis pour être les plus proches des meilleures déterminations disponibles (voir le tableau ci-dessous), et non comme les valeurs solaires définitives. Leur emploi systématique dans toutes les formules pertinentes et/ou dans les calculs de modèles est le garant d'une conversion uniforme en unités SI. Les symboles $L_{\odot}$ and $R_{\odot}$, par exemple, ne doivent être utilisés qu'en référence aux estimations existantes de la luminosité

solaire et du rayon solaire, y compris leurs incertitudes;

2. Que la même disposition soit prise pour exprimer les propriétés planétaires en unités des rayons équatoriaux et polaires de la Terre et de Jupiter (c'est-à-dire en adoptant les valeurs nominales notées $\mathcal{R}^{\rm N}_{e\rm E}$, $\mathcal{R}^{\rm N}_{p\rm E}$, $\mathcal{R}^{\rm N}_{e\rm J}$, and $\mathcal{R}^{\rm N}_{p\rm J}$, exprimés en mètres), et les paramètres de masse terrestre et jovien $(\mathcal{GM})^{\rm N}_{\rm E}$ and $(\mathcal{GM})^{\rm N}_{\rm J}$,respectivement (exprimés en $\rm m^3\,s^{-2}$). Les symboles tels que $\rm GM_E$, employés dans le système de constantes astronomiques IAU 2009 (Luzum *et al.* 2011) ne doivent être utilisés qu'en référence aux estimations existantes, y compris leurs incertitudes;

3. Que le Système des Constantes de Conversion Nominales Solaire et Planétaire soit adopté, comme indiqué dans les Tables ci-dessous:

SOLAR CONVERSION CONSTANTS		
$1\mathcal{R}^{\rm N}_\odot$	=	6.957×10^8 m
$1\mathcal{S}^{\rm N}_\odot$	=	$1361\,\rm W\,m^{-2}$
$1\mathcal{L}^{\rm N}_\odot$	=	3.828×10^{26} W
$1\mathcal{T}^{\rm N}_{\rm eff\odot}$	=	5772 K
$1(\mathcal{GM})^{\rm N}_\odot$	=	$1.327\,124\,4 \times 10^{20}\ \rm m^3 s^{-2}$

PLANETARY CONVERSION CONSTANTS		
$1\,\mathcal{R}^{\rm N}_{e\rm E}$	=	6.3781×10^6 m
$1\,\mathcal{R}^{\rm N}_{p\rm E}$	=	6.3568×10^6 m
$1\,\mathcal{R}^{\rm N}_{e\rm J}$	=	7.1492×10^7 m
$1\,\mathcal{R}^{\rm N}_{p\rm J}$	=	6.6854×10^7 m
$1\,(\mathcal{GM})^{\rm N}_{\rm E}$	=	$3.986\,004 \times 10^{14}\,\rm m^3\,s^{-2}$
$1\,(\mathcal{GM})^{\rm N}_{\rm J}$	=	$1.266\,865\,3 \times 10^{17}\,\rm m^3\,s^{-2}$

4. Que la masse d'un objet soit donnée en unités de masse solaire nominale $\mathcal{M}^{\rm N}_\odot$, en prenant le rapport $(GM)_{\rm object}/(\mathcal{GM})^{\rm N}_\odot$, ou en unités correspondantes jovienne et terrestre, $\mathcal{M}^{\rm N}_{\rm J}$ and $\mathcal{M}^{\rm N}_{\rm E}$, respectivement, en divisant par $(\mathcal{GM})^{\rm N}_{\rm J}$ et $(\mathcal{GM})^{\rm N}_{\rm E}$. Si les masses doivent être données en unités SI, elles doivent être exprimées en termes du rapport $(GM)_{\rm object}/G$, où la valeur de la constante newtonienne G doit être explicitement spécifiée dans la publication (par exemple, la valeur CODATA 2014 est $G = 6.67408\,(\pm 0.00031) \times 10^{-11}\,\rm m^3\,kg^{-1}\,s^{-2}$),

5. Que dans la mesure où des volumes nominaux sont requis, le volume terrestre nominal soit donné par $4\,\pi\,\mathcal{R}^{\rm N}_{e\rm E}{}^2\,\mathcal{R}^{\rm N}_{p\rm E}/3$, et le volume jovien nominal par $4\,\pi\,\mathcal{R}^{\rm N}_{e\rm J}{}^2\,\mathcal{R}^{\rm N}_{p\rm J}/3$.

Explications

1. La nécessité d'une précision accrue a conduit à introduire une distinction entre le Temps Barycentrique Coordonné (TBC), et le Temps Barycentrique Dynamique (TBD). C'est pour cette raison que la valeur du *pamamètre de masse solaire nominal* $(\mathcal{GM})^{\rm N}_\odot$ est adoptée comme nombre exact, à la précision assurant que

TBC et TBD ont des valeurs identiques (Luzum *et al.* 2011). Cette précision est considérée comme suffisante pour la plupart des applications en recherche stellaire et exoplanétaire actuellement envisageables.

2. Le *rayon solaire nominal* $\mathcal{R}^{\rm N}_{\odot}$ correspond au rayon solaire photométrique mesuré par Haberreiter *et al.* (2008),[15], qui ont résolu le désaccord subsistant depuis longtemps entre les rayons solaires sismique et photosphérique. Cette valeur de $\mathcal{R}^{\rm N}_{\odot}$ est en accord avec celle adoptée par Torres *et al.* (2010) dans leur récente compilation des mises à jour des rayons de systèmes binaires bien étudiés.

3. L'*irradiance solaire totale nominale* $\mathcal{S}^{\rm N}_{\odot}$ correspond à l'énergie électromagnétique totale moyenne du Soleil, intégrée sur toutes les longueurs d'onde, par unité de surface et par unité de temps, arrivant à une distance de 1 au — également mesurée de nos jours comme *l'irradiance solaire totale*("TSI": Total Solar Irradiance; e.g., Willson 1978) et connue historiquement sous le nom de *constante solaire* (Pouillet 1838). $\mathcal{S}^{\rm N}_{\odot}$ correspond à la TSI moyennée sur le cycle solaire 23 ($S_{\odot} = 1361\,(\pm 1)$ W m^{-2}; incertitude de 2σ ; Kopp *et al.*, in prep.)[16].

4. La *luminosité solaire nominale* $\mathcal{L}^{\rm N}_{\odot}$ correspond à la luminosité radiative solaire moyenne, arrondie au nombre de chiffres significatifs adéquat. La meilleure estimation actuelle (2015) de la luminosité solaire moyenne $L_{\odot}$ a été calculée à partir de la TSI moyennée sur les cycles solaires, et de la définition de l'unité astronomique donnée par l'UAI en 2012.[17].

5. La *température effective solaire nominale* $\mathcal{T}^{\rm N}_{\rm eff\odot}$ correspond à la température effective calculée à partir de la meilleure estimation actuelle (2015) de la luminosité radiative et du rayon photosphérique solaires, et de la valeur CODATA de 2014 pour la constante de Stefan-Boltzmann[18], arrondie au nombre de chiffres significatifs adéquat.

6. Les paramètres $\mathcal{R}^{\rm N}_{e\rm E}$ and $\mathcal{R}^{\rm N}_{p\rm E}$ correspondent respectivement aux rayons équatorial et polaire terrestres "zéro marée", tels qu'adoptés d'après les conventions IERS 2003 et 2010 (McCarthy & Petit 2004; Petit & Luzum 2010) et le système IAU 2009 des constantes astronomiques (Luzum *et al.* 2011), et adoptés par le Groupe de Travail de l'UAI "Cartographic Coordinates and Rotational Elements" (Archinal *et al.* 2011). Si le rayon équatorial et le rayon polaire ne sont pas explicitement spécifiés, il faut comprendre que le *rayon terrestre nominal* se rapporte suivant l'usage à $\mathcal{R}^{\rm N}_{e\rm E}$.

7. Les paramètres $\mathcal{R}^{\rm N}_{e\rm J}$ and $\mathcal{R}^{\rm N}_{p\rm J}$ correspondent respectivement aux rayons équatorial et polaire joviens au niveau de pression de 1 bar adoptés par le Groupe de Travail de l'UAI "Cartographic Coordinates and Rotational Elements" (Archinal *et al.* 2011). Si le rayon équatorial et le rayon polaire ne sont pas explicitement spécifiés, il faut comprendre que le *rayon jovien nominal* se rapporte suivant l'usage à $\mathcal{R}^{\rm N}_{e\rm J}$.

8. Le *paramètre de masse terrestre nominal* $(\mathcal{GM})^{\rm N}_{\rm E}$ est adopté d'après le système IAU 2009 des constantes astronomiques (Luzum *et al.* 2011), mais arrondi à la précision pour laquelle les valeurs de TCB et TCD sont en accord. Le *paramètre de masse jovienne nominal* $(\mathcal{GM})^{\rm N}_{\rm J}$ est calculé sur la base du paramètre de masse pour le système de Jupiter d'après le système IAU 2009 des constantes astronomiques

(Luzum *et al.* 2011), après soustraction de la contribution des satellites galiléens (Jacobson *et al.* 2000). La valeur obtenue est arrondie à la précision pour laquelle les valeurs de TCB et TCD sont en accord, et les incertitudes sur les masses des satellites sont considérées comme négligeables.

9. La valeur nominale d'une quantité Q peut être transcrite en LaTeX à l'aide des définitions ci-après dans les textes et les équations:

```
\newcommand{\Qnom}{\hbox{$\mathcal{Q}^{\rm N}_{\odot}$}}
\newcommand{\Qn}{\mathcal Q^{\rm N}_{\odot}}
```

Références

Archinal, B. A., A'Hearn, M. F., Bowell, E., *et al.* 2011, Celestial Mechanics and Dynamical Astronomy 109, 101

Haberreiter, M., Schmutz, W., Kosovichev, A. G. 2008, ApJ, 675, L53

Harmanec, P., Prša, A. 2011, PASP, 123, 976

Jacobson, R. A., Haw, R. J., McElrath, T. P., & Antreasian, P. G. 2000, J. Astronaut. Sci. 48(4), 495

Kopp, G. 2014, Journal of Space Weather and Space Climate, 4, A14

Kopp, G., Lawrence, G., Rottman, G., 2005, Solar Physics, 230, 129

Kopp, G., & Lean, J. L. 2011, Geophys. Res. Letters, 38, L01706

Luzum, B., Capitaine, N., Fienga, A., *et al.* 2011, Celestial Mechanics and Dynamical Astronomy, 110, 293

McCarthy, D. D. & Petit, G. 2004 IERS Technical Note No. 32, 1

Meftah, M., Irbah, A., Hauchecorne, A., *et al.* 2015, Solar Physics, 290, 673

Petit, G., Luzum, B. (Eds.) 2010 IERS Technical Note No. 36

Pouillet, C. S. M. 1838, *Memoire sur le chaleur solaire*, Paris, Bachelier

Prša, A. & Harmanec, P. 2012, Proc. IAU Symp. 282, Cambridge Univ., Press, 339

Schmutz W., Fehlmann A., Finsterle W., *et al.* 2013, AIP Conf. Proc. 1531, p. 624???627, doi:10.1063/1.4804847

Torres, G., Andersen, J., Giménez, A. 2010, A&A Rev., 18, 67

Willson, R. C. 2014, Astrophysics & Space Science, 352, 341

Willson, R. C. 1978, Journal of Geophysical Research, 83, 4003

Notes

[14]Il est à noter, par exemple, que puisque les vitesses de rotation projetées des étoiles ($v \sin i$) sont mesurées en unités SI, l'usage de valeurs différentes pour le rayon solaire peut conduire à des différences mesurables des périodes de rotation des étoiles géantes (voir Harmanec & Prša 2011).

[15] Haberreiter *et al.* (2008) ont mesuré pour le rayon photosphérique solaire une valeur de 695 658 ($\pm$ 140) km. La valeur adoptée $\mathcal{R}^{\rm N}_{\odot}$ est basée sur cette quantité, avec le nombre de chiffres significatifs approprié compte tenu de cette incertitude, et diffère légèrement de la valeur du rayon solaire nominal suggérée par Harmanec & Prša (2011) et Prša & Harmanec (2012).

[16] La TSI est variable au niveau de $\sim$0.08% ($\sim$1 W m^{-2}) et il est possible qu'elle soit variable

avec une amplitude légèrement plus grande au fil des siècles. Les instruments spatiaux mesurant la TSI ont une calibration absolue au niveau de 0.03% (Kopp 2014). L'expérience TIM/SORCE a obtenu une valeur de TSI inférieure à celle mesurée précédemment sur la base de l'instrument complètement caractérisé TIM (Kopp *et al.* 2005, Kopp & Lean 2011). Cette échelle de TSI révisée a été confirmée par la suite par PREMOS/PICARD, le premier radiomètre TSI spatial calibré dans le vide au Laboratoire de Radiométrie TSI (TSI Radiometer Facility, TRF), incluant une traçabilité SI préalable au lancement (Schmutz *et al.* 2013). Les expériences DIARAD/PREMOS (Meftah *et al.* 2015), ACRIM3/ACRIMSat (Willson 2014), VIRGO/SoHO, et TCTE/STP-Sat3 (http://lasp.colorado.edu/home/tcte/) ont donné des résultats compatibles avec cette nouvelle échelle compte tenu des incertitudes instrumentales, DIARAD, ACRM3 et VIRGO ayant été l'objet de corrections après le lancement, et TCTE ayant été validé au TRF avant son lancement en 2013. Les observations effectuées au cours du Cycle 23 avec ces expériences donnent des valeurs pour la constante solaire (arrondies au nombre de chiffres significatifs appropriés) et une incertitude (2σ) de: $S_\odot = 1361\ \mathrm{W\,m^{-2}}$ ($\pm 1\ \mathrm{W\,m^{-2}}$;. L'incertitude inclut les contributions sur la précision absolue des instruments TSI les plus récents, ainsi que les incertitudes sur l'extrapolation au Cycle 23 des valeurs trouvées par les instruments plus anciens.

[17] La Résolution B2 adoptée par la XXVIIIe Assemblée Générale en 2012 a défini l'unité astronomique (au) par convention comme *une unité de longueur exactement égale à 149 597 870 700 m*. La meilleure estimation actuelle de la TSI (voir la Note 3 ci-dessus) est alors en accord avec la meilleure estimation actuelle de la luminosité radiative solaire moyenne, $L_\odot = 4\pi\,(1\,\mathrm{au})^2\, S_\odot = 3.8275\,(\pm\,0.0014) \times 10^{26}\ \mathrm{W}$.

[18] La valeur CODATA 2014 de l a constante de Stefan-Boltzmann est $\sigma = 5.670\,367\,(\pm\,0.000\,013) \times 10^{-8}\ \mathrm{W\,m^{-2}\,K^{-4}}$. Le calcul de la meilleure estimation actuelle de la température effective solaire donne $T_{\mathrm{eff},\odot} = 5772.0\,(\pm\,0.8)\,\mathrm{K}$.

RÉSOLUTION B4

Protection des observations radioastronomiques dans l'intervalle de fréquences 76-81 GHz contre les brouillages causés par les radars automobiles

Proposée par la Commission 40 de l'UAI "Radioastronomie"

La XXIXe Assemblée Générale de l'Union Astronomique Internationale,

Reconnaissant

1. Que l'Union Astronomique Internationale est un membre sectoriel du Secteur "Radiocommunications" de l'Union Internationale des Télécommunications (UITR);

2. Que les observations radioastronomiques bénéficient d'une protection des législations nationales, dans les bandes qui leur sont allouées, contre les brouillages causés par les services utilisant les ondes radioélectriques, sur la base du Règlement des Radiocommunications (RR) de l'Union Internationale des Télécommunications (UIT);

3. Que les intervalles de fréquence 76-77,5 GHz et 79-81 GHz sont alloués au service de radioastronomie, qui est considéré comme un "service primaire" (RR, Article 5 Section II);

4. Que l'Article 29.9 (Section III) du RR stipule que "Lorsqu'elles assurent la protection du service de radioastronomie contre les brouillages à titre permanent ou temporaire, les administrations utilisent, selon le cas, des moyens tels que la séparation géographique, l'effet d'écran du terrain, la directivité de l'antenne, l'utilisation du partage dans le temps et de la plus faible puissance d'émission pratiquement réalisable."

Considérant

1. Que les observations radioastronomiques consistent en l'observation de signaux extrêmement faibles en provenance de sources cosmiques;

2. Que les récepteurs radioastronomiques sont d'une sensibilité exceptionnellement élevée, ce qui les rend hautement susceptibles d'être affectés par des brouillages en provenance de signaux anthropiques;

3. Que les radiofréquences sont une ressource limitée qui doit être partagée;

4. Que les fabricants automobiles projettent d'utiliser des radars dans le domaine des ondes millimétriques dans l'intervalle de fréquences 76-81 GHz en vue d'un certain nombre d'utilisations, y compris l'accroissement de la sécurité routière;

5. Que le point 1.18 inscrit à l'ordre du jour de la Conférence sur les Radiocommunications Mondiales 2015 (World Radiocommunication Conference WRC-15) de l'UIT envisage l'allocation de l'intervalle de fréquences 77,5-78 GHz aux applications radar dans le monde entier, cette allocation devant y être appliquée conjointement avec les allocations de fréquences radar existantes dans l'intervalle 76-81 GHz;

6. Que l'UIT n'a pas prévu de mesures pour protéger les observations radioastronomiques dans l'intervalle 76-81 GHz contre les brouillages causés par les radars automobiles;

Décide

1. De demander à ce que la Conférence WRC-15 prenne toutes les mesures en son pouvoir pour assurer la protection des observations radioastronomiques dans l'intervalle 76-81 GHz contre les brouillages causés par les radars automobiles;

2. De formuler un avis selon lequel la protection la plus efficace pour les observations radioastronomiques est la séparation géographique;

3. D'envoyer un exemplaire de la présente Résolution aux Administrations qui sont en charge ou qui abritent des observations radioastronomiques dans l'intervalle 76-81 GHz, là où les radars automobiles fonctionnent ou envisagent de fonctionner dans l'intervalle 76-81 GHz;

4. D'encourager les astronomes, et plus particulièrement ceux travaillant dans les pays concernés par la Décision 3 ci-dessus, à concourir activement à protéger les observations radio-astronomiques dans l'intervalle 76-81 GHz.

Transactions IAU, Volume XXIXB
Proceedings IAU Symposium No. Volume XXIXB, 2018
P. Benvenuti, ed.
© International Astronomical Union 2019
doi:10.1017/S174392131800399X

CHAPTER V

REPORT OF THE EXECUTIVE COMMITTEE

1. Introduction

The three years that have elapsed since the Beijing GA have been extremely rich and varied, and have seen major evolutions in the history of the IAU.

The actions of the Executive Committee can be highlighted in three categories:

- **Actions towards the community:**
  - Follow-up of the Division restructuring approved at the Beijing GA in 2012: reforming the Commissions;
  - Reorganization of the scientific meetings at GAs, with the merging of "Special Sessions" and "Joint Discussions" into "Focus Meetings", and with the introduction of "Division Meetings";
  - Frequent use of electronic voting (signing-up for Commissions, elections, etc.) after improving the reliability of the membership database;
  - Complete reorganization of the web site;
  - Start of the digitization of the IAU Archives (publications and documents)
- **Global actions towards Education and Development:**
  - Consolidation of the Office of Astronomy for Development (OAD) in Cape Town;
  - Consolidation of the Office for Astronomy Outreach (OAO) in Tokyo;
  - Creation of a new "Office for Young Astronomers" (OYA) in Oslo.
- **Actions towards the public:**
  - Participation in public naming of solar-system satellites and planetary features;
  - Worldwide contest for public naming of exoplanetary systems and their host stars ("NameExoWorlds");
  - Organizing a well-identified IAU "cornerstone" of the UNESCO 2015 "International Year of Light & Light Technologies" ("Cosmic Light").

These actions, and other EC activities, are described in the following Sections.

2. Executive Committee

2.1. *Executive Committee Members*

The composition of the Executive Committee (2012-2015), as approved by the General Assembly in Beijing, is as follows:

Officers

President	Norio	Kaifu	(Japan)
President-Elect	Silvia	Torres-Peimbert	(Mexico)
General Secretary	Thierry	Montmerle	(France)
Assistant General Secretary	Piero	Benvenuti	(Italy)

Vice-Presidents (First term)

Renée	Kraan-Korteweg	(South Africa)
Xiaowei	Liu	(China Nanjing)
Dina	Prialnik	(Israel)

Vice-Presidents (Second term)

Matthew	Colless	(Australia)
Jan	Palous	(Czech Republic)
Marta	Rovira	(Argentina)

Advisors

Robert	Williams	(past President, USA)
Ian	Corbett	(past General Secretary, UK)

To this list was added George Miley (the Netherlands), as Adviser on Astronomy for Development.

2.2. *Executive Committee meetings*

The Executive Committee met on five occasions:

EC92 (Beijing, XXVIII General Assembly): Sep.1, 2012
EC93 (Nara, hosted by N. Kaifu): May 9-11, 2013
EC94 (Canberra, hosted by M. Colless): Apr.30-May 2, 2014
EC95 (Padova, hosted by P. Benvenuti): Apr.15-17, 2015
EC96 (Honolulu, XXVIII General Assembly: Aug. 2 and Aug. 7, 2015

In addition, Officers' Meetings were held in Paris:

OM2013: Feb. 7-8, 2013
OM2014: Feb. 13-14, 2014
OM2015: Jan.15-16, 2015

This last meeting allowed the Officers to attend the Opening Ceremony of the International Year of Light & Light Technologies at UNESCO on Jan. 19-20 (see Section 7).

The IAU cocktail, traditionally held at Paris Observatory during the OM, is the opportunity to thank the Institut d'Astrophysique de Paris (IAP) and the Paris Observatory for their hospitality and assistance during the year.

Abridged minutes of the 2013 meeting have been published in the Information Bulletin, issues IB111 and IB112 (2013). (See also Section 12.)

2.3. *EC Working Groups*

During this triennium, the EC Working Groups were as follows:

"Continuing" Working Groups

- IAU General Assemblies (Chair: Gang Zhao): http://www.iau.org/science/scientific_bodies/working_groups/126/

- Women in Astronomy (Chair: Francesca Primas): http://www.iau.org/science/scientific_bodies/working_groups/122/

"Ad hoc" Working Groups

- Public Naming of Planets and Planetary Satellites (Chair: Thierry Montmerle): http://www.iau.org/science/scientific_bodies/working_groups/209/

- International Year of Light: Cosmic Light (Chair: Richard F. Green): http://www.iau.org/science/scientific_bodies/working_groups/212/

These two WGs are temporary, and are discussed in Section 7.

3. Secretariat

3.1. *Location and Staff*

The IAU Secretariat is located on the second floor of the "Institut d'Astrophysique de Paris" (IAP), 98bis, Bd Arago, 75014 Paris. A copper plaque has been installed near the street entrance of the IAP, courtesy of the new IAP Director, F. Bernardeau, to help visitors and publicize the IAU. The premises (three office rooms and an archive storage room) are rented under contract with the French CNRS (owner of the IAP building), which we thank for advantageous conditions, that include IAP help to maintain the Secretariat hardware and software systems. The current staff is composed of:

- Muriel Besson, Head of Administration (full time)
- Madeleine Smith-Spanier, Assistant and Database Manager (full time)
- Ginette Rude, Archivist (0.4 FTE, or two days/week).

M. Besson and M. Smith-Spanier each occupy an office room, while G. Rude has a desk in the Archive storage room. The General Secretary has an office room large enough for small meetings (4-5 persons). Note that there is no extra space available in the IAU "corner".

3.2. *History*

The current Secretariat staff listed above is entirely different from the Secretariat of the previous triennium, except for Archivist Ginette Rude (see below). It has been stabilised for less than a year, after a complicated history. This is only the third Secretariat in over two decades; M. Orine had been the IAU Secretary for about 20 years until she passed away suddenly in 2008, then Vivien Reuter took over the position (in March 2008) now known as "Head of Administration", with Jana Zilova recruited as "Database Manager" in July 2009 (hence just before the Rio 2009 GA). The "Database" in question is the "membership database", where the complete profiles of the IAU members (institutions, postal and email addresses, personal data, IAU body affiliations, etc.) are kept, the physical maintenance of the data being outsourced to ESO (see below). This task has now become a crucial activity to keep the contacts and circulate information between individual members, the Union as a whole, and the Executive (in particular because of the widespread use of electronic voting this triennium (see Section 4).

Then Jana Zilova went on sick leave just before the 2012 Beijing GA, leaving V. Reuter and GS I. Corbett handle a very difficult situation. Eventually J. Zilova didn't return to the IAU Office after Beijing, and for legal reasons her contract was terminated in March 2013. In parallel, V. Reuter had already announced before Beijing that she would leave the IAU towards the end of 2012, and a procedure was conducted that eventually led to the recruitment before Beijing of the current Head of Administration, Muriel Besson (actual start of contract in October 8, 2012), with V. Reuter training M. Besson until she left the IAU (end of January 2013).

On the other hand, finding a replacement for J. Zilova was not easy, but eventually a new Database Manager, Julia Nauer, was hired in January 2013. Unfortunately, after only a few months of excellent work, Julia decided to return to her native Switzerland, ending her contract in August 2013. Then a friend of hers, Madeleine Smith-Spanier, a US citizen living in Paris, agreed to succeed her - and again a period of a few weeks was necessary to train her to the job - and to the IAU as a whole. But the situation was unstable again, in that Madeleine was on a 1-yr, non-extendable, US-France Young Workers visa, with a contract starting in October 2013. For a permanent-track job allowing her to continue, it was necessary, according to the French law, to make the position public, and have Madeleine compete with other candidates. The procedure was run by M. Besson, and among the applying candidates, it was clear that Madeleine was the best qualified. So Madeleine, after having had to return temporarily to the US for working visa reasons, is now on a permanent position (French "indefinite duration contract", with a 10-yr working visa) since Oct. 20, 2014.

To summarize, over a period of three years since the pre-Beijing period, the Secretariat has had two successive Heads of Administration, and three different "Database managers".

Actually, the situation could have been worse, because Ginette Rude, a retired former assistant to the Paris Observatory President, who was hired previously to be a part-time "Archivist" for the IAU publications and internal documents (on a permanent contract since January 2009), had announced that she would probably leave the Secretariat after Beijing. Fortunately, she became very interested the "IAU Archive" digitizing project with Cambridge University Press (see Section 12), and decided to stay. An Archivist-like profile is clearly needed at the Secretariat, especially with the 100th Anniversary of the IAU in sight (2019 - celebrated at the 2018 GA in Vienna); it will be very difficult to replace Ginette when she eventually decides to leave for good.

3.3. *Internal Tasks*

The tasks of the Secretariat staff are briefly described in what follows. Note however that the staff is organized as a team, with constant communication so that some tasks may be shared or temporarily transferred in case of need.

Head of Administration

Day-to-day accounting (income: dues from National Members or "funds-in-transit", travel and other expenses, grants, etc.). Contacts with National Members and various administrations (French taxes, social security, etc.); contracts (staff hiring; photocopier, phone, etc.). Supervising of the Secretariat staff. Contacts with the bank (HSBC). Note that almost all expenditures are now done via electronic transfer, and each transfer is authorized individually by the GS. Very few bank checks are signed (by the GS) and each is photocopied. The HoA may however authorize transfers or sign checks when the GS is unavailable. The bank accounts can be monitored electronically at any time, with restricted access.

Assistant & Database Manager
Contacts with individual members. Updating the membership database, in collaboration with ESO (see next section), queries to National members and National Committees for Astronomy (NCA), etc. Organizing electronic consultations and elections. Help in the CUP editorial process ("ScholarOne" online submission of proceedings; see Section 12). Also, answering simple queries from the public (addressed to the "iaupublic@iap.fr" mailbox), or transferring them to competent astronomers (GS, Minor Planet Center, WGSBN, etc.).

Archivist
IAU and other publications: maintaining the IAU library, follow-up of scientific meetings (copyright forms, , etc.). Preparing and/or compiling documents to be digitized by CUP (see Section 12). Liaison with other libraries (observatories; Académie des Sciences, etc.). Storing and classification of "physical" IAU documents (EC meetings, etc.).

3.4. *Outsourced Tasks*

Essentially all high-level server tasks are outsourced under contract with ESO: general IAU web site maintenance and development, Press Office (web Announcements), maintenance of National and Individual membership and IAU bodies databases (Divisions, Commissions and Working Groups). This collaboration continues to prove extremely fruitful and dynamic developments and decisions are generally implemented almost in real time.

Another outsourced task in the present triennium has been professional accounting, consistent with the IAU legal status (non-profit organization based in France). (See Section 14)

4. Electronic Voting and Membership Database

4.1. *Electronic Voting*

The possibility for the whole IAU membership to vote electronically has been introduced during the previous triennium, with a subscription to the UK-based voting platform "Mi-Voice".

Any vote or consultation takes place as follows:

- the relevant mass-mailing list is created (from a limited group, like Commission members, to the whole IAU membership or nearly 10,000 "active" members);
- a message with the appropriate instructions is sent, including a "unique voting code" (the equivalent of a "polling card"), giving access to the ballot (list of options or slate of candidates);
- at the end of the vote, the results are compiled automatically, and then announced manually using certain criteria (e.g., publishing the names of the elected candidates, but not of the defeated ones).

Note that organizing a vote requires coordination between three different actors:

- Mi-Voice in Southampton (UK) for attributing a list of randomly generated voting codes (up to 10,000), then recording and sorting the votes;
- The IAU Secretariat in Paris to attribute a voting code to each voting member (i.e., to each email address);
- ESO in Garching (Germany) to establish an up-to-date customized mailing list and send the messages with voting codes inserted.

During the present triennium, this voting platform has been used extensively, and very successfully, for various issues related to the follow-up of the Division restructuring (see Section 6):

- Joining Divisions (1-9) (winter 2012-2013)
- Establishing Division Steering Committees, with 6 "at-large" members (Spring 2013)
- Expression of interest for proposed new Commissions (Winter 2013-2014)
- Sign-up for the selected new Commissions (May 2014)
- Voting for the new Division Steering Committees: Division Presidents and Vice-Presidents, and replacement of 3 at-large candidates (June 2014)
- Establishing the Organizing Committees of the selected new Commissions (June-July 2014).

As an example of the effort made by the Secretariat and actors mentioned above, the Division and Commission elections were run in parallel in a time span of two months. There were 102 candidates for the 9 Divisions, and 289 for the 35 Commissions (in all 391 candidates). This translated into 23271 unique voting codes being sent to the 9245 "active members" (see below), for the 44 quasi-simultaneous elections, when taking into account the multiple individual affiliations to Divisions (1 to 9) and Commissions (1 to 3).

4.2. *Membership Database*

A key to the validity of the electronic voting is to have a reliable membership database, in other words of "IAU citizens" having the ability to vote.

An extensive search has been conducted by the Secretariat to locate "missing members" (typically, no or invalid email addresses) by contacting the NCAs, the National Members, Commissions and Divisions, even Google. Over the past triennium we have recovered almost 500 members.

Especially in view of the electronic voting, we have defined as "active members" those filling two conditions: (i) to have a valid email address, (ii) to be affiliated to at least one Division.

For each National Member, the "IAU Directory":
(`http://www.iau.org/administration/membership/national/`) now lists "active members" (active link to their full profile), and "inactive members" (in black) - meaning that not enough information is available (this may include undeclared deceased members, etc.). With these definitions, the IAU Directory now lists 9245 "active members" (i.e., having the availability to vote), and 11277 members in total. The lists evolve in real time as new information is communicated to the Secretariat (iauinfos@iap.fr).

5. Membership of the Union

5.1. *New National Members*

Only one country applied for National Membership in 2015: Colombia (Category I). This was a strong application, with 25 Individual Members proposed for admission by the Membership Committee. The application followed a number of highly visible initiatives (proposed LARIM in 2016, organization of IAU Symposium 327 on "Fine Structure and Dynamics of the Solar Atmosphere", approval of an "Andean OAD Regional Node"; see Section 8), well coordinated by the various institutions of Colombia (Observatorio Nacional, Universidad de los Andes, etc.).

This brings the total number of IAU National Members to 74 (see `http://www.iau.org/administration/membership/national/`).

5.2. *Individual Members*

The Individual Members of the IAU are recorded in the "IAU Member Directory", which is now updated in real time as new information comes in (change of address, decessions, etc.). The information is normally provided by the NCAs (or equivalent if there is none, especially in small countries), with which the Secretariat is frequently in contact. As explained in Section 4, for each country the Directory now shows a list of members, differentiated according to their status:

- "active members", having a valid email address and affiliated to at least one Division: these are equivalently the "voting members" to whom mass emails are being sent on various occasions, either globally (messages from the Executive, e-Newsletters, etc.), or partially (messages from Divisions or Commissions); the names are linked to their profiles (underlined in blue);
- "inactive members", having some missing information (in general an invalid email address or not being affiliated to a Division in spite of requests; this may also include deceased members not notified to the Secretariat). Note that as soon as the correct information is received by the Secretariat, the "inactive" status can be immediately restored to "active".

At the time of writing, there are currently 9245 active members, out of a total of 11276 members in the IAU database. (See `http://www.iau.org/administration/membership/individual/`)

In Honolulu, 1189 new Individual Members will be admitted, as compared with 1008 in Beijing and 887 in Rio. The number of deceased members reported at Honolulu will be close to 400 (including yet unrecorded previous members).

6. Divisions, Commissions and Working Groups

6.1. *New Divisional structure*

Following the adoption of Resolution B5 at the Beijing GA concerning the new Divisional structure proposed by the EC, the Division Presidents and the GS, on behalf of the EC, started to work together on solidifying the Division Steering Committees (by organizing elections), and on preparing a deep reform of the Commissions, itself having an impact on existing Working Groups. In particular, the DPs were invited at all the EC meetings since 2013.

The list of the new Divisions (with their final names) is as follows:

Division A Fundamental Astronomy
Division B Facilities, Technologies and Data Science
Division C Education, Outreach and Heritage
Division D High Energy Phenomena and Fundamental Physics
Division E Sun and Heliosphere
Division F Planetary Systems and Bioastronomy
Division G Stars and Stellar Physics
Division H Interstellar Matter and Local Universe
Division J Galaxies and Cosmology

In Beijing, the GA approved the list of Division Presidents and Vice-Presidents presented by the EC. Being therefore appointed, and not elected, and together, the "continuity tradition", according to which Vice-Presidents would normally succeed the Presidents, was broken for the first time in the history of the IAU. This created a "virgin ground" to complete the Division restructuring by establishing the "Division Steering Committees" (replacing the "Organizing Committees"), with the main task of implemented the second recommendation of Resolution B5, i.e., to undertake a reform of the Commissions.

As a result, the DPs, with the approval of the EC, "invented" a new composition for the Division Steering Committees (DSC), by: (i) including ex-officio the Presidents of the existing Commissions (which was crucial given that the Commissions were to be reformed), and (ii) adding 6 elected "at-large" members, i.e., members drawn from their respective communities. Therefore, the very first step was to establish these communities themselves by asking all IAU members to sign-up for any number of new Divisions (i.e., from 1 to 9).

This original sign-up process (the main goal of which was thus to establish the "electorate" of each Division), launched in December 2012 (see `http://www.iau.org/news/announcements/detail/ann12026/`) lasted for three months, and turned out to be the first of a long series of electronic consultations using the Mi-Voice platform.

On average, about 25% of the IAU members chose "voluntarily" to be members of the new Divisions. Then what about the remaining 75% who didn't sign up? At the time, all IAU members had to belong to at least one Commission, which was one of the conditions for being accepted as a new individual member. Since each Commission was assigned to a Division, the non-voting members were assigned the Division(s) corresponding to the Commission(s) they were members of. The Division "constituencies" being established in this way, elections of the at-large DSC members could be organized. Potential candidates were thus encouraged to apply, after consultation with Commission Presidents, and the Mi-Voice poll set-up by the Secretariat. (The current Commission affiliation for each Division can be found via the corresponding links of the IAU web page, e.g., Division A: `http://www.iau.org/science/scientific_bodies/divisions/A/commissions/`)

Therefore, a few weeks after the start of the current triennium, the DSCs and their Presidents were ready to start working on the Commission reform.

6.2. *Commission Reform*

The main milestones of the Commission reform were planned and articulated around the EC Meetings, where the DPs were invited to present their conclusions to the EC members.

The reform was thus broken down in three phases:

1) Nara (EC93): "Commission Reset"

Following discussions between the GS and DPs, the DPs proposed to the EC to reset the Commissions, i.e., to terminate all existing Commissions at the Honolulu GA, and organize a "Call for proposals for new Commissions". It was expected, in particular, that Commissions would be created on science areas currently absent from the present list of Commissions.

In the subsequent months, many discussions took place between DPs (with feedback from the DSCs) and the GS to reach an appropriate definition of Commissions adapted to each specific subject areas covered by the Divisions, and to set up the Call itself. An inspiration was the Call for Proposals for Symposia, which features a first deadline for submitting Letters of Intent, and a second deadline for submitting Full Proposals.

2) Canberra (EC94): Call for Proposals

Guidelines for submitting proposals, proposal forms, etc. were presented to the EC. A major topic of discussion between DPs had been to promote multidisciplinarity, i.e., Commissions supported by more than one Division. (See `http://www.iau.org/news/announcements/detail/ann14014/`)

Three categories of Commissions were therefore defined:

- "Regular" Commissions (supported by one Division only);
- "Inter-Division" Commissions (supported by two or more Divisions, but featuring a "parent" Division, with the Commission President an ex-officio member of its DSC);
- "Cross-Division" Commissions (supported by two or more Divisions, but this time on equal grounds, i.e., with one Commission OC member being appointed ex-officio member of the DSC of each Division).

A timeline was agreed to implement the Call for Proposals for Commissions, with a deadline of Oct. 15 for the Letter of Intent, and Jan. 31 for the Full Proposals, so that the community would have ample time to reflect on the proposed Commissions, interact with the proposers and exchange opinions with the Executive on a dedicated web Forum (which turned out to be used rather sparingly), and for the proposers to possibly merge proposals on neighbouring topics (which has indeed been the case). The timeline also included milestones for organizing OC elections for the new Commissions, with the goal that the Commissions and their OC would be approved by the EC at the Honolulu GA. (See `http://www.iau.org/administration/events/past/`)

In all, 48 Letters of Intent were received by the Secretariat (see `http://www.iau.org/news/announcements/detail/ann14029/`). The submission indicated that the Commission reform was on the right track: in round numbers, compared with the existing 40 Commissions, about a third of the existing Commissions would not be re-proposed, a third would be re-proposed with sometimes significant changes, and a third were proposed on entirely new topics (such as Gravitational Waves, Astrostatistics, Computational Astrophysics, etc.).

During the 3.5 months of web posting of the Letters of Intent, electronic "Expressions of interest" were organized on three occasions, as a preview of the potential interest of the community for the proposed Commissions (see the last poll: `http://www.iau.org/news/announcements/detail/ann15002/`)

Then the Full Proposals entered a "recommendation phase", in which the Division Presidents would recommend a number of new Commissions for approval by the EC.

3) Padua (EC95): Commission approval and OC elections

In total, 47 proposals were reviewed (one was not confirmed), this time both, and independently, by the Division Presidents (committee chaired by the GS), and by the EC Vice-Presidents (committee chaired by the President-Elect). Then the reviews were combined, and individual proposals re-discussed if necessary.

As a result of this "dual selection", a list of 35 Commissions was submitted to the EC for approval, with a complementary list of proposals recommended for temporary "Division Working Groups" in the next triennium.

These approved "post-GA Commissions" (i.e., to take effect only at the Honolulu GA) can be found at: `http://www.iau.org/science/scientific_bodies/postga_commissions/`.

6.3. *Commissions and Divisions Elections*

6.3.1. *Commissions Elections*

Once the new Commissions were approved by the EC, to be established at the GA, it was necessary to organize the sign-up of their members, followed by the elections of their OC. Given the exceptional nature of the exercise (including for Commissions that never existed before), the DPs and EC endorsed the following election procedure:

a) The total number of position was adjusted to the size of the Commissions membership (see `http://www.iau.org/news/announcements/detail/ann15013/`), in line with the Working Rules: if N is the number of members, and P the number of positions: $P = 6$ for $N < 99$, $P=7$ for $100 < N < 199$, and $P = 8$ for $N > 200$.

b) The proposers of approved Commissions would not stand for election, but be ex-officio members of the OC, the lead proposer being future President by default (or else another proposer);

c) Separate elections would be held for the position on Commission Vice-President (normally future President);

d) "Open" elections would be held, i.e., self-appointed at-large candidates could stand for OC elections, including for the position of VP;

e) No candidate could stand for more than one position;

f) In order to encourage candidates for the VP position (the Working Rules state that there should be at least two candidates), the decision was made to opt for a "soft" VP election, in which the VP candidates would also stand for regular member position, so that if defeated as VP they would have a second chance to be elected as regular member.

The voters were asked to vote for at least one candidate (or abstain), up to a number defined by the number of positions available (different in each Commission depending on how many proposers would stay: some left to be elected elsewhere, or resigned, etc.)

The candidates were asked to write a statement and provide a photo; the Secretariat added a link to the IAU profile in the member Directory, for each of the 289 candidates.

In spite of the complexity of the poll, the elections worked extremely well. The Secretariat was able to handle the 35 elections successfully, which was quite an achievement. The results were published on July 15, in conformity with the timeline that was posted on the web site. (See `http://www.iau.org/news/announcements/detail/ann15024/`)

6.3.2. *Divisions Elections*

However, in parallel with the Commissions elections, the Divisions had to organize their own elections: Presidents and Vice-Presidents at the same time, an unprecedented situation (since the current ones were appointed together in Beijing), and half (3) of the previously elected at-large DSC members (which implied that the other half was requested to step down).

Here also an "open" vote was held, but with rules different than for the Commissions:

a) The elections for President and Vice-President were "hard": the defeated candidates would simply lose the corresponding election;
b) Inasmuch as possible, there should be at least two P and two VP candidates (there were exceptions, discussed and agreed by all DPs and endorsed by the EC);
c) Only three at-large positions were offered to the vote, a simple feature compared to the Commissions elections;
d) Here also, no candidate could stand for more than one election;
e) This rule was also extended to the Commissions elections ("one person-one function" rule).

Here also, the candidates were asked to write a statement and provide a photo; the Secretariat added a link to the IAU profile in the member Directory, for the 102 candidates.

The main difficulty was to manage the 9 simultaneous Divisions elections in parallel with the 35 Commissions elections.

The process worked extremely well also (and was much simpler than for Commissions). The results of the Divisions DSC elections were posted on July 31, as planned. (See `http://www.iau.org/news/announcements/detail/ann15021/`.)

6.4. *Working Groups*

Normally, Working Groups have to be approved at each GA, to continue existing (if they ask for it).

Since the current Commissions will be terminated at Honolulu, their WG will also be terminated. Following the Bye-Laws, new WG will have to be created, or re-created, by the relevant Commissions after Honolulu, with the approval of the relevant DSCs.

Division WGs are not concerned by the Commission reform, but they will be reviewed too for renewal, since some of them may now fall scientifically under the purview of new Commissions.

7. EC Actions towards the Public

7.1. *Creation of "ad hoc" EC Working Groups*

During this triennium, the EC was directly engaged into two highly visible actions towards the public, which were conducted under the purview of two EC Working Groups:

a) inviting the public to participate in the naming of solar-system planetary satellites and surface features, and of exoplanetary systems: WG "Public Naming of Planets and Planetary Satellites", chaired by the GS (http://www.iau.org/science/scientific_bodies/working_groups/209/)

b) participate in the UNESCO 2015"International Year of Light & Light Technologies" ("IYL2015" for short): WG "Cosmic Light", chaired by R. Green, with P. Benvenuti as EC representative (http://www.iau.org/science/scientific_bodies/working_groups/212/).

7.2. *WG "Public Naming of Planets and Planetary Satellites"*

The major actors of this Working Group are the "Working Group on Planetary Surface Nomenclature" (WGPSN, Division F; chaired by R. Schulz), and the "Working Group on Small Bodies Nomenclature" (WGSBN, Division F, Secretary G. Williams). Their decade-long experience in dealing with the public (e.g., for proposing asteroid names), and with space agencies (ESA, NASA) has been invaluable.

The actions of this WG have been twofold:

a) Solar-system objects: The year 2015 is exceptional for the space exploration of the solar system. Aside from missions ongoing for several years (Mars rovers and orbiters, Cassini), a number of highlights have been achieved (apart from the ESA Rosetta-Philae mission, which doesn't involve name approval by the IAU):

(1) Participate in the naming of craters on Mercury (the last discovered before the end of the Messenger mission); (http://www.iau.org/news/pressreleases/detail/iau1506/)

(2) Similarly, participate in the naming of topological features on Ceres as a dwarf planet (Dawn mission)

(3) The public naming of Pluto's surface features (fly-by of the New Horizons mission on July 14) is however still under debate at the time of writing (see http://www.iau.org/news/pressreleases/detail/iau1502/)

Previously, in collaboration with NASA, the IAU had sanctioned in 2013 the naming of Pluto's four latest, HST-discovered satellites (naming criteria and official approval of names: http://www.iau.org/news/pressreleases/detail/iau1303/).

b) Public Naming of Exoplanets and their Host Stars: A contest was organized to engage the public in naming exoplanetary systems and their host stars (nicknamed "ExoWorlds"). This contest, named "NameExoWorlds", is the first to be initiated by the IAU from the start. The goal being to submit naming proposals to the vote of the largest possible public worldwide, a cooperation with the well-known "citizen-science" organisation Zooniverse was set up. The contest was originally organized in several phases:

(1) Selection by the IAU of a set of 305 exoplanets offered for naming (making up 260 exoplanetary systems with 1 to 5 exoplanets), having reliable data (typically monitored for more than 5 years); (`http://www.iau.org/news/pressreleases/detail/iau1404/`)
(2) In parallel, creation of an "IAU Directory for World Astronomy", open to clubs and associations interested in astronomy ("clubs" for short), to record registrations to the contest; (`http://www.iau.org/news/pressreleases/detail/iau1406/`)
(3) Voting by the registered clubs to select 20 "most popular" ExoWorlds to be named (`http://www.iau.org/news/pressreleases/detail/iau1505/`);
(4) Submitting naming proposals (one per club per ExoWorld) to the IAU;
(5) Posting of the proposals on the web and having the public worldwide vote for the best proposals;
(6) Announcing the "winners" and the winning names for each of the 20 ExoWorlds at a special ceremony during the GA in Honolulu (after due consideration to cultural and geographical balance).

At the time of writing, 601 clubs are registered with the Directory, and 225 naming proposals have been received. Unfortunately, due to unexpected difficulties with Zooniverse, phase (5) had to be delayed, and will be the one announced at the GA, as an official "Public Event" (`http://www.iau.org/news/pressreleases/detail/iau1507/`). At the time of writing, how to announce the winning names (Phase (6)) is still under discussion, in coordination with the WG.

7.3. *WG "Cosmic Light"*

The name of the WG is that of a "cornerstone" of large projects submitted to the IYL2015 Organizing Committee (`http://www.iau.org/iyl/`). It was created by the EC at the beginning of 2014, and quickly capitalized on the experience gained by the IAU during IYA2009. The priorities of the IAU were soon established, as:

a) "the Right to Starlight" (implementation of Resolution B5 approved at the Rio 2009GA): dark skies and astronomical sites protection (both professional and public)
b) light from the Universe, in particular contribute to celebrating of the 100th anniversary of Einstein's Theory of Relativity

The WG issued a Call for Proposals for world-class projects covering item (a), with a special seed funding of 30 kE (`http://www.iau.org/iyl/proposals/`). The selected projects are:

- Cosmic Light Awareness (`http://www.iau.org/iyl/cornerstones/cosmiclightawareness/`
- Galileoscope (`http://www.iau.org/iyl/cornerstones/galileoscope/`)
- Light: Beyond the Bulb (`http://www.iau.org/iyl/cornerstones/lightbeyondthebulb/`)

7.4. *Outlook for the future*

Normally these two WG, which are time-limited, will be dissolved in early 2016, after due reporting and review. Perhaps some links should be maintained between the EC and the

WGPSN and WGSPN, and the WG "Public Naming of Planets and Planetary Satellites" kept operating in the next triennium after the end of the NameExoWorld contest.

8. Executive Summary of the OAD Activities in the triennium

The activities of the Office of Astronomy for Development were reviewed by the Executive Committee in all its annual meetings. In particular, the overall performance of the OAD spanning the period 2010-2014 was analyzed in depth together with the Report prepared by the ad hoc Review panel (see doc EC97-8-b OAD Review report-final) during the EC95 (Padova, April 15-17, 2015).

Therefore this document will only highlights in a concise form the main achievements of the OAD in the past triennium and concentrate on the negotiations between IAU and NRF that produced the revised version of the IAU-NRF Agreement on OAD, to be approved by the EC during the EC96-1 session.

For a more extended and comprehensive description of the OAD activities in the period 2010-2014, please refer to the Document OAD Self Evaluation which was prepared by the Director of OAD, Kevin Govender, for the Review Panel.

The IAU Office of Astronomy for Development was established in March 2011 as a joint venture between the IAU and the NRF to implement the IAU decadal Strategic Plan (SP), Astronomy for Development (AfD): 2010-2020.

Despite its small size (3 FTE) the OAD has been successful in implementing key elements of the SP during the first 3 years of its existence. Major accomplishments, as envisaged in the SP include:

- Establishment of its infrastructure and tools to communicate with an associated large global network of committed participants
- Establishment of 3 AfD Task Forces (TFs) covering (i) Universities and Research, (ii) Children and Schools and (iii) Public Outreach
- Development of a dedicated web portal, including the on-line submission procedures for the annual call for projects (`http://www.astro4dev.org`)
- Establishment of 4 OAD Regional Nodes (East Asia, South East Asia, East Africa, Southern Africa), with 2 additional ones approved more recently (West Africa, Andean South America) and others in the approval or planning stage (Portuguese Language Expertise Centre, Jordan ROAD, Armenia ROAD) .
- Recruitment of over 550 OAD volunteers
- Stewardship of the IAU development and education activities with the institution of an annual (highly oversubscribed) global call for proposals in each of the 3 Task Force sectors and coordination and management of the selection procedures
- Management and monitoring of contracts, projects and financial reporting for 42 projects selected by the TFs and approved by the IAU Extended Development Oversight Committee over the first 2 years.

During its first 3 years, the accomplishments and visibility of the OAD have brought substantial benefits and visibility both to the IAU and to South Africa for a tiny increment of the national and international expenditures on astronomy and space sciences.

Some important ambitions of the decadal Strategic Plan and the OAD have not been realised during the first triennium due to a lack of human resources. The most obvious of these is fund raising and another is the organisation of flagship "Astronomy for Development" projects.

9. Educational and Outreach activities: "IAU Offices"

In addition to the "Office of Astronomy for Development" (OAD) in Cape Town (see Section 8), the IAU has created an "Office for Astronomy Outreach" in Tokyo (not its original name), and this year a third one, the "Office for Young Astronomers" in Oslo. These initiatives demonstrate the continuous will of the IAU to be "global", and not restricted to its Secretariat in Paris.

9.1. *Office for Astronomy Outreach ("OAO": Tokyo)*

The present OAO was actually founded during the previous triennium (2009-2012), in the aftermath of the 2009 International Year of Astronomy, with the creation of the position of "Public Outreach Coordinator"(POC) at NOAJ (Tokyo), in charge to "resurrect" the network of "National Outreach Contacts" set up during IYA2009, and also to be the Editor of the "CAP Journal" (see Section 10). The position was filled by Sarah Reed just before the Beijing GA. Unfortunately, due to health reasons, Sarah could not stay in Japan and had to return to the UK. This happened in Spring 2013, just before EC93 in Nara. Given the projects initiated towards the public by the EC (see Section 7), it was decided to offer a position again, this time perhaps more ambitiously, since the position was to be the head of an "Office of Astronomy Outreach" (with a one-person staff to begin with), as "International Outreach Coordinator" (IOC). A call was issued internationally, which was quite successful since nearly 30 applications were received.

The selected candidate was Sze-leung Cheung, a Chinese citizen from Hong-Kong, who had in particular an excellent record of initiatives taken during IYA2009. Sze-leung was hired in early 2014, and presented to the EC at EC94 in Canberra. Sze-leung is now deeply involved in the "NameExoWorld" and "Cosmic Light" projects, and regularly publishes the "Astronomy Outreach Newsletter (latest Issue: `http://www.iau.org/news/announcements/detail/ann15022/`).

This time the OAO was on a sound basis, and the EC approved the hiring of a second person to the OAO, with the position of "International Outreach Assistant" (IOA). A new call was issued, with the full participation of Sze-leung. Again the call was very successful, and after selection and negotiation within a short list of three high-level candidates, a young lady from Portugal, Lina Canas, was hired just before EC95 in Padua, and presented to the EC via telecon.

At the time of writing, the OAO is very dynamic and in excellent shape, ready for new projects for the next triennium.

The OAO is co-funded by NAOJ and JAXA (Japan), Academia Sinica (China Taipeh), NOAC (China Nanjing), NARIT (Thailand), IIA (India), and KASI (Korea).

9.2. *Office fro Young Astronomers ("OYA": Oslo)*

This is a newly created structure, approved by the EC at EC95 in Padova, and recently signed by the IAU and the Norwegian Academy of Sciences and Letters (NASL). It is based on the existing "International School for Young Astronomers" (ISYA), which have been existing for many years, and have been funded since 2006 by NASL as a result of a first Agreement set up by past GS Oddbjorn Engvold.

The new Agreement, called "Office for Young Astronomers", is an enhancement of the previous one, allowing for development not only of ISYAs (with the establishment of a joint IAU-NASL Steering Committee), but providing a flexible framework for future projects.

The text of the Agreement, although including no staffing, was strongly inspired by the existing Agreements for the OAD and OAO, and this logically refers now to an "Office for Young Astronomers" in Oslo, home to the NASL.

10. CAP Journal

CAPjournal (`http://www.capjournal.org`), is published by the IAU Division C Commission 55, "Communicating Astronomy with the Public" (CAP). It is an open-access, peer-reviewed journal for astronomy communicators that was launched in 2007. The current publication frequency is three issues per year. 2100 copies printed copied of each issue are distributed, with several thousand additional copies distributed as PDFs. The publication of the CAPjournal is a core function of C55.

During the change of leadership of the Office of Astronomy Outreach from Sarah Reed to Sze-leung Cheung (Section 9), it was decided to change the Editor in Chief, CAPjournal to Georgia Bladon on a contractual basis. She has done a sterling job of continuing the publication. The IAU funds the printing costs and the cost of the Editor in Chief. ESO funds the distribution, the proofreading and other miscellaneous costs.

The last issue Issue 17 was published in June 2015. The next issues are planned for September 2015 and December 2015. The editors are working on a special issue about the exciting science communication done around the Rosetta mission, but it is not clear if the team can commit to this due to the ongoing mission itself.

11. Scientific meetings

11.1. *List of Meeetings (2012-2015)*

As usual, during the present triennium the IAU organized 9 Symposia each year, and 3 Regional IAU Meetings ("RIMs": APRIM, Asia-Pacific; LARIM: Latin America; MEARIM: Middle-East and Africa).

However, at the Beijing GA 8 Symposia were organized (compared to the usual 6), because of the high quality of the proposals and the impact of their topics on the general program.

The list of IAU Symposia (all published by Cambridge University Press, see Section 12) is as follows (more details can be found at `http://www.iau.org/publications/iau/symposia/`; Symposia without Editors have not been published yet):

2012:

IAUS 287 **Cosmic Masers - from OH to Ho**
Stellenbosch, South Africa, 29 Jan -3 Feb 2012
Eds. R.S. Booth, E.M.L. Humphreys, W.H.T. Vilemmings

IAUS 288 **Astrophysics from Antarctica**
Beijing, China, August 20-24, 2012
Eds. Michael Burton, Xiangqun Cui, and Nicholas Tothill

IAUS 289 **Advancing the Physics of Cosmic Distances**
Beijing, China, August 20-24, 2012
Ed. Richard de Grijs

IAUS 290 **Feeding Compact Objects: Accretion on All Scales Beijing**
China, August 20-24, 2012
Eds. Chenmin Zhang, Tomaso Belloni, Mariano Méndez, Shuangnan Zhang

IAUS 291 **Neutron Stars and Pulsars: Challenges and Opportunities after 80 years**
Beijing, China, August 20-24, 2012
Ed. Joeri van Leeuwen

IAUS 292 **Molecular Gas, Dust, and Star Formation**
Beijing, China, August 20-24, 2012
Eds. Tony Wong, Jürgen Ott

IAUS 293 **Formation, Detection, and Characterization of Extrasolar Habitable Planets**
Beijing, China, August 20-24, 2012
Ed. Nader Haghighipour

IAUS 294 **Solar and Astrophysical Dynamos and Magnetic Activity**
Beijing, China, August 20-24, 2012
Eds. Alexander G. Kosovichev, Yihua Yan, Elisabete de Gouveia Dal Pino, Lidia van Driel- Gesztelyi

IAUS 295 **The intriguing life of massive galaxies**
Beijing, China, August 20-24, 2012
Eds. Daniel Thomas, Anna Pasquali, Ignacio Ferreras

2013:

IAUS 296 **Supernova Environmental Impacts**
Kolcata, India, January 2013
Eds. Alak Ray, Richard A. McCray

IAUS 297 **The Diffuse Interstellar Bands**
Noordwijkerhout, The Netherlands, May 2013
Eds. Jan Cami, Nick L. J. Cox

IAUS 298 **Setting the scene for Gaia and LAMOST**
Lijiang, China, May 2013
Eds. Sofia Feltzing, Gang Zhao, Nicholas A. Walton, Patricia A. Whitelock

IAUS 299 **Exploring the Formation and Evolution of Planetary Systems**
Victoria, Canada, June 2013
Eds. Mark Booth, Brenda C. Matthews, James R. Graham

IAUS 300 **Nature of Prominences and their role in Space Weather**
Paris, France, June 2013
Eds. Brigitte Schmieder, Jean-Marie Malherbe, S. T. Wu

IAUS 301 **Precision Asteroseismology**
Wroclaw, Poland, August 2013
Eds. Joyce A. Guzik, William J. Chaplin, Gerald Handler, Andrzej Pigulski

IAUS 302 **Magnetic Fields throughout Stellar Evolution**
Biarritz, France, August 2013
Eds. Pascal Petit , Moira Jardine and Hendrik C. Spruit

IAUS 303 **The Galactic Center: Feeding and Feedback in a Normal Galactic Nucleus**
Santa Fe, USA, October 2013
Eds. Lorant Sjouwerman, Cornelia Lang, Jürgen Ott

IAUS 304 **Multiwavelength AGN Surveys and Studies**
Byurakan, Armenia, October 2013
Eds. Areg M. Mickaelian, David B. Sanders

2014:

IAUS 305 **Polarimetry: The Sun to Stars and Stellar Environments**
Punta Leona, Costa Rica, December 2014

IAUS 306 **Statistical challenges in 21st Century Cosmology (SCCC 21)**
Lisbon, Portugal, May 2014

IAUS 307 **New windows on massive stars: asteroseismology, interferometry, and spectropolarimetry**
Geneva, June 2014
Eds. Georges Meynet , Cyril Georgy , José Groh and Philippe Stee

IAUS 308 **The Zeldovich Universe: Genesis and Growth of the Cosmic Web**
Tallinn, Estonia, June 2014

IAUS 309 **Galaxies in 3D across the Universe**
Vienna, Austria, July 2014

IAUS 310 **Complex planetary systems Namur**
Belgium, July 2014

IAUS 311 **Galaxies Masses as Constraints of Formation Models**
Oxford, United Kingdom, July 2014

IAUS 312 **Star Clusters and Black Holes in Galaxies across Cosmic Time**
Beijing, China Nanjing, August 2014

IAUS 313 **Extragalactic Jets from every angle**
Puerto Ayora, Galapagos Islands, Ecuador, September 2014

There was no Symposium organized in 2015 before the GA. In addition, three Regional Meetings were organized (not published by the IAU):

LARIM 2013 XIV Latin American Regional IAU Meeting
Florianópolis, SC, Brazil, November 2013
Contact: Zulema Abraham

APRIM 2014 **12th Asia-Pacific IAU Regional Meeting**
Daejeon, Korea, Rep of., August 2013
Contact: Young Chol Minh

MEARIM 2014 **Third Middle-East and Africa IAU Regional Meeting**
Beirut, Lebanon, September 2014
Contact: Roger Hajjar

11.2. *Changes introduced for General Assembly Meetings*

At the EC93 meeting in Nara, important changes were approved by the EC and Division Presidents, concerning the scientific meetings at the General Assembly:

(1) Merging of the "Special Sessions" and "Joint Discussions" into single, medium- length topical meetings, named "Focus Meetings" (1-2.5 days)
(2) Creation of "Division Meetings", to highlight the scientific activities of Divisions independently of the topics of Symposia and Focus Meetings; for some Divisions this would also include business sessions (9 parallel sessions of 2 days).

Note that Commissions and Working Groups may also organize their own business sessions during a GA. Given the Commission reform undertaken during the present triennium (see Section 6), it will be important to review the respective roles of Division Meetings and business sessions of their Commissions and Working Groups.

12. IAU web site, Publications & Archives

12.1. *IAU web site*

The IAU web site, as a means of communication with the member community and with the public, has been completely redesigned by the Press Office (Press Officer: Lars Christensen; webmaster: Raquel Yumi Shida, both at ESO), and presented and approved by the Executive Committee at EC93 in Nara. (See `http://www.iau.org/news/announcements/detail/ann14012/`)

The old web site is now completely forgotten, and has given way to a modern, dynamic presentation with catch-eye images, "Featured News" carousel, etc. New pull-down menus allow quick access to internal information (meetings, committees, Divisions, Commissions, etc.). "Private" information, i.e., reserved for IAU members (such as statements of election candidates, access to proposals, etc.), are protected by a specific login (IAU membership code) and personal password.

The IAU home page (iau.org) now has several features, designed to enhance and clarify communication:

- Top: "Featured News", designed as a carousel to attract attention on the last five important news: these can be "Announcements" (i.e., news from the General Assembly, new Commissions, etc.), or "Press Releases" (sent to Press Offices around the world, thus more oriented towards the public, such as a public naming campaigns, prize announcements, etc.).
- Below: the latest Press Releases and Announcements, with access to the whole list.

To date, since the publication of the new web site, 72 Announcements have been published, as well as 16 Press Releases, since EC93.

Then in the bottom half of the page are regular news of general interest and deadlines (generally administrative), news of planetary surface feature names, and standing "themes" for the public (Selling Star Names, Pluto, How to become an Astronomer, The Constellations, etc.).

In addition, when a Press Release is issued, a recap "e-Newsletter" (sometimes featuring a message from the GS if appropriate) is sent, hence on an irregular basis, to all the "active" members of the IAU Member Directory (see Section 5).

To the right is a column listing easy-to-reach standing information (General Assembly, OAD, OAO, past publications, etc.).

12.2. *Publications*

12.2.1. *Contract with Cambridge University Press*

The Publisher for all IAU publications is Cambridge University Press (CUP). The current contract has been renewed until the end of 2018, i.e., the end of the year of the GA in Vienna. Therefore, a continuation or termination has to be decided at the 2018 EC meeting preceding the GA.

The present status is that the collaboration with CUP has been truly exceptional, in the best "academic spirit" (CUP is actually a non-profit company), demonstrating a lot of competence and flexibility, and it is hard to imagine a reason why the current contract should not be renewed in 2018.

12.2.2. *Symposia*

The "IAU Symposia Series" is the flagship of all IAU publications, bringing in fact a significant revenue to the budget of the Union. The debate "paper vs. electronic" is now obsolete, as no more than 10 % of the Proceedings are now printed on paper. Most of the revenue comes from subscriptions of libraries - most of which now lack shelf space to store all the books and publications.

Technically, the Symposia Series are handled like a journal (issues are numbered, and there must be in principle 9 issues per year). A new system is being tested (called "ScholarOne") by which contributions to Symposia will be accepted in real time to complete a volume, instead of having to wait for the whole volume to start production. This new procedure is expected to relieve pressure on the Editors, and speed up publication (ultimately down to a few weeks after the last contribution has been received).

The list of Symposia published during the present triennium is as follows:

2012:

IAUS 280 **The Molecular Universe**
Toledo, Spain, June 2011
Eds. J. Cernicharo, R.Bachiller

IAUS 281 **Binary Paths to Type Ia Supernovae Explosions**
Padova Italy, July 2011
Eds. Rosanne Di Stefano, Marina Orio, Maxwell Moe

IAUS 282 **From Interacting Binaries to Exoplanets: Essential Modeling Tools**
Tatranska Lomnica, Slovakia July 2011
Eds. Mercedes Richards, Ivan Hubeny

IAUS 283 **Planetary Nebulae: An Eye to the Future**
Puerto de la Cruz, Tenerife, July 2011
Eds. A Manchado, L Stanghellini, D Schönberner

IAUS 284 **The Spectral Energy Distribution of Galaxies**
Preston, United Kingdom, September 2011
Eds. Richard J Tuffs, Cristina C Popescu

IAUS 285 **New Horizons in Time-Domain Astronomy**
Oxford, UK, September 2011
Eds. Elizabeth Griffin, Robert Hanisch, Rob Seaman

IAUS 286 **Comparative Magnetic Minima: Characterizing quiet times in the Sun and Stars**
Mendoza, Argentina October 2011
Eds. David Webb and Cristina Mandrini

IAUS 287 **Cosmic Masers - from OH to Ho**
Stellenbosch, South Africa, Feb 2012
Eds. R.S. Booth, E.M.L. Humphreys, W.H.T. Vilemmings

2013:

IAUS 288 **Astrophysics from Antarctica**
Beijing, China, August 20-24, 2012
Eds. Michael Burton, Xiangqun Cui, and Nicholas Tothill

IAUS 289 **Advancing the Physics of Cosmic Distances**
Beijing, China, August 20-24, 2012
Ed. Richard de Grijs

IAUS 290 **Feeding Compact Objects: Accretion on All Scales**
Beijing, China, August 20-24, 2012
Eds. Chenmin Zhang, Tomaso Belloni, Mariano Méndez, Shuangnan Zhang

IAUS 291 **Neutron Stars and Pulsars: Challenges and Opportunities after 80 years**
Beijing, China, August 20-24, 2012
Ed. Joeri van Leeuwen

IAUS 292 **Molecular Gas, Dust, and Star Formation**
Beijing, China, August 20-24, 2012
Eds. Tony Wong, Jürgen Ott

2014:

IAUS 293 **Formation, Detection, and Characterization of Extrasolar Habitable Planets**
Beijing, China, August 20-24, 2012
Ed. Nader Haghighipour

IAUS 294 **Solar and Astrophysical Dynamos and Magnetic Activity**
Beijing, China, August 20-24, 2012
Eds. Alexander G. Kosovichev, Yihua Yan, Elisabete de Gouveia Dal Pino, Lidia van Driel- Gesztelyi

IAUS 295 **The intriguing life of massive galaxies**
Beijing, China, August 20-24, 2012
Eds. Daniel Thomas, Anna Pasquali, Ignacio Ferreras

IAUS 296 **Supernova Environmental Impacts**
Kolcata, India, January 2013
Eds. Alak Ray, Richard A. McCray

IAUS 297 **The Diffuse Interstellar Bands**
Noordwijkerhout, The Netherlands, May 2013
Eds. Jan Cami, Nick L. J. Cox

IAUS 298 **Setting the scene for Gaia and LAMOST**
Lijiang, China, May 2013
Eds. Sofia Feltzing, Gang Zhao, Nicholas A. Walton, Patricia A. Whitelock

IAUS 299 **Exploring the Formation and Evolution of Planetary Systems**
Victoria, Canada, June 2013
Eds. Mark Booth, Brenda C. Matthews, James R. Graham

IAUS 300 **Nature of Prominences and their role in Space Weather**
Paris, France, June 2013
Eds. Brigitte Schmieder, Jean-Marie Malherbe, S. T. Wu

IAUS 301 **Precision Asteroseismology**
Wroclaw, Poland, August 2013
Eds. Joyce A. Guzik, William J. Chaplin, Gerald Handler, Andrzej Pigulski

IAUS 302 **Magnetic Fields throughout Stellar Evolution**
Biarritz, France, August 2013
Eds. Pascal Petit , Moira Jardine and Hendrik C. Spruit

IAUS 303 **The Galactic Center: Feeding and Feedback in a Normal Galactic Nucleus**
Santa Fe, USA, October 2013
Eds. Lorant Sjouwerman, Cornelia Lang, Jürgen Ott

IAUS 304 **Multiwavelength AGN Surveys and Studies**
Byurakan, Armenia, October 2013
Eds. Areg M. Mickaelian, David B. Sanders

2015:

IAUS 307 **New windows on massive stars: asteroseismology, interferometry, and spectropolarimetry**
Geneva, June 2014
Eds. Georges Meynet, Cyril Georgy , José Groh and Philippe Stee

12.2.3. *General Assembly Proceedings*

There are three publications (also with CUP) associated with General Assemblies. If the GA is held in year N:

- Transactions A (TrA: Reports of Astronomy), compiling mainly triennial reports of Commissions and Working Groups and published in year N;
- Transactions B (TrB: Proceedings of the General Assembly), compiling the reports of the GA Business Sessions, the separate business meetings of Commissions and WGs, the membership of Divisions, Commissions, countries, etc. ; normally published in year N+1;
- Highlights of Astronomy, containing the proceedings for the scientific meetings at the GA (other than GA Symposia), i.e., "Special Sessions" and "Joint Discussions", and also Invited Discourses; normally published in year N+2.

This scheme could not be followed for this triennium for the Beijing volumes, essentially because of the Commission Reform:

- *TrA* was not requested, but is now being turned into a "legacy" volume, since the present Commissions will disappear in Honolulu. The EC has made the decision (at EC94 in Canberra) to suppress this volume, but in fact it still needs to be published to mark the end of the Commission regime that has existed since the creation of the IAU in 1919. Therefore it will be published after the GA (with Oct. 15 as a deadline for submitting triennal, even hexennial reports for Commissions not having sent their report for the Beijing GA; and also the Honolulu GA business meetings of these Commissions). It will be edited by the current GS, as would have been the case in the past;

- *TrB* has been delayed for a long time but is now in production;
- *Highlights of Astronomy* (#15) has been published earlier this year.

This issue will be the last of the series. Because of the merging of the GA "Special Sessions" and "Joint Discussions" into "Focus Meetings" (see Section 11), a new series will be launched: "Astronomy in Focus" (agreed by CUP).

TrB will host the new Division Meetings (Section 11) proceedings, in the form of summaries of the meetings, with contributions accessible on-line only.

12.2.4. *Information Bulletin*

Because of the revamping of the web site, and the new, dynamic way of announcing IAU news, the traditional bi-annual Information Bulletin (IB) has been temporarily suspended. The last issues published were IB111 (February 2013), and IB112 (July 2013).

It is expected that the IB will resume publication in the next triennium, but its contents and editorial policy have to be rediscussed in view of the increasing importance of the web site and its fast turnaround time for news (in contrast, the IB is normally published every six months). Turning it into an Archive of selected documents having appeared on the web site, and/or EC documents and records, are possibilities to be explored in the next triennium.

12.3. *IAU Archives*

As a result of an obvious necessity, and triggered by the proximity of the IAU centennial (it was founded in 1919 so celebrations are expected on the occasion of the GA2018 in Vienna), a project to digitize the IAU publications and documents has been started with CUP.

The project has two parts:

- the "IAU Book Archive" project, which concerns all IAU publications since the founding of the IAU (Transactions, Colloquia, Symposia, etc.);
- the "IAU Document Archive" project, which concerns physical archives currently stored at the French Académie des Sciences (corresponding to A. Blaauw's book: "History of the IAU, the first 50 years"; Kluwer), and more recently at the IAU Secretariat. The selection of these archives has started by G. Rude, but it needs supervision by a historian (who should be D. Baneke, from Leiden University).

Concerning the "Book Archive" project, difficult legal issues have been encountered, notably with Springer, who bought two of the previous IAU publishers (Reidel and Kluwer). After more than two years, no agreement has been reached between the IAU (advised by CUP) and Springer to agree on copyrights, but Springer has nevertheless unilaterally decided to put their IAU publications part of their own "Book Archive" collection (for a fee), and CUP is about to do the same (with free access, thus at a cost for the IAU): to date CUP has digitized about 180 volumes (Symposia and Colloquia since 1955). The story may go on for some time until the whole issue is finally settled legally.

Apart from the CUP archive digitization project, with the help of Ginette Rude, the web site now offers access to all IAU Executive Committees since 1919: (http://www.iau.org/administration/executive_bodies/past_committees/executive_co mmittee/), and to all GA Newspapers since the first issue (Moscow, 1958): (http://www.iau.org/publications/iau/ga_newspapers/).

13. Relations with other scientific organizations

13.1. *ICSU*

The IAU is a member of the International Council of Science (ICSU). Through ICSU, the IAU participates in highly relevant data activities under CODATA (The Committee on Data for Science and Technology) and WDS (The World Data System). Other activities include COSPAR (Space Research), SCOSTEP (Solar-Terrestrial Physics), IUCAF (Radio Astronomy and Space Science) and SCAR (Antarctic Research). Through these bodies the IAU is able to influence worldwide developments and strategy in fields of interest, but is increasingly concerned that the strategic priorities of ICSU are of little direct relevance to the IAU, and indeed to many of the "basic science" unions. This concern has been raised to the attention of the newly appointed Executive (2015) and constructive discussion on possible improvements of the IAU - ICSU relationship has been started. In particular the activities of the IAU-OAD on promotion of science through the international network of Regional Nodes represent a possibility for joint sponsored initiatives.

13.2. *UNESCO*

The IAU and UNESCO have renewed in 2013 a 5-yr MoU concerning the "Astronomy and World Heritage" initiative (see http://www.iau.org/news/pressreleases/detail/iau1302/).

In this document, the IAU commits itself to enhance its advisory role for preserving astronomical sites (including modern facilities) in the framework of the "World Heritage" in science. (This has been achieved recently with the creation of Commission C4 "Astronomy and World Heritage", under Division C.)

A number of actions have taken place in the present triennium leading to nominations to World Heritage (Pulkovo Obervatory, etc., but also proposed "High-Altitude Observatories"). The IAU also maintains the very informative "IAU-UNESCO Portal to the Heritage of Astronomy (http://www2.astronomicalheritage.net/).

It is expected that a scientific committee will be established by UNESCO, including the IAU and ICOMOS (International Committee on Monuments and Sites), to advise and select proposals for nominations according to UNESCO rules (which are quite complex).

13.3. *UN-COPUOS (The United Nations Committee on the Peaceful Uses of Outer Space)*

The IAU is an active contributor in this body, and particularly on the issue of forecasting, and potentially mitigating, future impacts of Near Earth Objects (NEOs; comets: NECs; asteroids: NEAs) on Earth. In particular in 2013 and 2014 the General Assembly of the United Nations (UN) has agreed that an international response is necessary to coordinate and develop mitigation measures to address the Near-Earth Object (NEO) impact threat. Mitigation includes detection, follow-up, and characterization of NEO impact threats as well as the development of possible deflection techniques. Consequently the International Asteroid Warning Network (IAWN) and Space Mission Planning Advisory Group (SMPAG) was recommended by the Working Group and formally endorsed by COPUOS at its 56th session in June 2013, by the 68th session of the UN General

Assembly in December 2013, and by the 69th session of the UN General Assembly in December 2014. IAU, through the IAU Minor Planet Center, is actively participating in the activities of the IAWN network.

13.4. *COSPAR*

The IAU is collaborating with COSPAR on the organization of their "Capacity Building Workshops" (`https://cosparhq.cnes.fr/events/cb-workshops`), the GS being an ex-officio member of the SOC, for those of astronomical interest. The last ones took place in Mexico (2014: X-ray astrophysics), China (2013: High-Energy Astrophysics), and Buenos Aires (2012: IR and mm Astronomy). The next one will be in Brazil (2015: Planetary Mission Data Analysis).

14. Financial Matter

The Accounts for the present triennium (2012-2014) and the proposed budget for the next triennium (2016-2018) are given in detail as a separate document, that has also been sent to the Finance Committee for review.

Here we give comments about the most important issues.

14.1. *Accounts 2012-2014*

14.1.1. *Income*

By far the main income of the IAU comes from the dues paid by National Members according to their Dues Category, each Category being associated with a certain number of "Units of Contribution" (Bye-Laws, Chap. VII).

The Unit of Contribution (UC) is approved by the General Assembly. Its value for the year 2015 is 2860 €. Adding up all Categories, but not taking into account suspended countries, yields a total of 305 UC, representing a nominal income of $305 \times 2860 =$ 872300 €. Another important contribution comes from royalties of IAU publications with Cambridge University Press (variable, but for this triennium around 100 k€). So in round members, in a normal year the IAU should have an income close to 1 M€.

Unfortunately, we are constantly struggling with unpaid dues, which regularly represent about 150 k€for a given calendar year. But the year 2014 has been particularly bad, with over 300 k€missing in dues. While some averaging effect does take place (dues can be paid late, or partly, in a given year covering a previous year), we have observed a steady decline (of 2-3 % /year). The situation can be serious because a large part of the expenditures is essentially fixed (salaries for the 2.4 FTE of the Secretariat, running costs of the EC and other administration, etc.). The only flexibility comes from "external" activities such as educational programs, which are strategic to the IAU actions, and could be in peril, whereas these educational programs (in the widest sense, i.e., including "capacity building" actions such as grants for Symposia) add up to more than 50 % of the IAU budget.

14.1.2. *Expenditures*

The budget approved by the Beijing GA did not include (temporary) additional expenditures that reflected actions undertaken by the EC during the present triennium:

(a) As explained in Section 6,, to implement Resolution B5 about Commission reform, the Division Presidents were invited to all EC meetings, starting with EC93 in Nara. The resulting additional cost was about 30 kE/year (or about 3 kE/DP/yr).

(b) The Agreement over the Office of Astronomy Outreach (OAO) includes a contribution of 25 kE/year for the OAO staff (see Section 9)
(c) To participate in the International Year of Astronomy (IYL2015) cornerstone "Cosmic Light", seed funding of 30 kE has been allotted in 2015 to a Call for Proposals for global projects (see Section 7)

An increasingly important expenditure is the IAU contribution to the OAD: in 2015 this contribution was 50 kE/yr for running costs, and 110 kE/yr for (seed) funding of projects selected by the Task Forces of the OAD. Overheads a not negligible (a Steering Committee meeting costs about 10 kE), so, with about 170 kE/yr, the OAD alone represents about 20 % of the IAU budget.

On the other hand, the GS resided in Paris, so the IAU studio could be assigned to G. Rude as part of her salary, represented a saving of about 20 kE/yr.

14.2. *Proposed budget 2016-2018*

The main differences with the present budget are as follows:

14.2.1. *Income*

We propose to increase the UC by 2 % /year, in continuation of the increase approved in Beijing. This increase in meant to take inflation into account, which is significantly variable from country to country.

Of course we hope that the National Members will pay their dues more regularly, and that some suspended countries will come back to the community.

14.2.2. *Expenditure*

No "special operations" comparable to the previous triennium are foreseen at the moment.

- The Division Presidents will be invited to one EC meeting only (other meetings will be organized by telecon): 30 kE for the whole triennium;
- The rental of the studio must be re-incorporated (20 kE/yr);
- There is an additional contribution to the OAO for the second staff hired in 2015 (25 kE/yr; see Section 9);
- Additional funding is proposed for the OAD, at the level of 20 kE/yr. This is mostly dedicated to implementing the OAD Review Committee recommendation to seek a part-time staff (or contractor) for fund-raising activities (see Section 8);
- Additional funding (20 kE the first yr, 10 kE/yr the following years) is proposed for outsourcing centralized IT developments for Divisions and Commissions (web page maintenance, databases, etc.: ESO, see Section 3), so that the access doesn't change from triennium to triennium when Presidents change.

14.3. *Accounting and Auditing*

During the previous triennium, the accounts were maintained by the Head of Administration V. Reuter, who used the SAGE professional accounting system. The accounts were then converted by a professional accountant at the end of each year into the legally required format, and the annual accounts were submitted to an external auditor. The 2012 accounts were approved in this way.

Starting in 2013, the new Head of Administration, being unfamiliar with SAGE, opted (with the GS agreement) for a complete outsourcing of the accounting, i.e., providing data to the same accountant, who then did both the SAGE accounting and the legal

BUDGET OF INCOME 2016 - 2018 (Euros, rounded):

		2016	2017	2018	Total
A	Adhering Organizations (National Member dues)	883 851	901 425	919 605	2 704 881
C	Total Royalties (CUP)	60 000	60 000	60 000	180 000
D	Funds in Transit (OAO- ISYA-Gruber Fellowship)	89 960	89 960	89 960	269 880
F	Bank Revenue	5 000	5 000	5 000	15 000
	TOTAL INCOME	1 038 811	1 056 385	1 074 565	3 169 761

BUDGET OF EXPENDITURE 2016-2018 (Euros, rounded):

		2016	2017	2018	Total
M	General Assembly	15 000	15 000	410 000	440 000
N	Scientific Activities	207 800	227 920	66 960	502 680
O	Educational Projects	252 100	252 100	252 100	756 300
P	Funds in Transit (OAO- ISYA-Gruber Fellowship)	89 960	89 960	89 960	269 880
Q	Cooperation with the other Unions	30 030	30 660	31 290	91 980
R	Executive Committee	79 493	79 683	79 874	239 050
S	Publications (IB-Archives)	38 000	13 000	13 000	64 000
T	Staff Salaries, Taxes & Charges	149 843	153 487	157 220	460 550
T3	Outsourced tasks	89 850	79 870	79 890	249 610
T4	General Office Expenditures	25 400	25 400	25 400	76 200
	TOTAL EXPENDITURE	980 976	970 580	1 209 194	3 160 750
	INCOME over EXPENDITURE	57 835	85 805	-134 629	9 011

Figure 1. Proposed budget 2016-2018

presentation at an additional cost of about 6kE/yr. To save money (about 6 kE/yr), it was then decided to drop the external auditor. The accounts for 2013 and 2014 were thus approved by the outsourced accountant.

It should be noted that, because the bank accounts (at HSBC) are accessible electronically at any time, and all transfers are also electronic (each transfer being authorized by the GS), the day-to-day accounting is quite robust and transparent. Very few checks are signed nowadays, and most of them for administration (French taxes, etc.).

However, at a Officers meeting held during the EC95 in Padova, with I. Corbett (past GS) as consultant, it was decided that, following the recommendation of the Finance Committee, the procedure stated in Working Rule 24.b., will be re-introduced with annual external audits starting from the year 2015.

14.4. *MBM Lawsuit*

I. Corbett informed the EC at the EC95 meeting in Padova that the IAU action against the previous auditors MBM (in charge of the IAU accounts until 2008) had been dismissed (to his surprise and disappointment) by the Court and that it had been decided not to appeal. This action is therefore terminated.

14.5. *Accounts 2012-2015 and Budget 2016-2018*

The corresponding tables are provided in a separate document for EC96-1 (identical to that sent to the Finance Committee).

15. Gruber Foundation

The IAU is involved in two awards funded by the Gruber Foundation:

- the Cosmology Prize (worth 500 k$);
- the Fellowship (stipend of 50 k$ for a young astronomer).

The Cosmology Prize is awarded each year. The IAU nominates one member of the selection committee (currently Wendy Freeman, Chair). The IAU issues a "Featured News" Announcement when the recipient(s) are announced by the Gruber Foundation. When the selection falls in a GA year, the Prize is presented during the Opening Ceremony (see http://www.iau.org/news/announcements/detail/ann15016/).

In contrast, the IAU is entirely responsible for managing the "TGF Fellowship": candidate selection and management of the award in negotiation with the recipient's institute.

The rules have evolved at the initiative of the IAU and in agreement with the Gruber Foundation. Indeed, the first two years of this triennium showed a modest interest for the award, frequently considered as a post-doc fellowship, however not competitive with the best post-docs. In addition, the closing date for the applications (Nov.1) was too early to make credible plans for a post-doc. And lastly, there were administrative problems of various sorts linked with overheads on the amount of the fellowship.

Nevertheless, the first two years (and the last of the previous triennium) showed that there were excellent candidates at the ending-PhD/starting-post-doc stage. Therefore, the suggestion was to target more precisely this young astronomer career segment, and to move the application deadline to March 15 to be closer to the end of a normal PhD (http://www.iau.org/science/grants_prizes/gruber_foundation/fellowships/)

This change met with great success, since there were 2 candidates for the 2013 Fellowship, 7 for 2014, and 26 for 2015 (i.e., after the modification of the rules). This year's recipient hasn't even finished his PhD (http://www.iau.org/science/grants_prizes/gruber_foundation/fellowships/recipients/fellow2015/).

16. NASL & Kavli Prize in Astrophysics

The relations with the Norwegian Academy of Sciences and Letters are excellent. The latest episode has been the signature of the Agreement creating the "Office for Young Astronomers", to enhance the current agreement concerning the International Schools for Young Astronomers (see Section 9).

This document includes a supplementary section referring to the Kavli Prize in Astrophysics, that reinforces the role of the IAU in establishing the selection committee. (For now this role is limited to selecting members out of a pre-defined list of names proposed by the Royal Society, the Max-Planck Society, and the US Academy of Sciences.)

Transactions IAU, Volume XXIXB
Proceedings IAU Symposium No. Volume XXIXB, 2018
P. Benvenuti, ed.
© International Astronomical Union 2019
doi:10.1017/S1743921318004003

EXECUTIVE COMMITTEE REPORT: APPENDIX I

I. IAU – PROPOSED INCOME for 2015 - 2018 (EURO, rounded)
[XXIX GA August 2015]

I	INCOME	2016 (EUR)	2017 (EUR)	2018 (EUR)	2016 - 2018 (EUR)
	unit of contribution	2 917	2 975	3 035	
	adjustment for inflation	2.0%	2.0%	2.0%	
	number of units	303	303	303	
A	**ADHERING ORGANIZATIONS**				
A1	National Member dues	*883 851*	*901 425*	*919 605*	*2 704 881*
B	**GRANTS**				
B4	NASL grant for ISYA (US$30,000)	23 610	23 610	23 610	70 830
C	**ROYALTIES**				
C2	CUP	60 000	60 000	60 000	180 000
D	**FUNDS IN TRANSIT**				
D3	TGF Fellowship (USD 50 000/yr)	41 350	41 350	41 350	124 050
F	**BANK REVENUE**	5 000	5 000	5 000	15 000
G	**TOTAL INCOME**	**1 038 811**	**1 056 385**	**1 074 565**	**3 169 761**

II. IAU – PROPOSED BUDGET OF EXPENDITURE 2016 - 2018 (EURO, rounded)

II	BUDGET OF EXPENDITURE	2016 (EUR)	2017 (EUR)	2018 (EUR)	2016 - 2018 (EUR)
M	**GENERAL ASSEMBLIES**				
M1	PREPARATION COSTS	15 000	15 000	25 000	55 000
M2	GRANTS GA INCL. 6 SYMPOSIA	- -	- -	385 000	385 000
	Total GENERAL ASSEMBLIES	*15 000*	*15 000*	*410 000*	*440 000*
N	**SCIENTIFIC ACTIVITIES**				
N1	SPONSORED MEETINGS				
N1.1	GRANTS IAU SYMPOSIA OUTSIDE GA	181 080	181 080	60 360	422 520
N1.2	GRANTS REGIONAL IAU MEETINGS (RIMs)	20 120	40 240	0	60 360
N1.3	CO-SPONSORED MEETINGS	1 600	1 600	1 600	4 800
	Sub-total Sponsored Meetings	*202 800*	*222 920*	*61 960*	*487 680*
N2	WORKING GROUPS				
N2.1.2	COMMISSION WORKING GROUPS	5 000	5 0000	5 000	15 000
	Total SCIENTIFIC ACTIVITIES	*207 800*	*227 920*	*66 960*	*502 680*

	Expenditure (cont'd)	**2016** (EUR)	**2017** (EUR)	**2018** (EUR)	**2016 - 2018** (EUR)
O	**EDUCATIONAL ACTIVITIES**				
O1	IAU/OYA/ISYA	5 000	5 000	5 000	150 000
O5	OAD Project Costs	110 000	110 000	110 000	330 000
	OAD Operating Costs	50 000	50 000	50 000	150 000
	OAD Steering Committee	10 100	10 100	10 100	30 300
	OAD provision for future operations	20 000	20 000	20 000	60 000
O11	co-sponsoring COSPAR Capacity Buiding Workshops	3 100	3 100	3 100	9 300
O5.1	OAO Project Costs	50 000	50 000	50 000	150 000
	OAO Operating Costs	2 000	2 000	2 000	6 000
O6	COSPAR	5 000	5 000	5 000	15 000
	total EDUCATIONAL ACTIVITIES	*252 100*	*252 100*	*252 100*	*756 300*
P	**FUNDS IN TRANSIT**				
P1	OAO-NAOJ	25 000	25 000	25 000	75 000
P2	Grüber Fellowship (US$ 50 000)	41 350	41 350	41 350	124 050
P3	OYA / NASL / ISYA	23 610	23 610	23 610	70 830
	total FUNDS IN TRANSIT	*89 960*	*89 960*	*89 960*	*269 880*
Q	**COOP. with OTHER UNIONS**				
Q2	dues to other unions				
Q2.1	icsu	21 200	21 800	22 400	65 400
Q2.2	iucaf	5 030	5 060	5 090	15 180
Q2.3	icsti	800	800	800	2 400
	Other: unesco	3 000	3 000	3 000	9 000
	total COOP. with OTHER UNIONS	*30 030*	*30 660*	*31 290*	*91 980*

	Budget of Expenditure (cont'd)	**2016** (EUR)	**2017** (EUR)	**2018** (EUR)	**2016 - 2018** (EUR)
R	**EXECUTIVE COMMITTEE**				
R1	Executive Committee Meetings	35 000	35 000	35 000	105 000
R1.1	EC expenses other than meetings	0	0	0	0
R2	Officers' Meetings	13 422	13 503	13 584	40 509
R3	Officers' expenditure (other)				
R3.1	General Secretary expenditure				
R3.1.1	GS Paris duty	17 571	17 680	17 790	53 041
R3.1.2	GS Other expenses	3 000	3 000	3 000	9 000
R3.5	Assistant General Secretary	1 500	1 500	1 500	4 500
R3.2	President	5 000	5 000	5 000	15 000
R3.3	President-elect	1 000	1 000	1 000	3 000
R4	Press-Office	3 000	3 000	3 000	9 000
	total EXECUTIVE COMMITTEE	*79 493*	*79 683*	*79 874*	*239 050*
S	**PUBLICATIONS**				
S1	IAU Information Bulletin	3 000	3 000	3 000	9 000
	Other: CAP Journal	5 000	5 000	5 000	15 000
	Archives + Centannial Book	30 000	5 000	5 000	40 000
	total PUBLICATIONS	*38 000*	*13 000*	*13 000*	*64 000*

	Budget of Expenditure (cont'd)	**2016** (EUR)	**2017** (EUR)	**2018** (EUR)	**2016 - 2018** (EUR)
T	**SECRETARIAT / ADMIN.**				
T1	Salaries & Charges	149 843	153 487	157 220	460 550
T3	Outsourced Tasks				
T3.1	Web/DB Development at ESO	17 600	17 600	17 600	52 800
T3.2	Data Base management at ESO	30 000	30 000	30 000	90 000
T3.2.0	Web Divisions management at ESO	20 000	10 000	10 000	40 000
T3.2.1	Mi-voice subscription	2 800	2 820	2 840	8 460
T3.3	IT Assistance in Paris	2 000	2 000	2 000	6 000
T3.4	Personnel Administration	2 000	2 000	2 000	6 000
T3.5	Accounting	7 000	7 000	7 000	21 000
T3.5	SAGE subscription	950	950	950	2 850
T3.6	Auditing	6 500	6 500	6 500	19 500
T3.7	Legal Fees	1 000	1 000	1 000	3 000
	sub-total office support costs	*89 850*	*79 870*	*79 890*	*249 610*
T4	General Office expenditure				
T4.1	Post	1 500	1 500	1 500	4 500
T4.2	Telephone and Internet	3 000	3 000	3 000	9 000
T4.3	Rent (INSU/IAP)	4 600	4 600	4 600	13 800
T4.4	IT Software & hardware	1 500	1 500	1 500	4 500
T4.5	Copier/Printer, rental,	7 000	7 000	7 000	14 000
T4.6	Office Supplies	1 000	1 000	1 000	3 000
T4.7	Miscellaneous items books, posters, etc	2 800	2 800	2 800	8 400
T4.9	Bank Charges	4 000	4 000	4 000	12 000
U4	Taxes on revenue	3 500	3 500	3 500	10 500
	sub-total General Office expenditure	*28 900*	*28 900*	*28 900*	*76 200*
	total SECRETARIAT / ADMIN.	*268 593*	*262 257*	*266 010*	*786 360*

Budget of Expenditure (cont'd)	2016 (EUR)	2017 (EUR)	2018 (EUR)	2016 - 2018 (EUR)
total EXPENDITURE	**980 976**	**970 580**	**1 209 194**	**3 160 750**
total INCOME	**1 038 811**	**1 056 385**	**1 074 565**	**3 169 761**
INCOME OVER EXPENDITURE	**+ 57 835**	**+ 85 805**	**– 134 629**	**+9 011**

Transactions IAU, Volume XXIXB
Proceedings IAU Symposium No. Volume XXIXB, 2018
P. Benvenuti, ed.
© International Astronomical Union 2019
doi:10.1017/S1743921318004015

CHAPTER VI

Commissions Reports

Traditionally, Chapter VI of the *Transactions B* Volume contains the Reports by the IAU Commissions. However the old Commissions ceased to exist at the XXIX General Assembly and were replaced by the new ones (see the Preface). To mark this important milestone in the history of the IAU, it was decided to dedicate the Volume XXIXA of the *Transactions A* of the IAU to the Reports by the old Commissions. The Volume, edited by Thierry Montmerle, is entitled "Commissions legacy Reports": because of this, the Chapter VI of the current *Transactions B* Volume has been left empty, but it was kept as a chapter not alter the traditional order.

Piero Benvenuti
General Secretary

Transactions IAU, Volume XXIXB
Proceedings IAU Symposium No. Volume XXIXB, 2018
Piero Benvenuti, ed.
© International Astronomical Union 2019
doi:10.1017/S1743921318004027

CHAPTER VII

TWENTY NINETH GENERAL ASSEMBLY

IAU STATUTES, BYE-LAWS, AND WORKING RULES

Honolulu, 21 August 2015

1. IAU Statutes

I. OBJECTIVE

1. The International Astronomical Union (hereinafter referred to as the Union) is an international nongovernmental organization. Its objective is to promote the science of astronomy in all its aspects.

II. DOMICILE AND INTERNATIONAL RELATIONS

2. The legal domicile of the Union is Paris, France.

3. The Union adheres to, and co-operates with the body of international scientific organizations through the International Council for Science (ICSU). It supports and applies the policies on the Freedom, Responsibility, and Ethics in the Conduct of Science defined by ICSU.

III. COMPOSITION OF THE UNION

4. The Union is composed of:

4.a. National Members (adhering organizations)

4.b. Individual Members (adhering persons)

IV. NATIONAL MEMBERS

5. An organization representing a national professional astronomical community, desiring to promote its participation in international astronomy and supporting the objective of the Union, may adhere to the Union as a National Member.

6. An organization desiring to join the Union as a National Member while developing professional astronomy in the community it represents may do so:

6a. on an interim basis, on the same conditions as above, for a period of up to nine years. After that time, it must apply to become a National Member on a permanent basis or its membership in the Union will terminate;

6b. on a prospective basis for a period of up to six years if its community has less than six Individual Members. After that time it must apply to become a National Member on either an interim or permanent basis or its membership in the Union will terminate.

7. A National Member is admitted to the Union on a permanent, interim, or prospective basis by the General Assembly. It may resign from the Union by so informing the General Secretary in writing.

8. A National Member may be either:

8.a. the organization by which scientists of the corresponding nation or territory adhere to ICSU or:

8.b. an appropriate National Society or Committee for Astronomy, or:

8.c. an appropriate institution of higher learning.

9. The adherence of a National Member is automatically suspended if its annual contributions, as defined in Articles 23c and 23e below have not been paid for five years; it resumes, upon the approval of the Executive Committee, when the arrears in contributions have been paid in full. After five years of suspension of a National Member, the Executive Committee may recommend to the General Assembly to terminate the Membership.

10. A National Member is admitted to the Union in one of the categories specified in the Bye-Laws.

V. INDIVIDUAL MEMBERS

11. A professional scientist who is active in some branch of astronomy may be admitted to the Union by the Executive Committee as an Individual Member. An Individual Member may resign from the Union by so informing the General Secretary in writing.

VI. GOVERNANCE

12. The governing bodies of the Union are:

12.a. The General Assembly;

12.b. The Executive Committee; and

12.c. The Officers.

VII. GENERAL ASSEMBLY

13. The General Assembly consists of the National Members and of Individual Members. The General Assembly determines the overall policy of the Union.

13.a. The General Assembly approves the Statutes of the Union, including any changes therein.

13.b.The General Assembly approves Bye-Laws specifying the Rules of Procedure to be used in applying the Statutes.

13.c. The General Assembly elects an Executive Committee to implement its decisions and to direct the affairs of the Union between successive ordinary meetings of the General Assembly. The Executive Committee reports to the General Assembly.

13.d. The General Assembly appoints a standing Finance Committee to advise the Executive Committee on its behalf on budgetary matters between General Assemblies, and to advise the General Assembly on the approval of the budget and accounts of the Union. The Finance Committee consists of not more than 8 members of different national affiliations, including a Chairperson, proposed by the National Members, and remains in office until the end of the next General Assembly.

13.e. The General Assembly appoints a Special Nominating Committee to prepare a suitable slate of candidates for election to the incoming Executive Committee.

13.f. The General Assembly appoints a standing Membership Committee to advise the Executive Committee on its behalf on matters related to the admission of Individual Members. The Membership Committee consists of not more than 8 members of different national affiliations, including a Chairperson, proposed by the National Members, and remains in office until the end of the next General Assembly.

14. Voting at the General Assembly on issues of a primarily scientific nature, as determined by the Executive Committee, is by Individual Members. Voting on all other matters is by National Member. Each National Member authorises a representative to vote on its behalf.

14.a. On questions involving the budget of the Union, the number of votes for each National Member is one greater than the number of its category, referred to in article 10. National Members with interim status, or which have not paid their dues for years preceding that of the General Assembly, may not participate in the voting.

14.b. On questions concerning the administration of the Union, but not involving its budget, each National Member has one vote, under the same condition of payment of dues as in §14.a.

14.c.National Members may vote by correspondence on questions concerning the agenda for the General Assembly.

14.d. A vote is valid only if at least two thirds of the National Members having the right to vote by virtue of article §14.a. participate in it by either casting a vote or signalling an abstention. An abstention. An abstention is not considered a vote cast.

15. The decisions of the General Assembly are taken by an absolute majority of the votes cast. However, a decision to change the Statutes requires the approval of at least two thirds of all National Members having the right to vote by virtue of article §14.a. Where there is an equal division of votes, the President determines the issue.

15.a. To enable the widest possible participation of Individual Members the Executive Committee may decide that voting on certain issues of a primarily scientific nature, as determined by the Executive Committee, shall be open for electronic voting for not more than 31 days counting from the close of the General Assembly at which the issue was raised.

15.b. The Executive Committee shall give Members not less than 3 months notice before the opening of the General Assembly of the intention to open certain issues to electronic voting after the General Assembly.

16. Changes in the Statutes or Bye-Laws can only be considered by the General Assembly if a specific proposal has been duly submitted to the National Members and placed on the Agenda of the General Assembly by the procedure and deadlines specified in the Bye-Laws.

VIII. EXECUTIVE COMMITTEE

17. The Executive Committee consists of the President of the Union, the President-Elect, six Vice-Presidents, the General Secretary, and the Assistant General Secretary, elected by the General Assembly on the proposal of the Special Nominating Committee.

IX. OFFICERS

18. The Officers of the Union are the President, the General Secretary, the President-Elect, and the Assistant General Secretary. The Officers decide short-term policy issues within the general policies of the Union as decided by the General Assembly and interpreted by the Executive Committee.

X. SCIENTIFIC DIVISIONS

19. As an effective means to promote progress in the main areas of astronomy, the scientific work of the Union is structured through its Scientific Divisions. Each Division covers a broad, well-defined area of astronomical science, or deals with international

matters of an interdisciplinary nature. As far as practicable, Divisions should include comparable fractions of the Individual Members of the Union.

20. Divisions are created or terminated by the General Assembly on the recommendation of the Executive Committee. The activities of a Division are organized by an Organizing Committee chaired by a Division President. The Division President and a Vice-President are elected by the General Assembly on the proposal of the Executive Committee, and are ex officio members of the Organizing Committee.

XI. SCIENTIFIC COMMISSIONS

21. Within Divisions, the scientific activities in well-defined disciplines within the subject matter of the Division may be organized through scientific Commissions. In special cases, a Commission may cover a subject common to two or more Divisions and then becomes a Commission of all these Divisions.

22. Commissions are created or terminated by the Executive Committee upon the recommendation of the Organizing Committee(s) of the Division(s) desiring to create or terminate them. The activities of a Commission are organized by an Organizing Committee chaired by a Commission President. The Commission President and a Vice-President are appointed by the Organizing Committee(s) of the corresponding Division(s) upon the proposal of the Organizing Committee of the Commission.

XII. BUDGET AND DUES

23. For each ordinary General Assembly the Executive Committee prepares a budget proposal covering the period to the next ordinary General Assembly, together with the accounts of the Union for the preceding period. It submits these to the Finance Committee for advice before presenting them to the vote of the General Assembly.

23.a.The Finance Committee examines the accounts of the Union from the point of view of responsible expenditure within the intent of the previous General Assembly, as interpreted by the Executive Committee. It also considers whether the proposed budget is adequate to implement the policy of the General Assembly. It submits reports on these matters to the General Assembly before its decisions concerning the approval of the accounts and of the budget.

23.b. The amount of the unit of contribution is decided by the General Assembly as part of the budget approval process.

23.c. Each National Member pays annually a number of units of contribution corresponding to its category. The number of units of contribution for each category shall be specified in the Bye-Laws.

23.d. A vote on matters under article 23 is valid only if at least two thirds of the National Members having the right to vote by virtue of article §14.a. cast a vote. In all cases an abstention is not a vote, but a declaration that the Member declines to vote.

23.e. National Members having interim status pay annually one half unit of contribution.

23.f. National Members having prospective status pay no contribution.

23.g. The payment of contributions is the responsibility of the National Members. The liability of each National Members in respect of the Union is limited to the amount of contributions due through the current year.

XIII. EMERGENCY POWERS

24. If, through events outside the control of the Union, circumstances arise in which it is impracticable to comply fully with the provisions of the Statutes and Bye-Laws of the Union, the Executive Committee and Officers, in the order specified below, shall take such actions as they deem necessary for the continued operation of the Union. Such action shall be reported to all National Members as soon as this becomes practicable, until an ordinary or extraordinary General Assembly can be convened. The following is the order of authority: The Executive Committee in meeting or by correspondence; the President of the Union; the General Secretary; or failing the practicability or availability of any of the above, one of the Vice-Presidents.

XIV. DISSOLUTION OF THE UNION

25. A decision to dissolve the Union is only valid if taken by the General Assembly with the approval of three quarters of the National Members having the right to vote by virtue of article §14.a. Such a decision shall specify a procedure for settling any debts and disposing of any assets of the Union.

XV. FINAL CLAUSES

26. These Statutes enter into force on 21 August 2012.

27. The present Statutes are published in French and English versions. For legal purposes, the French version is authoritative.

2. Statuts de l'UAI, Honolulu, 21 août 2015

I. OBJECTIF

1. L'Union Internationale Astronomique (dénommée ci-après Â« l'Union Â») est une organisation non gouvernementale internationale. Son objectif est de promouvoir la science de l'astronomie sous tous ses aspects.

II. DOMICILIATION ET RELATIONS INTERNATIONALES

2. Le domicile légal de l'Union est situé à Paris, en France.

3. L'Union adhère à, et coopère avec, l'ensemble des organisations scientifiques internationales à travers le Conseil International pour la Science (ICSU). Elle soutient et applique les directives sur la Liberté, la Responsabilité, et l'Ethique pour la bonne conduite des sciences telles que définies par l'ICSU.

III. COMPOSITION DE l'UNION

4. L'Union se compose de :

4.a. Membres Nationaux (organisations),

4.b. Membres Individuels (personnes physiques).

IV. MEMBRES NATIONAUX

5. Une organisation représentant une communauté astronomique professionnelle nationale, désireuse de développer sa participation sur la scène de l'astronomie internationale et soutenant les objectifs de l'Union, peut adhérer à l'Union en qualité de Membre National.

6. Une organisation désireuse de rejoindre l'Union en qualité de Membre National tout en développant l'astronomie professionnelle dans la communauté qu'elle représente peut le faire :

6.a. De manière temporaire, selon les conditions précitées, pour une période maximale de neuf ans. Passé ce délai, elle doit demander à devenir Membre National de manière permanente, à défaut son adhésion à l'Union sera résiliée.

6.b. De manière prospective, pour une période maximale de six ans si sa communauté compte moins de six Membres Individuels. Passé ce délai, elle devra demander à devenir Membre National de manière temporaire ou permanente, à défiaut de quoi son adhésion sera résiliée.

7. Un Membre National est admis dans l'Union de manière permanente, temporaire ou prospective par l'Assemblée Générale. Il peut se retirer de l'Union en informant le Secrétaire Général de son retrait par écrit.

8. Un Membre National peut être :

8.a. l'organisation par laquelle les scientifiques de la nation correspondante ou du territoire correspondant, adhérant à l'ICSU ou :

8.b. Une Société ou Comité National(e) compétent(e) d'Astronomie, ou :

8.c. Un établissement compétent d'enseignement supérieur.

9. L'adhésion d'un Membre National est automatiquement suspendue si ses cotisations annuelles, telles que définies aux Articles 23c et 23e ci-dessous n'ont pas été payées pendant cinq ans ; elle sera rétablie, sur approbation du Comité Exécutif, lorsque les arriérés relatifs à ses cotisations auront été payés en totalité. Après cinq ans de suspension d'un Membre National, le Comité Exécutif peut recommander à l'Assemblée Générale l'exclusion de ce Membre.

10. Un Membre National est admis dans l'Union au travers de l'une des catégories spécifiées dans le Règlement Intérieur.

V. MEMBRES INDIVIDUELS

11. Un scientifique professionnel exerçant son activité dans un des domaines de l'astronomie peut être admis dans l'Union par le Comité Exécutif en qualité de Membre Individuel. Un Membre Individuel peut quitter de l'Union en informant de son retrait le Secrétaire Général, par écrit.

VI. GOUVERNANCE

12. La gouvernance de l'Union est constituée par :

12.a. L'Assemblée Générale ;

12.b. Le Comité Exécutif ;

12.c. Les Membres du Bureau du Comité Executif.

VII. ASSEMBLEE GENERALE

13. L'Assemblée Générale se compose des Membres Nationaux et des Membres Individuels. L'Assemblée Générale détermine les grandes orientations de l'Union.

13.a. L'Assemblée Générale ratifie les Statuts de L'Union, y compris tous changements apportés à ces Statuts.

13.b. L'Assemblée Générale ratifie le Règlement Intérieur spécifiant les Règles de Procédure devant être suivies lors de l'application des Statuts.

13.c. L'Assemblée Générale élit un Comité Exécutif afin de mettre en œuvre ses décisions et le charger de diriger les affaires de l'Union entre d'une Assemblée Générale à l'autre. Le Comité Exécutif présente ses comptes-rendus à l'Assemblée Générale.

13.d. L'Assemblée Générale nomme un Comité des Finances pour conseiller le Comité Exécutif en son nom sur les questions budgétaires entre les Assemblées Générales et pour proposer ses recommandations à l'Assemblée Générale concernant l'approbation du budget et des comptes de l'Union. Le Comité des Finances se compose au plus de huit Membres de différentes représentations nationales et comprend un Président, proposé par les Membres Nationaux, qui conservera ses fonctions jusqu'à la fin de l'Assemblée Générale suivante.

13.e. L'Assemblée Générale nomme un Comité Spécial des Nominations afin de préparer un éventail de candidats en vue de l'élection du Comité Exécutif suivant.

13.f. L'Assemblée Générale nomme un Comité des Admissions pour permettre au Comité Exécutif de se prononcer sur l'admission de nouveaux Membres Individuels. Le Comité des Admissions se compose au plus de huit Membres de différentes représentations nationales et comprend un Président, proposé par les Membres Nationaux, qui conservera ses fonctions jusqu'à la fin de l'Assemblée Générale suivante.

14. Les votes lors de l'Assemblée Générale portant sur des questions de nature essentiellement scientifique, telles que déterminées par le Comité Exécutif, se font par les Membres Individuels. Les votes portant sur toutes les autres questions se font par les Membres Nationaux. Chaque Membre National nomme un représentant pour voter en son nom.

14.a. Concernant les questions ayant trait au budget de l'Union, le nombre de voix pour chaque Membre National est supérieur d'une unité au nombre définissant sa catégorie, telle que définie à l'Article 10 des Statuts. Les Membres Nationaux dont le statut est temporaire ou prospectif, ou qui ne sont pas à jour de leur cotisation au moment de l'Assemblée Générale, ne peuvent pas prendre part au vote.

14.b. Concernant les questions relatives à l'administration de l'Union, mais n'ayant pas trait au budget, chaque Membre National dispose d'une voix, selon les mêmes conditions relatives au paiement des cotisations mentionnées à l'Article 14.a.

14.c. Les Membres Nationaux peuvent voter par correspondance sur des questions mises à l'ordre du jour de l'Assemblée Générale.

14.d. Un vote n'est valide que si au moins deux tiers des Membres Nationaux ayant le droit de voter en vertu de l'Article 14.a. y participent, soit en votant, soit en se prononçant par une abstention. Une abstention n'est pas considérée comme l"expression d'un vote.

15. Les décisions de L'Assemblée Générale sont prises à la majorité absolue des voix. Cependant, la décision de modifier les Statuts exige l'approbation d'au moins deux tiers de tous les Membres Nationaux ayant droit de voter en vertu de l'Article 14.a. Lorsqu'il y a égalité des voix, il revient au Président de statuer sur l'issue du vote.

15.a. Afin de favoriser la plus large participation possible des Membres Individuels, le Comité Exécutif pourra décider que le vote portant sur certaines questions de nature essentiellement scientifique, tel que le déterminera le Comité Exécutif, pourront faire l'objet d'un vote électronique 31 jours au plus à partir de la clÃ´ture de l'Assemblée Générale à laquelle la question a été soulevée.

15.b. Le Comité Exécutif enverra un préavis d'au moins trois mois aux Membres avant l'ouverture de l'Assemblée Générale les informant des questions devant être soumises à un vote électronique après l'Assemblée Générale.

16. Les changements apportés aux Statuts ou au Règlement Intérieur ne peuvent être examinés par l'Assemblée Générale que si une proposition spécifique a été dûment soumise aux Membres Nationaux et mentionnée à l'ordre du jour de l'Assemblée Générale par la procédure et dans des délais spécifiés dans le Règlement Intérieur.

VIII. COMITE DE DIRECTION

17. Le Comité Exécutif se compose du Président de l'Union, du Président désigné (son successeur), de six Vice-Présidents, du Secrétaire Général, et du Secrétaire Général Adjoint, élus par l'Assemblée Générale sur proposition du Comité Spécial des Nominations.

IX. MEMBRES DU BUREAU

18. Les Membres de l'Union sont le Président, le Secrétaire Général, le Président désigné et le Secrétaire Général Adjoint. Les Membres du Bureau prennent des décisions sur le court terme dans le cadre des orientations générales de l'Union telles que décidées par l'Assemblée Générale et appliquées par le Comité Exécutif.

X. DIVISIONS SCIENTIFIQUES

19. En tant que moyen efficace pour favoriser les progrès dans les principaux domaines de l'astronomie, le travail scientifique de l'Union est structuré autour de ses Divisions Scientifiques. Chaque Division recouvre un large domaine bien défini de l'astronomie, ou traite de sujets internationaux de nature interdisciplinaire. Autant que possible, les Divisions comprendront chacune un nombre comparable de Membres Individuels de l'Union.

20. Les Divisions sont créées ou supprimées par l'Assemblée Générale sur recommandation du Comité Exécutif. Les activités d'une Division sont organisées par un Comité d'Organisation présidé par le Président de Division. Le Président de Division et le Vice-Président sont élus par l'Assemblée Générale sur proposition du Comité Exécutif, et sont Membres ès-qualité du Comité d'Organisation.

XI. COMMISSIONS SCIENTIFIQUES

21. A l'intérieur des Divisions, les activités scientifiques réparties en disciplines bien définies peuvent être organisées à travers des Commissions Scientifiques. Dans certains cas spécifiques, une Commission peut recouvrir une discipline commune à une ou plusieurs Divisions et devient alors une Commission de l'ensemble de ces Divisions.

22. Les Commissions sont créées ou supprimées par le Comité Exécutif sur recommandation de la ou des Division(s) désirant les créer ou les supprimer. Les activités d'une Commission sont organisées par un Comité d'Organisation présidé par le Président de Commission. Le Président de Commission et un Vice-Président sont nommés par le ou les Comité(s) d'Organisation de la ou des Division(s) correspondantes sur proposition du Comité d'Organisation de la Commission.

XII. BUDGET ET COTISATIONS

23. En vue de chaque Assemblée Générale, le Comité Exécutif prépare une proposition de budget couvrant la période jusqu'à l'Assemblée Générale suivante, accompagnée des comptes de l"Union pour la période précédente. Il les soumet au Comité des Finances en vue de leur examen avant leur soumission au vote de L'Assemblée Générale.

23.a. Le Comité des Finances examine les comptes de l'Union au regard des dépenses approuvées par l'Assemblée Générale précédente, selon l'exécution qu'en a fait le Comité Exécutif. Il se penche aussi sur l'adéquation ou non du budget proposé en vue de la mise en œuvre des orientations de l'Assemblée Générale. Il soumet son rapport à l'Assemblée Générale avant ses décisions concernant l'approbation des comptes et du budget.

23.b. Le montant des unités de cotisation est décidé par l'Assemblée Générale au titre du processus d'approbation du budget.

23.c. Chaque Membre National verse annuellement un nombre d'unités de cotisation correspondant à sa catégorie. Le nombre d'unités de cotisation pour chaque catégorie est spécifié dans le Règlement Intérieur.

23.d. Un vote sur les questions du ressort de l'Article 23 n'est valable que si les deux tiers des représentants des Membres Nationaux ayant le droit de vote en vertu de l'Article 14.a. prennnent part à ce vote. Dans tous les cas, une abstention ne constitue par un vote, mais une déclaration selon laquelle que le Membre National décide de ne pas voter.

23.e. Les Membres Nationaux ayant un statut d'intérim payent annuellement une demi-unité de cotisation.

23.f. Les Membres Nationaux ayant un statut prospectif ne payent aucune cotisation.

23.g. Le paiement des contributions relève de la responsabilité des Membres Nationaux. La responsabilité de chaque Membre National relative à l'Union se limite au montant des cotisations dues au cours de l'année.

XIII. PLEINS POUVOIRS

24. Si, en cas d'événements hors du contrÃ´le de l'Union, les circonstances font qu'il est impossible de respecter pleinement les dispositions des Statuts et du Règlement Intérieur de l'Union, le Comité Exécutif et les Membres du Bureau, dans l'ordre spécifié ci-après, agiront comme ils l'estimeront nécessaire pour assurer la continuité du fonctionnement de l'Union. De telles décisions seront rapportées aux Membres Nationaux dès que cela sera possible, jusqu'à ce qu'une Assemblée Générale Ordinaire ou Extraordinaire soit convoquée.

L'ordre hiérarchique est le suivant : le Comité Exécutif, en réunion ou par correspondance ; le Président de L'Union ; le Secrétaire Général ; ou à défaut de possibilité ou de disponibilité de ce qui précède, un des Vice-Présidents.

XIV. DISSOLUTION DE L'UNION

25. La décision de dissoudre l'Union n'est recevable que si elle est prise par l'Assemblée Générale avec l'approbation des trois quarts des Membres Nationaux ayant le droit de voter en vertu de l'Article 14.a. Une telle décision spécifiera une procédure visant à apurer les comptes et à disposer des actifs de l'Union.

XV. CLAUSES FINALES

26. Les présents Statuts entrent en vigueur le 21 août 2012.

27. Ils devront être publiés en français et en anglais. A toute fin juridique, la version française fera autorité.

3. IAU Bye-Laws

I. MEMBERSHIP

1. An application for admission to the Union as a National Member shall be submitted to the General Secretary by the proposing organization at least eight months before the next ordinary General Assembly.

2. The Executive Committee shall examine the application and resolve any outstanding issues concerning the nature of the proposed National Member and the category of membership (§ VII.25). Subsequently, the Executive Committee shall forward the application to the General Assembly for decision, with its recommendation as to its approval or rejection.

3. The Executive Committee shall examine any proposal by a National Member to change its category of adherence to a more appropriate level. If the Executive Committee is unable to approve the request, either party may refer the matter to the next General Assembly.

4. Individual Members are admitted by the Executive Committee upon the nomination of a National Member or the President of a Division. The Executive Committee shall publish the criteria and procedures for membership, and shall consult the Membership Committee before admitting new Individual Members.

II. GENERAL ASSEMBLY

5. The ordinary General Assembly meets, as a rule, once every three years. Unless determined by the previous General Assembly, the place and date of the ordinary General Assembly shall be fixed by the Executive Committee and be communicated to the National Members at least one year in advance.

6. The President may summon an extraordinary General Assembly with the consent of the Executive Committee, and must do so at the request of at least one third of the

National Members. The date, place, and agenda of business of an extraordinary General Assembly must be communicated to all National Members at least two months before the first day of the Assembly.

7. Matters to be decided upon by the General Assembly shall be submitted for consideration by those concerned as follows, counting from the first day of the General Assembly:

7.a. A motion to amend the Statutes or Bye-Laws may be submitted by a National Member or by the Executive Committee. Any such motion shall be submitted to the General Secretary at least nine months in advance and be forwarded, with the recommendation of the Executive Committee as to its adoption or rejection, to the National Members at least six months in advance.

7.b. The General Secretary shall distribute the draft budget prepared by the Executive Committee to the National Members at least eight months in advance. Any motion to modify this budget, or any other matters pertaining to it, shall be submitted to the General Secretary at least six months in advance. The Executive Committee shall consider whether or not to adopt any such motion in a modified budget, which shall be distributed to the National Members at least four months in advance. Should the Executive Committee decide to reject the motion it shall also be submitted to the General Assembly with the reasons for its rejection.

7.c. Any motion or proposal concerning the administration of the Union, and not affecting the budget, by a National Member, or by the Organizing Committee of a Scientific Division of the Union, shall be placed on the Agenda of the General Assembly, provided it is submitted to the General Secretary, in specific terms, at least six months in advance.

7.d. Any motion of a scientific character submitted by a National Member, a Scientific Division of the Union, or by an ICSU Scientific Committee or Program on which the Union is formally represented, shall be placed on the Agenda of the General Assembly, provided it is submitted to the General Secretary, in specific terms, at least six months in advance.

7.e. The complete agenda, including all such motions or proposals, shall be prepared by the Executive Committee and submitted to the National Members at least four months in advance.

8. The President may invite representatives of other organizations, scientists in related fields, and young astronomers to participate in the General Assembly. Subject to the agreement of the Executive Committee, the President may authorise the General Secretary to invite representatives of other organizations, and the National Members or other appropriate IAU bodies to invite scientists in related fields and young astronomers.

III. SPECIAL NOMINATING COMMITTEE

9. The Special Nominating Committee consists of the President and past President of the Union, a member proposed by the retiring Executive Committee, and four members

selected by the representatives of the National Members from up to twelve candidates proposed by Presidents of Divisions, with due regard to an appropriate distribution over the major branches of astronomy.

9.a. Except for the President and immediate past President, present and former members of the Executive Committee shall not serve on the Special Nominating Committee. No two members of the Special Nominating Committee shall belong to the same nation or National Member.

9.b. The General Secretary and the Assistant General Secretary participate in the work of the Special Nominating Committee in an advisory capacity, and the President-Elect may participate as an observer.

10. The Special Nominating Committee is appointed by the General Assembly, to which it reports directly. It assumes its duties immediately after the end of the General Assembly and remains in office until the end of the ordinary General Assembly next following that of its appointment, and it may fill any vacancy occurring among its members.

IV. OFFICERS AND EXECUTIVE COMMITTEE

11. Terms of office:

11.a. The President of the Union remains in office until the end of the ordinary General Assembly next following that of election. The President-Elect succeeds the President at that moment.

11.b. The General Secretary and the Assistant General Secretary remain in office until the end of the ordinary General Assembly next following that of their election. Normally the Assistant General Secretary succeeds the General Secretary, but both officers may be re-elected for another term.

11.c. The Vice-Presidents remain in office until the end of the ordinary General Assembly following that of their election. They may be immediately re-elected once to the same office.

11.d. The elections take place at the last session of the General Assembly, the names of the candidates proposed having been announced at a previous session.

12. The Executive Committee may fill any vacancy occurring among its members. Any person so appointed remains in office until the end of the next ordinary General Assembly.

13. The past President and General Secretary become advisers to the Executive Committee until the end of the next ordinary General Assembly. They participate in the work of the Executive Committee and attend its meetings without voting rights.

14. The Executive Committee shall formulate Working Rules to clarify the application of the Statutes and Bye-Laws. Such Working Rules shall include the criteria and procedures by which the Executive Committee will review applications for Individual Membership; standard Terms of Reference for the Scientific Commissions of the Union;

rules for the administration of the Union's financial affairs by the General Secretary; and procedures by which the Executive Committee may conduct business by electronic or other means of correspondence. The Working Rules shall be published electronically and in the Transactions of the Union.

15. The Executive Committee appoints the Union's official representatives to other scientific organizations.

16. The Officers and members of the Executive Committee cannot be held individually or personally liable for any legal claims or charges that might be brought against the Union.

V. SCIENTIFIC DIVISIONS

17. The Divisions of the Union shall pursue the scientific objects of the Union within their respective fields of astronomy. Activities by which they do so include the encouragement and organization of collective investigations, and the discussion of questions relating to international agreements, cooperation, or standardization. They shall report to each General Assembly on the work they have accomplished and such new initiatives as they are undertaking.

18. Each Scientific Division shall consist of:

18.a. An Organizing Committee, normally of 6-12 persons, including the Division President and Vice-President, and a Division Secretary appointed by the Organizing Committee from among its members. The Committee is responsible for conducting the business of the Division.

18.b. Members of the Union accepted by the Organizing Committee in recognition of their special experience and interests.

19. Normally, the Division President is succeeded by the Vice-President at the end of the General Assembly following their election, but both may be re-elected for a second term. Before each General Assembly, the Organizing Committee shall organize an election from among the membership, by electronic or other means suited to its scientific structure, of a new Organizing Committee to take office for the following term. Election procedures should, as far as possible, be similar among the Divisions and require the approval of the Executive Committee.

20. Each Scientific Division may structure its scientific activities by creating a number of Commissions. In order to monitor and further the progress of its field of astronomy, the Division shall consider, before each General Assembly, whether its Commission structure serves its purpose in an optimum manner. It shall subsequently present its proposals for the creation, continuation or discontinuation of Commissions to the Executive Committee for approval.

21. With the approval of the Executive Committee, a Division may establish Working Groups to study welldefined scientific issues and report to the Division. Unless specifically re-established by the same procedure, such Working Groups cease to exist at the

next following General Assembly.

VI. SCIENTIFIC COMMISSIONS

22. A Scientific Commission shall consist of:

22.a. A President and an Organizing Committee consisting of 4-8 persons elected by the Commission membership, subject to the approval of the Organizing Committee of the Division;

22.b. Members of the Union, accepted by the Organizing Committee, in recognition of their special experience and interests, subject to confirmation by the Organizing Committee of the Division.

23. A Commission is initially created for a period of six years. The parent Division may recommend its continuation for additional periods of three years at a time, if sufficient justification for its continued activity is presented to the Division and the Executive Committee. The activities of a Commission are governed by Terms of Reference, which are based on a standard model published by the Executive Committee and are approved by the Division.

24. With the approval of the Division, a Commission may establish Working Groups to study well-defined scientific issues and report to the Commission. Unless specifically re-appointed by the same procedure, such Working Groups cease to exist at the next following General Assembly.

VII. ADMINISTRATION AND FINANCES

25. Each National Member pays annually to the Union a number of units of contribution corresponding to its category as specified below. National Members with interim status pay annually one half unit of contribution, and those with prospective status pay no dues.

Categories as defined in Statutes, §10:	I	II	III	IV	V	VI	VII	VIII	IX	X	XI	XII
Number of units of contribution:	1	2	4	6	10	14	20	27	35	45	60	80

26. The income of the Union is to be devoted to its objects, including:

26.a. the promotion of scientific initiatives requiring international co-operation;

26.b. the promotion of the education and development of astronomy world-wide;

26.c. the costs of the publications and administration of the Union.

27. Funds derived from donations are reserved for use in accordance with the instructions of the donor(s). Such donations and associated conditions require the approval of the Executive Committee.

28. The General Secretary is the legal representative of the Union. The General Secretary is responsible to the Executive Committee for not incurring expenditure in excess of the amount specified in the budget as approved by the General Assembly.

29. The General Secretary shall consult with the Finance Committee (cf. Statutes § 13.d.) in preparing the accounts and budget proposals of the Union, and on any other matters of major importance for the financial health of the Union. The comments and advice of the Finance Committee shall be made available to the Officers and Executive Committee as specified in the Working Rules.

30. An Administrative office, under the direction of the General Secretary, conducts the correspondence, administers the funds, and preserves the archives of the Union.

31. The Union has copyright to all materials printed in its publications, unless otherwise arranged.

VIII. FINAL CLAUSES

32. These Bye-Laws enter into force on 21 August 2012.

33. The present Bye-Laws are published in French and English versions. For legal purposes, the French version is authoritative.

4. Bye-laws de l'UAI, Honolulu, 21 août 2015

I. I. QUALITÉ DE MEMBRE

1. Une demande d'admission à l'Union en qualité de Membre National devra être soumise au Secrétaire Général par l'organisation demandeuse au moins huit mois avant l'Assemblée Générale ordinaire suivante.

2. Le Comité Exécutif devra examiner la demande et résoudre toute question en suspens concernant la nature du Membre National proposé et le type d'adhésion (§ VII.25). Ensuite, le Comité Exécutif transmettra la demande à l'Assemblée Générale afin que celle-ci prenne sa décision, en l'accompagnant de sa recommandation quant à son approbation ou son rejet.

3. Le Comité Exécutif doit examiner toute proposition d'un Membre National de changement de son type d'adhésion à un niveau plus approprié. Si le Comité Exécutif ne peut approuver la demande, l'une ou l'autre partie peut référer la question à l'Assemblée Générale suivante.

4. Les Membres Individuels sont admis par le Comité Exécutif lors de la nomination d'un Membre National ou du Président d'une Division. Le Comité Exécutif doit publier les critères et procédures d'adhésion, et doit consulter le Comité d'Adhésion avant d'admettre de nouveaux Membres Individuels.

II. I. ASSEMBLÉE GÉNÉRALE

5. L'Assemblée Générale ordinaire se réunit, en règle générale, une fois tous les trois ans. Sauf si l'Assemblée Générale précédente l'a décidé, le lieu et la date de l'Assemblée Générale ordinaire seront fixés par le Comité Exécutif et communiqués aux Membres Nationaux au minimum un an à l'avance.

6. Le Président peut convoquer une Assemblée Générale extraordinaire avec l'autorisation du Comité Exécutif, et doit le faire à la demande d'au moins un tiers des Membres Nationaux. La date, le lieu et l'ordre du jour d'une Assemblée Générale extraordinaire doivent être communiqués à tous les Membres Nationaux au minimum deux mois avant le premier jour de l'Assemblée.

7. Les points devant être traités par l'Assemblée Générale doivent être soumis à sa considération par les personnes concernées comme suit, en comptant à partir du premier jour de l'Assemblée Générale :

7.a. une proposition de modification des Statuts peut être soumise par un Membre National ou par le Comité Exécutif. Toute proposition doit être soumise au Secrétaire Général au minimum neuf mois au préalable et transmise, avec la recommandation du Comité Exécutif quant à son adoption ou son rejet, aux Membres Nationaux au minimum six mois au préalable.

7.b. Le Secrétaire Général distribuera le budget prévisionnel préparé par le Comité Exécutif aux Membres Nationaux au moins huit mois au préalable. Toute proposition de modification de ce budget, ou toute autre question y afférente, sera soumise au Secrétaire Général au minimum six mois au préalable. Le Comité Exécutif doit considérer si adopter ou non une telle proposition de modification du budget, laquelle sera distribuée aux Membres Nationaux au minimum quatre mois au préalable. Si le Comité Exécutif décide de rejeter la proposition, celle-ci doit également être soumise à l'Assemblée Générale accompagnée des motifs de son rejet.

7.c. Toute motion ou proposition concernant l'administration de l'Union, et n'affectant pas le budget, par un Membre National, ou par le Comité Organisateur d'une Division Scientifique de l'Union, sera mise à l'Ordre du jour de l'Assemblée Générale, à condition qu'elle soit soumise au Secrétaire Général, en des termes spécifiques, au minimum six mois au préalable.

7.d. Toute proposition à caractère scientifique soumise par un Membre National, une Division Scientifique de l'Union, ou par un Comité ou Programme Scientifique ICSU dans lequel l'Union est formellement représentée, doit être mise à l'Ordre du jour de l'Assemblée Générale, à condition qu'elle soit soumise au Secrétaire Général, en des termes spécifiques, au minimum six mois au préalable.

7.e. L'ordre du jour complet, incluant toutes ces propositions, sera préparé par le Comité Exécutif et soumis aux Membres Nationaux au minimum quatre mois au préalable.

8. Le Président peut inviter des représentants d'autres organisations, scientifiques dans des domaines liés, et jeunes astronomes à participer à l'Assemblée Générale. Avec l'accord du Comité Exécutif, le Président peut autoriser le Secrétaire Général à inviter des

représentants d'autres organisations, et les Membres Nationaux ou autres organes IAU appropriés à inviter des scientifiques dans des domaines liés et jeunes astronomes.

III. COMITÉ DE NOMINATION SPÉCIAL

9. Le Comité de Nomination Spécial est constitué du Président et de l'ancien Président de l'Union, d'un membre proposé par le Comité Exécutif sortant, et quatre membres sélectionnés par les représentants des Membres Nationaux jusqu'à douze candidats proposés par les Présidents des Divisions, en tenant bien compte d'une répartition appropriée dans les principaux rameaux d'astronomie.

9.a. À l'exception du Président et de l'ancien Président immédiat, les membres actuels et anciens membres du Comité Exécutif ne peuvent siéger au Comité de Nomination Spécial. Deux membres du Comité de Nomination Spécial ne peuvent provenir de la même nation ou du même Membre National.

9.b. Le Secrétaire Général et le Secrétaire Général Adjoint participent au travail du Comité de Nomination Spécial en qualité de conseillers, et le Président Élu peut participer en qualité d'observateur.

10. Le Comité de Nomination Spécial est nommé par l'Assemblée Générale dont il dépend directement. Il assume ses fonctions immédiatement après la fin de l'Assemblée Générale et demeure en poste jusqu'à la fin de l'Assemblée Générale ordinaire suivant celle de sa nomination, et il peut combler toute vacance se produisant parmi ses membres

IV. DIRIGEANTS ET COMITÉ EXÉCUTIF

11. Mandat :

11.a. Le Président de l'Union demeure en poste jusqu'à la fin de l'Assemblée Générale ordinaire suivant celle de son élection. Le Président Élu succès au Président actuel.

11.b. Le Secrétaire Général et le Secrétaire Général Adjoint demeurent en poste jusqu'à la fin de l'Assemblée Générale ordinaire suivant celle de leur élection. Normalement, le Secrétaire Général Adjoint succède au Secrétaire Général, mais les deux dirigeants peuvent être réélus pour un autre mandat.

11.c. Les Vice-Présidents demeurent en poste jusqu'à la fin de l'Assemblée Générale ordinaire suivant celle de leur élection. Ils peuvent être immédiatement réélus une fois à la même fonction.

11.d. Les élections ont lieu au cours de la dernière session de l'Assemblée Générale, les noms des candidats proposés ayant été annoncés au cours d'une session préalable.

12. Le Comité Exécutif peut combler toute vacance se produisant entre ses membres. Toute personne ainsi nommée demeure en poste jusqu'à la fin de l'Assemblée Générale ordinaire suivante.

13. Les anciens Président et Secrétaire Général deviennent conseillers du Comité Exécutif jusqu'à la fin de l'Assemblée Générale ordinaire suivante. Ils participent au travail du Comité Exécutif et assistent à ses réunions sans droits de vote.

14. Le Comité Exécutif formulera des Règles de Travail afin de clarifier l'application des Statuts. De telles Règles de Travail doivent inclure les critères et procédures selon lesquels le Comité Exécutif étudiera les demandes d'Adhésion Individuelle ; les Termes de Référence standards des Commissions Scientifiques de l'Union ; les règles d'administration des affaires financières de l'Union par le Secrétaire Général et les procédures selon lesquelles le Comité Exécutif peut conduire ses activités par des moyens de correspondance électroniques ou autres. Les Règles de Travail seront publiées électroniquement et dans les Transactions de l'Union.

15. Le Comité Exécutif nomme les représentants officiels de l'Union dans d'autres organisations scientifiques.

16. Les Dirigeants et membres du Comité Exécutif ne peuvent être tenus individuellement ou personnellement responsables pour toute réclamation ou toutes poursuites pouvant être engagées à l'encontre de l'Union.

V. DIVISIONS SCIENTIFIQUES

17. Les Divisions de l'Union doivent poursuivre les objets scientifiques de l'Union dans leurs domaines d'astronomie respectifs. Les activités y afférentes incluent l'encouragement et l'organisation d'investigations collectives, et la discussion de questions en rapport avec les accords internationaux, la coopération, ou la normalisation. Ils doivent rendre compte à chaque Assemblée Générale du travail qu'ils ont accompli et des initiatives qu'ils ont prises.

18. Chaque Division Scientifique est constituée de :

18.a. Un Comité de Pilotage, normalement 6 à 12 personnes, incluant le Président de la Division et le Vice-président, et un Secrétaire de Division nommé par le Comité de Pilotage parmi ses membres. Le Comité de Pilotage est responsable de conduire l'activité de la Division.

18.b. Les membres de l'Union acceptés par le Comité de Pilotage en reconnaissance de leurs expériences et intérêts particuliers.

19. Normalement, le Vice-président succède au Président de la Division à la fin de l'Assemblée Générale suivant leur élection, mais tous les deux peuvent être réélus pour un second mandat. Avant chaque Assemble Générale, le Comité de Pilotage organisera une élection parmi les membres, par des moyens électroniques ou autres adaptés à sa structure scientifique, d'un nouveau Comité de Pilotage devant prendre ses fonctions pour le mandat suivant. Les procédures d'élection doivent, dans la mesure du possible, être similaires parmi les Divisions et requièrent l'approbation du Comité Exécutif.

20. Chaque Division Scientifique peut structurer ses activités scientifiques en créant plusieurs Commissions. Afin de suivre l'évolution de son domaine d'astronomie, la Divi-

sion doit considérer, avant chaque Assemblée Générale, si la structure de sa Commission atteint son objectif de manière optimum. Elle doit ensuite présenter ses propositions pour la création, le maintien ou la suppression de Commissions au Comité Exécutif pour approbation.

21. Avec l'approbation du Comité Exécutif, une Division peut constituer des Groupes de Travail chargés d'étudier des questions scientifiques bien définies et d'en rendre compte à la Division. Sauf s'ils sont spécifiquement ré-établis par la même procédure, ces Groupes de Travail cesseront d'exister lors de l'Assemblée Générale qui suit immédiatement.

VI. COMMISSIONS SCIENTIFIQUES

22. Une Commission Scientifique est constituée de :

22.a. Un Président, un Vice-président et un Comité Organisateur constitué de 4 à 8 personnes élues par les membres de la Commission, sous réserve de l'approbation du Comité de Pilotage de la Division. De plus, un Secrétaire de Commission sera nommé par le Comité Organisateur parmi ses membres ;

22.b. Les membres de l'Union acceptés par le Comité Organisateur en reconnaissance de leurs expériences et intérêts particuliers.

23. Une Commission est initialement créée pour une période de six ans. La Division mère peut recommander sont maintien durant des périodes additionnelles de trois ans à un moment donné, si une justification suivante de son maintien est présentée à la Division et au Comité Exécutif. Les activités d'une Commission sont régies par les Termes de Référence, lesquels sont basés sur un modèle standard publié par le Comité Exécutif et approuvés par la Division.

24. Avec l'approbation de la Division, une Commission peut constituer des Groupes de Travail chargés d'étudier des questions scientifiques bien définies et d'en rendre compte à la Commission. Sauf s'ils sont spécifiquement renommés par la même procédure, ces Groupes de Travail cesseront d'exister lors de l'Assemblée Générale qui suit immédiatement.

VII. ADMINISTRATION AND FINANCES

25. Chaque Membre National paie chaque année à l'Union un nombre d'unités de contribution correspondant à sa catégorie telle que spécifiée ci-après. Les Membres Nationaux ayant un statut provisoire paient chaque année une demi-unité de contribution, et ceux ayant un statut potentiel ne paient aucun droit.

Catégories telles que définies à l'article 10 des Statuts, §10 :	I	II	III	IV	V	VI	VII	VIII	IX	X	XI	XII
Nombre d'unités de contribution :	1	2	4	6	10	14	20	27	35	45	60	80

26. Le revenu de l'Union doit être utilisé pour atteindre ses objectifs, notamment :

26.a. la promotion d'initiatives scientifiques requérant une coopération internationale ;

26.b. la promotion de l'éducation et du développement de l'astronomie dans le monde entier ;

26.c. les coûts des publications et d'administration de l'Union.

27. Les fonds provenant de donations doivent être strictement utilisés conformément aux instructions du/des donateur(s). De telles donations et conditions associées requièrent l'approbation du Comité Exécutif.

28. Le Secrétaire Général est le représentant légal de l'Union. Le Secrétaire Général est responsable envers le Comité Exécutif de ne pas engager de dépenses supérieures au montant spécifié dans le budget tel qu'approuvé par l'Assemblée Générale.

29. Le Secrétaire Général doit consulter les Comités Financiers (cf. Statuts, § 13.d.) lors de la préparation des comptes et propositions de budget de l'Union, et sur toute autre question importante pour la santé financière de l'Union. Les commentaires et avis du Comité Financier doivent être mis à disposition des Dirigeants et du Comité Exécutif comme il est spécifié dans les Règles de Travail.

30. Un Responsable administratif, sous la direction du Secrétaire Général, conduit la correspondance, administre les fonds, et préserve les archives de l'Union.

31. L'Union a un droit d'auteur sur tous les éléments imprimés dans ses publications, sauf disposition contraire.

VIII. CLAUSES FINALES

32. Les présents Statuts entrent en vigueur le 21 août 2012.
**

33. Les présents Statuts sont publiés en français et en anglais. À des fins légales, la version française prévaut.

5. IAU Working Rules

INTRODUCTION AND RATIONALE

The Statutes of the International Astronomical Union (IAU) define the goals and organizational structure of the Union, while the Bye-Laws specify the main tasks of the various bodies of the Union in implementing the provisions of the Statutes. The Working Rules are designed to assist the membership and governing bodies of the Union in carrying out these tasks in an appropriate and effective manner. Each of the sections below is preceded by an introduction outlining the goals to be accomplished by the procedures specified in the succeeding paragraphs. The Executive Committee updates the Working Rules as necessary to reflect current procedures and to optimize the services of the IAU to its membership.

I. NON-DISCRIMINATION

The International Astronomical Union (IAU) follows the regulations of the International Council for Science (ICSU) and concurs with the actions undertaken by their Standing Committee on Freedom in the Conduct of Science on non-discrimination and universality of science (cf. § 22 below)

II. NATIONAL MEMBERSHIP

The aim of the rules for applications for National Membership is to ensure that the proposed National Member adequately represents an astronomical community not already represented by another Member, and that such membership will be of maximum benefit for the community concerned (cf. Statutes §IV).

1. Applications for National Membership should therefore clearly describe the following essential conditions :

1.a. the precise definition of the astronomical community to be represented by the proposed Member ;

1.b. the present state and expected development of that astronomical community ;

1.c. the manner in which the proposed National Member represents this community ;

1.d. whether the application is for membership on a permanent, interim, or prospective basis ; and

1.e. the category in which the prospective National Member wishes to be classified (cf. Bye-Laws §25).

1.f. the process by which the National Membership annual dues will be paid promptly and in full.

2. Applications for National Membership shall be submitted to the General Secretary, who will forward them to the Executive Committee for review as provided in the Statutes.

III. INDIVIDUAL MEMBERSHIP

Professional scientists whose research is directly relevant to some branch of astronomy are eligible for election as Individual Members of the Union (cf. Statutes §V). Individual Members are normally admitted by the Executive Committee on the proposal of a National Member. However, Presidents of Divisions may also propose individuals for membership in cases when the normal procedure is not applicable or practicable (cf. Bye-Laws §4). The present rules are intended to ensure that all applications for membership are processed on a uniform basis, and that all members are fully integrated in and contributing to the activities of the Union.

3. The term "Professional Scientist" shall normally designate a person with a doctoral degree (Ph.D.) or equivalent experience in astronomy or a related science, and whose professional activities have a substantial component of work related to astronomy.

4. National Members and Division Presidents may propose Individual Members who fall outside the category of professional scientist but who have made major contributions to the science of astronomy, e.g., through education or research related to astronomy. Such proposals should be accompanied by a detailed motivation for what should be seen as exceptions to the rule.

5. Eight months before an ordinary General Assembly, National Members and Presidents of Divisions will be invited to propose new Individual Members; these proposals should reach the General Secretary no later than five months before the General Assembly. Late proposals will normally not be taken into consideration. Proposals from Presidents of Divisions will be communicated by the IAU 3 months before the General Assembly to the relevant National Members, if any, who may add the person(s) in question to their own list of proposals.

6. National Members shall promptly inform in writing the General Secretary of the death of any Individual Member represented by them. National Members are also urged to propose the deletion of Individual Members who are no longer active in astronomy by including a written agreement of the member concerned. Such proposals should be submitted to the General Secretary at the same time as proposals for new Individual Members.

7. Proposals for membership shall include the full name, date of birth, and nationality of the candidate, postal and electronic addresses, the University, year, and subject of the M.Sc./Ph.D. or equivalent degree, current affiliation and occupation, the proposing National Member or Division, the Division(s) and/or Commission(s) which the candidate wishes to join, and any further detail that might be relevant.

8. The standing Membership Committee advises the Executive Committee on its behalf on matters related to the admission of Individual Members (Statute 13f.).

8.a Three months before a General Assembly, the Membership Committee shall prepare a list of at least 10 Individual Members of the Union who accept to serve on the Committee for the next triennium if elected, including a nominee for Chair. The General Secretary shall forward this list to the National Members and invite additional nominations from them with a deadline of one month before the General Assembly. The Membership Committee shall verify that the resulting slate complies with the rules in Statutes 13 and with general principles of scientific, geographical and gender balance. Members shall not normally serve more than two consecutive terms, and it is desirable that roughly half of the members are replaced at each election. The Chair of the Membership Committee shall present the resulting slate of nominations to the National Members together with the report of the Committee on the previous triennium at the beginning of the General Assembly, for final election at its closing session.

9. The General Secretary shall submit all proposals for Individual Membership to the Membership Committee for review, consolidated into two lists :

9.a one containing all proposals by National Members ; and

9.b one containing all proposals by Presidents of Divisions, in accordance with Bye-Law §4 .

10. The Membership Committee shall examine all proposals for individual membership and advise the Executive Committee on the proposals for individual membership.

11. In exceptional cases, the Executive Committee may, on the proposal of a Division, admit an Individual Member between General Assemblies. Such proposals shall be prepared as described above (cf. § 2) and submitted with a justification of the request to bypass the normal procedure. The Executive Committee shall consult the Nominating Committee or relevant National Member before approving such exceptions to the normal procedure.

12. The General Secretary shall maintain updated lists of all National and Individual Members, and shall make these available to the membership in electronic form. The procedures for dissemination of these lists shall be set by the General Secretary in such a way that the membership directory be properly protected against unintended or inappropriate use.

IV. RESOLUTIONS OF THE UNION

Traditionally, the decisions and recommendations of the Union on scientific and organizational matters of general and significant importance are expressed in the Resolutions of the Union. In order for such Resolutions to carry appropriate weight in the international community, they should address astronomical matters of significant impact on the international society, or matters of international policy of significant importance for the international astronomical community as a whole.

Resolutions should be adopted by the Union only after thorough preparation by the relevant bodies of the Union. The proposed resolution text should be essentially complete before the beginning of the General Assembly, to allow Individual and National Members time to study them before discussion and debate by the General Assembly. The following procedures have been designed to accomplish this :

13. Proposals for Resolutions to be adopted by the Union may be submitted by a National Member, by the Executive Committee, a Division, a Commission or a Working Group. They should address specific issues of the nature described above, define the objectives to be achieved, and describe the action(s) to be taken by the Officers, Executive Committee, or Divisions to achieve these objectives.

14. Resolutions proposed for vote by the Union fall in three categories as set out in Article 14 of the Union's Statutes :

14.a. Resolutions with implications for the budget of the Union (Statute 14a) ; or

14.b. Resolutions affecting the administration of the Union but without financial implications (Statute 14b).

14.c. Resolutions of a primarily scientific nature (Statute 14).

Proposals for Resolutions should be submitted on standard forms appropriate for each type, which are available from the IAU Secretariat. They may be submitted in either English or French and will be discussed and voted upon in the original language. Upon submission each proposed Resolution is posted on the Union web site. When the approved Resolutions are published, a translation to the other language will be provided.

15. Resolutions with implications for the budget of the Union must be submitted to the General Secretary at least nine months before the General Assembly in order to be taken into account in the budget for the impending triennium. All other Resolutions must be submitted to the General Secretary six months (Bye-Laws 7c and 7d) before the beginning of the General Assembly. The Executive Committee may decide to accept late proposals in exceptional circumstances.

16. Before being submitted to the vote of the General Assembly, proposed Resolutions will be examined by the Executive Committee, Division Presidents, and by a Resolutions Committee, which is nominated by the Executive Committee. The Resolutions Committee consists of at least three members of the Union, one of whom should be a member of the Executive Committee, and one of whom should be a continuing member from the previous triennium. It is appointed by the General Assembly during its final session and remains in office until the end of the following General Assembly.

17. The Resolutions Committee will examine the content, wording, and implications of all proposed Resolutions promptly after their submission. In particular, it will address the following points :

i. suitability of the subject for an IAU Resolution;

ii. correct and unambiguous wording;

iii. consistency with previous IAU Resolutions.

The Resolutions Committee may refer a Resolution back to the proposers for revision or withdrawal if it perceives significant problems with the text, but can neither withdraw nor modify its substance on its own initiative. The Resolutions Committee advises the Executive Committee whether the subject of a proposed Resolution is primarily a matter of policy or primarily scientific. The Resolutions Committee will also notify the Executive Committee of any perceived problems with the substance of a proposed Resolution.

18. The Executive Committee will examine the substance and implications of all proposed Resolutions. Proposed Resolutions shall be published in the General Assembly Newspaper before the final session, and shall state if the Resolution is open to electronic voting after the General Assembly. The Resolutions Committee will present the proposals during a plenary session of the General Assembly with its own recommendations, and those of the Executive Committee, if any, for their approval or rejection. A representative of the body proposing the Resolution will be given the opportunity to defend the Resolution in front of the General Assembly, after which a general discussion shall take place

19. Resolutions with implications for the budget of the Union are voted upon by the National Members during the final plenary session of the General Assembly. Other resolutions may be voted upon by the National Members or by Individual Members as

appropriate according to the Statutes of the Union by correspondence after the General Assembly. The Union will facilitate electronic discussion of all Resolutions on the Union website in advance of a vote either at the General Assembly or electronically (Statutes 15a. and 15.b.).

V. EXTERNAL RELATIONS

Contacts with other international scientific organizations, national and international public bodies, the media, and the public are increasing in extent and importance. In order to maintain coherent overall policies in matters of international significance, clear delegation of authority is required. Part of this is accomplished by having the Union's representatives in other scientific organization appointed by the Executive Committee (cf. Bye-Laws §15). Supplementary rules are given in the following.

20. Representatives of the Union in other scientific organizations are appointed by the Executive Committee upon consultation with the Division(s) in the field(s) concerned.

21. In other international organizations, e.g. in the United Nations Organization, the Union is normally represented by the General Secretary or Assistant General Secretary, as decided by the Executive Committee.

22. The Union strongly supports the policies of the International Council for Science (ICSU) as regards the freedom and universality of science. Participants in IAU sponsored activities who feel that they may have been subjected to discrimination are urged, first, to seek clarification of the origin of the incident, which may have been due to misunderstandings or to the cultural differences encountered in an international environment. Should these attempts not prove successful, contact should be made with the General Secretary who will take steps to resolve the issue.

23. Public statements that are attributed to the Union as a whole can be made only by the President, the General Secretary, or the Executive Committee. The General Secretary may, in consultation with the relevant Division, appoint Individual Members of the Union with special expertise in questions that attract the attention of media and the general public as IAU spokespersons on specific matters.

VI. FINANCIAL MATTERS

The great majority of the Union's financial resources are provided by the National Members, as laid out in the Statutes §XII and Bye-Laws §VII. The purpose of the procedures described below is twofold : (i) to provide the best possible advice and guidance to the General Secretary and Executive Committee in planning and managing the Union ?s financial affairs, and (ii) to provide National Members with a mechanism for continuing input to and oversight over these affairs between and in preparation for the General Assemblies. The procedures adopted to accomplish this are as follows :

24. At the end of each of its final sessions the General Assembly appoints a Finance Sub-Committee of 5-6 members, including a Chair.

24.a. Three months before a General Assembly, the Finance Committee shall prepare a list of at least 10 Individual Members of the Union who accept to serve on the Committee for the next triennium if elected, including a nominee for Chair. The General Secretary shall forward this list to the National Members and invite additional nominations from them with a deadline of one month before the General Assembly. The Finance Committee shall verify that the resulting slate complies with the rules in Statutes 13 and with general principles of scientific, geographical and gender balance. Members shall not normally serve more than two consecutive terms, and it is desirable that roughly half of the members are replaced at each election. The Chair of the Finance Committee shall present the resulting slate of nominations to the National Members together with the report of the Committee on the previous triennium at the beginning of the General Assembly, for final election at its closing session. The Finance Committee remains in office until the end of the next General Assembly (cf. Statutes §13.d.) and cooperates with the National Members, Executive Committee and General Secretary in the following manner :

24.b. After the end of each year the General Secretary will call for a legal audit of the accounts by a properly licensed, external auditor. The auditor will make a report addressed to the General Assembly. The General Secretary provides the Finance Sub-Committee with the auditor's report and summary reports covering the financial performance of the Union as compared to the approved budget, together with an analysis of any significant departures, and information on any Executive Committee approvals of budget changes. Upon receipt of the above reports from the General Secretary, the Finance Sub-Committee examines the accounts of the Union in the light of the corresponding budget and any relevant later decisions by the Executive Committee. It reports its findings and recommendations to the Executive Committee at its next meeting. The Finance Sub-Committee may at any time, at the request of the Executive Committee or the General Secretary, or on its own initiative, advise the General Secretary and/or the Executive Committee on any aspect of the Union's financial affairs.

24.c. Towards the end of the year preceding that of a General Assembly, the General Secretary shall submit a preliminary draft of the budget for the next triennium to the Finance Committee for review. The draft budget, updated as appropriate following the comments and advice of the Finance Committee, shall be submitted to the Executive Committee for approval together with the report of the Finance Committee and shall be sent to the National Members as a draft budget as stipulated in Bye-Law 7b. The final budget proposal as approved by the Executive Committee shall be submitted to the National Members with a statement of the views of the Finance Committee on the proposal.

24.d. Before the first session of a General Assembly, the Finance Sub-Committee shall submit a report, including the auditor's reports, to the Executive Committee on its findings and recommendations concerning the development of the Union's finances over the preceding triennium. The Finance Sub-Committee shall also prepare, in consultation with the National Members, a slate of candidates for the composition of the Finance Sub-Committee in the next triennium, preferably providing a balance between new and continuing members.

24.e. The report of the Finance Committee, together with the audited detailed accounts and the earlier comments on the proposed budget for the next triennium, will form a suitable basis for the discussions of the Finance Committee leading to its recommendations to the General Assembly concerning the approval of the accounts for the previous triennium and the budget for the next triennium, as well as the new Finance Sub-Committee to serve during that period.

25. The General Secretary is responsible for managing the Union's financial affairs according to the approved budget (cf. Bye-Laws §28).

25.a. In response to changing circumstances, the Executive Committee may approve such specific changes to the annual budgets as are consistent with the intentions of the General Assembly when the budget was approved.

25.b. Unless authorized by the Executive Committee, the General Secretary shall not approve expenses exceeding the approved budget by more than 10% of any corresponding major budget line or 2% of the total budget in a given year, whichever is larger. This restriction does not apply in cases when external funding has been provided for a specific purpose, e.g. travel grants to a General Assembly.

25.c. Unless specifically identified in the approved budget, contractual commitments in excess of €50,000, or with performance terms in excess of 3 years require the additional approval of the Union President.

26. The National Representatives, in approving the accounts for the preceding triennium, discharge the General Secretary and the Executive Committee of liability for the period in question.

VII. RULES OF PROCEDURE FOR THE EXECUTIVE COMMITTEE

The Executive Committee must respond quickly to events and thus it needs to be able to have discussions and take decisions on a relatively short timescale and without meeting in person. The following rules, as required by Bye-Law 14, are designed to facilitate EC action in a flexible manner, while giving such decisions the same legal status as those taken at actual physical meetings.

27. The Executive Committee should meet in person at least once per year. In years of a General Assembly it should meet in conjunction with and at the venue of the General Assembly. In other years, the Executive Committee decides on the date and venue of its regular meeting. The meetings of the Executive Committee are chaired by the President or, if the President is unavailable, by the President Elect or by one of the Vice-Presidents chosen by the Executive Committee to serve in this capacity.

28. The date and venue of the next regular meeting of the Executive Committee shall be communicated at least six months in advance to all its members and the Advisors, and to all Presidents of Divisions. Any of these persons may then propose items for inclusion in the Draft Agenda of the meeting before the date posted on the IAU Deadlines page.

29. Outgoing and incoming Presidents of Divisions are invited to attend all non-confidential sessions of the outgoing and incoming Executive Committee, respectively, in the year of a General Assembly. The President will invite Presidents of Divisions to attend the meetings of the Executive Committee in the years preceding a General Assem-

bly. Division Presidents attend these sessions with speaking right, but do not participate in any voting.

30. The Executive Committee may take official decisions if at least half of its members participate in the discussion and vote on an issue. Decisions are taken by a simple majority of the votes cast. In case of an equal division of votes, the Chair's vote decides the issue. Members who are unable to attend may, by written or electronic correspondence with the President before the meeting, authorize another member to vote on her/his behalf or submit valid votes on specific issues.

31. If events arise that require action from the Executive Committee between its regular meetings, the Committee may meet by teleconference or by such electronic or other means of correspondence as it may decide. In such cases, the Officers shall submit a clear description of the issue at hand, with a deadline for reactions. If the Officers propose a specific decision on the issue, the decision shall be considered as approved unless a majority of members vote against it by the specified deadline. In case of a delay in communication, or if the available information is considered insufficient for a decision, the deadline shall be extended or the decision deferred until a later meeting at the request of at least two members of the Executive Committee.

32. The Officers of the IAU should, as a rule, meet once a year at the IAU Secretariat in order to discuss all matters of importance to the Union. The other members of the Executive Committee and the Division Presidents shall be invited to submit items for discussion at the Officers' Meetings and shall receive brief minutes of these Meetings.

33. Should any member of the Executive Committee have a conflict of interest on a matter before the Executive Committee that might compromise their ability to act in the best interests of the Union, they shall declare their conflict of interest to the Executive Committee, and such conflict shall be recorded by the Secretariat. The remaining members of the Executive Committee determine the appropriate level of participation in such issues for members with a potential conflict of interest.

VIII. SCIENTIFIC MEETINGS AND PUBLICATIONS

Meetings and their proceedings remain a major part of the activities of the Union. The purpose of scientific meetings is to provide a forum for the development and dissemination of new ideas, and the proceedings are a written record of what transpired.

34. The General Secretary shall publish in the Transactions and on the IAU web site rules for scientific meetings organized or sponsored by the Union.

35. The proceedings of the General Assemblies and other scientific meetings organized or sponsored by the Union shall, as a rule, be published. To ensure prompt publication of Proceedings of IAU Symposia and Colloquia, the Assistant General Secretary is authorized to oversee the production of the material for the Proceedings. The Union shall publish an Information Bulletin at regular intervals to keep Members informed of current and future events in the Union. The Union shall also publish a more informal, periodic Newsletter which it distributes electronically to its members. The Executive Committee decides on the scope, format, and production policies for such publications, with due

regard to the need for prompt publication of new scientific results and to the financial implications for the Union. At the present time, publications are in printed and in electronic form.

36. Divisions, Commissions, and Working Groups shall, with the approval of the Executive Committee, be encouraged to issue Newsletters or similar publications addressing issues within the scope of their activity.

IX. TERMS OF REFERENCE FOR DIVISIONS

The Divisions are the scientific backbone of the IAU. They have a main responsibility for monitoring the scientific and international development of astronomy within their subject areas, and for ensuring that the IAU will address the most significant issues of the time with maximum foresight, enterprising spirit, and scientific judgment. To fulfill this role IAU Divisions should maintain a balance between innovation and continuity. The following standard Terms of Reference have been drafted to facilitate that process, within the rules laid down in the Statutes §X and the Bye-Laws §V.

37. As specified in Bye-Law 18, the scientific affairs of the Division are conducted by an Organizing Committee of up to 12 members of the Division, headed by the Division President, Vice-President, and Secretary. Thus, all significant decisions of the Division require the approval of the Organizing Committee, and the President and Vice-President are responsible for organizing the work of the Committee so that its members are consulted in a timely manner. Contact information for the members of the Organizing Committee shall be maintained at the Division web site.

Unless agreed otherwise by the Executive Committee on a case by case basis, the President of a Division cannot be President of another Division or of a Commission, or be Chair of a Working Group.

38. Individual Members of the Union are admitted to membership in a Division by its Organizing Committee (cf. Bye-Laws §18). Individual Members active within the field of activity of the Division and interested in contributing to its development should contact the Division Secretary, who will consult the Organizing Committee on the admission of the candidates.

38.a. The Division Secretary shall maintain a list of Division members for ready consultation by the community, including their Commission memberships if any. Updates to the list shall be provided to the IAU Secretariat on a running basis.

38.b. Members may resign from a Division by so informing the Division Secretary.

38.c. In the event of a Division being newly formed, Individual Members can themselves elect to join the Division. Before the General Assembly following that at which the new Division was created its Organising Committee shall scrutinise and confirm the Division membership.

39. The effectiveness of the Division relies strongly on the scientific stature and dedication of its President and Vice-President to the mission of the Division. The Executive

Committee, in proposing new Division Presidents and Vice-Presidents for election by the General Assembly, will rely heavily on the recommendations of the Organizing Committee of the Division. In order to prepare a strong slate of candidates for these positions, and for the succession on the Organizing Committee itself, the following procedures shall normally apply :

39.a. Candidates are proposed and selected from the membership of the Division on the basis of their qualifications, experience, and stature in the fields covered by the Division. In addition, the Organizing Committees should have proper gender balance and broad geographical representation.

39.b. At least six months before a General Assembly, the Organizing Committee submits to the membership of the Division a list of candidates for President, Vice-President (for which there should be at least two persons willing to serve), Secretary, and the Organizing Committee for the next triennium. The Organizing Committee requests nominations from the entire membership in preparing this list, and then conducts a vote, normally electronically, among all Division members for the above offices, the results of which are reported to the General Secretary at least three months before the General Assembly. The Vice-President is normally nominated to succeed the President. The outgoing President participates in the deliberations of the new Organizing Committee in an advisory capacity.

39.c. If more names are proposed than there are positions to be filled on the new Organizing Committee, the outgoing Organizing Committee devises the procedure by which the requisite number of candidates is elected by the membership. The resulting list is communicated to the General Secretary at least two months before the General Assembly. The General Secretary may allow any outstanding issues to be resolved at the business meeting of the Division during the General Assembly. If for any reason the Organizing Committee has not been able to arrange for the election of new officers and an Organizing Committee by two months before the GA, the EC will nominate a VP and Organizing Committee at its first General Assembly meeting.

39.d. A member of the Organizing Committee normally serves a maximum of two terms, unless elected Vice-President of the Division in her/his second term. Presidents may serve for only one term.

39.e. The Organizing Committee decides on the procedures for designating the Division Secretary, who maintains the web site, records of the business and membership of the Division, and other rules for conducting its business by physical meetings or by correspondence.

39.f. In the event of a newly formed Division, paragraphs 39a - 39c do not apply. The Executive Committee shall consult the Organizing Committees of the relevant predecessor Divisions on possibl candidates for President and Vice-President of the new Division for the next triennium. The Executive Committee shall select the names to be proposed to the Generals Assembly for election.

39.g. As soon as possible after their election at a General Assembly, the President and Vice-President of the new Division shall request nominations to the Organising Committee from the membership of the Division and then conduct a vote among Division members, the results of which are reported to the General Secretary. The Organising Committee elects a Secretary from its membership.

40. A key responsibility of the Organizing Committee is to maintain an internal organization of Commissions and Working Groups in the Division which is conducive to the fulfillment of its mission. The Organizing Committee shall take the following steps to accomplish this task in a timely and effective manner :

40.a. Within the first year after a General Assembly - with the business meeting of the Commission at the General Assembly itself as a natural starting point - the Organizing Committee shall discuss with its Commissions, and within the Organizing Committee itself, if changes in its Commission and Working Group structure may enable it to accomplish its mission better in the future. As a rule, Working Groups should be created (following the rules in Bye-Law 21 and Bye-Law 23) for new activities that are either of a known, finite duration or are exploratory in nature. If experience, possibly from an existing Working Group, indicates that a major section of the Division's activities require a coordinating body for a longer period (a decade or more), the creation of a new Commission may be in order.

40.b. Whenever the Organizing Committee is satisfied that the creation of a new Working Group or Commission is well motivated, it may take immediate action as specified in Bye-Law 21 or Bye-Law 23. In any case, the Organizing Committee submits its complete proposal for the continuation, discontinuation, or merger of its Commissions and Working Groups to the General Secretary at least three months before the next General Assembly.

40.c. The President and Organizing Committee maintain frequent contacts with the other IAU Divisions to ensure that any newly emerging or interdisciplinary matters are addressed appropriately and effectively.

X. TERMS OF REFERENCE FOR COMMISSIONS

The role of the Commissions is to organize the work of the Union in specialized subsets of the fields of their parent Division(s), when the corresponding activity is judged to be of considerable significance over times of a decade or more. Thus, new Commissions may be created by the Executive Committee with the agreement of all the Divisions when fields emerge that are clearly in sustained long-term development and where the Union may play a significant role in promoting this development at the international level. Similarly, Commissions may be discontinued by the Executive Committee upon the recommendation of the parent Division when their work can be accomplished effectively by the parent Division. In keeping with the many-sided activities of the Union, Commissions may have purely scientific as well as more organizational and/or interdisciplinary fields. They will normally belong and report to one of the IAU Divisions, but may be common to two or more Divisions. The following rules apply if a Division has more than one Commission.

41. The activities of a Commission are directed by an Organizing Committee of 4-8 members of the Commission, headed by a Commission President and Vice-President (cf. Bye-Laws §22). A member of the Organizing Committee normally serves a maximum of two terms, unless elected Vice-President of the Commission in her/his second term.

Presidents may serve for only one term. All members of the Organizing Committee are expected to be active in this task, and are to be consulted on all significant actions of the Commission. The Organizing Committee appoints a Commission Secretary who maintains the records of the membership and activities of the Commission in co-operation with the Division Secretary and the IAU Secretariat. Contact information for the members of the Organizing Committee shall be maintained at the Commission web site.

Unless agreed otherwise by the Executive Committee on a case by case basis, the President of a Commission cannot be President of a Division or of another Commission, or be Chair of a Working Group.

42. Individual Members of the Union, who are active in the field of the Commission and wish to contribute to its progress, are admitted as members of the Commission by the Organizing Committee. Interested Members should contact the Commission Secretary, who will bring the request before the Organizing Committee for decision. Members may resign from the Commission by notifying the Commission Secretary. Before each General Assembly, the Organizing Committee may also decide to terminate the Commission membership of persons who have not been active in the work of the Commission; the individuals concerned shall be informed of such planned action before it is put into effect. The Commission Secretary will report all changes in the Commission membership to the Division Secretary and the IAU Secretariat.

43. At least six months before a General Assembly, the Organizing Committee submits to the membership of the Commission a list of candidates for President, Vice-President (for which there should be the names of two persons willing to serve), the Organizing Committee, and heads of Program Groups for the next triennium. The Organizing Committee requests nominations from the entire membership in preparing this list, and then conducts a vote, normally electronically, among all the members for the above offices, the results of which are reported to the General Secretary at least three months before the General Assembly. The Vice-President is normally nominated to succeed the President. The outgoing President participates in the deliberations of the new Organizing Committee in an advisory capacity. If more names are proposed than available elective positions, the outgoing Organizing Committee devises the procedure by which the requisite number of candidates is elected by the membership. The resulting list is submitted to the Organizing Committee of the parent Division(s) for approval before the end of the General Assembly. Members of the Organizing Committee normally serve a maximum of two terms, unless elected Vice-President of the Commission. Presidents may serve for only one term.

44. At least six months before each General Assembly, the Organizing Committee shall submit to the parent Division(s) a report on its activities during the past triennium, with its recommendation as to whether the Commission should be continued for another three years, or merged with one or more other Commissions, or discontinued. If a continuation is proposed, a plan for the activities of the next triennium should be presented, including those of any Working Groups which the Commission proposes to maintain during that period.

45. The Organizing Committee decides its own rules for the conduct of its business by physical meetings or (electronic) correspondence. Such rules require approval by the Organizing Committee of the parent Division(s).

46. The procedural rules applying to the establishment of a new Division shall also apply to the establishment of a new Commission. Where there is no 'relevant predecessor Commission(s)' the parent Division(s) shall submit to the potential membership of the new Commission a list of candidates for President, Vice-President and Organising Committee for the next triennium.

XI. SPECIAL NOMINATING COMMITTEE

47. Approximately six months before the start of the General Assembly the General Secretary shall invite Division Presidents and the members of the Executive Committee to nominate potential members of the SNC with a deadline of 3 months before the General Assembly. The Executive Committee shall prepare a list of candidates in consultation with the Membership Committee for appointment at the final Business Session of the General Assembly. The SNC, once appointed, shall elect its own Chair.

Transactions IAU, Volume XXIXB
Proceedings IAU Symposium No. Volume XXIXB, 2018
Piero Benvenuti, ed.
© International Astronomical Union 2019
doi:10.1017/S1743921318004039

CHAPTER VIII

NEW MEMBERS AND DECEASED MEMBERS AT THE GENERAL ASSEMBLY

Report of the IAU Membership Committee on the last Triennium (2012–2015)

Committee Members:

Sara R. Heap (Sara.R.Heap@nasa.gov)
Christian Henkel (chenkel@mpifr-bonn.mpg.de)
Myungshin Im (mim@astro.snu.ac.kr)
Lex Kaper (L.Kaper@uva.nl)
Rene A. Méndez Bussard (rmendez@u.uchile.cl)
Helmut O. Rucker (helmut.rucker@oeaw.ac.at)
Nikolay N. Samus (samus@sai.msu.ru/)
Ramotholo R. Sefako (rrs@saao.ac.za)

Time Frame

Activities started in early 2015. Individuals applying for nomination for IAU Individual Membership should have contacted their NCA or Adhering Organization before December 1, 2014. The deadline for approved nominations for Individual Members to be submitted to the IAU Secretariat by NCAs or Adhering Organizations was February 3, 2015. Therefore, the focus of this report are the last months prior to the IAU XXIX General Assembly (August 2015) in Honolulu.

Basic Statistics

For application period 2014/5, there were a total of 1275 applications from 49 countries. 1189 of these applications, i.e. 93.3%, could be accepted. To the rare exceptions: More than half of the Indian applications have been rejected by their NCA, mainly due to insufficient experience of the candidates. Seven candidates went to Morocco, which is suspended. Having applied through the US form, attempts to contact them personally remained unsuccessful. Several Australian, Belgian, and in one case also a UK application have been rejected by their NCAs because of a too recent Ph.D., reflecting different rules for acceptance which are defined to a significant part by the National Committees. Most Argentinian and a few Egypt applications did not provide a minimum of the information needed for an acceptance. Only few applications were coming from students. There was notably one such application also from Australia. In addition, a few double and one triple application could be identified. Overall, less than 100 applications were not successful.

Intrinsic Quality of the Applications

A large amount of applications, about 30%, was incomplete, lacking either birthday, year of acquired Ph.D, present professional status or the requested three papers. Many even lacked a combination of these few but very fundamental parts of general information. Under such conditions, it is very hard to judge the quality of a given application. Checking, for example, the missing publication list with ADS always implies the danger of including other people with the same last name and initial of the given name. Furthermore, quite often, topics or titles were given, while a detailed reference, including year, journal, volume and page number were all absent.

A related issue refers to the autentity of the given publications. Checking them in a few cases, they always turned out to be real. However, it would be fairly easy to just invent a few references and paper titles. One candidate chose a clever way. He listed his publications on ADS and attached the web-link in the first of the three boxes for the papers. This immediately revealed his entire bibliography. Nevertheless, providing such an exhausting information requires to have a name which is not too commonly found.

To the time of the application: According to the IAU working rules on individual membership, paragraph 5, late proposals will normally not be taken into consideration. This time, Columbians could apply at a later time than five months prior to the General Assembly. Since this is related to astro-political considerations (Columbia applies for National Membership), it is definitely not the business of the membership committee to comment on it except that we think that such delayed actions (where the IAU has little influence on timing) should be part of the report. There were also individual late applications from other countries, but their number was very small and did not require much additional work.

Another special case was Iran, which is currently suspended, but is on its way back to the IAU. So there are some listed new Iranian members. These will become official once Iran has recovered from suspended status.

Finally, it is important to state that the membership committee should have some people being member for more than one three year period. The leave of the very well informed previous chairman, Johannes Andersen, left a gap of knowledge. Furthermore, the collaboration between the different members of the membership committee could be improved, which will be an important point of discussion, when four of the old (and an unknown number of new) committee members will meet during the IAU GA in Hawaii.

Suggestions for Potential Improvements

(1) The question arises, in how far incomplete applications should be treated in the future. When papers or other important information are missing, should it really be the task of the IAU and its membership committee to try to find out about the missing facts? Or should the applications be screened superficially (either by the committee or the IAU

Secretariat) soon after their arrival? So that incomplete applications can be sent back right away – with the comment, that if the application will be resubmitted in a complete form within a reasonable amount of time, it will still be considered.

(2) In the application document, everyone can choose to provide either his/her private or professional address. Both is apparently considered to be equivalent. It may be a good idea to emphasize that a professional address is preferable. And to request a good explanation in case such an address can eventually not be given.

(3) A postdoc, having to leave his position after a few years and not yet being certain to find another job in the field, may not be a suitable candidate. Somehow the professional future of a given person should be secured. Otherwise, what to do with the many postdocs, who become IAU members but have to leave the field after their second or third postdoc occupation? While the IAU has certainly the possibilities to deal with such complexities, it requires working time which could perhaps be better used for other issues. To give an example: I joined the IAU soon after getting tenure. And always felt that this was about the right time.

The List

Below the names of the new members are given, recommended by the Membership Committee for admission at the GA. Those nominated by the Division Presidents are presented first, followed by the majority of successful applicants, which were nominated by National Members.

1. New members admitted at the General Assembly

The following lists give the names of the 1008 new Individual Members admitted at the XXIXth General Assembly, ordered by National Member.

For a complete list of IAU members, please consult the IAU Directory:
`http://www.iau.org/administration/membership/individual/`.

For a complete list of IAU National Members (member countries), please visit:
`http://www.iau.org/administration/membership/national/`.

NEW MEMBERS ADMITTED BY THE XXIXTH GENERAL ASSEMBLY HONOLULU, 3-14 AUG. 2015

[23 Oct. 2015]

1. New Individual Members nominated by Division Presidents†

Division B:
Hussain, Shahid (Pakistan)

† The names of the new Individual Members are repeated under the relevant National Member where applicable.

Division C:
Cheung, Sze-leung (Japan)
Mauduit, Jean-Christophe (France)
Mozaffari, Seyyed Mohammad (Iran)
Sandu, Oana (Germany)

Division E:
Baraka, Suleiman (Gaza, Palestine)
Vargas Dominguez, Santiago (Colombia)

Division F:
Guliyev, Ayyub (Azerbaijan)
Holman, Matthew (USA)
Nabiyev, Shaig (Azerbaijan)
Spoto, Federica (Italy)

Division G:
Showman, Adam P. (USA)

Division J:
Khan, Fazeel Mahmood (Pakistan)

2. New Individual Members of non-Member countries nominated by neighbouring National Members‡

Serbia
Kuzmanovska, Olgica (Macedonia, Former Rep of Yugoslavia)

South Africa
Beeharry, Girish (Mauritius)
Backes, Michael (Namibia)
Steenkamp, Riaan (Namibia)

Malaysia
Banks, Timothy (Singapore)

3. New Individual Members listed by National Member

Argentina

Cardaci, Monica
De Vito, Maria
Maffione, Nicolas
Miller, Marcelo
Pacheco, Ana
Podesta, Ricardo
Torres, Andrea
Weidmann, Walter

‡ The list of non-Member countries and their new Individual Members is given at the end (with country names in italics).

Armenia

Andreasyan, Ruben
Gyulzadyan, Marietta
Hakobyan, Artur
Hakobyan (Akopian), A.
Movsessian, Tigran
Ter-Kazarian, Gagik
Yeghikyan, Ararat

Australia

Allison, James
Bell, Martin
Bolejko, Krzysztof
Breen, Shari
Casagrande, Luca
Catinella, Barbara
Cole, Andrew
Cooke, Jeff
Cortese, Luca
Crighton, Neil
Curran, Peter
d'Orgeville, Céline
Davis, Tamara
Dawson, Joanne
Duffy, Alan
Ernest, Allan
Federrath, Christoph
Fenner, Yeshe
For, Bi-Qing
Hamacher, Duane
Howell, Eric
Huber, Daniel
Jacob, Andrew
Johnson, Megan
Kapinska, Anna
Keane, Evan
Kim, Hansik
Konstantopoulos, Iraklis
Lowe, Vicki
Marshall, Jonathan
Martell, Sarah
McClelland, David
Meurer, Gerhardt
O'Sullivan, Shane
Obreschkow, Danail
Parkinson, David
Popping, Attila
Reichardt, Christian
Rigaut, Francois
Robotham, Aaron
Shannon, Ryan
Shelyag, Sergiy
Sun, Xiaohui
Tescari, Edoardo
Tonini, Chiara
Trott, Cathryn
Zucker, Daniel

Austria

Baldovin, Carla
Baumjohann, Wolfgang
Goerdt, Tobias
Ratzka, Thorsten

Belgium

Buitink, Stijn
Clesse, Sébastien
Hughes, Thomas
Paladini, Claudia
Raskin, Gert
Sluse, Dominique
Van Grootel, Valérie
van Marle, Allard Jan

Brazil

Akras, Stavros
Aleman, Isabel
Alvarez Candal, Alvaro
Araujo, Rosana
Borges Fernandes, M.
Braga Ribas, Felipe
Charbonnier, Aldée
Cherman, Alexandre
Chies Santos, Ana
Cypriano, Elysandra
de Souza, Rafael
Fraga, Luciano
Gaspar, Helton
Ghezzi, Luan
Gonçalves, Thiago
Izidoro F. da Costa, A.
Maia, Francisco
Makler, Martin
Marcolino, Wagner
Overzier, Roderik
Rabello-Soares, M. C.
Riguccini, Laurie
Sfair, Rafael
Spinelli, Patrícia
Trevisan, Marina
Vasconcelos, Maria

Bulgaria

Belcheva, Maya
Bozhilov, Vladimir
Koleva, Kostadinka
Kopchev, Valentin
Manusheva, Mariyana
Peneva, Stoyanka
Petrov, Georgi
Tomova, Mima
Tsvetkova, Svetla

Canada

Afshordi, Niayesh
Albert, Justin
Albert, Loïc
Artigau, Etienne
Boley, Aaron
Broderick, Avery
Cartledge, Stefan
Chastenay, Pierre
Frolov, Andrei
Gallo, Luigi
Heinke, Craig
Hlavacek-Larrondo, J.
Holder, Gilbert
Khalack, Viktor
Kothes, Roland
Langill, Philip
LeBlanc, Francis
Lovekin, Catherine
Malo, Lison
Mandy, Margot
Martimbeau, Nathalie
Matzner, Christopher
McConnachie, Alan
Moon, Dae-Sik
Morsink, Sharon
Murray, Norman
Nelson, Lorne
Normandeau, Magdalen
Parker, Laura
Pfeiffer, Harald
Pogosian, Levon
Pogosyan, Dmitri
Pudritz, Ralph
Robishaw, Timothy
Sato, Takashi
Schieven, Gerald
Scott, William
Spekkens, Kristine
Spencer, Locke
Thacker, Robert
Vanderlinde, Keith
Wadsley, James

Chile

Anguita, Timo
Bustos, Ricardo
Cortes, Cristian
Galazutdinov, Gazinur
Gallenne, Alexandre
Gomez, Matias
Padilla, Nelson
Pignata, Giuliano
Roman-Lopes, Alexandre
Schleicher, Dominik
Valcarce, Aldo

China Nanjing

CHEN, Xuepeng
DONG, Ruifang
FENG, Hua
HOU, Yonghui
HUANG, Jing
HUANG, Yong
JIA, Xiaolin
JIANG, Haifeng
JIN, Shuanggen
LI, Cheng
LI, Di
LI, Zhiyuan
LI, Zhuo
LIANG, En-Wei
LIU, Qinghui
LIU, Tiexin
QIU, Keping
SHI, Yong
SONG, Shuli
WANG, Jiasong
WANG, Ran
WANG, Xin
WANG, Zhongxiang
WU, Bobing
WU, Jian
XU, BO
XU, Haiguang
XUE, Yanjie
XUE, Yongquan
YUAN, Qirong
ZHAN, Hu
ZHANG, Haiying
ZHANG, Sijiong
ZHAO, Bing
ZHAO, Wen
ZHOU, Aiying
ZHU, Zong-Hong

China Taipei

ASADA, Keiichi
CHEN, Ming-Tang
CHIANG, Lung-Yih
KINOSHITA, Daisuke
LEE, Chin-Fei
Lehner, Matthew
LIN, Zhong-Yi
SUYU, Sherry
TSENG, Wei-Ling
WANG, Ming-Jye

Colombia

Arenas-Salazar, Jose R.
Arias Callejas, Veronica
Casas Miranda, R.
Chaparro, Germán
Cuartas-Restrepo, Pablo
Forero-Romero, Jaime
Garcia Varela , Jose
Gonzalez, Guillermo
Higuera Garzon, Mario

Larranaga, Eduard
López, Framsol
Montoya Martinez, J.
Munoz Cuartas, Juan
Nunez, Luis
Pachon, Leonardo
Pinzon, Giovanni
Portilla, José
Poveda Tejada, Nicanor
Reina, Jerson
Sabogal Martinez, B.
Sanabria Gomez, José
Silva-Villa, Esteban
Vargas Dominguez, Santiago
Vera Villamizar, Nelson
Zuluaga, Jorge

Croatia, the Republic of

Calogovic, Jasa
Delhaize, Jacinta
Delvecchio, Ivan
Miettinen, Oskari
Skokic, Ivica
Smolcic, Vernesa
Zic, Tomislav

Czech Republic

Bezdek, Aleš
Bohacova, Martina
Civis, Svatopluk
Ebrova, Ivana
Ferus, Martin
Haas, Jaroslav
Jilkova, Lucie
Krizek, Michal
Kubatova, Brankica
Kunneriath, Devaky
Orlitova, Ivana
Ripa, Jakub
Schee, Jan
Svoboda, Jifi
Taylor, Rhys

Denmark

Antoci, Victoria
Gall, Christa
Grillo, Claudio
Hansen, Camilla
Hsiao, Eric
Koay, Jun Yi
Silva Aguirre, Victor

Egypt

Abdel Rahman, Helal
Ahmed, Nasr
Ali, Gamal
Attia, Gamal
Awad, Zainab
Dwidar, Hany
Edris, Khaled
Elkhateeb, Magdy
Elnagahy, Farag
Elsanhoury, Waleed
Gadallah, Kamel
Ghoneim, Rasha
Hanna, Yousry
Hendy, Yasser
Ibrahim, Makram
Khattab, El Amira Hend
Metwally, Hayman
Mohamed, Ahmed E.
Nouh, Mohamed
Owis, Ashraf
Rahoma, Walid
Samwel, Susan
Selim, Ibrahim
Shaker, Ashraf
Takey, Ali
Youssef, Manal

Estonia

Kama, Mihkel
Tuvikene, Taavi
Vurm, Indrek

Ethiopia

Gidelaw, Amare

Finland

Finoguenov, Alexis
Hovatta, Talvikki
Lindfors, Elina
Pelkonen, Veli-Matti
Penttila, Antti
Schmidt, Juergen
Vainio, Rami

France

Alves, Marta
Amari, Tahar
André, Philippe
Arzoumanian, Doris
Audit, Edouard
Bacmann, Aurore
Ballot, Jérôme
Bazin, Cyrille
Bonnefoy, Mickaël
Bouret, Jean-Claude
Carry, Benoît
Caux, Emmanuel
Chardonnet, Pascal
Chassande-Mottin, Eric
Chiavassa, Andrea
Commerçon, Benoît
Coulais, Alain
Cox, Nick
Cuillandre, Jean-Charles
Dartois, Emmanuel
Dubernet, Marie-Lise
Ferrari, Chiara
Foglizzo, Thierry
Gabici, Stefano
Gavazzi, Raphael
Gonzalez, Matthias
Gratier, Pierre
Groussin, Olivier
Guillot, Tristan
Hassler, Donald
Hathi, Nimish
Haubois, Xavier
Hugot, Emmanuel
Janvier, Miho
Knödlseder, Jürgen
Kordopatis, Georges
Lainey, Valéry
Lamy, Laurent
langlois, Maud
Lavaux, Guilhem
Leauthaud, Alexie
Le Brun, Amandine
Lenain, Jean-Philippe
Leroy, Nicolas
Lo Faro, Barbara
Lopez Ariste, Arturo
Marcowith, Alexandre
Marin, Frédéric
Mauduit, J.-Christophe
Maury, Anaëlle
Mei, Simona
Melchior, Anne-Laure
Meyer-Vernet, Nicole
Micelotta, Elisabetta
Morin, Julien
Oksala, Mary
Padovani, Marco
Pety, Jérôme
Pignatale, Francesco
Pratt, Gabriel
Quantin-Nataf, Cathy
Rahmani, Hadi
Renner, Stéfan
Robert, Vincent
Rodriguez, Sebastien
Rodriguez Ferreira, J. G.
Roukema, Boudewijn
Santerne, Alexandre
Schultheis, Mathias
Scribano, Yohann
Starck, Jean-Luc
Strugarek, Antoine
Szopa, Cyril
Talbi, Dahbia
Tavabi, Ehsan
Tisserand, Patrick
Venot, Olivia
Zanda, Brigitte

Germany

Allen, Bruce
Astraatmadja, Tri
Ball, Warrick
Bonafede, Annalisa
Cescutti, Gabriele
Danzmann, Karsten
Davies, Richard
Denker, Carsten
Elsaesser, Dominik
Fohlmeister, Janine
Geier, Stephan
Gillessen, Stefan
Hussain, Gaitee
Kauffmann, Jens
Kitaura, Francisco
Krabbe, Alfred
Kronberg, Elena
Kuckein, Christoph
Kunder, Andrea
Labadie, Lucas
Lacerda, Pedro
Liefke, Carolin
Maturi, Matteo
Minchev, Ivan
Mueller, Michael
Murphy, Simon
Oberst, Jürgen
Pillai, Thushara
Polsterer, Kai
Pössel, Markus
Rafferty, David
Roa Garzon, Maximo A.
Sandin, Christer
Sandu, Oana
Schartmann, Marc
Schinzel, Frank
Schrabback, Tim
Schunker, Hannah
Schwarz, Dominik
Seitz, Manuela
Shapiro, Alexander
Simon, Patrick
Soja, Rachel
Srama, Ralf
Tautz, Robert
Urrutia, Tanya
Vazza, Franco
Verbiest, Joris
Walcher, Jakob
Weller, Jochen

Greece

Balasis, Georgios
Bitsakis, Theodoros
Bratsolis, Emmanuel
Gazeas, Kosmas
Karampelas, Antonios
Katsanikas, Matthaios
Kontogiannis, Ioannis
Koulouridis, Elias
Leonidaki, Ioanna
Liakos, Alexios
Manousakis, Antonios
Meli, Athina
Petropoulou, Vasiliki

Hungary

Bognar, Zsofia
Borkovits, Tamas
Csizmadia, Szilárd
Dobos, Lászlo
Horvath, Istvan
Kovacs, Tamás
Marton, Gábor
Molnar, Lászlo
Murakozy, Judit
Regaly, Zsolt
Sandor, Zsolt
Szalai, Tamas
Vida, Krisztián

Iceland

Johannesson, Gudlaugur

India

Ajith, Parameswaran
Arun, K.G.
Chandra, Poonam
Dewangan, Gulab
Gupta, Alok
Gupta, Neeraj
Hanasoge, Shravan
Joshi, Yogesh
Kharb, Preeti
Malali, Sampoorna
Misra, Ranjeev
Naik, Sachindranatha
Oberoi, Divya
Pai, Archana
Pandey, Jeewan
Pandey, Shashi
Puravankara, Manoj
Sivarani, Thirupathi
Srivastava, Abhishek
Sule, Aniket
Sutaria, Firoza
Vadawale, Santosh

Iran, Islamic Rep. of

Baghram, Shant
Dadashi, Neda
Javadi, Atefeh
Miraghaei, Halime
Mosleh, Moein
Mozaffari, Seyyed Mohammad
Sheikhnezami, Somayeh
Taghizadeh T., Javad
Tavasoli, Saeed

Ireland

Bloomfield, David
Bracken, Colm
Moriarty, Patrick

Israel

Cohen, Ariel
Dar, Arnon
Granot, Jonathan
Gurfil, Pini
Manulis, Ilan
Sagiv, Ilan
Sarid, Gal
Sternberg, Amiel
Verbin, Yosef

Italy

Alessi, Elisa Maria
Bacciotti, Francesca
Bastieri, Denis
Bedin, Luigi
Boccato, Caterina
Bolli, Pietro
Botticella, Maria Teresa
Brucato, John
Buson, Sara
Campana, Sergio
Canestrari, Rodolfo
Cappelluti, Nico
Carini, Roberta
Cassaro, Pietro
Cignoni, Michele
Colpi, Monica
Covino, Stefano
Dalessandro, Emanuele
De Angelis, Ilaria
Fabjan, Dunja
Ferrari, Valeria
Ferraro, Francesco
Ferruzzi Caruso, Debora
Fioretti, Valentina
Fontani, Francesco
Franzetti, Paolo
Gardiol, Daniele
Grieco, Valentina
Gualtieri, Leonardo
Guidi, Gianluca
Lamia, Livio
Lanzoni, Barbara
Liuzzo, Elisabetta
Losurdo, Giovanni
Maio, Umberto
Massi, Fabrizio
Micheli, Marco
Molinari, Sergio
Molinaro, Marco
Mucciarelli, Alessio
Nascimbeni, Valerio
Natalucci, Lorenzo
Orosei, Roberto
Palomba, Ernesto
Pancino, Elena
Paolillo, Maurizio
Piranomonte, Silvia
Pizzone, Rosario
Razzano, Massimiliano
Re Fiorentin, Paola
Romano, Donatella
Rondoni, Lamberto
Rueda Hernandez, Jorge Armando
Scandariato, Gaetano
Spavone, Marilena
Spitoni, Emanuele
Stella, Luigi
Stratta, Giulia
Tarchi, Andrea
Tortora, Crescenzo
Turrini, Diego
Vito, Fabio

Japan

AKAHORI, Takuya
ANDO, Masaki
Antolin, Patrick
ARAKAWA, Masahiko
Cheung, Sze-leung
EGUSA, Fumi
FUJII, Michiko
FUJIWARA, Hideaki
FURUSAWA, Hisanori
HATSUKADE, Bunyo
HAYASHI, Yoshi-Yuki
HIGUCHI, Aya
HOSHINO, Masahiro
IKOMA, Masahiro
IKUTA, Chisato
ISHII, Shun
ISHIYAMA, Tomoaki
ITOH, Yousuke
KAJITA, Takaaki
KAKAZU, Yuko
KATSUKAWA, Yukio
KAWAGOE, Shio
KAWASHIMA, T.
KOMIYA, Yutaka
KOMUGI, Shinya
KOUZUMA, Shinjirou
KUBO, Yuki
KUSAKABE, Nobuhiko
LEE, Shiu-Hang
MAEDA, Yoshitomo
MIYAMOTO, Yusuke
MIZUTA, Akira
MOTOYAMA, K.
NAGAI, Hiroshi
NAGAI, Makoto
NAGAYAMA, Takahiro
NAKAJIMA, Motoki
NISHIMURA, Nobuya
NOBUKAWA, M.
NODA, Manabu
NOZAWA, Takaya
OKAMOTO, Sakurako
SAITO, Masao
SAKAI, Nami
SATO, Shuichi
SHIMAJIRI, Yoshito
SHINKAI, Hisa-aki
SOTANI, Hajime
TAKAHASHI, H.
TAKAHASHI, Hiroyuki
TAKEI, Dai
TAMURA, Yoichi
TANAKA, Masaomi
TODA, Kouichi
TORII, Kazufumi
TSUKAGOSHI, Takashi
TSUMURA, Kohji
UJIHARA, Hideki
URAKAWA, Seitaro
WAKITA, Shigeru
WATANABE, Y.
WU, Ronin
YAMANAKA, Masayuki
YASUI, Chikako
YASUTAKE, Nobutoshi

Kazkhstan

Shestakova, Lyubov
Denissyuk, Eduard
Dubovichenko, Sergey
Kondratyeva, Ludmila
Kusakin, Anatoly
Makukov, Maxim
Valiullin, Rashit

Korea, Rep. of

BACH, Kiehunn
CHOI, Young-Jun
CHUN, Moo-Young
HWANG, Ho Seong
HWANG, Narae
JUNG, Taehyun
KIM, Chang-Goo
KIM, Hak-Sub
KIM, Hyosun
KIM, Jihyun
KO, Jongwan
KWAK, Kyujin
LEE, Ho-Gyu
LEE, Seong-Kook
MIHN, Byeong-Hee
OH, Sang Hoon
PARK, Kyung Sun
PARK, Sun Mie
SHIM, Hyunjin
SHIN, Junho
SHINN, Jong-Ho
SOHN, Bong Won
SUNG, Suk-Kyung
YUN, Youngjoo

Malaysia

Yusof, Norhasliza

Mexico

Calcaneo-Roldan, Carlos
de la Fuente Acosta, E.
Garcia-Diaz, Maria
Gomez, Gilberto
Gonzalez Dominguez, R.
Gonzalez-Lopezlira, R.
Jack, Dennis
Luna, Abraham
Morisset, Christophe
Nagel, Erick
Navarro-Gonzalez, R.
Nunez Lopez, R.
Olguin, Lorenzo
Peimbert, Antonio
Pichardo, Barbara
Raga, Alejandro
Rodriguez, Lino
Rodriguez-Rico, Carlos
Sanchez, Leonardo
Segura, Antigona
Velazquez, Miguel
Watson, Alan
Zapata, Luis

Netherlands

Ando, Shin'ichiro
Berge, David
Bertone, Gianfranco
Cimo, Giuseppe
Désert, Jean-Michel
Deul, Erik
Janssen, Gemma
Klein Wolt, Marc
Koerding, Elmar
Kruithof, Gert
McCaughrean, Mark
McNamara, Paul

New Zealand

Rattenbury, Nicholas

Norway

Jafarzadeh, Shahin
Mendes Domingos P., T.

Poland

Balucinska-Church, M.
Bejger, Michał
Pollo, Agnieszka
Soida, Marian
Szuszkiewicz, Ewa
Wlodarczyk, Ireneusz
Wyrzykowski, Lukasz

Portugal

Afonso, Jose
Alves de Oliveira, C.
Barbosa, Domingos
Bonifacio, Vitor
Brandao, Isa
Campante, Tiago
Cardoso, Cátia
Correia, Alexandre
da Cunha, Elisabete
Fernandes, Cristina
Figueira, Pedro
Garcia Hernandez, A.
Grave, Jorge
Krone Martins, Alberto
Kumar, Nanda
Lopes, Ilidio
Machado, Pedro
Martins, Zita
Messias, Hugo
Mimoso, José
Nunes, Nelson
Papaderos, Polychronis
Rojas-Ayala, Bárbara
Sobral, David
Sousa, Sérgio
Tereno, Ismael

Romania

Moldovan, Dan Nelu
Szucs-Csillik, Iharka-M.

Russian Federation

Agafonov, Mikhail
Akimkin, Vitaly
Baluev, Roman
Belyaev, Andrey
Bondarenko, Yuri
Burenkov, Alexander
Chupina, Natalia
Demidov, Mikhail
Demidova, Tatiana
Efimova, Natalia
Egorov, Oleg
Goranskij, Vitaly
Gorbovskoy, Evgeny
Gvaramadze, Vasilii
Ibrahimov, Mansur
Ivanchik, Alexandre
Kaisina, Elena
Khabarova, Olga
Khamitov, Irek
Khoperskov, Sergey
Kondratyev, Boris
Kononov, Dmitry
Kurdubov, Sergei
Larchenkova, Tatiana
Pilipenko, Sergey
Popova, Olga
Sotnikova, Yulia
Sych, Robert
Tarakanov, Peter
Tsvetkov, Alexander
Zhelenkova, Olga

Serbia, Republic of

Bogosavljevic, Milan
Cseki, Attila
Jovanovic, Bora
Latkovic, Olivera
Milic, Ivan
Milisavljević, Slaviša
Nina, Aleksandra
Onic, Dusan
Pavlovic, Rade
Prodanovic, Tijana
Todorovic, Nataša
Vidojevic, Sonja

Slovakia

Hambalek, Lubomir
Kundra, Emil
Sekeras, Matej
Veres, Peter
Vilagi, Jozef
Zigo, Pavel

South Africa

Allie, Muhammad
Blyth, Sarah-Louise
Boettcher, Markus
Brink, Jeandrew
Chen, Andrew
Cluver, Michelle
Dave, Romeel
de Witt, Aletha
Deane, Roger
Hilton, Matthew
Holbrook, Jarita
Kotze, Marissa
Libert, Yannick
Manxoyi, Sivuyile
Mohamed, Shazrene
Ray, Subharthi
Sievers, Â Jonathan
Skelton, Rosalind
Townsend, Lee
van Soelen, Brian
Weltman, Amanda

Spain

Altieri, Bruno
Angulo, Raul
Armano, Michele
Bonoli, Silvia
Cervino, Miguel
Diego, José
Esteban Pinillos, Héctor
Galindo Mendoza, F. J.
Gil de Paz, Armando
Gutierrez, Jordi
Kuulkers, Erik
Madiedo, Jose
Marco Castillo, F.
Martinez Uso, M. J.
Mas-Hesse, Jose Miguel
Munoz-Darias, Teodoro
Ness, Jan-Uwe
O'Mullane, William
Orsi Moyano, Alvaro
Osorio, Mayra
Prada, Francisco
Rodriguez-Pacheco, J.
Rosenberg Gonzalez, A.
Sanchez Contreras, C.
Sanchez-Portal, Miguel
Solano, Enrique
Suarez, Juan Carlos
Vaduvescu, Ovidiu
Verde, Licia
Yepes, Gustavo

Sweden

Amanullah, Rahman
Becherini, Yvonne
Conrad, Jan
Fuglesang, Christer
Geppert, Wolf
Janson, Markus
Kero, Johan
Larsson, Bengt
Larsson, Josefin
Marti-Vidal, Ivan
Naslund, Magnus
Ramstedt, Sofia
Stenberg Wieser, G.
Wirstrom, Eva
Yang, Jun
Zackrisson, Erik

Switzerland

Atek, Hakim
Battaglia, Marina
Cava, Antonio
Falanga, Maurizio
Kleint, Lucia
Paraficz, Danuta
Schawinski, Kevin

Thailand

Irawati, Puji
Jaroenjittichai, Phrudth
Maithong, Wiraporn
Poshyachinda, Saran
Rujopakarn, Wiphu
Sanguansak, Nuanwan
Sawangwit, Utane
Surina, Farung
Tummuangpak, P.
Wongwathanarat, A.
Yuma, Suraphong

Turkey

Afsar, Melike
Akyuz, Aysun
Alis, Sinan
Bakis, Hicran
Bozkurt, Zeynep
Devlen, Ahmet
Devlen, Ebru
Eksi, Kazim
Ergin, Tülün
Erkut, Mehmet
Ertan , Unal
Gogus, Ersin
Inam, Sitki
Isik, Emre
Özel, Mehmet

Özişik, Tuncay
Ulubay Siddiki, Ayse
Yakut, Kadri
Yerli, Sinan
Yildiz, Mutlu
Yildiz, Umut

Ukraine

Babyk, Iurii
Ishchenko, Maryna
Kleshchonok, Valerii
Kolesnyk, Yuriy
Krushevska, Viktoriia
Kulinich, Yurij
Kulyk, Iryna
Melekh, Bohdan
Petruk, Oleh
Stanislavsky, Aleksander
Tishkovets, Victor
Zakharenko, Vyacheslav
Zakhozhay, Olga
Zinchenko, Igor

United Kingdom

Avison, Adam
Baker, Deborah
Bernard-Salas, Jeronimo
Bolton, James
Calabrese, Erminia
Casewell, Sarah
Casey, Morag
Chapman, Emma
Clements, David
Coppin, Kristen
Cotter, Garret
Darnley, Matthew
Dominik, Martin
Fossey, Stephen
Green, Lucie
Gregory, Scott
Hammond, Neil
Hermes, James
Jarvis, Johanna
Joachimi, Benjamin
Jones, Geraint
Jones, Hugh
Kerins, Eamonn
Kreplin, Alexander
Long, David
Lyman, Joseph
Malik, Karim
Marchetti, Lucia
Morton, Richard
Muldrew, Stuart
Nelmes, Susan
Perez-Suarez, David
Pickering, Juliet
Prema, Paresh
Ridpath, Ian
Sargent, Mark
Sathyaprakash, B.
Sekhar, Aswin
Simoes, Paulo
Smith, Alexis
Snodgrass, Colin
Stanway, Elizabeth
Starling, Rhaana
Unruh, Yvonne
Valori, Gherardo
Veras, Dimitri
Watkins, Laura
Williams, David
Zhang, Keke

United States

Alatalo, Katherine
Alcock, Charles
Allen, Branden
Andrews, Sean
Argon, Alice
Atreya, Prakash
Babu, G. Jogesh
Baker, John
Barger, Kat
Barron, Eric
Becker, Andrew
Becker, George
Bellini, Andrea
Bellm, Eric
Bennett, Jeffrey
Beresnyak, Andrey
Berger, Edo
Berukoff, Steven
Bischoff, Colin
Blanton, Elizabeth
Blundell, Raymond
Borucki, William
Braatz, James
Bradac, Marusa
Bradley, Richard
Brady, Patrick
Brown, Duncan
Bryden, Geoffrey
Bulbul, Esra
Cadonati, Laura
Caldwell, Douglas
Cara, Mihai
Centeno, Rebecca
Chakrabarty, Deepto
Chambers, Kenneth
Cheung, Chi (Teddy)
Chiar, Jean
Chomiuk, Laura
Cisewski, Jessi
Cornish, Neil
Cucchiara, Antonino
D'Angelo, Gennaro
Dalle Ore, Cristina
Dame, Thomas
Damodaran, Anish
Dcruz, Noella
Debes, John
Demorest, Paul
Di Stefano, Rosanne
Dickinson, Mark
Dixon, William
Dore, Olivier
Dorland, Bryan
Doty, Steven

Dudik, Rachel
Ebeling, Harald
Farnocchia, Davide
Farrington, Christopher
Fassett, Caleb
Ferguson, Henry
Fish, Vincent
Fisher, Robert
Flanagan, Kathryn
Fleming, Scott
Flohic, Helene
Ford, Alyson
Frayer, David
Frebel, Anna
Frinchaboy, Peter
Garcia, Javier
Gilbert, Karoline
Gordon, David
Gosain, Sanjay
Goudfrooij, Paul
Goulding, Andy
Green, Joel
Grogin, Norman
Gullberg, Steven
Gunther, Hans
Gurwell, Mark
Hahn, Michael
Hamaguchi, Kenji
Hennessy, Gregory
Hewitt, Jacqueline
Hickox, Ryan
Hodge, Jacqueline
Hoffman, Jennifer
Holley-Bockelmann, J.
Holman, Matthew
Howell, Dale
Hussain, Shahid
Jain, Kiran
Jenness, Timothy
Jirdeh, Hussein
Johns Krull, Christopher
Joyce, Richard
Kalogera, Vicky
Kashyap, Vinay
Kassin, Susan
Katsavounidis, E.
Keating, Brian
Keto, Eric
Keyes, Charles (Tony)
Kiessling, Alina
Kite, Edwin
Konopacky, Quinn
Kopp, Greg
Kovac, John
Kupper, Andreas
Laine, Seppo
Lawrence, Stephen
Lawton, Brandon
Laycock, Silas
Lazzarini, Albert
Le Bail, Karine
Lee, Janice
Lehner, Nicolas
Levesque, Emily
Lewis, Nikole
Livas, Jeffrey
Lloyd, James
Loeb, Abraham
Looney, Leslie
Lotz, Jennifer
Luo, Bin
MacKenty, John
Maddalena, Ronald
Madura, Thomas
Mahabal, Ashish
Mandelbaum, Rachel
Mandic, Vuk
Marchis, Franck
McCollough, Michael
McDowell, Jonathan
McGrath, Elizabeth
McQuinn, Matthew
Melendez, Marcio
Mills, Elisabeth
Mioduszewski, Amy
Moro Martin, Amaya
Morse, Jon
Mroczkowski, Tony
Mueller, Guido
Muench, August
Munn, Jeffrey
Murphy, Nicholas
Nagai, Daisuke
Nations, Harold
Nitta, Nariaki
Norton, Aimee
Nugent, Peter
Oberg, Karin
Oesch, Pascal
Oliversen, Nancy
Palo, Scott
Park, Ryan
Parriott, Joel
Peek, Joshua
Peeples, Molly
Pena-Guerrero, Maria
Penny, Laura
Pepper, Joshua
Petitpas, Glen
Pillepich, Annalisa
Postman, Marc
Prestage, Richard
Puatua, Wendy
Qi, Chunhua
Rabinowitz, David
Rafelski, Marc
Randall, Scott
Rao, Ramprasad
Rao, Sandhya
Reed, Michael
Reed, Phillip
Ressler, Michael
Rest, Armin
Ritchey, Adam
Ryan, Erin
Saar, Steven
Sabbi, Elena
Salviander, Sarah
Sana, Hugues
Sarajedini, Vicki
Saulson, Peter
Schaefer, Gail
Schlaufman, Kevin
Scowen, Paul
Shawhan, Peter
Sheppard, Scott
Sheth, Kartik
Shoemaker, Deirdre
Showman, Adam P.
Shporer, Avi
Sjouwerman, Lorant
Skibba, Ramin
Slater, Stephanie
Slavin, Jonathan
Smith, Denise
Smutko, Michael
Soderberg, Alicia
Soderblom, Jason
Sonnentrucker, Paule
Soummer, Remi

Staguhn, Johannes
Stahl, H.
Stassun, Keivan
Stern, Daniel
Street, Rachel
Strolger, Louis-Gregory
Tegmark, Max
Thompson, William
TIAN, Hui
Tobin, John
Tombesi, Francesco
Tremblay, Grant
Treuthardt, Patrick
Troja, Eleonora
Tsai, Chao-Wei
Turnshek, Diane
Vestrand, W. Thomas
Weinstein, Alan
Weintroub, Jonathan
Williams, Roy
Wolk, Scott
Wood-Vasey, William
Woodward, Charles
Wright, Shelley
Young, Peter
Zakamska, Nadia
Zhao, Ping

Vatican City State

Macke, Robert

Venezuela

Gatuzz, Efrain
Hernandez, Fabiola
Lacruz, Elvis
Vieira Villarreal, Rosa

Viet Nam

Cao, Tuan
Nguyen, Thao
Pham, Nhung
Pham, Tuan-Anh

4. New Individual Members listed by non-Member countries

Azerbaijan, Republic of

Guliyev, Ayyub
Nabiyev, Shaig

Gaza, Palestine

Baraka, Suleiman

Macedonia, Former Rep. of Yugoslavia

Kuzmanovska, Olgica

Mauritius

Beeharry, Girish

Namibia

Backes, Michael
Steenkamp, Riaan

Pakistan

Hussain, Shahid
Khan, Fazeel Mahmood

Singapore

Banks, Timothy

2. Deceased members (2012-2015)

Abdelkawi, M.
Abulazm, Mohamed
Adam, Madge
Afanasjeva, Praskovja
Ahmed, Imam
Ahmed, Mostafa
Aiad, A.
AKABANE , Kenji
Albers, Henry
Ambruster, Carol
Anderson, Kinsey
Andrienko, Dmitry
Andrillat, Yvette
Antalova, Anna
Arnaud, Jean-Paul
Arnold, James
Arp, Halton
Arsenijevic, Jelisaveta
Aslanov, I.
Asseo, Estelle
Auman, Jason
Aurass, Henry
Backer, Donald
Bajaja, Esteban
Banos, George
Barbaro, Guido
Bartaya, R.
Barth, Charles
Batrakov, Yurij
Batson, Raymond
Baum, William
Bautz, Laura
Beggs, Denis
Bell, Roger
Benevolenskaya, Elena
Berger, Christiane
Bernacca, Pierluigi
Bernstein, Hans-Heinrich
Bettis, Dale
Bhat, Chaman
Bhattacharyya, J.
Billingham, John
Blow, Graham
Bochonko, D.
Bohm, Karl-Heinz
Boley, Forrest
Bonnor, W.
Bottinelli, Lucette
Boulon, Jacques
Bouska, Jiří
Bowen, George
Boyarchuk, Alexander
Braccesi, Alessandro
Breahna, Iulian
Brosterhus, Elmer
Broten, Norman
Brown, Robert
Buchler, J.
Burgess, Alan
Burton, William
Bystrova, Natalja
Campbell, Belva
Cardona, Octavio
Carr, Thomas
Casanovas, Juan
Castelli, John
Caswell, James
Cayrel de Strobel, Giusa
Celis, Leopoldo
Cester, Bruno
Chamberlain, Joseph
Chopinet, Marguerite
CHOU, Kyong-Chol
Christy, Robert
Chubb, Talbot
Chupp, Edward
Climenhaga, John
Clutton-Brock, Martin
Colgate, Stirling
Collins, George
Considere, Suzanne
Coradini, Angioletta
Couteau, Paul
Cox, Arthur
Dalgarno, Alexander
Davis, Morris
Davis, Richard
Deeming, Terence
de Jonge, J.
Delsemme, Armand
Denishchik, Yurii
Denisse, Jean-Francois
Dewhirst, David
de Young, David
Divan, Lucienne
Djurovic, Dragutin
Dolidze, Madona
Dominko, Fran
Dommanget, Jean
Donn, Bertram
Doroshenko, Valentina
Dufay, Maurice
Duncombe, Raynor
Dzhapiashvili, Victor
Dzigvashvili, R.
Eddy, John
Efimov, Yuri
El Basuny, Ahmed

Elford, William
Elliot, James
El Shahawy, Mohamad
Elste, Günther
Elston, Wolfgang
Eminzade, T.
Eshleman, Von
Evans, J.
Ewen, Harold
FANG, Li-Zhi
Felten, James
Finzi, Arrigo
Fitch, Walter
Fleischer, Robert
Fletcher, J.
Florsch, Alphonse
Fomenko, Alexandr
Fomin, Piotr
Fong, Richard
Fox, Kenneth
Frost, Kenneth
Frye, Glenn
FUJITA, Yoshio
Gadsden, Michael
Galal, A.
Galletto, Dionigi
Galt, John
Gamaleldin, Abdulla
Garlick, George
Genkin, Igor
Gerlei, Otto
Gietzen, Joseph
Gil, Janusz
Godfrey, Peter
Godoli, Giovanni
Gordon, Kurtiss
Goudas, Constantine
Gould, Robert
Grasberg, Ernest
Grasdalen, Gary
Grebenikov, Evgenij
Grishchuk, Leonid
Grossmann-Doerth, Ulrich
Grudler, Pierre
Guest, John
Gulmen, Ömür
Gurzadyan, Grigor
Guseinov, O.
Gusejnov, Ragim
Hack, Margherita
Hadjidemetriou, John
Hall, Douglas
Hamid, S.
Hamm, Monique
Hansen, Carl
Hansen, Richard
Heddle, Douglas
Hein, Righini
Hellali, Yhya
Henon, Michel
Herbig, George
Herczeg, Tibor
Heyvaerts, Jean
Hobbs, Robert
Horsky, Jan
Hovhannessian, Rafik
HUANG, Runqian
HUANG, Yinn-Nien
Hulth, Per
Hunten, Donald
Hysom, Edmund
Iannini, Gualberto
Ill, Marton
INAGAKI, Shogo
Ipatov, Alexander
ISHIDA, Keiichi
Issa, Issa
Jarrell, Richard
Jones, Albert
Jones, Barrie
Jones, Frank
Journet, Alain
Juszkiewicz, Roman
Kahlmann, Hans
Kalenichenko, Valentin
Kaluzny, Janusz
KANG , Gon-Ik
Kapisinsky, Igor
Karygina, Zoya
Kasumov, Fikret
KAWARA , Kimiaki
Khare, Bishun
Kharin, Arkadiy
Khromov, Gavriil
Khrutskaya, Evgenia
KIM, Bong-Kyu
Kiselev, Aleksej
Kislyuk, Vitaliy
KITAMURA, Masatoshi
Kleczek, Josip
Koch, David
Kokurin, Yurij
Korchak, Alexander
Kostina, Lidiya
Kostyakova, Elena
Kraft, Robert
Kramer, Kh
Kraushaar, William
Krogdahl, W.
Kultima, Johannes
Kumajgorodskaya, Raisa
Kuril'chik, Vladimir
Kurochka, Lev
Kurpinska-Winiarska, M.
Labeyrie, Jacques
Laclare, Francis
Lafon, Jean-Pierre
Lal, Devendra
Lapushka, Kazimirs
Lavrov, Mikhail
Lavrukhina, Augusta
Lecar, Myron
Lehmann, Marek
Leikin, Grigerij
Leushin, Valerij
Lin, Chia
LING, Chih-Bing
LI, Qibin
Lloyd Evans, Thomas
Locke, Jack
Loden, Kerstin
Lominadze, Jumber
Loskutov, Viktor
Loucif, Mohammed
Lovell, Bernard
Lumme, Kari
LU, Ruwei
LU, Tan
Major, John
Malumian, Vigen
Mammano, Augusto
Mancuso, Santi
Marie, Mohamed
Marlborough, Michael
Martin, William
Masani, A.
Mathez, Guy
MA, Xingyuan
Mayfield, Earle
Mazure, Alain

McCrosky, Richard
McGee, Richard
McIntosh, Bruce
McNamara, Delbert
Mead, Jaylee
Men, Anatolij
Message, Philip
Mezger, Peter
Michard, Raymond
Michel, F.
Mihalas, Dimitri
Mikesell, Alfred
Mikhailov, Alexander
Mikisha, Anatoly
Millet, Jean
Minakov , Anatoliy
Mioc, Vasile
Mogilevskij, Emmanuil
Moore, Patrick
Morbey, Christopher
Morris, Stephen
Morton, G.
Motz, Lloyd
Murray, Stephen
Namba, Osamu
Nather, R.
Nedbal, Dalibor
Nefedeva, Antonina
Nelson, Robert
Neugebauer, Gerry
Neukum, Gerhard
Nicholls, Ralph
Nikoloff, Ivan
Nikolov, Nikola
NISHI , Keizo
Novikov, Sergej
OBI, Shinya
O'Donoghue, Darragh
Ogelman, Hakki
Oliver, John
Omarov, Tuken
Opolski, Antoni
OWAKI, Naoaki
Pacholczyk, Andrzej
Page, Arthur
Parise, Ronald
PARK, Hong-Seo
Pascoal, Antonio
Peale, Stanton
Peery, Benjamin
Pellas, Paul
Perez, Maria
Petford, Alfred
Petrov, Gennadij
Pillinger, Colin
Pneuman, Gerald
Polyachenko, Valerij
Poulle, Emmanuel
Prevot-Burnichon, Marie-Louise
Price, Stephan
Proisy, Paul
Pronik, Iraida
Pryce, Maurice
Pushkin, Sergej
Rachkovsky, D.
Rashkovskij, Sergey
Rawlings, Steven
Rayrole, Jean
Razin, Vladimir
Rense, William
Richards, Mercedes
Rizvanov, Naufal
Robinson, Leif
Robley, Robert
Rohlfs, Kristen
Romanchuk, Pavel
Romano, Giuliano
Roman, Rodica
Rosen, Edward
Roslund, Curt
Rouse, Carl
Routly, Paul
Roy, Archie
Rubin, Robert
Rudnicki, Konrad
Ruffle, Paul
Rule, Bruce
Rusu, I.
Ruzdjak, Vladimir
Saad, Nadia
Sahade, Jorge
SAITO, Kuniji
SAITO , Takao
SAKASHITA , Shiro
Sargent, Wallace
Satterthwaite, Gilbert
Savage, Ann
Scarrott, Stanley
Schmidt, Hermann
Schmitt, Dieter
Schnell, Anneliese
Schultz, Georg
Schutz, Bob
Searle, Leonard
Semel, Meir
Shakeshaft, John
Shakhbazian, Romelia
Shakhbazyan, Yurij
Shaltout, Mosalam
Shapiro, Maurice
Sheffer, Evgenij
Shefov, Nikolaj
Shobbrook, Robert
Shukla, K.
Sinnerstad, Ulf
Sitarski, Grzegorz
Smit, J.
Sokolov, Viktor
Sokolsky, Andrej
Soliman, Mohamed
Sotirovski, Pascal
Souriau, Jean-Marie
Stachnik, Robert
Stahr-Carpenter, M.
Stankevich, Kazimir
Stefl, Stanislav
Steinberg, Jean-Louis
Stoeger, William
Stone, R.
SUGAWA, Chikara
Sultanov, G.
Svestka, Zdenek
Sykora, Július
Szeidl, Bela
TAKASE, Bunshiro
TAKENOUCHI , Tadao
TAKEUTI, Mine
Talon, Raoul
TANAKA , Riichiro
TANAKA , Yasuo
Tatevyan, Suriya
Tempesti, Piero
Thiry, Yves
Thomas, Hans
Thomas, Roger
Tofani, Gianni
TOTSUKA , Yoji
Townes, Charles
Trefftz, Eleonore

Tsap, Teodor
TSUBAKI , Tokio
UESUGI , Akira
Upton, E.
Valeev, Sultan
van Bueren, Hendrik
van Houten-Groeneveld, Ingrid
Vardya, M.
Vaughan, Arthur
Velkov, Kiril
Veron, Philippe
Vicente, Raimundo
Volland, Hans
Voroshilov, Volodymyr
Walker, Helen
Walker, Robert
Warwick, James
Weekes, Trevor
Wehinger, Peter
WEI, Wenren
Wellgate, G.
Wentzel, Donat
Westerhout, Gart
Whitaker, Ewen
Wijnbergen, Jan
Wilson, Albert
Wilson, Peter
Wolfe, Arthur
Wood, John
YAMASHITA , Kojun
YANG, Fumin
YANG, Shijie
Yoss, Kenneth
Young, Judith
Young, Louise
Yuldashbaev, Taimas
Zahn, Jean-Paul
Zander, Rodolphe
Zhitnik, Igor
Ziznovsky, Jozef

Transactions IAU, Volume XXIXB
Proceedings IAU Symposium No. Volume XXIXB, 2018
P. Benvenuti, ed.
© International Astronomical Union 2019
doi:10.1017/S1743921318004040

CHAPTER IX

DIVISIONS, COMMISSIONS, & WORKING GROUPS

NOTE. This chapter lists the main IAU scientific bodies: Divisions, Commissions and their Working Groups.

For the new 9 Divisions which were introduced at the XXVIIIth General Assembly (August 2012) we report both the Organizing Committees in force during the triennium 2012-2015 and those elected for the triennium 2015-2018.

The old Commissions were all dismissed at the XXIXth General Assembly and replaced by the new ones: the final report about the activities of the old Commissions are reported in a separate Volume (Transaction XXIX A - Legacy Report). This Chapter lists the new Commissions and their respective Organizing Committees.

The Working Groups in force during the 2012-2015 triennium and those appointed at the XXIXth General Assembly for the triennium 2015-2018 are also reported.

The members of the Divisions and their (old) Commissions are listed in the following Chapters (X and XI respectively).

Divisions, Commissions and Working Groups for the triennium 2012-2015

Division A Fundamental Astronomy

President:
Sergei A. Klioner
Technische Universität Dresden
Lohrmann Observatory
Mommsenstr 13
01062 - Dresden
Germany

Vice-President:
Jacques Laskar
Observatoire de Paris
IMCCE
77 Av Denfert Rochereau, Bat. A
75014 - Paris
France

Secretary:
Susan Gessner Stewart
United States Naval Observatory
Astronomical Applications
3450 Massachusetts Ave., NW
20392-5420 Washington
United States

Steering Committee Members:

Nicole Capitaine (France), Cheng-Li Huang (China Nanjing), Catherine Y. Hohenkerk (United Kingdom), Mizuhiko HOSOKAWA (Japan), Dennis D. McCarthy (United States), François Mignard (France), Alessandro Morbidelli (France), Dimitri Pourbaix (Belgium), Michael H. Soffel (Germany), Norbert Zacharias (United States), Richard Gross (United States), Sergei A. Klioner (Germany), Jacques Laskar (France), Anthony G.A. Brown (Netherlands), Sylvio Ferraz-Mello (Brazil), Susan Gessner Stewart (United States)

Commissions:

Commission 4 Ephemerides
Commission 7 Celestial Mechanics & Dynamical Astronomy
Commission 8 Astrometry
Commission 19 Rotation of the Earth
Commission 31 Time
Commission 52 Relativity in Fundamental Astronomy

Division B Facilities, Technologies and Data Science

President:
David Richard Silva
National Optical Astronomy Observatory
950 North Cherry Avenue
AZ 85719 - Tucson
United States

Vice-President:
Pietro Ubertini
INAF
IAPS - Institute for Space Astrophysics and Planetology
Via Fosso del Cavaliere 100
00133 - Roma
Italy

Secretary:
Michael G. Burton
Armagh Observatory and Planetarium
College Hill
BT61 9DG - Armagh
United Kingdom

Steering Committee Members:

Michael G. Burton (United Kingdom), Malcolm G. Smith (Chile), Françoise Genova (France), Anthony James Beasley (United States), Gloria M. Dubner (Argentina), Lisa Storrie-Lombardi (United States), Robert J. Hanisch (United States), Hitoshi YAMAOKA (Japan), Lyudmila I. Mashonkina (Russian Federation), Jessica Mary Chapman (Australia), Richard F. Green (United States), Gerard T. van Belle (United States), Simon F. Portegies Zwart (Netherlands), David Richard Silva (United States), Pietro Ubertini (Italy), Alistair Walker (Chile)

Commissions:

Commission 5 Documentation & Astronomical Data
Commission 6 Astronomical Telegrams
Commission 14 Atomic & Molecular Data
Commission 25 Astronomical Photometry and Polarimetry
Commission 30 Radial Velocities
Commission 40 Radio Astronomy
Commission 50 Protection of Existing & Potential Observatory Sites
Commission 54 Optical & Infrared Interferometry

Division C Education, Outreach and Heritage

President:
Mary Kay M. Hemenway
University of Texas Austin
McDonald Observatory
2515 Speedway, C1400
TX 78712 - Austin
United States

Vice-President:
Hakim Luthfi Malasan
Faculty of Mathematics & Natural Sciences
Institut Teknologi Bandung
Astronomy Division & Bosscha Observatory
Lembang - West Java
40391 - Bandung
Indonesia

Secretary:
Rajesh Kochhar
Panjab University
Mathematics Department
Sector14
160014 - Chandigarh
India

Steering Committee Members:

Michele Gerbaldi (France), Xiaochun Sun (China, Nanjing), Clive Ruggles (United Kingdom), Juan Antonio Belmonte Aviles (Spain), Lars Lindberg Christensen (Germany), Beatriz Elena Garcia (Argentina), Raymond P. Norris (Australia), Pedro Russo (Netherlands), Mary Kay M. Hemenway (United States), Hakim Luthfi Malasan (Indonesia), Jean-Pierre de Greve (Belgium), Kazuhiro SEKIGUCHI (Japan)

Commissions:

Commission 41 History of Astronomy
Commission 46 Astronomy Education & Development
Commission 55 Communicating Astronomy with the Public

Division D High Energy Phenomena and Fundamental Physics

President:
Diana Mary Worrall
University of Bristol
HH Wills Physics Laboratory
Tyndall Avenue
BS8 1TL - Bristol
United Kingdom

Vice-President:
Felix Albert Aharonian
Dublin Institute for Advanced Studies
School of Cosmic Physics
31 Fitzwilliam Place
2 - Dublin
Ireland

Secretary:
Elena Pian
INAF
Institute of Space Astrophysics and Cosmic Physics
Via P. Gobetti 101
40129 - Bologna
Italy

Steering Committee Members:
Christine Jones (United States), Chryssa Kouveliotou (United States), Tadayuki TAKAHASHI (Japan), Diana Mary Worrall (United Kingdom), Felix Albert Aharonian (Ireland), Xavier Barcons (Germany), John G. Kirk (Germany), Anna Wolter (Italy), Elena Pian (Italy)

Commissions:
Commission 44 Space & High Energy Astrophysics

Division E Sun and Heliosphere

President:
Lidia van Driel-Gesztelyi
University College London
Space and Climate Physics, Mullard Space Science Laboratory
London
United Kingdom

Vice-President:
Yihua YAN
National Astronomical Observatories
20A Datun Rd
Chaoyang District
100012- Beijing
China Nanjing

Secretary:
Gianna Cauzzi
INAF - Osservatorio Astrofisico di Arcetri
Largo E Fermi 5
50125- Firenze
Italy

Steering Committee Members:
Sarah Gibson (United States), Marc L DeRosa (United States), Peng-Fei CHEN (China, Nanjing), Arnab Rai Choudhuri (India), Lyndsay Fletcher (United Kingdom), Ingrid Mann (Norway), Karel J. Schrijver (United States), Rudolf von Steiger (Switzerland), Lidia van Driel-Gesztelyi (United Kingdom), Yihua YAN (China, Nanjing)

Commissions:

Commission 10 Solar Activity
Commission 12 Solar Radiation & Structure
Commission 49 Interplanetary Plasma & Heliosphere

Division F Planetary Systems and Bioastronomy

President:
Giovanni B. Valsecchi
INAF- IAPS
Via Fosso del Cavaliere 100
Tor Vergata
00133 - Roma
Italy

Vice-President:
Nader Haghighipour
University of Hawaii
Institute for Astronomy
2680 Woodlawn Dr
HI 96822- Honolulu
United States

Secretary:
Gonzalo Tancredi
Facultad de Ciencias
Departamento de Astronomía
Igua 4225
11400- Montevideo
Uruguay

Steering Committee Members:
Pamela Kilmartin (New Zealand), Alberto Cellino (Italy), Dominique Bockelee-Morvan (France), Mark T Lemmon (United States), Didier Queloz (United Kingdom), William M. Irvine (United States), Steven R. Chesley (United States), Petrus Matheus Marie Jenniskens (United States), Pascale Ehrenfreund (United States), Alain Lecavelier des Etangs (France), Jiří Borovička (Czech Republic), Paul Winchester Chodas (United States), Giovanni B. Valsecchi (Italy), Nader Haghighipour (United States)

Commissions:

Commission 15 Physical Study of Comets & Minor Planets
Commission 16 Physical Study of Planets & Satellites
Commission 20 Positions & Motions of Minor Planets, Comets & Satellites
Commission 22 Meteors, Meteorites & Interplanetary Dust
Commission 51 Bio-Astronomy
Commission 53 Extrasolar Planets (WGESP)

Division G Stars and Stellar Physics

President:
Ignasi Ribas
Institut de Ciencies de l'Espai
IEEC-CSIC
Cami de Can Magrans, s/n
08193 - Bellaterra, Barcelona
Spain

Vice-President:
Corinne Charbonnel
Geneva Observatory (Switzerland)
& CNRS (France)
Chemin des Maillettes 51
1290 - Versoix
Switzerland

Secretary:
Virginia Trimble
University of California Irvine
Physics-Astronomy Department
4575 Physics
CA 92697-4575 - Irvine
United States

Steering Committee Members:
Conny Aerts (Belgium), Karen Pollard (New Zealand), Richard O. Gray (United States), Marco Limongi (Italy), Katia Cunha (Brazil), Beatriz Barbuy (Brazil), Mercedes T. Richards (United States), David R. Soderblom (United States), Virginia Trimble (United States), Corinne Charbonnel (Switzerland), Birgitta Nordström (Denmark), Francesca D'Antona (Italy), Brian D. Mason (United States), Joachim Puls (Germany), Ignasi Ribas (Spain)

Commissions:
Commission 26 Double & Multiple Stars
Commission 27 Variable Stars
Commission 29 Stellar Spectra
Commission 35 Stellar Constitution
Commission 36 Theory of Stellar Atmospheres
Commission 42 Close Binary Stars
Commission 45 Stellar Classification

Division H Interstellar Matter and Local Universe

President:
Ewine F. van Dishoeck
Leiden University
Leiden Obsevatory
PO Box 9513
2300 RA - Leiden
Netherlands

Vice-President:
Jonathan Bland-Hawthorn
AAO
PO Box 296
NSW 1710 - Epping
Australia

Secretary:
Bruce G. Elmegreen
IBM
T J Watson Res Center

1101 Kitchawan Road
10598 - Yorktown Heights
United States

Steering Committee Members:
Thomas Henning (Germany), Sun Kwok (China, Nanjing), Giovanni Carraro (Chile), Eileen D. Friel (United States), Annie C. Robin (France), Holger Baumgardt (Australia), Bruce G. Elmegreen (United States), Diego Mardones (Chile), Michael R. Meyer (Switzerland), Birgitta Nordström (Denmark), Ewine F. van Dishoeck (Netherlands)

Commissions:
Commission 33 Structure & Dynamics of the Galactic System
Commission 34 Interstellar Matter
Commission 37 Star Clusters & Associations

Division J Galaxies and Cosmology

President:
Françoise Combes
Observatoire de Paris
LERMA
61 Av de l'Observatoire, Bat. A
75014 - Paris
France

Vice-President:
Thanu Padmanabhan
IUCAA
Post Bag 4 Ganeshkhind
Pune University Campus
411 007 - Pune
India

Secretary:
Thaisa Storchi-Bergmann
UFRGS
Inst de Física
Campus do Vale
CP 15051
91501-970 - Porto Alegre, RS
Brazil

Steering Committee Members:
Ofer Lahav (United Kingdom), Brian P. Schmidt (Australia), John S. Gallagher III (United States), Marijn Franx (Netherlands), Jayant Murthy (India), Elaine M. Sadler (Australia), Thaisa Storchi-Bergmann (Brazil), Françoise Combes (France), Thanu Padmanabhan (India), Andrew J. Bunker (United Kingdom), Stéphane J. Courteau (Canada), Monica Rubio (Chile)

Commissions:
Commission 21 Galactic and Extragalactic Background Radiation
Commission 28 Galaxies
Commission 47 Cosmology

Division Working Groups Inter-Division Working Groups Commission Working Groups

Division	*Working Group Name*	*Chair*
D A	Astrometry by Small Ground-Based Telescopes	M. Assafin
D A	Multi-waveband Realisations of International Celestial Reference System	F. Mignard
D A	Numerical Standards in Fundamental Astronomy (NSFA)	B. Luzum
D A	Pulsar-based Timescales	G. Hobbs
D A	Redefinition of UTC	D. D. McCarthy
D A	Standards of Fundamental Astronomy (SOFA)	C. Y. Hohenkerk
D A	Third Realisation of International Celestial Reference Frame	C. S. Jacobs
D B	Future Large Scale Facilities	G. F. Gilmore
D B	Site Testing Instruments	A.i A. Tokovinin
D E	Impact of Magnetic Activity on Solar and Stellar Environments	D. Nandi
D E	International Collaboration on Space Weather	L. Damé
D E	International Data Access	R., D. Bentley
D F	Exoplanets for the Public	A. Lecavelier des Etangs
D F	Near Earth Objects	A. W. Harris
D F	Planetary System Nomenclature (WGPSN)	R. M. Schulz
D F	Small Bodies Nomenclature (SBN)	J. Tichá
D G	Abundances in Red Giants	
D G	Active B Stars	C. E. Jones
D G	Ap & Related Stars	G. Mathys
D G	Massive Stars	A. H. Davó

Divisions	*Working Group Name*	*Chair*
D A-F	Cartographic Coordinates & Rotational Elements	B. A. Archinal
D A-G	Nominal Units for Stellar & Planetary Astronomy	P. Harmanec
D C-E	Solar Eclipses	J. M. Pasachoff
D E-G	Solar-Type Stars	P. Petit

Inter-Commission Working Groups

Commission	*Working Group Name*	*Chair*
C 4	Standardizing Access to Ephemerides	James Lindsay Hilton
C 5	Astrostatistics and Astroinformatics	Eric D. Feigelson
C 5	Designations	Marion Schmitz
C 5	FITS	Lucio Chiappetti
C 5	Libraries	Marsha Bishop
C 5	TG on Preservation & Digitization of Photographic Plates	
C 5	Virtual Observatories, Data Centers & Networks	
C 12	Coordination of Synoptic Observations of the Sun	Alexei A. Pevtsov
C 14	Atomic Data	
C 14	Molecular Data	John Harry Black
C 14	Solids & Their Surfaces	Gianfranco Vidali
C 15	Asteroid Families	Alberto Cellino
C 15	Data Bases for Comet Spectroscopy	Irakli Simonia

Commission	*Working Group Name*	*Chair*
C 15	Physical Studies of Asteroids	David J. Tholen
C 15	Physical Studies of Comets	Daniel Craig Boice
C 15	TG On Asteroid Magnitudes	Karri Muinonen
C 15	TG On Asteroid Polarimetric Albedo Calibration	Ricardo Alfredo Gil-Hutton
C 19	Theory of Earth Rotation	Jose Manuel Ferrándiz
C 22	Meteor Shower Nomenclature	Tadeusz Jan Jopek
C 25	Infrared Astronomy	Kevin Volk
C 26	Catolog of Orbital Elements of Spectroscopic Binary Systems	Dimitri Pourbaix
C 26	Maintenance of the Visual Double Star Database	William I. Hartkopf
C 30	Radial-Velocity Standard Stars	Gerard Jasniewicz
C 30	Stellar Radial Velocity Bibliography	Orlando Hugo Levato
C 34	Astrochemistry	Thomas J. Millar
C 34	Planetary Nebulae	Letizia Stanghellini
C 40	Astrophysically Important Spectral Lines	Masatoshi OHISHI
C 40	Historic Radio Astronomy	Kenneth I. Kellermann
C 40	The Radio Astronomy Definition of Continuum Flux Density in Broad-Band SED Observations	Ronald D. Ekers
C 41	Archaeoastronomy and Astronomy in Culture	Raymond P. Norris
C 41	Archives	Ileana Chinnici
C 41	Astronomy and World Heritage	Gudrun B. E. Wolfschmidt
C 41	Historical Instruments	Yunli Shi
C 41	Johannes Kepler	Terence J. Mahoney
C 41	Transits of Venus	Wayne Orchiston
C 46	The IAU Dictionary of Astronomical Concepts	Mohammad Heydari-Malayeri
C 55	CAP Conferences	Ian E. Robson
C 55	CAP Journal	Sarah Jane Reed
C 55	Communicating Heliophysics	Carine Briand
C 55	New Media	Pamela L. Gay
C 55	New Ways of CAP	Michael John West
C 55	Outreach Professionalization & Accreditation	Richard Tresch Fienberg
C 55	Public Outreach Information Management	Pedro Russo
C 55	Washington Charter For CAP	Dennis Crabtree

Executive Committee Working Groups

Commission	*Working Group Name*	*Chair*
C4,C7,C8,C16,C20	Natural Planetary Satellites	Jean-Eudes Arlot

Working Group Name	*Chair*
IAU General Assemblies	
International Year of Light: Cosmic Light	Richard F. Green
Public Naming of Planets and Planetary Satellites	Thierry Montmerle
Women in Astronomy	Francesca Primas

Commissions (until 2015)

Commission 4 - Ephemerides

President: Catherine Y. Hohenkerk
UKHO
HM Nautical Almanac Office
Admiralty Way
TA1 2DN - Taunton
United Kingdom

Vice-President: Jean-Eudes Arlot
Observatoire de Paris
IMCCE
77 Av Denfert Rochereau
75014 - Paris
France

Organizing Committee Members William Thuillot (France), Mitsuru Sôma (Japan), Agnès Fienga (France), Marina V. Lukashova (Russian Federation), Jose Manuel Ferrándiz (Spain), John A. Bangert (United States), William M. Folkner (United States), Elena V. Pitjeva (Russian Federation), Steven A. Bell (United Kingdom), Sean E. Urban (United States), Jean-Eudes Arlot (France), George H. Kaplan (United States)

Commission 5 - Documentation & Astronomical Data

President:
Robert J. Hanisch
National Institute of Standards and Technology
Office of Data and Informatics
100 Bureau Drive - Stop 8300
MD 20899-8300 - Gaithersburg
United States

Vice-President:
Michael W. Wise
ASTRON
Netherlands Institute for Radio Astronomy
P.O. Box 2
7990 AA - Dwingeloo
Netherlands

Organizing Committee Members:
Daniel Egret (France), Anja C. Schröder (South Africa), Douglas Tody (United States), R. Elizabeth M. Griffin (Canada), Ajit K. Kembhavi (India), Fabio Pasian (Italy), Tara Murphy (Australia), Marsha Bishop (United States), Heinz J Andernach (Mexico)

Commission 6 - Astronomical Telegrams

President:
Hitoshi YAMAOKA
National Astronomical Observatory of Japan
Public Relations Center
2-21-1, Osawa,
181-8588 - Mitaka
Japan

Vice-President:
Daniel W. E. Green
Harvard University
Department of Earth and Planetary Sciences
Hoffman Lab 209
20 Oxford St.
MA 02138 - Cambridge
United States

Organizing Committee Members:
Alan C. Gilmore (New Zealand), Nikolay N. Samus (Russian Federation), Gareth V. Williams (United States), Timothy B. Spahr (United States), Kaare Aksnes (Norway), Jana Tichá (Czech Republic), Syuichi NAKANO (Japan)

Commission 7 - Celestial Mechanics & Dynamical Astronomy

President:
Alessandro Morbidelli
Observatoire de la Côte d'Azur
Bd de l'Observatoire
BP 4229
06304 - Nice Cedex 4
France

Vice-President:
Cristian Beaugé
Observatorio Astronómico
Laprida 854
Barrio Observatorio
5000 - Córdoba
Argentina

Organizing Committee Members:
Piet Hut (United States), Alessandra Celletti (Italy), Nader Haghighipour (United States), Fernando Virgilio Roig (Brazil), Jacques Laskar (France), Seppo Mikkola (Finland)

Commission 8 - Astrometry

President:
Norbert Zacharias
US Naval Observatory
Astrometry Department
3450 Massachusetts Ave NW
DC 20392-5420 - Washington
United States

Vice-President:
Anthony G.A. Brown
Leiden University
Leiden Observatory
PO Box 9513
2300 RA - Leiden
Netherlands

Organizing Committee Members:
Ramachrisna Teixeira (Brazil), Oleksandr V. Shulga (Ukraine), Li CHEN (China, Nanjing), Valeri Makarov (United States), Jean Souchay (France), Stephen C. Unwin (United States), Naoteru GOUDA (Japan)

Commission 10 - Solar Activity

President:
Karel J. Schrijver
Lockheed Martin Advanced
Technology Center
Solar and Astrophysics Laboratory
Dept. A021S Bldg. 252
3251 Hanover Street
CA 94304-1191 - Palo Alto
United States

Vice-President:
Lyndsay Fletcher
University of Glasgow
School of Physics and Astronomy
Kelvin Bldg
G12 8QQ - Glasgow
United Kingdom

Organizing Committee Members:

Daniel O. Gómez (Argentina), Paul S. Cally (Australia), Paul Charbonneau (Canada), S. Sirajul Hasan (India), Yihua YAN (China, Nanjing), Astrid Veronig (Austria), Ayumi ASAI (Japan), Sarah Gibson (United States)

Commission 12 - Solar Radiation & Structure

President:
Gianna Cauzzi
INAF
Osservatorio Astrofisico di Arcetri
Largo E Fermi 5
50125 - Firenze
Italy

Vice-President:
Nataliia Gennadievna Shchukina
Main Astronomical Observatory
Solar Physics Department
27 Acad. Zabolotnoho St
03680 - Kyiv
Ukraine

Organizing Committee Members:

Mingde DING (China, Nanjing), Stuart M. Jefferies (United States), Natalie Krivova (Germany), Michele Bianda (Switzerland), Sergio Dasso (Argentina), Dean-Yi CHOU (China, Taipei), Axel Brandenburg (Sweden), Fernando Moreno-Insertis (Spain)

Commission 14 - Atomic & Molecular Data

President:
Lyudmila I. Mashonkina
Inst of Astronomy RAS
Pyatnitskaya ul 48
119017 - Moscow
Russian Federation

Vice-President:
Farid Salama
NASA- Ames Research Center
Space Science & Astrobiology Division
Building N245
MS 245 6
CA 94035-1000 - Moffett Field
United States

Organizing Committee Members:
France Allard (France), Paul S. Barklem (Sweden), Helen J. Fraser (United Kingdom), Gillian Nave (United States), Peter Beiersdorfer (United States), Hampus Nilsson (Sweden)

Commission 15 - Physical Study of Comets & Minor Planets

President:
Dominique Bockelee-Morvan
Observatoire Paris-Site de Meudon
LESIA
5 Pl Jules Janssen
92195 - Meudon Cedex
France

Vice-President:
Ricardo Alfredo Gil-Hutton
Complejo Astronómico El Leoncito
Conicet
Av España 1512 Sur
J5402DSP - San Juan
Argentina

Secretary:
Daniel Hestroffer
Observatoire de Paris
IMCCE
77 Av Denfert Rochereau - Bat. A
75014 - Paris
France

Organizing Committee Members:
Irina N. Belskaya (Ukraine), Hideyo KAWAKITA (Japan), Javier Licandro (Spain), Hajime YANO (Japan), Björn J.R. Davidsson (Sweden), Elisabetta Dotto (Italy), Thais Mothé-Diniz (Norway), Diane H. Wooden (United States), Alan Fitzsimmons (United Kingdom)

Commission 16 - Physical Study of Planets & Satellites

President:
Mark T Lemmon
Texas A & M University
Atmospheric Sciences Department
TX 77843-3150 - College Station
United States

Organizing Committee Members:
Sang Joon KIM (Korea, Rep of), Leonid V. Ksanfomality (Russian Federation), Luisa M. Lara (Spain), David Morrison (United States), Victor G. Tejfel (Kazakhstan), Padma A. Yanamandra-Fisher (United States)

Commission 19 - Rotation of the Earth

President:
Cheng-Li Huang
Shanghai Astronomical Observatory CAS
80 Nandan Rd
200030 - Shanghai
China Nanjing

Vice-President:
Richard Gross
Jet Propulsion Laboratory
Geodynamics and Space Geodesy Group
Mail Stop 238-600
4800 Oak Grove Drive
91109 - Pasadena, CA
United States

Secretary:
Florian Seitz
Technische Universität München
Deutsches Geodätisches Forschungsinstitut (DGFI)
Arcisstr. 21
D-80333 - Munich
Germany

Organizing Committee Members:
Vladimir E. Zharov (Russian Federation), Daniela Thaller (Germany), Harald Schuh (Germany), Oleg A. Titov (Australia), Zinovy M. Malkin (Russian Federation), Bernd Richter (Germany), Benjamin F. Chao (China, Taipei), David A. Salstein (United States), Wiesław Kosek (Poland), Christian Bizouard (France)

Commission 20 - Positions & Motions of Minor Planets, Comets & Satellites

President:
Steven R. Chesley
CALTECH
JPL - MS 301 150
4800 Oak Grove Dr
CA 91109 - Pasadena
United States

Vice-President:
Daniela Lazzaro
Observatorio Nacional
Coordenadoria de Astronomia e Astrofisica
R. Gal. Jose Cristino 77
São Cristóvão
20921-400 - Rio de Janeiro, RJ
Brazil

Secretary:
Andrea Milani Comparetti
Università di Pisa
Dipartimento di Matematica
Piazzale B. Pontecorvo 5
56127 - Pisa
Italy

Organizing Committee Members:
Robert Jedicke (United States), Mikael Granvik (Finland), Shinsuke ABE (Japan), Jana Tichá (Czech Republic), Timothy B. Spahr (United States), Jin ZHU (China, Nanjing), Alan C. Gilmore (New Zealand), Petr Pravec (Czech Republic)

Commission 21 - Galactic and Extragalactic Background Radiation

President:
Jayant Murthy
Indian Institute of Astrophysics
II Block Koramangala
560 034 - Bangalore
India

Vice-President:
Jayant Murthy
Indian Institute of Astrophysics
II Block Koramangala
560 034 - Bangalore
India

Organizing Committee Members:
Ingrid Mann (Norway), William J. Baggaley (New Zealand), Kalevi Mattila (Finland), Anny-Chantal Levasseur-Regourd (France), Junichi WATANABE (Japan), Eli Dwek (United States)

Commission 22 - Meteors, Meteorites & Interplanetary Dust

President:
Petrus Matheus Marie Jenniskens
SETI Institute
189 Bernardo Ave
CA 94043 - Mountain View
United States

Vice-President:
Jiří Borovička
Astronomical Institute of the Academy of Sciences of the Czech Republic
Department of Interplanetary Matter
Fričova 298
251 65 - Ondřejov
Czech Republic

Secretary:
Tadeusz Jan Jopek
A. M. University
Astronomical Observatory
Ul. Sloneczna 36
60-286 - Poznan
Poland

Organizing Committee Members:
Jin ZHU (China, Nanjing), Galina O. Ryabova (Russian Federation), Shinsuke ABE (Japan), Diego Janches (United States), Masateru ISHIGURO (Korea, Rep of), Jérémie J. Vaubaillon (France), Guy Joseph Consolmagno (Vatican City State)

Commission 25 - Astronomical Photometry and Polarimetry

President:
Alistair Walker
AURA/CTIO/NOAO
Casilla 603
La Serena
Chile

Vice-President:
Saul J. Adelman
The Citadel
Physics Dept
171 Moultrie St
SC 29409-0270 - Charleston
United States

Organizing Committee Members:
Wen Ping CHEN (China, Taipei), J. Allyn Smith (United States), Steve B. Howell (United States), Kevin Volk (United States), Antonio Mario Magalhaes (Brazil), Barbara J. Anthony-Twarog (United States), Jens Kirkeskov Knude (Denmark), Donald W. Kurtz (United Kingdom), John W. Menzies (South Africa), W. Schoeneich (Germany)

Commission 26 - Double & Multiple Stars

President:
Brian D. Mason
US Naval Observatory
Astrometry Dept
3450 Massachusetts Ave NW
DC 20392-5420 - Washington
United States

Vice-President:
Yurij Yu Balega
Special Astrophysical Observatory RAS
Nizhnij Arkhyz
Zelenchukskaya
369167 - Karachaevo-Cherkesia
Russian Federation

Organizing Committee Members:
Bo Reipurth (United States), Andrei A. Tokovinin (Chile), Patricia J C Lampens (Belgium), Jose-Angel Docobo (Spain), Vakhtang S. Tamazian (Spain), Marco Scardia (Italy), Frédéric Arenou (France), Edouard Oblak (France), Theo A. ten Brummelaar (United States)

Commission 27 - Variable Stars

President:
Karen Pollard
University of Canterbury
Department of Physics-Astronomy
Private Bag 4800
8020 - Christchurch
New Zealand

Vice-President:
Christopher Simon Jeffery
Armagh Observatory
College Hill
BT61 9DG - Armagh
United Kingdom

Organizing Committee Members:
Klaus G. Strassmeier (Germany), Dennis Stello (Australia), S. O. Kepler (Brazil), Saskia Hekker (Germany), Katrien Uytterhoeven (Spain), Katrien Kolenberg (Belgium), David E. Mkrtichian (Thailand), Laurent Eyer (Switzerland), Márcio Catelan (Chile)

Commission 28 - Galaxies

President:
John S. Gallagher III
University of Wisconsin Madison
Department of Astronomy
475 N Charter St
WI 53706 - Madison
United States

Organizing Committee Members:
Stéphane J. Courteau (Canada), Monica Rubio (Chile), Marijn Franx (Netherlands), Avishai Dekel (Israel), Elena Terlevich (United Kingdom), Chanda J. Jog (India), Naomasa NAKAI (Japan), Linda J. Tacconi (Germany), Shardha Jogee (United States)

Commission 29 - Stellar Spectra

President:
Katia Cunha
Observatorio Nacional
R. Gal. Jose Cristino 77
São Cristóvão
20921-400 - Rio de Janeiro, RJ
Brazil

Vice-President:
David R. Soderblom
Space Telescope Science Institute
3700 San Martin Dr
MD 21218-2410 - Baltimore
United States

Organizing Committee Members:
Paul A Crowther (United Kingdom), Vanessa M. Hill (France), Martin Asplund (Australia), Jorge Melendez (Brazil), Nikolai E. Piskunov (Sweden), Kim A. Venn (Canada), David Yong (Australia), Wako AOKI (Japan), Kenneth G. Carpenter (United States)

Commission 30 - Radial Velocities

President:
Dimitri Pourbaix
Université Libre de Bruxelles
IAA
CP 226
Bd du Triomphe
1050 - Bruxelles
Belgium

Vice-President:
Tomaž Zwitter
University of Ljubljana
Faculty of Mathematics & Physics
Jadranska 19
1000 - Ljubljana
Slovenia, the Republic of

Organizing Committee Members:
Alain Jorissen (Belgium), David A. Katz (France), Matthias Steinmetz (Germany), Alceste Z. Bonanos (Greece), Tsevi Mazeh (Israel), Dante Minniti (Chile), Francesco A. Pepe (Switzerland)

Commission 31 - Time

President:
Mizuhiko HOSOKAWA
National Institute of Information and Communications Technology
Senior Executive Director
4-2-1 Nukuikitamachi
184-8795 - Koganei, Tokyo
Japan

Vice-President:
Elisa Felicitas Arias
BIPM
Time Department
Pavillon de Breteuil
92312 - Sèvres Cedex France

Organizing Committee Members:
Demetrios N. Matsakis (United States), Vladimir E. Zharov (Russian Federation), Shougang ZHANG (China, Nanjing), Philip Tuckey (France), William Markowitz (United States)

Commission 33 - Structure & Dynamics of the Galactic System

President:
Birgitta Nordström
University of Copenhagen
Niels Bohr Institute
Juliane Maries Vej 30
2100 - Copenhagen
Denmark

Vice-President:
Jonathan Bland-Hawthorn
AAO
PO Box 296
NSW 1710 - Epping
Australia

Organizing Committee Members:
Evangelie (Lia) Athanassoula (France), Felix J. Lockman (United States), Chanda J. Jog (India), Annie C. Robin (France), Dante Minniti (Chile), Sofia Feltzing (Sweden), Jonathan Bland-Hawthorn (Australia)

Commission 34 - Interstellar Matter

President:
Sun Kwok
University of Hong Kong
Faculty of Science
Chong Yuet Ming Physics Bldg
Pokfulam Rd
Hong Kong
China Nanjing

Vice-President:
Bon-Chul KOO
Seoul National University
Astronomy Dept
San 56-1 Shillimdong
Kwanak-gu
151-742 - Seoul
Korea, Rep of

Organizing Committee Members:
Thomas Henning (Germany), Paola Caselli (Germany), Ji YANG (China, Nanjing), Mika J. Juvela (Finland), Laszlo Viktor Toth (Hungary), Sylvie Cabrit (France), Elisabete M. de Gouveia Dal Pino (Brazil), Michal Różyczka (Poland), Dieter Breitschwerdt (Germany), Masato TSUBOI (Japan), Michael G. Burton (United Kingdom), Susana Lizano (Mexico)

Commission 35 - Stellar Constitution

President:
Marco Limongi
INAF
Osservatorio Astronomico di Roma
Via Frascati 33
Monte Porzio Catone
00040 - Roma
Italy

Vice-President:
John C. Lattanzio
Monash University
School of Physics and Astronomy
10 College Walk
Clayton Campus
VIC 3800 - Victoria
Australia

Organizing Committee Members:
Inma Dominguez (Spain), Giora Shaviv (Israel), Amanda I. Karakas (Australia), Marcella Marconi (Italy), Jacco Th. van Loon (United Kingdom), Claus Leitherer (United States), Jordi Isern (Spain)

Commission 36 - Theory of Stellar Atmospheres

President:
Joachim Puls
LMU München
Universitäts-Sternwarte
Scheinerstr 1
81679 - München
Germany

Vice-President:
Ivan Hubeny
University of Arizona
Steward Observatory and Department of Astronomy
933 N. Cherry Avenue
AZ 85721 - Tuscon
United States

Organizing Committee Members:
Rolf-Peter Kudritzki (United States), Mats Carlsson (Norway), Tatiana A. Ryabchikova (Russian Federation), France Allard (France), Martin Asplund (Australia), Carlos Allende Prieto (Spain), Thomas R. Ayres (United States), Bengt Gustafsson (Sweden)

Commission 37 - Star Clusters & Associations

President:
Giovanni Carraro
European Southern Observatory
Science Operations
Alonso de Cordova 3107
19001 - Santiago de Chile
Chile

Vice-President:
Richard de Grijs
Peking University
Kavli Institute for Astronomy and Astrophysics
Yi He Yuan Lu 5
Hai Dian District
100871 - Beijing
China Nanjing

Organizing Committee Members:
Peter B. Stetson (Canada), Douglas Paul Geisler (Chile), Simon P. Goodwin (United Kingdom), Barbara J. Anthony-Twarog (United States), Bruce G. Elmegreen (United States), Dante Minniti (Chile)

Commission 40 - Radio Astronomy

President:
Jessica Mary Chapman
CSIRO Astronomy and Space Science
PO Box 76
NSW 1710 - Epping
Australia

Vice-President:
Gabriele Giovannini
INAF
Istituto di Radioastronomia
Via P Gobetti 101
40129 - Bologna
Italy

Organizing Committee Members:
Monica Rubio (Chile), Raffaella Morganti (Netherlands), Justin L. Jonas (South Africa), Ren-Dong NAN (China, Nanjing), Joseph Lazio (United States), Prajval Shastri (India), Christopher L. Carilli (United States), Hisashi HIRABAYASHI (Japan), Richard E. Hills (United Kingdom)

Commission 41 - History of Astronomy

President:
Brian Warner
University Cape Town
Astronomy Department
Private Bag
7700 - Rondebosch
South Africa

Vice-President:
Xiaochun Sun
Inst History Natural Science, CAS
55 Zhongguancun East Road
100190
Beijing
China Nanjing

Secretary:
David Valls-Gabaud
Observatoire de Paris
LERMA
61 Avenue de l'Observatoire
75014 - Paris

France

Organizing Committee Members:
Eugene F. Milone (Canada), Juan Antonio Belmonte Aviles (Spain), Raymond P. Norris (Australia), Brenda G. Corbin (United States), Mitsuru SOMA (Japan), Luisa Pigatto (Italy)

Commission 42 - Close Binary Stars

President:
Mercedes T. Richards
Pennsylvania State University
Department of Astronomy & Astrophysics
525 Davey Lab
PA 16802 - University Park
United States

Vice-President:
Theodor Pribulla
Slovak Academy of Sciences
059 60
Tatranska Lomnica
Slovakia

Organizing Committee Members:
Horst Drechsel (Germany), Carla Maceroni (Italy), Andrej Prša (United States), John Southworth (United Kingdom), Shay Zucker (Israel), Tomaž Zwitter (Slovenia, the Republic of), Colin David Scarfe (Canada), Ulisse Munari (Italy), Joanna Mikołajewska (Poland), David H. Bradstreet (United States)

Commission 44 - Space & High Energy Astrophysics

President:
Christine Jones
Harvard Smithsonian Center for Astrophysics
High Energy Astrophysics Division
MS 2
60 Garden Str
MA 02138-1516 - Cambridge
United States

Vice-President:
Noah Brosch
Tel Aviv University
Department of Physics & Astronomy/ Wise Observatory
Ramat Aviv
PO Box 39040
69978 - Tel Aviv
Israel

Organizing Committee Members:
João Braga (Brazil), Hernan Quintana (Chile), Diana Mary Worrall (United Kingdom), Hideyo KUNIEDA (Japan), Mattheus de Graauw (Chile), Haruyuki OKUDA (Japan), Marco Salvati (Italy), Thierry Montmerle (France), Kulinder Pal Singh (India), Matthew G. Baring (United States), Martin Adrian Barstow (United Kingdom), Eugene M. Churazov (Russian Federation), Jean Eilek (United States), Jayant Murthy (India), Isabella Pagano (Italy)

Commission 45 - Stellar Classification

President:
Birgitta Nordström
University of Copenhagen
Niels Bohr Institute
Juliane Maries Vej 30
2100 - Copenhagen
Denmark

Vice-President:
Caroline Soubiran
Laboratoire d'Astrophysique de Bordeaux
LAB (CNRS UMR 5804)
2 r de l'Observatoire
BP 89
33270 - Floirac
France

Organizing Committee Members:
Sandy K. Leggett (United States), Harinder P. Singh (India), Werner W. Weiss (Austria), Carlos Allende Prieto (Spain), Richard O. Gray (United States), Adam J. Burgasser (United States), Ranjan Gupta (India), Margaret Murray Hanson (United States)

Commission 46 - Astronomy Education & Development

President:
Jean-Pierre de Greve
Vrije Universiteit Brussel
Dept of Physics-DNTK
Pleinlaan 2
1050 - Brussel
Belgium

Vice-President:
Beatriz Elena García
Instituto de Tecnologías en Detección y Astropartículas (ITeDA)
ITeDA Mendoza
Azopardo 313
Godoy Cruz
5501 - Mendoza
Argentina

Organizing Committee Members:
Magda G. Stavinschi (Romania), Roger Ferlet (France), John B. Hearnshaw (New Zealand), Silvia Torres-Peimbert (Mexico), Rosa M. Ros (Spain), Jay M. Pasachoff (United States), Barrie W. Jones (United Kingdom), George Kildare Miley (Netherlands), Laurence A. Marschall (United States), Edward F. Guinan (United States), Rajesh Kochhar (India)

Commission 47 - Cosmology

President:
Brian P. Schmidt
ANU RSAA
Mount Stromlo Observatory
Cotter Rd
Weston Creek

ACT 2611 - Canberra
Australia

Organizing Committee Members:
Andrew J. Bunker (United Kingdom), Ofer Lahav (United Kingdom), Douglas Scott (Canada), Yipeng JING (China, Nanjing), Anton M. Koekemoer (United States), Olivier Le Fèvre (France), Benedetta Ciardi (Germany), David C. Koo (United States)

Commission 49 - Interplanetary Plasma & Heliosphere

President:
Ingrid Mann
EISCAT Scientific Association
P. O. Box 812
98128 - Kiruna
Sweden

Vice-President:
P. K. Manoharan
Tata Institute of Fundamental Research (TIFR)
Radio Astronomy Centre (NCRA)
Post Box 8
643 001 - Udhagamandalam (Ooty)
India

Organizing Committee Members:
Igor V. Chashei (Russian Federation), Olga Malandraki (Greece), Natchimuthuk Gopalswamy (United States), Sarah Gibson (United States), Yoichiro Hanaoka (Japan), Carine Briand (France), David Lario (United States)

Commission 50 - Protection of Existing & Potential Observatory Sites

President:
Richard F. Green
University of Arizona
Steward Observatory
933 North Cherry Ave
AZ 85721-0065 -Tucson
United States

Vice-President:
Constance Elaine Walker
National Optical Astronomy Observatory
950 North Cherry Avenue
85719 - Tucson United States

Organizing Committee Members:
Beatriz Elena Garcia (Argentina), Ramotholo R. Sefako (South Africa), Ferdinando Patat (Germany), Elizabeth M. Alvarez del Castillo (United States), Masatoshi OHISHI (Japan), Anastasios Tzioumis (Australia), Margarita Metaxa (Greece)

Commission 51 - Bio-Astronomy

President:
Pascale Ehrenfreund
Space Policy Institute
Elliott School of International Affairs
1957 E Street NW
Suite 403
20052 - Washington DC
United States

Vice-President:
Sun Kwok
University of Hong Kong
Faculty of Science
Chong Yuet Ming Physics Bldg
Pokfulam Rd
Hong Kong
China Nanjing

Organizing Committee Members:
Nils G. Holm (Sweden), Ray Jayawardhana (Canada), Charles H. Lineweaver (Australia), Muriel Gargaud (France), Nader Haghighipour (United States), Masatoshi OHISHI (Japan), Douglas Galante (Brazil), Anny-Chantal Levasseur-Regourd (France)

Commission 52 - Relativity in Fundamental Astronomy

President:
Michael H. Soffel
Lohrmann Observatory TU Dresden
Planetary Geodesy
Mommsenstr 13
01062 - Dresden
Germany

Vice-President:
Sergei M Kopeikin
Department of Physics and Astronomy
University of Missouri
322 Physics Bldg
65211 - Columbia
United States

Organizing Committee Members:
Daniel Hestroffer (France), William M. Folkner (United States), Neil Ashby (United States), Jin-he TAO (China, Nanjing), Gérard Petit (France)

Commission 53 - Extrasolar Planets (WGESP)

President:
Alain Lecavelier des Etangs
Institut d'Astrophysique de Paris
98bis Bd Arago
75014 - Paris
France

Vice-President:
Dante Minniti
Universidad Andres Bello
Department of Physics
Fernandez Concha 700
Las Condes
Santiago
Chile

Organizing Committee Members:

Gang ZHAO (China, Nanjing), Ray Jayawardhana (Canada), Heike Rauer (Germany), Didier Queloz (United Kingdom), Eiichiro KOKUBO (Japan), Rosemary A. Mardling (Australia), Andrew Collier Cameron (United Kingdom), Peter Bodenheimer (United States)

Commission 54 - Optical & Infrared Interferometry

President:
Gerard T. van Belle
Lowell Observatory
1400 West Mars Hill Road
86001- Flagstaff
United States

Vice-President:
Denis Mourard
Observatoire de la Côte d'Azur
Laboratoire Lagrange
BP4229
06304 - Nice
France

Secretary:
Fabien Malbet
Université Joseph Fourier / Centre National de la Recherche Scientifique
IPAG - Institut de Planétologie et d'Astrophysique de Grenoble
BP 53
38041 - Grenoble cedex 9
France

Organizing Committee Members:

Michael James Ireland (Australia), Theo A. ten Brummelaar (United States), John Thomas Armstrong (United States), Markus Wittkowski (Germany), Bruno Lopez (France), Karine Perraut (France), Éric M. Thiébaut (France), John Stephen Young (United Kingdom)

Commission 55 - Communicating Astronomy with the Public

President:
Lars Lindberg Christensen
ESO
Karl-Schwarzschild-Str 2
85748- Garching
Germany

Vice-President:
Pedro Russo
Leiden Observatory
Universe Awareness International Office
P.O. Box 9513
NL-2300 RA
Leiden
Netherlands

Secretary:
FRichard Tresch Fienberg
American Astronomical Society

43 Everett Ave.
MA 02472-1881
Watertown
United State

Organizing Committee Members:
Zi ZHU (China, Nanjing), Ian E. Robson (United Kingdom), Carolina Johanna Ödman (South Africa), Kimberley Kowal Arcand (United States), Sze-leung Cheung (Japan), Kazuhiro SEKIGUCHI (Japan), Jin ZHU (China, Nanjing)

Divisions, Commissions and Working Groups for the triennium 2015-2018

Division A Fundamental Astronomy

President:
Anne Lemaître
University of Namur
Namur Center of Complex Systems (naXys)
Rempart de la Vierge 8
5000 - Namur
Belgium

Vice-President:
Daniel Hestroffer
Observatoire de Paris
IMCCE
77 Av Denfert Rochereau - Bat. A
75014 - Paris
France

Secretary:
Susan Gessner Stewart
United States Naval Observatory
Astronomical Applications
3450 Massachusetts Ave., NW
20392-5420 Washington
United States

Steering Committee Members:
Elisa Felicitas Arias (France), Fernando Virgilio Roig (Brazil), Cristian Beaugé (Argentina), William M. Folkner (United States), Anne Lemaître (Belgium), Sergei A. Klioner (Germany), Anthony G.A. Brown (Netherlands), Richard Gross (United States), Susan Gessner Stewart (United States), Nicole Capitaine (France), Catherine Y. Hohenkerk (United Kingdom), Daniel Hestroffer (France), Sylvio Ferraz-Mello (Brazil), Ralph A. Gaume (United States)

Commissions:
Commission A1 Astrometry
Commission A2 Rotation of the Earth

Commission A3 Fundamental Standards
Inter-Division A-F Commission Celestial Mechanics and Dynamical Astronomy
Cross-Division A-F Commission Solar System Ephemerides

Division B Facilities, Technologies and Data Science

President:
Pietro Ubertini
INAF
IAPS - Institute for Space Astrophysics and Planetology
Via Fosso del Cavaliere 100
00133 - Roma
Italy

Vice-President:
Michael G. Burton
University of New South Wales
School of Physics
NSW 2052 - Sydney
Australia

Secretary:
Michael G. Burton
University of New South Wales
School of Physics
NSW 2052 - Sydney
Australia

Steering Committee Members:

Eric D. Feigelson (United States), Ana I. Gómez de Castro (Spain), Peter Joseph Quinn (Australia), Wenwu TIAN (China Nanjing), Gabriele Giovannini (Italy), Farid Salama (United States), Saul J. Adelman (United States), David Richard Silva (United States), Pietro Ubertini (Italy), Michael G. Burton (Australia), Simon F. Portegies Zwart (Netherlands), Richard F. Green (United States), Lisa Storrie-Lombardi (United States), Michael W. Wise (Netherlands)

Commissions:

Commission B1 Computational Astrophysics
Commission B2 Data and Documentation
Commission B3 Astroinformatics and Astrostatistics
Commission B4 Radio Astronomy
Commission B5 Laboratory Astrophysics
Commission B6 Astronomical Photometry and Polarimetry
Inter-Division B-C Commission Protection of Existing and Potential Observatory Sites
Inter-Division B-H-J Commission Intergalactic Medium

Division C Education, Outreach and Heritage

President:
John B. Hearnshaw
University of Canterbury
Physics-Astronomy Department
Private Bag 4800
8020 - Christchurch
New Zealand

Vice-President:
Susana E. Deustua
Space Telescope Science Inst
Instruments Division
3700 San Martin Dr
MD 21218 - Baltimore
United States

Secretary:
Pamela L. Gay
Southern Ilinois University Edwardsville
Center for STEM Research, Education, & Outreach
Center for STEM REO
SIUE, Campus Box 2224
62025 - Edwardsville
United States

Steering Committee Members:

Katrien Kolenberg (Belgium), Saeko S. Hayashi (United States), Mary Kay M. Hemenway (United States), John B. Hearnshaw (New Zealand), Susana E. Deustua (United States), Pamela L. Gay (United States), Beatriz Elena García (Argentina), Pedro Russo (Netherlands), Xiaochun Sun (China Nanjing), Clive Ruggles (United Kingdom), Michele Gerbaldi (France), Kazuhiro SEKIGUCHI (Japan), Linda Elisabeth Strubbe (Canada)

Commissions:

Inter-Division B-C Commission Protection of Existing and Potential Observatory Sites
Commission C1 Astronomy Education and Development
Commission C2 Communicating Astronomy with the Public
Commission C3 History of Astronomy
Commission C4 World Heritage and Astronomy

Division D High Energy Phenomena and Fundamental Physics

President:
Chryssa Kouveliotou
George Washington University
Physics Department
Corcoran Hall
725 21st Street, NW
DC 20052 - Washington
United States

Vice-President:
Elena Pian
INAF
Institute of Space Astrophysics and Cosmic Physics
Via P. Gobetti 101
40129 - Bologna
Italy

Secretary:

Anna L Watts
University of Amsterdam
Astronomical Institute 'Anton Pannekoek'
Postbus 94249
1090 GE - Amsterdam
Netherlands

Steering Committee Members:

Anna L Watts (Netherlands), Tadayasu DOTANI (Japan), Isabelle Grenier (France), Neil Gehrels (United States), Diana Mary Worrall (United Kingdom), Chryssa Kouveliotou (United States), William Richard Forman (United States), Xavier Barcons (Spain), John G. Kirk (Germany), Anna Wolter (Italy), Elena Pian (Italy)

Commissions:

Commission D1 Gravitational Wave Astrophysics
Inter-Division D-G-H-J Commission Galaxy Spectral Energy Distributions
Cross-Division D-J Commission Supermassive Black Holes, Feedback and Galaxy Evolution

Division E Sun and Heliosphere

President:

Yihua YAN
National Astronomical Observatories
20A Datun Rd
Chaoyang District
100012- Beijing
China Nanjing

Vice-President:

Sarah Gibson
NCAR/HAO
p.o. box 3000
80301-3000 - Boulder
United States

Secretary:

Marc L DeRosa
Lockheed Martin Solar and Astrophysics Laboratory
3251 Hanover St. B/252
94304 - Palo Alto
United States

Steering Committee Members:

Toshifumi Shimizu (Japan), Eduard P. Kontar (United Kingdom), Nandita Srivastava (India), Natalie Krivova (Germany), Lidia van Driel-Gesztelyi (United Kingdom), Yihua YAN (China Nanjing), Lyndsay Fletcher (United Kingdom), Sarah Gibson (United States), Ingrid Mann (Sweden), Marc L DeRosa (United States), Arnab Rai Choudhuri (India), Rudolf von Steiger (Switzerland)

Commissions:

Commission E1 Solar Radiation and Structure
Commission E2 Solar Activity
Commission E3 Solar Impact Throughout the Heliosphere

Division F Planetary Systems and Bioastronomy

President:
Nader Haghighipour
University of Hawaii
Institute for Astronomy
2680 Woodlawn Dr
HI 96822- Honolulu
United States

Vice-President:
Gonzalo Tancredi
Facultad de Ciencias
Departamento de Astronomía
Igua 4225
11400 - Montevideo
Uruguay

Secretary:
Jean-Luc Margot
University of California
595 Charles E Young Drive East
CA 90095 - Los Angeles
United States

Steering Committee Members:

Maria A. Barucci (France), Sun Kwok (China Nanjing), Daniela Lazzaro (Brazil), Andrea Milani Comparetti (Italy), Nader Haghighipour (United States), Gonzalo Tancredi (Uruguay), Giovanni B. Valsecchi (Italy), Jiří Borovička (Czech Republic), Alain Lecavelier des Etangs (France), Jean-Luc Margot (United States), Paul Winchester Chodas (United States), Athena Coustenis (France), William M. Irvine (United States), Didier Queloz (United Kingdom)

Commissions:

Inter-Division A-F Commission Celestial Mechanics and Dynamical Astronomy
Commission F1 Meteors, Meteorites and Interplanetary Dust
Commission F2 Exoplanets and the Solar system
Commission F3 Astrobiology
Cross-Division A-F Commission Solar System Ephemerides

Division G Stars and Stellar Physics

President:
Corinne Charbonnel
Geneva Observatory (Switzerland)
& CNRS (France)
Chemin des Maillettes 51
1290 - Versoix
Switzerland

Vice-President:
David R. Soderblom
Space Telescope Science Institute
3700 San Martin Dr
MD 21218-2410 - Baltimore
United States

Secretary:
Tabetha Suzanne Boyajian
Louisiana State University
Department of Physics and Astronomy
Baton Rouge
United States

Steering Committee Members:
Francesca D'Antona (Italy), Tabetha Suzanne Boyajian (United States), Andrej Prša (United States), Artemio Herrero Davó (Spain), John C. Lattanzio (Australia), Christopher Simon Jeffery (United Kingdom), Ivan Hubeny (United States), Pierre Kervella (France), Geraldine Joan Peters (United States), Corinne Charbonnel (Switzerland), David R. Soderblom (United States), Ignasi Ribas (Spain), Martin Asplund (Australia)

Commissions:
Commission G1 Binary and Multiple Star Systems
Commission G2 Massive Stars
Commission G3 Stellar Evolution
Commission G4 Pulsating Stars
Commission G5 Stellar and Planetary Atmospheres
Inter-Division G-H-J Commission Stellar Clusters throughout Cosmic Space and Time
Inter-Division D-G-H-J Commission Galaxy Spectral Energy Distributions

Division H Interstellar Matter and Local Universe

President:
Bruce G. Elmegreen
IBM
T J Watson Res Center
1101 Kitchawan Road
10598 - Yorktown Heights
United States

Vice-President:
Leonardo Testi
INAF
Osservatorio Astrofisico di Arcetri
Largo E Fermi 5
50125 - Firenze
Italy

Secretary:
Francisca Kemper
Academia Sinica
Institute of Astronomy and Astrophysics
PO Box 23-141 10617 -Taipei
China Taipei

Organizing Committee Members:
Holger Baumgardt (Australia), Eva Schinnerer (Germany), Cristina Chiappini (Germany), Eva K. Grebel (Germany), Thomas J. Millar (United Kingdom), Letizia Stanghellini (United States), Richard de Grijs (China Nanjing), Bruce G. Elmegreen

(United States), Leonardo Testi (Italy), Francisca Kemper (China Taipei), Ewine F. van Dishoeck (Netherlands), Diego Mardones (Chile), Michael R. Meyer (Switzerland)

Commissions:

Commission H1 The Local Universe
Commission H2 Astrochemistry
Commission H3 Planetary Nebulae
Inter-Division G-H-J Commission Stellar Clusters throughout Cosmic Space and Time
Inter-Division D-G-H-J Commission Galaxy Spectral Energy Distributions
Inter-Division B-H-J Commission Intergalactic Medium

Division J Galaxies and Cosmology

President:
Claus Leitherer
Space Telescope Science Inst
3700 San Martin Dr
MD 21218-2410 - Baltimore
United States

Vice-President:
Matthew Arnold Malkan
UCLA
Physics-Astronomy Dept
Box 951547
405 Hilgard Ave
CA 90095-1547 - Los Angeles
United States

Secretary:
Jeremy R. Mould
Swinburne University
Centre for Astrophysics and Supercomputing
John St
Hawthorn - 3122
Melbourne
Australia

Organizing Committee Members:
Andrew J. Bunker (United Kingdom), Monica Rubio (Chile), Leslie Hunt (Italy), Marcella Carollo (Switzerland), Jeremy R. Mould (Australia), Denis Burgarella (France), Avery A. Meiksin (United Kingdom), Claus Leitherer (United States), Matthew Arnold Malkan (United States), Françoise Combes (France), Thaisa Storchi-Bergmann (Brazil), Stéphane J. Courteau (Canada)

Commissions:

Inter-Division G-H-J Commission Stellar Clusters throughout Cosmic Space and Time
Inter-Division D-G-H-J Commission Galaxy Spectral Energy Distributions
Inter-Division B-H-J Commission Intergalactic Medium
Cross-Division D-J Commission Supermassive Black Holes, Feedback and Galaxy Evolution

Division Working Groups

Division	*Working Group Name*	*Chair*	*Co-Chair(s)*
D A	Astrometry by Small Ground-Based Telescopes		William Thuillot, Marcelo Assafin
D A	Multi-waveband Realisations of International Celestial Reference System	François Mignard	
D A	Numerical Standards in Fundamental Astronomy (NSFA)	Brian Luzum	
D A	Standards of Fundamental Astronomy (SOFA)	C. Y. Hohenkerk	
D A	Third Realisation of International Celestial Reference Frame	Patrick Charlot	
D B	Time Domain Astronomy	Rob Seaman	
D B	UV Astronomy	A. I. Gómez de Castro	
D C	Star Names	Eric E. Mamajek	
D D	Supernovae	Avishay Gal-Yam	
D E	Impact of Magnetic Activity on Solar and Stellar Environments	Dibyendu Nandi	
D F	Exoplanets for the Public	A. Lecavelier des Etangs	
D F	Near Earth Objects	Patrick Michel	
D F	Planetary System Nomenclature (WGPSN)	Rita M. Schulz	
D F	Small Bodies Nomenclature (SBN)	Jana Tichá	
D G	Abundances in Red Giants	Jacco Th. van Loon	
D G	Active B Stars	Carol Evelyn Jones	
D G	Ap & Related Stars	Gautier Mathys	
D H	The Galactic Center		

Inter-Division Working Groups

Division	*Working Group Name*	*Chair*	*Co-Chair(s)*
D A-F	Cartographic Coordinates & Rotational Elements	Brent A. Archinal	
D B-E	Coordination of Synoptic Observations of the Sun		Frederic Clette, Alexei A. Pevtsov
D C-E	Solar Eclipses	Jay M. Pasachoff	

Commission Working Groups

Commission	*Working Group Name*	*Chair*	*Co-Chair(s)*
C A2	Theory of Earth Rotation and Validation (IAU / IAG Joint WG)	Jose M. Ferrándiz	
C B4	Historic Radio Astronomy	R. Wielebinski	R. T. Schilizzi
C B5	High-Accuracy Stellar Spectroscopy	Paul S. Barklem	
C B7	Site Protection	Richard F. Green	Saeko S. Hayashi
C B7	Technical Working Group	Diane E. Turnshek	

Commission	Working Group Name	Chair	Co-Chair(s)
C C1	Astronomy for Equity and Inclusion	Amelia Ortiz Gil	
C C1	Network for Astronomy School Education (NASE)	Rosa M. Ros	Beatriz E. García
C C1	Theory and Methods in Astronomy Education	Paulo S. Bretones	
C C2	CAP Conferences	Sze-leung Cheung	Oana M. Sandu
C C2	CAP Journal	Georgia R. Bladon	
C C2	Outreach Professionalization & Accreditation	Richard T. Fienberg	
C C2	Public Outreach Information Management	Lars L. Christensen	
C C2	Science Communication Research in Astronomy	Marta Entradas	
C C3	Historical Radio Astronomy (WGHRA)		
C C3	Johannes Kepler	Terence J. Mahoney	
C C4	Astronomical Heritage in Danger	Alejandro M. López	
C C4	Classical Observatories from the Renaissance to the 20th Century	Gudrun B. E. Wolfschmidt	
C C4	Heritage of Space Exploration	Mikhail Ya Marov	
C E1	Solar Irradiance	Greg Kopp	A. I. Shapiro
C F1	Meteor Shower Nomenclature (MSN-WG)	Diego Janches	

Inter-Commissions Working Groups

Commission	Working Group Name	Chair
CB7,CC1	Achieving Sustainable Development within a Quality Lighting Framework	
CB7,CC4	Windows to the Universe: High-Mountain Observatories, and Other Astronomical Sites of the Late 20th and Early 21st Centuries	Robert C. Smith
CC1,CF2,CF3,CH2	Education and Training in Astrobiology	Muriel Gargaud
CC1,CC4	Intangible Astronomical Heritage	Duane W. Hamacher
CC3,CC4	Archaeoastronomy and Astronomy in Culture	Raymond P. Norris

Executive Committee Working Groups

Working Group Name	Chair	Co-Chair(s)
Dark and Quiet Sky Protection	Richard F. Green	
Global Coordination of Ground and Space tAstrophysics	Roger L. Davies	David N. Spergel
International Year of Light: Cosmic Light	Richard F. Green	
Public Naming of Planets and Planetary Satellites	Thierry Montmerle	
Women in Astronomy	Francesca Primas	

Commissions (2015 - 2018)

Commission A1 - Astrometry

President:
Anthony G.A. Brown
Leiden University
Leiden Observatory
PO Box 9513
2300 RA - Leiden
Netherlands

Vice-President:
Jean Souchay
Observatoire de Paris
SYRTE
61 Av de l'Observatoire
75014 - Paris
France

Secretary:
Dafydd Wyn Evans
University of Cambridge
Institute of Astronomy
Madingley Road
CB3 0HA - Cambridge
United Kingdom

Organizing Committee Members:
Jean Souchay (France), Dafydd Wyn Evans (United Kingdom), Alexandre Humberto Andrei (Brazil), Stephen C. Unwin (United States), Yoshiyuki YAMADA (Japan), Norbert Zacharias (United States), Anthony G.A. Brown (Netherlands)

Commission A2 - Rotation of the Earth

President:
Richard Gross
Jet Propulsion Laboratory
Geodynamics and Space Geodesy Group
Mail Stop 238-600
4800 Oak Grove Drive
91109 - Pasadena, CA
United States

Vice-President:
Florian Seitz
Technische Universität München
Deutsches Geodätisches Forschungsinstitut
(DGFI)
Arcisstr. 21
D-80333 - Munich
Germany

Secretary:
Alberto Escapa
Universidad de Alicante
Matemática Aplicada. EPS
Campus de San Vicente
03690 - San Vicente de Raspeig - Alicante
Spain

Organizing Committee Members:
Jose Manuel Ferrándiz (Spain), Florian Seitz (Germany), Alberto Escapa (Spain), Daniela Thaller (Germany), Vladimir E. Zharov (Russian Federation), Richard Gross (United States)

Commission A3 - Fundamental Standards

President:
Catherine Y. Hohenkerk
UKHO
HM Nautical Almanac Office
Admiralty Way
TA1 2DN - Taunton
United Kingdom

Vice-President:
Brian Luzum
US Naval Observatory
3450 Massachusetts Ave NW
DC 20392 5420 - Washington
United States

Secretary:
Nancy A. Oliversen
United States Naval Observatory
Astronomical Applications Department
3450 Massachusetts Ave, N. W.
20392-5410 - Washington
United States

Organizing Committee Members:
Nicole Capitaine (France), Catherine Y. Hohenkerk (United Kingdom), Brian Luzum (United States), Charles H. Acton (United States), John A. Bangert (United States), Dennis D. McCarthy (United States), Nancy A. Oliversen (United States)

Commission B1 - Computational Astrophysics

President:
Simon F. Portegies Zwart
Leiden Observatory
Leiden University
PO Box 9513
NL2300 RA - Leiden
Netherlands

Vice-President:
Dmitrij V. Bisikalo
Inst of Astronomy RAS
Pyatnitskaya ul 48
119017 - Moscow
Russian Federation

Secretary:
Christian M. Boily
Observatoire Astronomique de Strasbourg
11 rue de l'Université
67000 - Strasbourg
France

Organizing Committee Members:
Dmitrij V. Bisikalo (Russian Federation), Sungsoo S. KIM (Korea, Rep of), Irina N. Kitiashvili (United States), Garrelt Mellema (Sweden), Michael Shara (United States), Christian M. Boily (France), Simon F. Portegies Zwart (Netherlands)

Commission B2 - Data and Documentation

President:
Michael W. Wise
ASTRON
Netherlands Institute for Radio Astronomy
P.O. Box 2
7990 AA - Dwingeloo
Netherlands

Vice-President:
Anja C. Schröder
SAAO
Observatory Road
P O Box 9
7935 - Cape Town
South Africa

Organizing Committee Members:
Michael W. Wise (Netherlands), Anja C. Schröder (South Africa), Chenzhou CUI (China Nanjing), R. Elizabeth M. Griffin (Canada), Robert J. Hanisch (United States), Arnold H. Rots (United States), Rob Seaman (United States)

Commission B3 - Astroinformatics and Astrostatistics

President:
Eric D. Feigelson
Pennsylvania State University
Astronomy & Astrophysics Dept
525 Davey Lab
PA 16802 - University Park
United States

Vice-President:
Prajval Shastri
Indian Institute of Astrophysics
Sarjapur Road
Koramangala
560 034 - Bengaluru
India

Secretary:
Saeqa Dil Vrtilek
Harvard Smithsonian CfA
HCO/SAO
MS 67
60 Garden Str
MA 02138-1516 - Cambridge
United States

Organizing Committee Members:
Eric D. Feigelson (United States), Saeqa Dil Vrtilek (United States), Eric B. Ford (United States), Alan Heavens (United Kingdom), Fionn Murtagh (United Kingdom), Yanxia ZHANG (China Nanjing), Prajval Shastri (India)

Commission B4 - Radio Astronomy

President:
Gabriele Giovannini
INAF
Istituto di Radioastronomia
Via P Gobetti 101
40129 - Bologna
Italy

Vice-President:
Anthony James Beasley
NRAO
520 Edgemont Rd
22903-2475 - Charlottesville
United States

Secretary:
Wim van Driel
Observatoire Paris-Site de Meudon
GEPI
5 Pl Jules Janssen
92195 - Meudon Cedex
France

Organizing Committee Members:
Anthony James Beasley (United States), Gabriele Giovannini (Italy), Wim van Driel (France), Xiaoyu HONG (China Nanjing), Joseph Lazio (United States), Nicholas Seymour (Australia), Jan Mathijs van der Hulst (Netherlands), Tony H. Wong (United States)

Commission B5 - Laboratory Astrophysics

President:
Farid Salama
NASA- Ames Research Center
Space Science & Astrobiology Division
Building N245
MS 245 6
CA 94035-1000 - Moffett Field
United States

Vice-President:
Helen J. Fraser
The Open Unviersity
Department of Physical Sciences, Astronomy Division
Walton Hall
MK7 6AA - Milton Keynes
United Kingdom

Secretary:
Gianfranco Vidali
Syracuse University
Department of Physics
221 Physics Building
NY 13244-1130 - Syracuse
United States

Organizing Committee Members:

Paul S. Barklem (Sweden), Farid Salama (United States), Gianfranco Vidali (United States), Thomas Henning (Germany), Harold Linnartz (Netherlands), Feilu WANG (China Nanjing), Helen J. Fraser (United Kingdom)

Commission B6 - Astronomical Photometry and Polarimetry

President:
Saul J. Adelman
The Citadel
Physics Dept
171 Moultrie St
SC 29409-0270 - Charleston
United States

Vice-President:
Antonio Mario Magalhaes
Universidade de São Paulo
Instituto de Astronomia, Geofisica e Ciencias Atmosfericas
Rua do Matão 1226
Cidade Universitaria
05508-090 - São Paulo
Brazil

Secretary:
Saul J. Adelman
The Citadel
Physics Dept
171 Moultrie St
SC 29409-0270 - Charleston
United States

Organizing Committee Members:

Antonio Mario Magalhaes (Brazil), Saul J. Adelman (United States), Pierre Bastien (Canada), Richard Ignace (United States), Carme Jordi (Spain), J. Allyn Smith (United States), Kevin Volk (United States)

Commission C1 - Astronomy Education and Development

President:
Beatriz Elena García
Instituto de Tecnologías en Detección y Astropartículas (ITeDA)
ITeDA Mendoza
Azopardo 313
Godoy Cruz
5501 - Mendoza
Argentina

Vice-President:
Paulo Sergio Bretones
Universidade Federal de São Carlos
Departamento de Metodologia de Ensino
13565-905 - São Carlos
Brazil

Secretary:
Kathleen DeGioia Eastwood
Northern Arizona University

Physics and Astronomy Department
Box 6010
AZ 86011-6010 - Flagstaff
United States

Organizing Committee Members:
Beatriz Elena García (Argentina), Paulo Sergio Bretones (Brazil), Jean-Pierre de Greve (Belgium), Christopher David Impey (United States), Nicoletta Lanciano (Italy), Amelia Ortiz Gil (Spain), Kathleen DeGioia Eastwood (United States)

Commission C2 - Communicating Astronomy with the Public

President:
Pedro Russo
Leiden Observatory
Universe Awareness International Office
P.O. Box 9513
NL-2300 RA - Leiden
Netherlands

Vice-President:
Richard Tresch Fienberg
American Astronomical Society
43 Everett Ave.
MA 02472-1881 - Watertown
United States

Secretary:
Kingsley C. Okpala
University of Nigeria
Physics & Astronomy
University Road, Nsukka, Enugu State, Nigeria
41001 - Nsukka, Enugu State
Nigeria

Organizing Committee Members:
Pedro Russo (Netherlands), Kingsley C. Okpala (Nigeria), Megan Kirsty Argo (United Kingdom), Lars Lindberg Christensen (Germany), Carol Ann Christian (United States), Sylvie D. Vauclair (France), Richard Tresch Fienberg (United States)

Commission C3 - History of Astronomy

President:
Xiaochun Sun
Inst History Natural Science, CAS
55 Zhongguancun East Road
100190 - Beijing
China Nanjing

Vice-President:
Wayne Orchiston
National Astronomical Research Institute of Thailand
191 Huay Kaew Road
Suthep District
50200 - Chiang Mai
Thailand

Organizing Committee Members:
Xiaochun Sun (China Nanjing), Wayne Orchiston (Thailand), Owen Gingerich (United States), Raymond P. Norris (Australia), Jay M. Pasachoff (United States), Christiaan L. Sterken (Belgium), David Valls-Gabaud (France)

Commission C4 - World Heritage and Astronomy

President:
Clive Ruggles
University of Leicester
School of Archaeology and Ancient History
University Road
LE1 7RH - Leicester
United Kingdom

Gudrun B. E. Wolfschmidt
Hamburg University, Center for history of science and technology
Department of Physics, Hamburg Observatory
Bundesstr 55 Geomatikum
D-20146 - Hamburg
Germany

Vice-President:

Secretary:
Roger Ferlet
UMR7095 CNRS
Institut d'Astrophysique de Paris
98bis Bd Arago
75014 - Paris
France

Organizing Committee Members:
Roger Ferlet (France), Clive Ruggles (United Kingdom), Gudrun B. E. Wolfschmidt (Germany), Siramas Komonjinda (Thailand), Mikhail Ya Marov (Russian Federation), Malcolm G. Smith (Chile)

Commission D1 - Gravitational Wave Astrophysics

President:
Neil Gehrels
NASA/GSFC
Code 661
MD 20771 - Greenbelt
United States

Vice-President:
Marica Branchesi
Università degli studi di Urbino "Carlo Bo"
Dipartimento di Scienze di Base e Fondamenti/Physics section
Via Santa Chiara, 27
61029 - Urbino
Italy

Organizing Committee Members:
Neil Gehrels (United States), Marica Branchesi (Italy), Pierre Binetruy (France), Federico Ferrini (Italy), Richard N. Manchester (Australia), David H. Shoemaker (United States), Robin Stebbins (United States)

Commission E1 - Solar Radiation and Structure

President:
Natalie Krivova
MPI für Sonnensystemforschung
Justus-von-Liebig-Weg 3
37077 - Göttingen Germany

Vice-President:
Alexander Kosovichev
New Jersey Institute of Technology
Department of Physics
Tiernan Hall, Room 463
University Heights
07102 - Newark
United States

Secretary:
Nicolas Labrosse
University of Glasgow
School of Physics and Astronomy
G12 8QQ - Glasgow
United Kingdom

Organizing Committee Members:
Michele Bianda (Switzerland), Nataliia Gennadievna Shchukina (Ukraine), Natalie Krivova (Germany), Alexander Kosovichev (United States), Nicolas Labrosse (United Kingdom), Gianna Cauzzi (Italy), Yoshinori SUEMATSU (Japan)

Commission E2 - Solar Activity

President:
Lyndsay Fletcher
University of Glasgow
School of Physics and Astronomy
Kelvin Bldg
G12 8QQ - Glasgow
United Kingdom

Vice-President:
Paul S. Cally
Monash University
School of Mathematical Sciences
and Monash Centre for Astrophysics
Wellington Road
Clayton
VIC 3800 - Victoria
Australia

Organizing Committee Members:
Manolis K. Georgoulis (Greece), Lyndsay Fletcher (United Kingdom), Karel J. Schrijver (United States), Paul S. Cally (Australia), Philippa K. Browning (United Kingdom), Jongchul CHAE (Korea, Rep of), Amy R. Winebarger (United States)

Commission E3 - Solar Impact Throughout the Heliosphere

President:
Ingrid Mann
EISCAT Scientific Association
P. O. Box 812
98128 - Kiruna
Sweden

Vice-President:
Carine Briand
Observatoire de Paris
LESIA/CNRS
5 place J. Janssen
92195 - Meudon
France

Organizing Committee Members:
Ingrid Mann (Sweden), Margit Haberreiter (Switzerland), Olga Malandraki (Greece), Dibyendu Nandi (India), Kanya KUSANO (Japan), Ilya G. Usoskin (Finland), Carine Briand (France)

Commission F1 - Meteors, Meteorites and Interplanetary Dust

President:
Jiří Borovička
Astronomical Institute of the Academy of Sciences of the Czech Republic
Department of Interplanetary Matter
Fričova 298
251 65 - Ondřejov
Czech Republic

Vice-President:
Diego Janches
Goddard Space Flight Center/NASA
Space Weather Lab
Code 674
20771 - Greenbelt
United States

Secretary:
Pavel Koten
Astronomical Inst. Academy of Sciences
Department of Interplanetary Matter
Fričova 298
251 65 - Ondřejov
Czech Republic

Organizing Committee Members:
Jiří Borovička (Czech Republic), Pavel Koten (Czech Republic), Margaret D. Campbell-Brown (Canada), Petrus Matheus Marie Jenniskens (United States), Galina O. Ryabova (Russian Federation), Jérémie J. Vaubaillon (France), David J. Asher (United Kingdom), Diego Janches (United States)

Commission F2 - Exoplanets and the Solar system

President:
Alain Lecavelier des Etangs
Institut d'Astrophysique de Paris
98bis Bd Arago
75014 - Paris
France

Vice-President:
Jack J. Lissauer
NASA/ARC
MS 245 3
CA 94035 1000 - Moffett Field
United States

Organizing Committee Members:

Alain Lecavelier des Etangs (France), Régis Courtin (France), Mark T Lemmon (United States), Patrick Michel (France), Alessandro Morbidelli (France), Feng TIAN (China Nanjing), Paul A. Wiegert (Canada), Jack J. Lissauer (United States)

Commission F3 - Astrobiology

President:
Sun Kwok
University of Hong Kong
Faculty of Science
Chong Yuet Ming Physics Bldg
Pokfulam Rd
Hong Kong
China Nanjing

Vice-President:
Masatoshi OHISHI
NAOJ
Astronomy Data Center
2-21-1 Osawa
Mitaka
181-8588 - Tokyo
Japan

Secretary:
Joseph A. Nuth
NASA/GSFC
Solar System Exploration Division
Code 690
MD 20771 - Greenbelt
United States

Organizing Committee Members:

Joseph A. Nuth (United States), Sun Kwok (China Nanjing), Masatoshi OHISHI (Japan), Muriel Gargaud (France), Jesus Martinez-Frias (Spain), Sergio Pilling (Brazil), Nils G. Holm (Sweden)

Commission G1 - Binary and Multiple Star Systems

President:
Andrej Prša
Villanova University
Dept. of Astrophysics and Planetary Science
800 E Lancaster Ave
19085 - Villanova
United States

Vice-President:
Virginia Trimble
University of California Irvine
Physics-Astronomy Department
4575 Physics
CA 92697-4575 - Irvine
United States

Secretary:
Christopher Adam Tout
University of Cambridge
Institute of Astronomy
The Observatories
Madingley Road
CB3 0HA - Cambridge
United Kingdom

Organizing Committee Members:
Andrej Prša (United States), Virginia Trimble (United States), Brian D. Mason (United States), Robert D. Mathieu (United States), Terry D. Oswalt (United States), John Southworth (United Kingdom), Tomaž Zwitter (Slovenia, the Republic of), Christopher Adam Tout (United Kingdom)

Commission G2 - Massive Stars

President:
Artemio Herrero Davó
Instituto de Astrofísica de Canarias
Research Division
C/ Vía Láctea s/n
38200 - La Laguna, Tenerife
Spain

Vice-President:
Jorick S. Vink
Armagh Observatory
College Hill
BT61 9DG - Armagh
United Kingdom

Secretary:
Nicole St-Louis
Université de Montréal
Département de Physique
CP 6128, Succ Centre-Ville
QC H3C 3J7 - Montréal
Canada

Organizing Committee Members:
You-Hua Chu (China Taipei), Artemio Herrero Davó (Spain), Jorick S. Vink (United Kingdom), Nicole St-Louis (Canada), Jose H. Groh (Switzerland), Gregor Rauw (Belgium), Asif ud-Doula (United States)

Commission G3 - Stellar Evolution

President:
John C. Lattanzio
Monash University
School of Physics and Astronomy
10 College Walk
Clayton Campus
VIC 3800 - Victoria
Australia

Vice-President:
Marc Howard Pinsonneault
Ohio State University
Astronomy Dept
140 W 18th Ave
OH 43210-1106 - Columbus
United States

Secretary:
Jacco Th. van Loon
Keele University
Astrophysics Group
Lennard-Jones Laboratories
ST5 5BG - Staffordshire
United Kingdom

Organizing Committee Members:
Jacco Th. van Loon (United Kingdom), John C. Lattanzio (Australia), Marc Howard Pinsonneault (United States), Zhanwen HAN (China Nanjing), Franz Kerschbaum (Austria), Marco Limongi (Italy), Marcella Marconi (Italy), Gražina Tautvaišiene (Lithuania)

Commission G4 - Pulsating Stars

President:
Christopher Simon Jeffery
Armagh Observatory
College Hill
BT61 9DG - Armagh
United Kingdom

Vice-President:
Jaymie Matthews
University of British Columbia
Dept of Physics and Astronomy
Hennings Bldg
6224 Agricultural Rd
BC V6T 1Z1 - Vancouver
Canada

Secretary:
Saskia Hekker
Max Planck Institute for Solar System Research
Justus von Liebigweg 3

37077 - Göttingen
Germany

Organizing Committee Members:
Christopher Simon Jeffery (United Kingdom), Jaymie Matthews (Canada), Saskia Hekker (Germany), Joyce Ann Guzik (United States), Karen Pollard (New Zealand), Hiromoto SHIBAHASHI (Japan), Dennis Stello (Australia)

Commission G5 - Stellar and Planetary Atmospheres

President:
Ivan Hubeny
University of Arizona
Steward Observatory and Department of Astronomy
933 N. Cherry Avenue
AZ 85721 - Tuscon
United States

Vice-President:
Carlos Allende Prieto
Instituto de Astrofisica de Canarias
Via Lactea s/n
38200 - La Laguna, Tenerife
Spain

Organizing Committee Members:
Ivan Hubeny (United States), France Allard (France), Katia Cunha (Brazil), John D. Landstreet (Canada), Thierry M Lanz (France), Lyudmila I. Mashonkina (Russian Federation), Adam P. Showman (United States), Carlos Allende Prieto (Spain)

Commission H1 - The Local Universe

President:
Eva K. Grebel
Astronomisches Rechen-Institut
Zentrum fuer Astronomie der Universitaet Heidelberg
Moenchhofstr. 12-14
University of Heidelberg
69120 - Heidelberg
Germany

Vice-President:
Dante Minniti
Universidad Andres Bello
Department of Physics
Fernandez Concha 700
Las Condes
Santiago
Chile

Organizing Committee Members:
Eva K. Grebel (Germany), Dante Minniti (Chile), Evangelie (Lia) Athanassoula (France), Sofia Feltzing (Sweden), Yasuo FUKUI (Japan), Vanessa M. Hill (France), Margaret Meixner (United States), Gang ZHAO (China Nanjing)

Commission H2 - Astrochemistry

President:
Thomas J. Millar
Queen's Univ Belfast
School of Maths-Physics
13 Stranmillis Rd
BT9 5FS - Belfast
United Kingdom

Vice-President:
Edwin A. Bergin
University of Michigan
Astronomy Dept
1085 S. University Ave
MI 48109-1107
Ann Arbor
United States

Secretary:
Maria R. Cunningham
UNSW
School of Physics
NSW 2052 - Sydney
Australia

Organizing Committee Members:
Maria R. Cunningham (Australia), Thomas J. Millar (United Kingdom), Satoshi YAMAMOTO (Japan), Edwin A. Bergin (United States), Yuri AIKAWA (Japan), Paola Caselli (Germany), Jes K. J. (Denmark)

Organizing Committee Members:
Eva K. Grebel (Germany), Dante Minniti (Chile), Evangelie (Lia) Athanassoula (France), Sofia Feltzing (Sweden), Yasuo FUKUI (Japan), Vanessa M. Hill (France), Margaret Meixner (United States), Gang ZHAO (China Nanjing)

Commission H3 - Planetary Nebulae

President:
Letizia Stanghellini
NOAO
950 North Cherry Ave
AZ 85726-6732 - Tucson
United States

Vice-President:
Albert Zijlstra
University of Manchester
School of Physics & Astronomy
Oxford Road
M13 9PL - Manchester
United Kingdom

Secretary:
Arturo Manchado
Instituto de Astrofísica de Canarias
Research
C/ Vía Láctea s/n
38200 - La Laguna, Tenerife
Spain

Organizing Committee Members:
Letizia Stanghellini (United States), Arturo Manchado (Spain), Orsola De Marco (Australia), Karen B. Kwitter (United States), Miriam Pena (Mexico), Albert Zijlstra (United Kingdom)

Inter-Divisions Commissions (2015 - 2018)

Inter-Division A-F Commission - Celestial Mechanics and Dynamical Astronomy

President:
Cristian Beaugé
Observatorio Astronómico
Laprida 854
Barrio Observatorio
5000 - Córdoba
Argentina

Vice-President:
Alessandra Celletti
Università di Roma Tor Vergata
Dipto di Matematica
Via della Ricerca Scientifica, 1
00133 - Roma
Italy

Secretary:
Bonnie Alice Steves
Glasgow Caledonian University
The Graduate School
Cowcaddens Rd
G4 0BA - Glasgow
United Kingdom

Organizing Committee Members:
Cristian Beaugé (Argentina), Bonnie Alice Steves (United Kingdom), Douglas P. Hamilton (United States), Eiichiro KOKUBO (Japan), Jacques Laskar (France), Daniel J. Scheeres (United States), Alessandra Celletti (Italy)

Inter-Division B-C Commission - Protection of Existing and Potential Observatory Sites

President:
Richard F. Green
University of Arizona
Steward Observatory
933 North Cherry Ave
AZ 85721-0065 -Tucson United States

Vice-President:
Constance Elaine Walker
National Optical Astronomy Observatory
950 North Cherry Avenue
85719 -Tucson
United States

Organizing Committee Members:
Richard F. Green (United States), David Galadí-Enríquez (Spain), Harvey Steven Liszt (United States), Ramotholo R. Sefako (South Africa), Yongheng ZHAO (China Nanjing), Constance Elaine Walker (United States)

Inter-Divisions G H J Commission - Stellar Clusters throughout Cosmic Space and Time

President:
Richard de Grijs
Peking University
Kavli Institute for Astronomy and Astrophysics
Yi He Yuan Lu 5
Hai Dian District
100871 - Beijing
China Nanjing

Vice-President:
Amanda I. Karakas
Monash University
Monash Centre for Astrophysics, School of Physics & Astronomy
School of Physics & Astronomy, Monash University
Clayton
3800 - Melbourne
Australia

Secretary:
Ernst Paunzen
Faculty of Sciences Masaryk University
Department of Theoretical Physics and Astrophysics
Kotlářská 2
611 37 - Brno
Czech Republic

Organizing Committee Members:

Ernst Paunzen (Czech Republic), Francesca D'Antona (Italy), Richard de Grijs (China Nanjing), Amanda I. Karakas (Australia), André Moitinho (Portugal), Jan Palouš (Czech Republic), Alison I. Sills (Canada)

Inter-Division D-G-H-J Commission - Galaxy Spectral Energy Distributions

President:
Denis Burgarella
Aix-Marseille University
Laboratoire d'Astrophysique de Marseille
Technopole de Chateau-Gombert
38, rue Joliot-Curie
13388 - Marseille Cedex 13
France

Vice-President:
Cristina Carmen Popescu
University of Central Lancashire
Jeremiah Horrocks Institute
PR1 2HE - Preston
United Kingdom

Secretary:
Daniel Schaerer
Université de Genève
Observatoire de Genève
51 chemin des Maillettes
1290 - Versoix
Switzerland

Organizing Committee Members:

Denis Burgarella (France), Cristina Carmen Popescu (United Kingdom), Amy J. Barger (United States), Asantha R. Cooray (United States), Robert C. Kennicutt (United Kingdom), Toru YAMADA (Japan), Daniel Schaerer (Switzerland)

Inter-Division B-H-J Commission - Intergalactic Medium

President:

Avery A. Meiksin
University of Edinburgh
Institute for Astronomy
Royal Observatory
Blackford Hill
EH9 3HJ - Edinburgh
United Kingdom

Vice-President:

Hsiao-Wen CHEN
Kavli Institute for Cosmological Physics
Department of Astronomy & Astrophysics
The University of Chicago
5640 S. Ellis Ave
Chicago, IL 60637
United States

Organizing Committee Members:

Avery A. Meiksin (United Kingdom), Hsiao-Wen CHEN (United States), Valentina D' Odorico (Italy), Nissim Kanekar (India), Jason X. Prochaska (United States), Joop Schaye (Netherlands)

Cross-Divisions Commissions (2015 - 2018)

Cross-Division D-J Commission - Supermassive Black Holes, Feedback and Galaxy Evolution

President:

William Richard Forman
Smithsonian Observatory
Harvard Smithsonian Center for Astrophysics
High Energy Astrophysics Division
60 Garden Street
MA 02138-1516 - Cambridge
United States

Vice-President:

Thaisa Storchi-Bergmann
UFRGS
Inst de Física
Campus do Vale
CP 15051
91501-970 - Porto Alegre, RS
Brazil

Organizing Committee Members:

William Richard Forman (United States), Thaisa Storchi-Bergmann (Brazil), Judith H. Croston (United Kingdom), Sebastian Heinz (United States), Roberto Maiolino (United Kingdom), Sera B. Markoff (Netherlands), Hagai Netzer (Israel), Marta Volonteri (France)

Cross-Division A-F Commission - Solar System Ephemerides

President:
Andrea Milani Comparetti
Università di Pisa
Dipartimento di Matematica
Piazzale B. Pontecorvo 5
56127 - Pisa
Italy

Vice-President:
William M. Folkner
JPL m/s 301-121
4800 Oak Grove Dr.
91109 - Pasadena
United States

Organizing Committee Members:

William M. Folkner (United States), Steven R. Chesley (United States), Andrea Milani Comparetti (Italy), Jean-Eudes Arlot (France), Elena V. Pitjeva (Russian Federation), Paolo Tanga (France)

Transactions IAU, Volume XXIXB
Proceedings IAU Symposium No. Volume XXIXB, 2018
P. Benvenuti, ed.
© International Astronomical Union 2019
doi:10.1017/S1743921318004052

CHAPTER X

DIVISIONS MEMBERSHIP
(until 31 August 2015)

NOTE. This chapter gives the membership of the IAU Divisions in force until the end of the XVIIIth General Assembly.

Division A Fundamental Astronomy

President: Sergei A. Klioner
Vice-President: Jacques Laskar

Steering Committee Members:

Nicole Capitaine
Cheng-Li Huang
Catherine Y. Hohenkerk
Mizuhiko HOSOKAWA
Dennis D. McCarthy
François Mignard
Alessandro Morbidelli
Dimitri Pourbaix
Michael H. Soffel
Norbert Zacharias
Richard Gross
Sergei A. Klioner
Jacques Laskar
Anthony G.A. Brown
Sylvio Ferraz-Mello
Susan Gessner Stewart

Members:

Abad Hiraldo, Carlos
Abad Medina, Alberto
Abalakin, Viktor
Abbas, Ummi
Abele, Màris
Abt, Helmut
Acton, Charles
Aghaee, Alireza
Aharonian, Felix
Ahmed, Abdel-aziz
Ahmed, Mostafa
AHN, SANG-HYEON
AHN, Youngsook
AK, Tansel
Aksnes, Kaare
Ali, Sabry
Allan, David
Alley, Carrol
Al-Malki, Mohammed
Al-Naimiy, Hamid
Ammons, Stephen
Andrade, Manuel
Andrei, Alexandre
Anosova, Joanna
Antonacopoulos, Gregory
Antonelli, Lucio Angelo
Aquilano, Roberto
Arabelos, Dimitrios
ARAKIDA, Hideyoshi
Archinal, Brent
Arenou, Frédéric
Argyle, Robert
Arias, Elisa
Arlot, Jean-Eudes
Arnold, Richard
Arribas Mocoroa, Santiago
Asanok, Kitiyanee
Ashby, Neil
Asher, David
Asplund, Martin
Assafin, Marcelo
Athanassoula, Evangelie (Lia)
Augereau, Jean-Charles
BABA, Junichi
Babusiaux, Carine
Badescu, Octavian
Badolati, Ennio
BAEK, Chang Hyun
Bailes, Matthew
Bakhtigaraev, Nail

Baldry, Ivan
Ballabh, Goswami
Balmino, Georges
Balona, Luis
Bandyopadhyay, A.
Bangert, John
Banni, Aldo
Barabanov, Sergey
Barberis, Bruno
Barbieri, Cesare
Barbosu, Mihail
Barkin, Yuri
Bartczak, Przemysław
Bartkevicius, Antanas
Bartlett, Jennifer
Bastian, Ulrich
Bates, Brian
Batten, Alan
Bazzano, Angela
Beasley, Anthony
Beaugé, Cristian
Beavers, Willet
Becker, Werner
Beers, Timothy
Belizon, Fernando
Bell, Steven
Benedict, George
Benest, Daniel
Benevides Soares, Paulo
Bergeat, Jacques
Bernardi, Fabrizio
Berthier, Jerôme
Bettis, Dale
Beutler, Gerhard
Beuzit, Jean-Luc
Bhatnagar, K.
Bien, Reinhold
Bietenholz, Michael
Birlan, Mirel
Bishop, Roy
Bizouard, Christian
Blair, David
Blakeslee, John
Boboltz, David
Boccaletti, Dino
Boehm, Johannes
Bois, Eric
Bolotin, Sergei
Bolotina, Olga
Bonanos, Alceste
Bongiovanni, Angel
Borczyk, Wojciech
Borderies, Nicole
Borisov, Borislav
Borisov, Galin
Borkowski, Kazimierz
Boroson, Bram
Boss, Alan
Bouchard, Antoine
Boucher, Claude
Bougeard, Mireille
Bourda, Geraldine
Boyajian, Tabetha
Boytel, Jorge
Bozis, George
Bradley, Arthur
Bradt, Hale
Branham, Richard
Breakiron, Lee
Breger, Michel
Breiter, Sławomir
Brentjens, Michiel
Brieva, Eduardo
Bromberg, Omer
Brookes, Clive
Brosche, Peter
Brouw, Willem
Brown, Anthony
Brož, Miroslav
Brozovic, Marina
Brumberg, Victor
Brunini, Adrian
Bruyninx, Carine
Bryant, John
Brzeziński, Aleksander
Bucciarelli, Beatrice
Buchhave, Lars
Bueno de Camargo, Julio
Buie, Marc
Burikham, Piyabut
Burki, Gilbert
Burns, Joseph
Bursa, Michal
Butkovskaya, Varvara
Butler, Paul
CAI, Michael
CAI, Yong
Calabretta, Mark
Campbell, Robert
Cannon, Kipp
Capitaine, Nicole
Caranicolas, Nicholas
Cardoso Santos, Nuno
Carlin, Jeffrey
Carney, Bruce
Carpino, Mario
Carpintero, Daniel
Carruba, Valerio
Carter, William
Casetti, Dana
Catanzaro, Giovanni
Cazenave, Anny
Cefola, Paul
Celletti, Alessandra
Chadid, Merieme
CHAE, Kyu Hyun
Chambers, John
Champion, David
Chao, Benjamin
Chapanov, Yavor
Chapront, Jean
Chapront-Touze, Michelle
Charlot, Patrick
Chemin, Laurent
CHEN, Alfred
CHEN, Li
CHEN, Li
CHEN, Linfei
CHEN, Wen Ping
CHEN, Yuqin
Chesley, Steven
CHOI, Kyu Hong
CHOU, Yi
Christou, Apostolos
Chung, Eduardo
CHUNG, Hyun-Soo
Cincotta, Pablo
Cionco, Rodolfo
Cioni, Maria-Rosa
Clarkson, Chris
Cochran, William
Colin, Jacques
Coma, Juan
Conrad, Albert
Contopoulos, George
Cooper, Nicholas
Corbin, Thomas
Corwin Jr, Harold
Costa, Edgardo
Couto da Silva, Telma

Crampton, David
Crézé, Michel
Crifo, Francoise
Crosta, Mariateresa
Csabai, Istvan
Cudworth, Kyle
Cvetkovi'c, Zorica
da Costa, Luiz
Dahn, Conard
Dalton, Gavin
Damljanovic, Goran
Danylevsky, Vassyl
da Rocha-Poppe, Paulo
Davis, Marc
Davis, Morris
Davis, Robert
Day-Jones, Avril
Debarbat, Suzanne
De Biasi, Maria
de Bruijne, Jos
De Cuyper, Jean-Pierre
de Felice, Fernando
Defraigne, Pascale
de Greiff, J.
Dehant, Véronique
Dejaiffe, Rene
de Jonge, J.
Deleflie, Florent
Dell'Antonio, Ian
Deller, Adam
Delmas, Christian
Del Santo, Melania
de Medeiros, Jose
DeNisco, Kenneth
Derekas, Aliz
Derman, Ethem
Dermott, Stanley
Descamps, Pascal
De Souza Pellegrini, Paulo
de Viron, Olivier
Devyatkin, Aleksandr
DI, Xiaohua
Dick, Steven
Dick, Wolfgang
Dickey, Jean
Dickey, John
Dickman, Steven
Dikova, Smilyana
Di Sisto, Romina
Domínguez, Mariano
do Nascimento, José
DONG, Shaowu
DONG, Xiaojun
Dorch, Søren Bertil
Dormand, John
Douglas, Nigel
Douglas, R.
Dourneau, Gerard
Downes, Turlough
Drake, Jeremy
Dravins, Dainis
Drimmel, Ronald
Drożyner, Andrzej
DU, Lan
Dubath, Pierre
Ducourant, Christine
Duma, Dmitrij
Dunham, David
Duriez, Luc
Dvorak, Rudolf
Dybczyński, Piotr
Eaton, Joel
Efroimsky, Michael
Eisner, Josh
El, Bakkali
Elipe, Antonio
Elkin, Vladimir
El Shahawy, Mohamad
Emelianov, Nikolaj
Emel'yanenko, Vacheslav
Emilio, Marcelo
Eppelbaum, Lev
Érdi, Bálint
Eroshkin, Georgij
Escapa, Alberto
Esguerra, Jose Perico
Espenak, Fred
Evans, Dafydd
Evans, Daniel
Fabricius, Claus
Fahey, Richard
Faller, James
Fallon, Frederick
FAN, Yu
FAN, Yufeng
Farnham, Tony
Faure, Cécile
Fedorov, Peter
Fedorova, Elena
Fekel, Francis
FENG, Chugang
Fernandes-Martin, Vera
Fernández, Laura
Fernández, Silvia
Ferrandiz, Jose
Ferrari, Fabricio
Ferraz-Mello, Sylvio
Ferrer, Martinez
Fey, Alan
Fienga, Agnès
Filacchione, Gianrico
Finch, Charlie
Firneis, Maria
Fletcher, J.
Fliegel, Henry
Floria Peralta, Luis
Florsch, Alphonse
Folgueira, Marta
Folkner, William
Foltz, Craig
Fomin, Valery
Fominov, Aleksandr
Forgács-Dajka, Emese
Fors, Octavi
Forveille, Thierry
Foschini, Luigi
Fouchard, Marc
Franz, Otto
Fredrick, Laurence
Fresneau, Alain
Froeschle, Claude
Froeschle, Michel
Fruchter, Andrew
Frutos-Alfaro, Francisco
FU, Yanning
FUJIMOTO, Masa-Katsu
FUJISHITA, Mitsumi
FUJITA, Mitsutaka
FUKUSHIMA, Toshio
Gai, Mario
Gallardo Castro, Carlos
Gambis, Daniel
Gamen, Roberto
Gandolfi, Davide
Gangadhara, R.T.
GAO, Buxi
GAO, Jian
GAO, Yuping
Gaposchkin, Edward
Garcia, Beatriz

Garcia-Barreto, José
Garrido, César
Gatewood, George
Gaume, Ralph
Gauss, Stephen
Gavras, Panagiotis
Gayazov, Iskander
Geffert, Michael
Geller, Aaron
GENG, Lihong
Georgelin, Yvon
Germain, Marvin
Getino Fernández, Juan
Giacaglia, Giorgio
GILLES, Dominique
Gilmore, Gerard
Giordano, Claudia
Giorgini, Jon
Giovanelli, Riccardo
Giuliatti Winter, Silvia
Glebova, Nina
Gnedin, Yurij
Goddi, Ciriaco
Goldreich, Peter
Gomes, Rodney
Gómez Fernández, Jose
Gontcharov, George
González, Gabriela
Gonzalez, Jorge
Goode, Philip
Goodwin, Simon
GOUDA, Naoteru
Goudas, Constantine
Gouguenheim, Lucienne
Gouliermis, Dimitrios
Goyal, A.
Goździewski, Krzysztof
Gozhy, Adam
Granveaud, Michel
Granvik, Mikael
Gray, David
Gray, Norman
Greenberg, Richard
Greenhouse, Matthew
Greisen, Eric
Griffin, Ian
Griffin, Roger
Gronchi, Giovanni
Gross, Richard
Groten, Erwin
Gudehus, Donald
Guibert, Jean
Guinot, Bernard
Gumjudpai, Burin
Gün, Gulnur
GUNJI, Shuichi
GUO, Ji
Gurshtein, Alexander
Gurzadyan, Vahagn
Gusev, Alexander
Guseva, Irina
Guzik, Joyce
HAAS, RÜDIGER
Hackman, Christine
Haghighipour, Nader
Hajian, Arsen
Hakopian, Susanna
Halbwachs, Jean-Louis
Hallan, Prem
Hamid, S.
Hamilton, Douglas
HAN, Inwoo
HAN, Tianqi
HAN, Yanben
HANADO, Yuko
Hanslmeier, Arnold
Hanson, Robert
Harper, David
Harris, Alan
Harris, Hugh
Hartkopf, William
Hau, George
HE, Miao-fu
Hefty, Jan
Heggie, Douglas
Hellali, Yhya
Helmer, Leif
Hemenway, Paul
Hering, Roland
Hernández-Monteagudo, Carlos
HERSANT, Franck
Hessels, Jason
Hestroffer, Daniel
Heudier, Jean-Louis
Hewett, Paul
Heyl, Jeremy
Hilditch, Ronald
Hill, Graham
Hilton, James
Ho, Wynn
Hobbs, David
Hobbs, George
Hobiger, Thomas
Høg, Erik
Hohenkerk, Catherine
Holmberg, Johan
Holz, Daniel
HONG, Zhang
HORI, Genichiro
Horn, Martin
Horner, Jonathan
HOU, Xiyun
Howard, Andrew
Howard, Sethanne
Hrivnak, Bruce
Hsieh, Henry
HSU, Rue-Ron
HU, Hongbo
HU, Xiaogong
HU, Yonghui
HU, Zhong wen
HUA, Yu
HUANG, Cheng
HUANG, Tianyi
Hube, Douglas
Hubrig, Swetlana
Hugentobler, Urs
Humphries, Colin
Hurley, Jarrod
Husárik, Marek
Hut, Piet
Hutter, Donald
Ianna, Philip
Ibata, Rodrigo
IIJIMA, Shigetaka
Ilyas, Mohammad
Imbert, Maurice
Inglis, Michael
INOUE, Takeshi
Iorio, Lorenzo
Ipatov, Sergei
Ireland, Michael
Irigoyen, Maylis
Irwin, Alan
Irwin, Michael
Ismail, Mohamed
ITO, Takashi
Ivanov, Dmitrii
Ivanov, Pavel

Ivanova, Violeta
Ivantsov, Anatoliy
Ivezic, Zeljko
Iyer, Balasubramanian
Izmailov, Igor
Izzard, Robert
Jackson, Paul
Jacobs, Christopher
Jacobson, Robert
Jäggi, Adrian
Jahreiss, Hartmut
Jakubík, Marian
Jalali, Mir Abbas
Janiczek, Paul
Jasniewicz, Gerard
Jassur, Davoud
Jefferys, William
JI, Jianghui
JIA, Lei
JIANG, Ing-Guey
Jiménez-Vicente, Jorge
JIN, WenJing
Johnson, Thomas
Johnston, Kenneth
Jones, Burton
Jones, Derek
Jordi, Carme
Jorissen, Alain
Jubier, Xavier
Kadouri, Talib
KAKUTA, Chuichi
Kalomeni, Belinda
Kalvouridis, Tilemachos
Kamal, Fouad
KAMEYA, Osamu
Kanayev, Ivan
Kaplan, George
Karachentsev, Igor
Karatekin, Özgür
Karttunen, Hannu
Kascheev, Rafael
Katz, David
Katz, Joseph
Kausch, Wolfgang
Kazantseva, Liliya
Kent, Brian
Kenworthy, Matthew
Keskin, Varol
Khalesseh, Bahram
Kharchenko, Nina
Khoda, Oleg
Kholshevnikov, Konstantin
Khrutskaya, Evgenia
Khumlemlert, Thiranee
Kim, Yoo Jea
KIM, Sungsoo
King, Ivan
KING, Sun-Kun
KINOSHITA, Hiroshi
Kitaeff, Vyacheslav
Kitiashvili, Irina
Klemola, Arnold
Klepczynski, William
Klioner, Sergei
Klock, Benny
Klocok, Lubomir
Klokocnik, Jaroslav
Knežević, Zoran
Knowles, Stephen
KOBAYASHI, Hideyuki
KOBAYASHI, Yukiyasu
Koivisto, Tomi
KOKUBO, Eiichiro
Kołaczek, Barbara
Konacki, Maciej
Kopeikin, Sergei
Korchagin, Vladimir
Kornoš, Leonard
Korsun, Alla
Koschny, Detlef
Kosek, Wiesław
Koshelyaevsky, Nikolay
Koshkin, Nikolay
Kostelecký, Jan
Kouba, Jan
Kouwenhoven, M.B.N. (Thijs)
Kovačević, Andjelka
Kovács, József
Kovalevsky, Jean
KOYAMA, Yasuhiro
KOZAI, Yoshihide
Krabbe, Angela
Kraft, Robert
Kramer, Michael
Krivov, Alexander
Kroll, Peter
Kudryavtseva, Nadezhda (Nadia)
Kuimov, Konstantin
Kumkova, Irina
KURAYAMA, Tomoharu
Kurzyńska, Krystyna
Kuznetsov, Eduard
Kwok, Sun
Lala, Petr
Lammers, Uwe
Landolt, Arlo
Lang, Mark
Lara, Martin
Laskar, Jacques
La Spina, Alessandra
Latham, David
Lattanzi, Mario
Latypov, A.
Lazorenko, Peter
Lazovic, Jovan
Lecavelier des Etangs, Alain
Lee, Hyun-chul
LEE, Jae Woo
LEE, Jun
LEE, Ki-Won
Lega, Elena
Lehmann, Marek
Lemaître, Anne
Lenhardt, Helmut
Lepine, Sebastien
Le Poncin-Lafitte, Christophe
Le Poole, Rudolf
Levato, Orlando
Levine, Stephen
Lewis, Brian
LI, Jinling
LI, Jinzeng
LI, Li-Xin
LI, Qi
LI, Qingkang
LI, xiaohui
LI, Yong
LI, Yuqiang
LI, Zhigang
LIAO, Dechun
LIAO, Xinhao
Libert, Anne-Sophie
Lieske, Jay
Lin, Douglas
Lindegren, Lennart

Lindgren, Harri
Lindqvist, Michael
Lissauer, Jack
LIU, Chengzhi
LIU, Ciyuan
LIU, Fukun
LIU, Tao
LIU, Wenzhong
LIU, Xiang
LIU, Yang
LIU, Yujuan
Livadiotis, George
Lo Curto, Gaspare
łokas, Ewa
Lopes, Paulo
Lopez, Carlos
Lopez, Jose
Lopez Moratalla, Teodoro
Lovis, Christophe
Lu, Phillip
LU, BenKui
LU, Chunlin
LU, Xiaochun
LU, Youjun
Lub, Jan
Lucchesi, David
Luck, John
Lukashova, Marina
Lundquist, Charles
LUO, Ali
LUO, Dingchang
Luri, Xavier
Luzum, Brian
Ma, Chopo
MA, Jingyuan
MA, Lihua
MA, Wenzhang
MA, Zhenguo
MacConnell, Darrell
Maciejewski, Andrzej
Maciesiak, Krzysztof
Maddison, Sarah
Madsen, Claus
Magnier, Eugene
Maigurova, Nadiia
Majid, Abdul
Makarov, Valeri
Makhlouf, Amar
Malhotra, Renu
Malkin, Zinovy
Mallamaci, Claudio
Mamajek, Eric
MANABE, Seiji
Manchanda, R.
Manchester, Richard
Mandel, Ilya
Mangum, Jeffrey
Manoharan, P.
Mapelli, Michela
Marchal, Christian
Marcialis, Robert
Marcy, Geoffrey
Margot, Jean-Luc
Marranghello, Guilherme
Marschall, Laurence
Marshalov, Dmitriy
Martín, Eduardo
Martinet, Louis
Martinez Fiorenzano, Aldo
Martins, Roberto
Marzari, Francesco
MASAKI, Yoshimitsu
Mason, Brian
Massaro, Francesco
Massey, Philip
Matas, Vladimir
Mathieu, Robert
Matsakis, Demetrios
Maurice, Eric
Mavraganis, Anastasios
Mayor, Michel
Mazeh, Tsevi
McAlister, Harold
McAteer, R. T. James
McLean, Brian
McMillan, Robert
Meibom, Soren
Melbourne, William
Melia, Fulvio
Melnick, Gary
Melnyk, Olga
Mendes, Virgilio
Méndez Bussard, Rene
Merman, Hirsh G.
Merriam, James
Métris, Gilles
Meylan, Georges
Michel, Patrick
Middleton, Christopher
Migaszewski, Cezary
MIKAMI, Takao
Mikkola, Seppo
Milani Comparetti, Andrea
Minazzoli, Olivier
Mink, Jessica
Minniti, Dante
Missana, Marco
MIZUNO, Yosuke
Mkrtichian, David
Modali, Sarma
Moffat, John
Mohd, Zambri
Monaco, Lorenzo
Monet, Alice
Monet, David
Montgomery, Michele
Moore-Weiss, John
Moorhead, James
Morabito, David
Morbey, Christopher
Morbidelli, Roberto
Moreira Morais, Maria
Morgan, Peter
MORI, Masao
Morrell, Nidia
Morrison, Leslie
Mota, David
Muanwong, Orrarujee
Mueller, Ivan
Müller, Andreas
Muiños Haro, José
Mukhopadhyay, Banibrata
Müller, Jürgen
Murray, Stephen
Muzzio, Juan
Mysen, Eirik
Nacozy, Paul
Naef, Dominique
NAGATAKI, Shigehiro
Nahar, Sultana
Najid, Nour-Eddine
NAKAGAWA, Akiharu
NAKAJIMA, Koichi
NAKASHIMA, Jun-ichi
Namouni, Fathi
Naoz, Smadar
Napolitano, Nicola
Näränen, Jyri
NARITA, Norio
Naroenkov, Sergey

Nastula, Jolanta
Navarro, Julio
Navone, Hugo
Nefedyev, Yury
Netzer, Nathan
Newhall, X.
Nice, David
Niell, Arthur
Niemi, Aimo
NIINUMA, Kotaro
Nikoloff, Ivan
Nilsson, Tobias
Ninković, Slobodan
NITTA, Atsuko
NIWA, Yoshito
Nobili, Anna
Nordström, Birgitta
Nothnagel, Axel
Novaković, Bojan
Noyelles, Benoît
Nunez, Jorge
Oetken, L.
Ofek, Eran
Ogando, Ricardo
O'Handley, Douglas
OHISHI, Masatoshi
OHNISHI, Kouji
Oja, Heikki
Oja, Tarmo
Okeke, Francisca
Okpala, Kingsley
Olive, Don
Olivier, Enrico
Ollongren, A.
Olsen, Hans
Oozeer, Nadeem
Orellana, Mariana
Orellana, Rosa
Orlov, Viktor
Osborn, Wayne
Osório, Isabel
Osório, José
Owen Jr, William
Page, Gary
Pakvor, Ivan
Pál, András
Panafidina, Natalia
Panessa, Francesca
Pannunzio, Renato
Pantoja, Carmen
Paquet, Paul
PARK, Pilho
Parv, Bazil
Pascu, Dan
Patten, Brian
Pauwels, Thierry
Pavlyuchenkov, Yaroslav
Peale, Stanton
Pedoussaut, André
Pejović, Nadezda
Penna, Jucira
Pepe, Francesco
Perets, Hagai
Perozzi, Ettore
Perrier-Bellet, Christian
Perryman, Michael
Pešek, Ivan
Peterson, Ruth
Petit, Gérard
Petit, Jean-Marc
Petrov, Sergey
Petrovskaya, Margarita
PHAN, Dong
Philip, A.G.
Picca, Domenico
Pilat-Lohinger, Elke
Pilkington, John
Pineau des Forets, Guillaume
PING, Jinsong
Pinigin, Gennadiy
Pireaux, Sophie
Pitjev, Nikolaj
Pitjeva, Elena
Pitkin, Matthew
Platais, Imants
Podolský, Jiri
Pollas, Christian
Polyakhova, Elena
Poma, Angelo
Popescu, Petre
Potapov, Vladimir
Pozanenko, Alexei
Poznanski, Dovi
Preston, George
Pribulla, Theodor
Protsyuk, Yuri
Proverbio, Edoardo
Prusti, Timo
Puetzfeld, Dirk
Pugliano, Antonio
Pushkin, Sergej
Puzia, Thomas
QIMING, Wang
Quintana, Hernan
Rafferty, Theodore
Ramírez, Jose
Ransom, Scott
Rastegaev, Denis
Rastorguev, Alexey
Ratnatunga, Kavan
Ray, James
Ray, Paul
Reasenberg, Robert
Reddy, Bacham
Reffert, Sabine
Reid, Warren
Reisenegger, Andreas
Reitze, David
Revaz, Yves
Reyes-Ruiz, Mauricio
Reynolds, John
Richardson, Derek
Richter, Bernd
Riles, Keith
Rizvanov, Naufal
Robe, Henri
Robertson, Douglas
Rodin, Alexander
Rodriguez-Eillamil, R.
Roemer, Elizabeth
Roeser, Siegfried
Rogister, Yves
Roig, Fernando
Romanov, Yuri
Romero Pérez, Maria
Ron, Cyril
Rosińska, Dorota
Rossello, Gaspar
Rossi, Alessandro
Rothacher, Markus
Rots, Arnold
Rovithis-Livaniou, Helen
Royer, Frédéric
Rubenstein, Eric
Rubin, Vera
Ruder, Hanns
Rushton, Anthony
Russell, Jane
Rusu, I.

Ryabov, Boris
Ryabov, Yurij
Rykhlova, Lidiya
Saad, Abdel-naby
Sachkov, Mikhail
Saemundsson, Thorsteinn
Saffari, Reza
Sage, Leslie
Sahai, Raghvendra
SAKAMOTO, Tsuyoshi
Salazar, Antonio
Salstein, David
Samarasinha, Nalin
Samus, Nikolay
Sanders, Gary
Sanders, Walter
Sansaturio, Maria
Sarasso, Maria
Sari, Re'em
SASAO, Tetsuo
SATO, Koichi
Scarfe, Colin
Schartel, Norbert
Scheeres, Daniel
Schilbach, Elena
Schildknecht, Thomas
Schillak, Stanisław
Schmitt, Henrique
Scholl, Hans
Scholz, Ralf-Dieter
Schramm, Thomas
Schreiber, Karl
Schrijver, Johannes
Schröder, Anja
Schubart, Joachim
Schuh, Harald
Schutz, Bob
Schwekendiek, Peter
Sconzo, Pasquale
Seaman, Rob
Segaluvitz, Alexander
šegan, Stevo
Ségransan, Damien
Seidelmann, P.
Seimenis, John
Sein-Echaluce, M.
Seitz, Florian
Seitzer, Patrick
SEKIDO, Mamoru
SEKIGUCHI, Naosuke
Selim, Hadia
Semerák, Oldrich
Serber, Alexander
Sese, Rogel Mari
Sevilla, Miguel
Seyed-Mahmoud, Behnam
Shaffer, David
Shakht, Natalia
Shankland, Paul
Shanklin, Jonathan
Shapiro, Irwin
Sharma, A.
Sharp, Nigel
Sheikh, Suneel
Shelus, Peter
SHEN, Kaixian
SHEN, Zhiqiang
Shevchenko, Ivan
SHI, Huli
Shinnaga, Hiroko
Shiryaev, Alexander
Shoemaker, David
SHU, Fengchun
Shulga, Oleksandr
Shuygina, Nadia
Sidlichovsky, Milos
Sidore'nkov, Nikolay
Siebert, Arnaud
Sigismondi, Costantino
Sima, Zdislav
Simkin, Susan
Simó, Carles
Simon, Jean-Louis
Sinachopoulos, Dimitris
Sinha, Krishnanand
Sivakoff, Gregory
Sivan, Jean-Pierre
Skripnichenko, Vladimir
Skuljan, Jovan
Skurikhina, Elena
Slade, Martin
Smart, Richard
Smirnova, Tatiana
Smith, Myron
Smylie, Douglas
Soderhjelm, Staffan
SOFIA LYKAWKA, Patryk
Sokolov, Leonid
Solarić, Nikola
Solivella, Gladys
Solovaya, Nina
Soltynski, Maciej
SOMA, Mitsuru
SONG, Jinan
Soria, Roberto
Sorokin, Nikolaj
Soubiran, Caroline
Souchay, Jean
Sovers, Ojars
Sozzetti, Alessandro
Spadaro, Daniele
Spagna, Alessandro
Spoljaric, Drago
Stadel, Joachim
Stairs, Ingrid
Standish, E.
Stappers, Benjamin
Stavinschi, Magda
Stefanik, Robert
Steigenberger, Peter
Stein, John
Steinmetz, Matthias
Stellmacher, Irène
Stephenson, F.
Stergioulas, Nikolaos
Sterken, Christiaan
Steves, Bonnie
Stickland, David
Storey, John
Strauss, Michael
SUGANUMA, Masahiro
Süli, Áron
Sultanov, G.
SUN, Fuping
SUN, Yisui
SUNADA, Kazuyoshi
SUNG, Hyun-Il
Suntzeff, Nicholas
Surkis, Igor
Sveshnikov, Mikhail
Sweatman, Winston
Szabados, Laszlo
Szenkovits, Ferenc
TAKAHASHI, Rohta
TAN, Baolin
TANG, Zheng-Hong
Tanga, Paolo
TAO, Jin-he
Tapley, Byron

Tarady, Vladimir
TARIS, François
Tarter, C.
Taş, Günay
Tatevyan, Suriya
Taylor, Donald
Taylor, Gregory
Teixeira, Paula
Teixeira, Ramachrisna
Tektunali, H.
ten Brummelaar, Theo
Tepper Garcia, Thorsten
Tessema, Solomon
Thaller, Daniela
Thiry, Yves
Thomas, Claudine
Thomas, Maik
Thomas, Peter
Thomsen, Bjarne
Thorsett, Stephen
Thorstensen, John
Thuillot, William
Thurston, Mark
TING, Yeou-Tswen
Tiscareno, Matthew
Titov, Oleg
Titov, Vladimir
Tokovinin, Andrei
Tomasella, Lina
Tommei, Giacomo
Tonry, John
Torres, Jesus Rodrigo
Tremaine, Scott
Trenti, Michele
Tripicco, Michael
TSAI, An-Li
Tsiganis, Kleomenis
Tsinganos, Kanaris
Tsuchida, Masayoshi
TSUJIMOTO, Takuji
Tuckey, Philip
Turck-Chièze, Sylvaine
Turon, Catherine
Tzioumis, Anastasios
UEDA, Haruhiko
Unwin, Stephen
Upgren, Arthur
Urban, Sean
Vahia, Mayank
Vallejo, Miguel
Valsecchi, Giovanni
Valtonen, Mauri
van Altena, William
van Dessel, Edwin
van Gent, Robert
Van Hoolst, Tim
van Langevelde, Huib
van Leeuwen, Floor
van Leeuwen, Joeri
Varvoglis, Harry
Vashkovyak, Sofja
Vashkov'yak, Mikhail
Vass, Gheorghe
Vassiliev, Nikolaj
Veillet, Christian
Vereshchagin, Sergej
Vernotte, François
Verschueren, Werner
Vicente, Raimundo
Vieira Neto, Ernesto
Vienne, Alain
Vilhena de Moraes, Rodolpho
Vilinga, Jaime
Vilkki, Erkki
Vinet, Jean-Yves
József Vinkó, Jozsef
Virtanen, Jenni
Vityazev, Veniamin
Vlahakis, Nektarios
Voges, Wolfgang
Vokrouhlicky, David
Volyanska, Margaryta
Vondrák, Jan
Walch, Jean-Jacques
Walker, Gordon
Wallace, Patrick
Walton, Nicholas
WANG, Guangli
WANG, Jia-Ji
WANG, Kemin
WANG, Na
WANG, Xiao-bin
WANG, Xiaoya
WANG, Yulin
WANG, Zhengming
Warren Jr, Wayne
Wasserman, Lawrence
WATANABE, Noriaki
Watts, Anna
Wayth, Randall
Weber, Robert
Wegner, Gary
WEI, Erhu
Weisberg, Joel
WEN, Linqing
Weratschnig, Julia
Whipple, Arthur
White, Graeme
Wicenec, Andreas
Wiegert, Paul
Wielen, Roland
Wilkins, George
Will, Clifford
Williams, Carol
Williams, James
Willstrop, Roderick
Wilson, P.
Winkler, Gernot
Winter, Othon
Wittenmyer, Robert
Wittmann, Axel
Wnuk, Edwin
Wooden, William
WU, Bin
WU, Guichen
WU, Haitao
WU, Jiun-Huei
WU, Lianda
WU, Shouxian
WU, Zhen-Yu
Wucknitz, Olaf
Wytrzyszczak, Iwona
XIA, Yifei
XIAO, Dong
XIAO, Naiyuan
XIAOSHENG, Wan
XIE, Yi
XIONG, Jianning
XU, Jiayan
Yahil, Amos
Yallop, Bernard
YAMADA, Yoshiyuki
YANG, Fumin
YANG, Stephenson
YANG, Tinggao
YANG, Xuhai
YANG, Zhigen
YANO, Taihei
YASUDA, Haruo

Yatsenko, Anatolij
Yatskiv, Yaroslav
YE, Shuhua
YI, Zhaohua
Yokoyama, Tadashi
YOKOYAMA, Koichi
Yoon, Suk-Jin
YOSHIDA, Haruo
YOSHIDA, Junzo
YOSHIZAWA, Masanori
Yseboodt, Marie
YU, Nanhua
YUASA, Manabu
YUTAKA, Shiratori
Zacchei, Andrea
Zacharias, Norbert
Zafiropoulos, Basil
Zagars, Juris
Zaggia, Simone
Zagretdinov, Renat
Zainal Abidin, Zamri
Zare, Khalil
ZHANG, Haotong
ZHANG, Sheng-Pan
ZHANG, Shougang
ZHANG, Wei
ZHANG, Wei
ZHANG, Weiqun
ZHANG, Xiaoxiang
ZHANG, Yang
ZHANG, Yong
ZHANG, Zhibin
ZHANG, Zhongping
ZHAO, Changyin
ZHAO, Haibin
ZHAO, You
Zharov, Vladimir
Zhdanov, Valery
ZHENG, Jia-Qing
ZHENG, Xuetang
ZHENG, Yong
ZHONG, Min
ZHOU, Hongnan
ZHOU, Ji-Lin
ZHOU, Li-Yong
ZHOU, Yonghong
ZHU, Liying
ZHU, Wenyao
ZHU, Yaozhong
ZHU, Zi
Zitrin, Adi
Zsoldos, Endre
Zucker, Shay
Zwitter, Tomaž

Division B Facilities, Technologies and Data Science

President: David Richard Silva
Vice-President: Pietro Ubertini

Organizing Committee Members:

Michael G. Burton
Malcolm G. Smith
Françoise Genova
Anthony James Beasley
Gloria M. Dubner
Lisa Storrie-Lombardi
Robert J. Hanisch
Hitoshi YAMAOKA
Lyudmila I. Mashonkina
Jessica Mary Chapman
Richard F. Green
Gerard T. van Belle
Simon F. Portegies Zwart
David Richard Silva
Pietro Ubertini
Alistair Walker

Members:

Abalakin, Viktor
Abdulla, Shaker
Ables, Harold
Ábrahám, Péter
Abraham, Roberto
Abraham, Zulema
Absil, Olivier
Abt, Helmut
Accomazzi, Alberto
Acharya, Bannanje
Acke, Bram
Ade, Peter
Adelman, Saul
Adler, David
Aerts, Conny
Afram, Nadine
Aggarwal, Kanti
Aghaee, Alireza
Agüeros, Marcel
A'Hearn, Michael
Ahumada, Javier
Aigrain, Suzanne
Aizenman, Morris
AK, Serap
AKABANE, Kenji
Akeson, Rachel
AKITAYA, Hiroshi
Aksnes, Kaare
Akujor, Chidi
Alberdi, Antonio
Albrecht, Rudolf
Alecian, Evelyne
Alexander, Joseph
Alexander, Paul
ALIBERT, Yann
Alissandrakis, Costas
Allan, Alasdair
Allard, France
Allard, Nicole

Allen, Lori
Allen, Ronald
Allende Prieto, Carlos
Allen Jr, John
Aller, Hugh
Aller, Margo
Altenhoff, Wilhelm
Altunin, Valery
Alvarez, Pedro
Alvarez del Castillo, Elizabeth
Ambrocio-Cruz, Silvia
Ambrosini, Roberto
AN, TAO
Anandaram, Mandayam
Andernach, Heinz
Andersen, Michael
Andersen, Torben
Andreon, Stefano
Andreuzzi, Gloria
Andruchow, Ileana
Angel, J.
Angione, Ronald
Anglada, Guillem
Anthony-Twarog, Barbara
Anton, Sonia
Antonelli, Lucio Angelo
Antonova, Antoaneta
Anupama, G.
Apostolovska, Gordana
Appenzeller, Immo
Arcidiacono, Carmelo
Ardeberg, Arne
Arduini-Malinovsky, Monique
Arenou, Frédéric
Aretxaga, Itziar
Argo, Megan
Argyle, Robert
Arion, Douglas
Arlot, Jean-Eudes
Armstrong, John
Arnal, Edmundo
Arons, Jonathan
Arribas Mocoroa, Santiago
Arsenijevic, Jelisaveta
Asanok, Kitiyanee
Asareh, Habibolah
Aschwanden, Markus
Ashley, Michael
Ashok, N.
Aspin, Colin
Assousa, George
Aubier, Monique
Audard, Marc
Aufdenberg, Jason
Augereau, Jean-Charles
Augusto, Pedro
Aungwerojwit, Amornrat
Aurass, Henry
Aussel, Herve
Avila Foucault, Remy
Baan, Willem
Baars, Jacob
Bååth, Lars
BABA, Naoshi
Babkovskaia, Natalia
Babu, G.S.D.
Bachiller, Rafael
Baddiley, Christopher
Badescu, Octavian
Baffa, Carlo
Bagla, Jasjeet
Bagri, Durgadas
BAI, Jinming
Bailer-Jones, Coryn
Bailes, Matthew
Baines, Ellyn
Bajaja, Esteban
Bajkova, Anisa
Baker, Andrew
Baker, Joanne
Bakker, Eric
Balanca, Christian
Balasubramanian, V.
Balasubramanyam, Ramesh
Baldinelli, Luigi
Baldry, Ivan
Baliyan, Kiran
Ball, Lewis
Ballester, Pascal
Bally, John
Balona, Luis
Balonek, Thomas
Balser, Dana
Balyshev, Marat
Bamford, Steven
Banerjee, Dipankar
Banhatti, Dilip
Baran, Andrzej
Baransky, Olexander
Barber, Robert
Barbieri, Cesare
Barbuy, Beatriz
Barden, Marco
Barklem, Paul
Barmby, Pauline
Barnbaum, Cecilia
Barnes, David
Barnes III, Thomas
Barone, Fabrizio
Barrett, Paul
Barrientos, Luis
Barrow, Colin
Bartaya, R.
Bartczak, Przemysław
Bartel, Norbert
Barthel, Peter
Bartkevicius, Antanas
Bartkiewicz, Anna
Baruch, John
Barvainis, Richard
Barway, Sudhashu
Baryshev, Andrey
Basart, John
Bash, Frank
Baskill, Darren
Bassett, Bruce
Bastien, Pierre
Basu, Kaustuv
Baudry, Alain
Bauer, Amanda
Baum, Stefi
Baume, Gustavo
Bautista, Manuel
Bayet, Estelle
Bazzano, Angela
Beasley, Anthony
Becciani, Ugo
Bechtold, Jill
Beck, Rainer
Beck, Sara
Beckmann, Volker
Behar, Ehud
Beiersdorfer, Peter
Bekki, Kenji
Beklen, Elif
Bellazzini, Michele
Bell Burnell, Jocelyn
Bely-Dubau, Francoise

Benacchio, Leopoldo
Benaglia, Paula
Bendjoya, Philippe
Benetti, Stefano
Benisty, Myriam
Benkhaldoun, Zouhair
Benn, Chris
Bennett, Charles
Bensammar, Slimane
Benson, James
Bentley, Robert
Benvenuti, Piero
Benz, Arnold
Berdyugin, Andrei
Berg, Richard
Berge, Glenn
Bergeat, Jacques
Berger, Jean-Philippe
Bergman, Per
Berkhuijsen, Elly
Bernat, Andrew
Bersier, David
Bertello, Luca
Berthier, Jerôme
Bertout, Claude
Bessell, Michael
Bhandari, Rajendra
Bhardwaj, Anil
Bhat, Ramesh
Bianda, Michele
Bieging, John
Biémont, Emile
Biermann, Peter
Bietenholz, Michael
Bigdeli, Mohsen
Biggs, James
Bignall, Hayley
Bignell, R.
Biraud, François
Biretta, John
Birkinshaw, Mark
Birkmann, Stephan
Bishop, Marsha
Bjorkman, Jon
Black, John
Blackwell-Whitehead, Richard
Blair, David
Blanco, Carlo
Blandford, Roger
Blazit, Alain
Blitz, Leo
Bloemhof, Eric
Blommaert, Joris
Blomme, Ronny
Blum, Robert
Blundell, Katherine
Boboltz, David
Bock, Douglas
Bockelee-Morvan, Dominique
Boden, Andrew
Bodewits, Dennis
Boechat-Roberty, Heloisa
Böker, Torsten
Bohlender, David
Boily, Christian
Bolatto, Alberto
Bommier, Veronique
Bonaccini, Domenico
Bonanno, Giovanni
Bond, Howard
Bond, Ian
Bondi, Marco
Bonneau, Daniel
Bontekoe, Romke
Boonstra, Albert
Booth, Roy
Bordé, Pascal
Borde, Suzanne
Borgman, Jan
Borisov, Galin
Borisova, Ana
Borka Jovanović, Vesna
Borkowski, Kazimierz
Borne, Kirk
Boroson, Bram
Borra, Ermanno
Bos, Albert
Bosch-Ramon, Valenti
Bosken, Sarah
Bot, Caroline
Bottinelli, Lucette
Bouchard, Antoine
Boumis, Panayotis
Bouton, Ellen
Bower, Geoffrey
Bowers, Phillip
Boyajian, Tabetha
Boyce, Peter
Boyle, Richard
Bradley, Arthur
Braithwaite, Jonathan
Brammer, Gabriel
Branchesi, Marica
Brandl, Bernhard
Brandner, Wolfgang
Brandt, William
Branduardi-Raymont, Graziella
Branscomb, L.
Braun, Robert
Breckinridge, James
Breger, Michel
Bregman, Jacob
Brentjens, Michiel
Brescia, Massimo
Breton, Rene
Bridle, Alan
Briggs, Franklin
Brinchmann, Jarle
Brinks, Elias
Britzen, Silke
Brocato, Enzo
Broderick, John
Bromage, Gordon
Bronfman, Leonardo
Brooks, Kate
Brosch, Noah
Broten, Norman
Brouw, Willem
Brown, Anthony
Brown, Douglas
Brown, Joanna
Brown, Jo-Anne
Brown, Michael
Brown, Robert
Brown, Thomas
Browne, Ian
Brunetti, Gianfranco
Brunner, Robert
Bruno, Roberto
Brunthaler, Andreas
Bryant, Julia
Bucciarelli, Beatrice
Buchlin, Eric
Buie, Marc
Bujarrabal, Valentin
Burderi, Luciano
Burke, Bernard

Burton, Michael
Buscher, David
Buser, Roland
Butler, Ray
Bzowski, Maciej
Cabanac, Remi
CAI, Mingsheng
Calabretta, Mark
Campana, Riccardo
Campbell, Robert
Campbell-Wilson, Duncan
Cantiello, Michele
Canto Martins, Bruno
Caon, Nicola
Cappellaro, Enrico
Capria, Maria
Caproni, Anderson
Caputi, Karina
Carbon, Duane
Carciofi, Alex
Caretta, Cesar
Carilli, Christopher
Carlberg, Raymond
Carlqvist, Per
Carney, Bruce
Caroubalos, Constantinos
Carpenter, Kenneth
Carramiñana, Alberto
Carrasco, Bertha
Carretti, Ettore
Carroll, P.
Carruba, Valerio
Carter, Brian
Carvalho, Joel
Casasola, Viviana
casassus, simon
Cash Jr, Webster
Casoli, Fabienne
Cassano, Rossella
Castelaz, Micheal
Castelletti, Gabriela
Castets, Alain
Caswell, James
Catelan, Márcio
Cayrel, Roger
cazaux, Stephanie
Cecconi, Baptiste
Cellino, Alberto
Cellone, Sergio
Celotti, Anna Lisa
Cenko, Stephen
Cepa, Jordi
Cernicharo, Jose
Cesetti, Mary
CH, ISHWARA CHANDRA
Chadid, Merieme
Champion, David
CHAN, Kwing
Chance, Kelly
Chandler, Claire
Chandrasekhar, Thyagarajan
CHANG, Hong
CHANG, Hsiang-Kuang
Chapanov, Yavor
Chapman, Jacqueline
Chapman, Jessica
Charles, Philip
Charlot, Patrick
Charmandaris, Vassilis
Chelli, Alain
Chelliah Subramonian, Stalin
CHEN, An-Le
CHEN, Guoming
CHEN, Huei-Ru
CHEN, Wen Ping
CHEN, Xinyang
CHEN, Xuefei
CHEN, Xuelei
CHEN, Yongjun
CHEN, Zhijun
CHEN, Zhiyuan
Chengalur, Jayaram
Cheung, Cynthia
Chiappetti, Lucio
CHIKADA, Yoshihiro
CHIN, Yi-nan
Chini, Rolf
CHO, Se Hyung
Choudhary, Debi Prasad
Choudhury, Tirthankar
Christiansen, Wayne
Christlieb, Norbert
CHU, Yaoquan
Chukwude, Augustine
CHUNG, Hyun-Soo
Chyży, Krzysztof
Ciardi, David
Cichowolski, Silvina
Ciliegi, Paolo
Cinzano, Pierantonio
Cioni, Maria-Rosa
Clark, Barry
Clark, David
Clark, David
Clark, Frank
Clarke, Tracy
Clayton, Geoffrey
Clegg, Andrew
Clem, James
Clemens, Dan
Clocchiatti, Alejandro
Coffey, Deirdre
Cohen, Marshall
Cohen, Richard
Colas, François
Coleman, Paul
Coletti, Donna
Colless, Matthew
Colomer, Francisco
Combes, Françoise
Combi, Jorge
Condon, James
Conklin, Edward
Connolly, Leo
Conti, Alberto
Contreras, Maria
Conway, John
Cook, Kem
Copin, Yannick
Corbally, Christopher
Corbel, Stéphane
Corbett, Ian
Corbin, Brenda
Cordes, James
Corliss, C.
Cornille, Marguerite
Corradi, Wagner
Corral, Luis
Cosentino, Rosario
Costa, Enrico
Costa, Joaquim
Costa, Marco
Costero, Rafael
Côte, Patrick
Côté, Stéphanie
Cotton Jr, William
Couch, Warrick

Couchman, Hugh
Coudé du Foresto, Vincent
Courbin, Frederic
Courtois, Helene
Coyne, S.J, George
Crabtree, Dennis
Cracco, Valentina
Crane, Patrick
Crause, Lisa
Crawford, David
Crawford, Fronefield
Crawford, Steven
Creech-Eakman, Michelle
Crézé, Michel
Cristiani, Stefano
Cristóbal, David
Croft, Steve
Croston, Judith
Crovisier, Jacques
Crutcher, Richard
Cruzalèbes, Pierre
Cruz-Gonzalez, Irene
Csabai, Istvan
Cuby, Jean-Gabriel
Cui, Wei
Cui, Xiangqun
CUI, Chenzhou
Cuillandre, Jean-Charles
Cullum, Martin
Cunniffe, John
Cunningham, Maria
Cuypers, Jan
Cvetkovi'c, Zorica
Dachs, Joachim
Dagkesamanskii, Rustam
Dahn, Conard
DAI, Zhibin
DAISAKU, Nogami
DAISHIDO, Tsuneaki
DAISUKE, Iono
Dalgarno, Alexander
Dalla, Silvia
Dallacasa, Daniele
Dalton, Gavin
Damé, Luc
D'Amico, Nicolo
Damljanovic, Goran
D'Ammando, Filippo
Danchi, William
Danford, Stephen
Dannerbauer, Helmut
Davidge, Timothy
Davies, Rodney
Davies, Roger
Davis, Christopher
Davis, Donald
Davis, Gary
Davis, Michael
Davis, Morris
Davis, Richard
Davis, Robert
Davis, Timothy
de Bergh, Catherine
De Bernardis, Paolo
de Boer, Klaas
de Bruijne, Jos
De Buizer, James
de Carvalho, Reinaldo
De Cuyper, Jean-Pierre
Deeg, Hans
de Frees, Douglas
Degaonkar, S.
de Gregorio-Monsalvo, Itziar
de Greiff, J.
de Jager, Cornelis
de Jager, Gerhard
de Jong, Roelof
de Kertanguy, Amaury
de Lange, Gert
Delannoy, Jean
de La Noë, Jerome
Deller, Adam
Del Olmo, Ascension
del Peloso, Eduardo
Delplancke, Francoise
Delsemme, Armand
de Medeiros, Jose
Demory, Brice-Olivier
den Herder, Jan-Willem
DeNisco, Kenneth
Denisse, Jean-Francois
Dennefeld, Michel
Dent, William
Depagne, Éric
de Petris, Marco
De Rossi, María
Derriere, Sebastien
de Ruiter, Hans
Désesquelles, Jean
Deshpande, Avinash
Deshpande, M.
Despois, Didier
de Vicente, Pablo
DeVorkin, David
Dewdney, Peter
Dhawan, Vivek
d'Hendecourt, Louis
Dhillon, Vikram
Diamond, Philip
Diaz, Marcos
Diaz, Ruben
Diaz Trigo, Maria
Dickel, Helene
Dickel, John
Dickey, John
Dickman, Robert
Di Cocco, Guido
Dieleman, Pieter
Diercksen, Geerd
Dieter Conklin, Nannielou
Dimitrijevic, Milan
Dintinjana, Bojan
Dionatos, Odysseas
Dipper, Nigel
Dixon, Robert
Djorgovski, Stanislav
Dluzhnevskaya, Olga
DOBASHI, Kazuhito
Dobrzycki, Adam
Dodd, Richard
D' Odorico, Valentina
Dodson, Richard
Dolan, Joseph
Domiciano de Souza, Armando
do Nascimento, José
DONG, Xiaojun
Doressoundiram, Alain
DOU, Jiangpei
Doubinskij, Boris
Dougados, Catherine
Dougherty, Sean
Downes, Dennis
Downes Wallace, Juan
Downs, George
Drake, Frank
Drake, Jeremy
Drake, Stephen
Dravins, Dainis

Dreher, John
Drew, Janet
Drimmel, Ronald
Drissen, Laurent
Driver, Simon
DU, Lan
Dubau, Jacques
Dubner, Gloria
Dubois, Pascal
Dubout, Renee
Ducati, Jorge
Ducourant, Christine
Dufay, Maurice
Duffett-Smith, Peter
Dukes Jr., Robert
Dulieu, Francois
Dulk, George
Dumke, Michael
Duorah, Hira
Durand, Daniel
Dutrey, Anne
Duvert, Gilles
Dwarakanath, K.
Dyson, Freeman
Eales, Stephen
Echevarria, Juan
Edelson, Rick
Ederoclite, Alessandro
Edwards, Paul
Edwards, Philip
Egret, Daniel
Ehle, Matthias
Eidelsberg, Michele
Eisner, Josh
Ekers, Ronald
Ekström García Nombela, Sylvia
Elia, Davide
Elias II, Nicholas
Elkin, Vladimir
Ellingsen, Simon
Ellis, Graeme
Elmhamdi, Abouazza
Elyiv, Andrii
Emerson, Darrel
Emerson, James
Emonts, Bjorn
Epstein, Eugene
Epstein, Gabriel
Ercan, Enise
Erickson, William
ESAMDIN, Ali
Esenoğlu, Hasan
Eshleman, Von
ESIMBEK, Jarken
Evans, Christopher
Evans, Dafydd
Evans, Ian
Ewing, Martin
EZAWA, Hajime
Fabiani, Sergio
Fabregat, Juan
Fabricius, Claus
Fabrika, Sergei
Facondi, Silvia
Fahlman, Gregory
Falcke, Heino
Faller, James
FAN, Yufeng
Fanaroff, Bernard
Fanti, Roberto
Farrell, Sean
Faulkner, Andrew
Faure, Cécile
Feain, Ilana
Feautrier, Nicole
Federici, Luciana
Federman, Steven
Fedotov, Leonid
Feigelson, Eric
Feinstein, Alejandro
Feldman, Paul
Feldman, Paul
Felli, Marcello
Feretti, Luigina
Ferland, Gary
Fernandes, Francisco
Fernández Lajús, Eduardo
Fernie, J.
Ferrarese, Laura
Ferrari, Attilio
Ferrari, Fabricio
Ferrari, Marc
Ferrini, Federico
Ferrusca, Daniel
Fey, Alan
Fich, Michel
Field, George
Figueiredo, Newton
Filacchione, Gianrico
Filipovic, Miroslav
Filippenko, Alexei
Fillion, Jean-Hugues
Fink, Uwe
Finn, Lee
Fisher, Richard
Fleck, Bernhard
Fleischer, Robert
Fletcher, Andrew
Florido, Estrella
Florkowski, David
Flower, David
Floyd, David
Fluke, Christopher
Fluri, Dominique
Focardi, Paola
Foing, Bernard
Foley, Anthony
Folgueira, Marta
Fomalont, Edward
Fornasier, Sonia
Fort, David
Forte, Juan
Forveille, Thierry
FOUCAUD, Sebastien
Fouqué, Pascal
Fox-Machado, Lester
Foy, Renaud
Frail, Dale
Fraix-Burnet, Didier
Franco, Gabriel Armando
Fraser, Helen
Frater, Robert
Frey, Sandor
Friberg, Per
Fridlund, Malcolm
Friedman, Scott
Fruchter, Andrew
Fuerst, Ernst
Fuhr, Jeffrey
FUKUI, Yasuo
FUKUSHIMA, Toshio
Fyfe, Duncan
Gabanyi, Krisztina
Gábor, Pavel
Gabriel, Alan
Gabriel, Carlos
Gabuzda, Denise
Gaensler, Bryan
Gai, Mario

Galadi-Enriquez, David
Galan, Maximino
Gallagher, Sarah
Gallagher III, John
Gallego, Jesús
Gallego, Juan Daniel
Gallimore, Jack
Galsgaard, Klaus
Gal-Yam, Avishay
Gammelgaard, Peter
GAN, Weiqun
Gangadhara, R.T.
GAO, Yu
Garay, Guido
Garcia, Beatriz
Garcia, Paulo
Garcia-Lorenzo, Maria
Gargaud, Muriel
Garrett, Michael
Garrido, César
Garrington, Simon
Garrison, Robert
Gary, Dale
Gasiprong, Nipon
Gastaldello, Fabio
Gaume, Ralph
Gawroński, Marcin
Gaylard, Michael
Gehrels, Neil
Gehrz, Robert
Geldzahler, Barry
Genet, Russell
GENG, Lihong
Genova, Françoise
Gentile, Gianfranco
Genzel, Reinhard
Gerard, Eric
Gerbaldi, Michele
Gergely, Tomas
Gervasi, Massimo
Gezari, Suvi
Ghigo, Francis
Ghosh, Swarna
Ghosh, Tapasi
Giani, Elisabetta
Gibson, David
Gibson, Sarah
Gilbank, David
Gilliland, Ronald
Gillon, Michaël
Gilmore, Alan
Giménez, Alvaro
Giménez de Castro, Carlos
Gioia, Isabella
Giorgi, Edgard
Giovannini, Gabriele
Girard, Julien
Giroletti, Marcello
Gitti, Myriam
Glagolevskij, Yurij
Glass, Ian
Glazebrook, Karl
Glindemann, Andreas
Glinski, Robert
Glover, Simon
Goddi, Ciriaco
Goebel, Ernst
Goedhart, Sharmila
Goicoechea, Luis
Golay, Marcel
Goldbach, Claudine
Golden, Aaron
Goldwire, Jr., Henry
Golev, Valeri
Golub, Leon
Gomez, Gonzalez
Gomez, Mercedes
Gomez, Monique
Gómez Fernández, Jose
Gómez Rivero, José
GONG, Biping
GONG, Xuefei
González, Gabriela
González, J.
Goode, Philip
Goodman, Alyssa
Goodrich, Robert
Gopalswamy, Natchimuthuk
Gordon, Chris
Gordon, Karl
Gordon, Mark
Gorschkov, Aleksandr
Gosachinskij, Igor
Goss, W. Miller
Gottesman, Stephen
Gouliermis, Dimitrios
Govender, Kevindran
Gower, Ann
Graae Jørgensen, Aleksandra
Grabowski, Bolesław
Graham, David
Graham, Eric
Graham, John
Grandi, Steven
Grant, Ian
Grauer, Albert
Gredel, Roland
Green, Anne
Green, Daniel
Green, David
Green, James
Green, Richard
Greenhill, Lincoln
Greenhouse, Matthew
Gregorini, Loretta
Gregorio-Hetem, Jane
Gregory, Philip
Greisen, Eric
Grenon, Michel
Grevesse, Nicolas
Grewing, Michael
Griffin, R. Elizabeth
Griffin, Roger
Grillmair, Carl
Grindlay, Jonathan
Groh, Jose
Groot, Paul
Grosbøl, Preben
Grothkopf, Uta
Grueff, Gavril
Grundahl, Frank
GU, Minfeng
GU, Xuedong
Gubchenko, Vladimir
Gudehus, Donald
Guelin, Michel
Guesten, Rolf
Guetter, Harry
Guglielmino, Salvatore
Guibert, Jean
Guidice, Donald
GUILLOTEAU, Stéphane
Guinan, Edward
Gulkis, Samuel
Gull, Stephen
Gulyaev, Sergei
GUO, Hongfang

GUO, Jianheng
Gupta, Yashwant
Gurvits, Leonid
Guseva, Irina
Guzik, Joyce
Gwinn, Carl
Hackman, Thomas
Haehnelt, Martin
Hänel, Andreas
Hakkila, Jon
Hall, Peter
Hallinan, Gregg
Hamuy, Mario
HAN, JinLin
Hanasz, Jan
HANDA, Toshihiro
Haniff, Christopher
Hanisch, Robert
Hankins, Timothy
Hanlon, Lorraine
HAO, Lei
Hardee, Philip
Harmer, Dianne
Harnett, Julienne
Harris, Alan
Harris, Daniel
Hartman, Henrik
Harvel, Christopher
Harvey-Smith, Lisa
Haschick, Aubrey
HASEGAWA, Tetsuo
Hatziminaoglou, Evanthia
Hau, George
Hauck, Bernard
Haverkorn, Marijke
Hayashi, Masahiko
Hayes, Donald
Haynes, Martha
Haynes, Raymond
HAZUMI, Masashi
HE, Jinhua
Heald, George
Heck, Andre
Heeralall-Issur, Nalini
Hefele, Herbert
Heidt, Jochen
Heiles, Carl
Heinz, Sebastian
Heiser, Arnold
Heiter, Ulrike
Helmer, Leif
Helmich, Frank
Helou, George
Hempel, Marc
Henkel, Christian
Henning, Thomas
Hensberge, Herman
Herbst, Eric
Herpin, Fabrice
Heske, Astrid
Hess, Kelley
Hessels, Jason
Hesser, James
Hessman, Frederic
Hestroffer, Daniel
Hetem Jr., Annibal
Hewish, Antony
Hibbard, John
Hickson, Paul
Hidayat, Bambang
Hidayat, Taufiq
Higgs, Lloyd
Hilditch, Ronald
Hillenbrand, Lynne
Hillier, John
Hills, Richard
HIRABAYASHI, Hisashi
HIRAMATSU, Masaaki
Hiriart, David
HIROTA, Tomoya
Hjalmarson, Ake
Hledík, Stanislav
Ho, Paul
Ho, Wynn
Hoang, Binh
Hobbs, George
Hochedez, Jean-François
Hodapp, Klaus
Hodge, Paul
Hönig, Sebastian
Hofmann, Wilfried
Hofner, Peter
Högbom, Jan
Hogg, David
Hojaev, Alisher
Hollis, Jan
Hollow, Robert
Homeier, Derek
HONG, Xiaoyu
HONMA, Mareki
Hopkins, Andrew
Hopp, Ulrich
Hora, Joseph
Horácek, Jiri
Hörandel, Jörg
Horiuchi, Shinji
Horne, Keith
Hornstrup, Allan
Horton, Anthony
Hotan, Aidan
Houde, Martin
House, Lewis
Howard III, William
Howell, Steve
Hron, Josef
HU, Zhong wen
HUANG, Hui-Chun
Huber, Martin
Hubrig, Swetlana
Huchtmeier, Walter
Hudec, Rene
Hudkova, Ludmila
Huebner, Walter
Hughes, David
Hughes, John
Hughes, Philip
Hummel, Christian
Humphreys, Elizabeth
Hunstead, Richard
Huovelin, Juhani
Hutter, Donald
Huynh, Minh
HWANG, Chorng-Yuan
Hyland, Harry
Ibrahim, Zainol Abidin
Ignjatovi'c, Ljubinko
IGUCHI, Satoru
IKEDA, Norio
Ikhsanov, Robert
Iliev, Ilian
Iliev, Ilian
Ilin, Gennadii
Ilyasov, Sabit
Ilyin, Ilya
IMAI, Hiroshi
INATANI, Junji
INOUE, Makoto
Ioannou, Zacharias
Ipatov, Aleksandr
Irbah, Abdanour

Ireland, Michael
Irvine, William
Irwin, Alan
Irwin, Patrick
ISHIDA, Manabu
ISHIGURO, Masato
Israel, Frank
Ivanov, Dmitrii
Ivantsov, Anatoliy
Ivezic, Zeljko
IWATA, Ikuru
IWATA, Takahiro
IYE, Masanori
Izzard, Robert
Jáchym, Pavel
Jackson, Carole
Jackson, Neal
Jackson, William
Jacoby, George
Jacq, Thierry
Jaffe, Walter
Jakobsen, Peter
Jamar, Claude
Jamrozy, Marek
Jannuzi, Buell
Janssen, Michael
Jassur, Davoud
Jauncey, David
Jeffers, Sandra
Jelić, Vibor
Jenkner, Helmut
JEONG, Woong-Seob
Jerzykiewicz, Mikołaj
Jewell, Philip
JIANG, Dongrong
JIANG, Zhibo
Jiménez-Vicente, Jorge
JIN, WenJing
JIN, Zhenyu
Joblin, Christine
Johnson, Donald
Johnson, Fred
Johnston, Helen
Johnston, Kenneth
Johnston-Hollitt, Melanie
Joly, Francois
Jonas, Justin
Jones, David
Jones, Dayton
Jones, Derek
Jones, Paul
Jordán, Andrés
Jordan, Carole
Jorden, Paul
Jordi, Carme
Jorgensen, Anders
Jørgensen, Uffe
Jorissen, Alain
Joshi, Umesh
Josselin, Eric
JUNG, Jae-Hoon
Kadiri, Samir
Käufl, Hans Ulrich
Kafka, Styliani (Stella)
Kaftan, May
Kaidanovski, Mikhail
KAIFU, Norio
KAKINUMA, Takakiyo
Kalberla, Peter
Kalomeni, Belinda
Kaltman, Tatyana
KAMAZAKI, Takeshi
KAMEGAI, Kazuhisa
KAMENO, Seiji
KAMEYA, Osamu
Kaminker, Alexander
Kamp, Inga
KANAMITSU, Osamu
Kanbach, Gottfried
Kandalyan, Rafik
Kanekar, Nissim
KANG, Gon-Ik
Kantharia, Nimisha
Kaňuchová, Zuzana
Kaplan, George
Karas, Vladimir
Kardashev, Nicolay
Karoff, Christoffer
Karouzos, Marios
Karpov, Nikolai
Kasiviswanathan, Sankarasubramanian
Kassim, Namir
Kastel, Galina
KASUGA, Takashi
KATO, Takako
Kaufmann, Pierre
Kausch, Wolfgang
Kavelaars, JJ.
Kaviraj, Sugata
KAWABATA, Koji
KAWABE, Ryohei
Kawada, Mitsunobu
KAWAGUCHI, Kentarou
KAWAMURA, Akiko
KAWARA, Kimiaki
Kazantseva, Liliya
Kazlauskas, Algirdas
Kebede, Legesse
Kedziora-Chudczer, Lucyna
Keeney, Brian
Kellermann, Kenneth
Kelly, Brandon
Kembhavi, Ajit
Kemper, Francisca
Kennedy, Eugene
Kent, Brian
KENTARO, Matsuda
Kentischer, Thomas
Kenworthy, Matthew
Kepler, S.
Kerber, Florian
Kern, Pierre
Kerschbaum, Franz
Kervella, Pierre
Keshet, Uri
Keskin, Varol
Kesteven, Michael
Khaikin, Vladimir
Khodachenko, Maxim
Khosroshahi, Habib
Khrutskaya, Evgenia
Kibblewhite, Edward
Kielkopf, John
Kijak, Jarosław
Kilborn, Virginia
Kilkenny, David
Killeen, Neil
Kilmartin, Pamela
Kim, Joo Hyeon
KIM, Hyun-Goo
KIM, Ji Hoon
KIM, Kwang tae
KIM, Sang Chul
KIM, Sang Joon
KIM, Seung-Lee
KIM, Tu Whan
KIM, Young-Soo
Kimball, Amy
Kinemuchi, Karen

King, Ivan
Kingston, Arthur
Kipper, Tonu
Kirby, Kate
Kislyakov, Albert
Kitaeff, Vyacheslav
Kitiashvili, Irina
Kjaergaard, Per
Klein, Karl
Klein, Ulrich
Knee, Lewis
Kneib, Jean-Paul
Knude, Jens
Knudsen, Kirsten
Ko, Hsien
KOBAYASHI, Hideyuki
Kocharovskij, Vladimir
Kocharovsky, Vitaly
KODA, Jin
Köhler, Rainer
Koen, Marthinus
Kohl, John
KOHNO, Kotaro
KOJIMA, Masayoshi
Kolb, Ulrich
Kolenberg, Katrien
Kolláth, Zoltan
Kolobov, Dmitri
Kolokolova, Ludmilla
Kołomański, Sylwester
Kolomiyets, Svitlana
Konacki, Maciej
Kondratiev, Vladislav
KONG, Xu
Konovalenko, Alexander
Kontizas, Evangelos
Kontizas, Mary
Koopmans, Leon
Kopatskaya, Evgenia
Kopylova, Yulia
Koribalski, Bärbel
Kornilov, Victor
Korpela, Eric
Korzhavin, Anatoly
Kóspál, Ágnes
KOSUGI, George
Kotulla, Ralf
Kouveliotou, Chryssa
Kovalev, Yuri
Kovalev, Yuri
Kovaleva, Dana
Kovalevsky, Jean
KOYAMA, Yasuhiro
KOZAI, Yoshihide
Kraan-Korteweg, Renée
Kraft, Ralph
Kramer, Busaba
Kramer, Michael
Kramida, Alexander
Kraus, Alexander
Kraus, Stefan
Kretschmar, Peter
Kreysa, Ernst
Krichbaum, Thomas
Krishna, Gopal
Krishnan, Thiruvenkata
Kriss, Gerard
Kroll, Peter
Kronberg, Philipp
Kroto, Harold
Krügel, Endrik
KUAN, Yi-Jehng
Kubát, Jiri
Kudryavtseva, Nadezhda (Nadia)
Kuijpers, H.
Kuin, Paul
Kuiper, Thomas
Kulkarni, Prabhakar
Kulkarni, Shrinivas
Kulkarni, Vasant
Kumkova, Irina
Kundt, Wolfgang
Kunert-Bajraszewska, Magdalena
Kunkel, William
Kuntschner, Harald
Kunz, Martin
Kupka, Friedrich
KURAYAMA, Tomoharu
Kurtanidze, Omar
Kurtz, Donald
Kurtz, Michael
Kurtz, Stanley
Kurucz, Robert
Kus, Andrzej
Kutner, Marc
Kuzkov, Volodymyr
Kwok, Sun
Labbe, Ivo
Labeyrie, Antoine
Lacour, Sylvestre
Lada, Charles
La Franca, Fabio
LAI, Shih-Ping
Laing, Robert
Lal, Dharam
Lambert, David
Landecker, Thomas
Landman, Donald
Landolt, Arlo
Landstreet, John
Landt, Hermine
Lang, Kenneth
Langer, William
Langhoff, Stephanie
Langston, Glen
Lapenta, Giovanni
LaRosa, Theodore
Larson, Stephen
Larsson, Stefan
Lasenby, Anthony
Laskarides, Paul
LAS VERGNAS, Olivier
Laugalys, Vygandas
Launay, Françoise
Launay, Jean-Michel
Laurent, Michel
Lawrence, Charles
Lawrence, G.
Lawrence, Jon
Lawson, Peter
Layzer, David
Lazauskaite, Romualda
Lazio, Joseph
Leach, Sydney
Leahy, J.
Lebohec, Stephan
Le Bouquin, Jean-Baptiste
Le Bourlot, Jacques
Lèbre, Agnes
Lebrón, Mayra
Le Contel, Jean-Michel
Lee, Hyun-chul
Lee, William
LEE, Chang Won
LEE, Jeong-Eun
LEE, Jung-Won
LEE, Sang-Sung
LEE, Yong Bok

LEE, Youngung
Lefloch, Bertrand
Le Floch, André
Leger, Alain
Lehnert, Matthew
Leibowitz, Elia
Leibundgut, Bruno
Leinert, Christoph
Leisawitz, David
Lemaire, Jean-louis
Lemaitre, Gérard
Lemke, Michael
Lena, Pierre
Lenc, Emil
Lenzen, Rainer
Leone, Francesco
Lépine, Jacques
Le Poole, Rudolf
Lequeux, James
Leroy, Jean-Louis
Lesch, Harald
Le Squeren, Anne-Marie
Lester, Daniel
Lesteven, Soizick
Lestrade, Jean-François
Leto, Giuseppe
Leung, Chun
Levreault, Russell
Lewis, Brian
Li, Hong-Wei
LI, Jinzeng
LI, Min
LI, Qingkang
LI, Shiguang
LI, Zhi
LIANG, Guiyun
Ligori, Sebastiano
Likkel, Lauren
Lilley, Edward
LIM, Jeremy
Linde, Peter
Lindgren, Harri
Lindqvist, Michael
Linnartz, Harold
Lintott, Chris
LipunOv, Vladimir
Lis, Dariusz
Liseau, René
Lisi, Franco
Lister, Matthew
Liu, Zhong
LIU, Chengzhi
LIU, Guoqing
LIU, Sheng-Yuan
LIU, Siming
LIU, Xiang
LIU, Xiaoqun
LIU, Yang
Lo, Fred K. Y.
Lo, Wing-Chi Nadia
Lobel, Alex
Locke, Jack
Lockman, Felix
Lockwood, G.
Lo Curto, Gaspare
Loiseau, Nora
Lomb, Nicholas
Long, Knox
Longair, Malcolm
Longo, Giuseppe
Lonsdale, Carol
Lopes, Paulo
Lopez, Bruno
Lopez, Sebastian
Loren, Robert
Loubser, Ilani
Loukitcheva, Maria
Loulergue, Michelle
Loup, Cecile
Louys, Mireille
Lovas, Francis
Love, Gordon
Lovell, James
Lowe, Robert
Lozinskaya, Tatjana
LU, Fangjun
Lub, Jan
Lubowich, Donald
Lucero, Danielle
Luks, Thomas
Luna, Homero
LUO, Ali
LUO, Xianhan
Luque-Escamilla, Pedro
Lutz, Barry
Lutz, Dieter
Luu, Jane
Lyne, Andrew
Lytvynenko, Leonid
MA, Guanyi
Maccarone, Maria Concetta
Macchetto, Ferdinando
MacDonald, Geoffrey
MacDonald, James
Machalski, Jerzy
Maciesiak, Krzysztof
Mack, Karl-Heinz
Mack, Peter
MacLeod, John
Macquart, Jean-Pierre
Madjarska, Maria
Madore, Barry
MAEDA, Keiichi
MAEHARA, Hideo
Magalhaes, Antonio Mario
Magnier, Eugene
Maillard, Jean-Pierre
Mainieri, Vincenzo
Maiolino, Roberto
Maitzen, Hans
Maíz Apellániz, Jesús
MAKINO, Fumiyoshi
MAKISHIMA, Kazuo
Malasan, Hakim
Malbet, Fabien
Malin, David
Malkov, Oleg
Malofeev, Valery
Manchanda, R.
Manchester, Richard
Mancìni, Dario
Mandel, Holger
Mandolesi, Nazzareno
Manfroid, Jean
Mann, Robert
Mannucci, Filippo
Manoharan, P.
Manset, Nadine
Mantovani, Franco
MAO, Rui-Qing
Maran, Stephen
Marano, Bruno
Marcaide, Juan-Maria
Marcy, Geoffrey
Mardones, Diego
Mardyshkin, Vyacheslav
Marecki, Andrzej
Marín-Franch, Antonio
Markkanen, Tapio
Markoff, Sera

Marleau, Francine
Marques, Dos
Marraco, Hugo
Marscher, Alan
Marsden, Stephen
Marshalov, Dmitriy
Martí, Josep
Martin, Christopher
Martín, Eduardo
Martin, Robert
Martín, Sergio
Martín Díaz, Carlos
Martinez, Peter
Martínez, Vicent
Martinez Roger, Carlos
Martin-Pintado, Jesus
Martins, Donald
Marvel, Kevin
Masciadri, Elena
Masheder, Michael
Mashonkina, Lyudmila
Masiero, Joseph
Maslennikov, Kirill
Masłowski, Józef
Mason, Brian
Mason, Helen
Mason, Paul
Massardi, Marcella
Massaro, Francesco
Massey, Philip
Masson, Colin
Masters, Karen
Mathys, Gautier
Matsakis, Demetrios
MATSUNAGA, Noriyuki
MATSUO, Hiroshi
MATSUSHITA, Satoki
Matthews, Brenda
Matthews, Henry
Mattig, W.
Mattila, Kalevi
Matveenko, Leonid
Matz, Steven
Mauersberger, Rainer
Mawet, Dimitri
Max, Claire
Maxwell, Alan
Mayer, Pavel
McAdam, Bruce
McAlister, Harold
McAteer, R. T. James
McCall, Benjamin
McCammon, Dan
McClure-Griffiths, Naomi
McConnell, David
McCulloch, Peter
McCutcheon, William
McDavid, David
McKean, John
McKenna-Lawlor, Susan
McLaren, Robert
McLean, Brian
McLean, Donald
McLean, Ian
McMahon, Richard
McMullin, Joseph
McNally, Derek
McWhirter, R.
Meadows, A.
Mebold, Ulrich
Meeks, M.
Meier, David
Mein, Pierre
Meištas, Edmundas
Melikidze, Giorgi
Mellier, Yannick
Melo, Claudio
Mendoza, V.
Mendoza-Torress, Jose-Eduardo
Menon, T.
Menten, Karl
Menzies, John
Merín Martín, Bruno
Messerotti, Mauro
Metaxa, Margarita
Metcalfe, Leo
Metcalfe, Travis
Meyer, Martin
Meyer, Michael
Mianes, Pierre
Michalec, Adam
Mickaelian, Areg
Mickelson, Michael
Middelberg, Enno
Migenes, Victor
Mihajlov, Anatolij
Mikhailov, Andrey
Miley, George
Milkey, Robert
Millan-Gabet, Rafael
Miller, Joseph
Miller, Neal
Millour, Florentin
Mills, Bernard
Milne, Douglas
Milone, Eugene
Minchin, Robert
Mingaliev, Marat
MINH, Young-Chol
Mink, Jessica
Minniti, Dante
Mintz Blanco, Betty
Mirabel, Igor
Miranda, Oswaldo
Mironov, Aleksey
Miroshnichenko, Alla
Mitchell, Kenneth
Mitton, Jacqueline
Mitton, Simon
Miyawaki, Ryosuke
MIYAZAKI, Atsushi
MIYOSHI, Makoto
MIZUNO, Akira
MIZUNO, Norikazu
Modali, Sarma
Moellenbrock III, George
Moffett, David
Moffett, Thomas
Mohd, Zambri
Mohr, Joseph
Moitinho, André
Molenda-żakowicz, Joanna
Molinari, Emilio
Momjian, Emmanuel
MOMOSE, Munetake
Monaco, Lorenzo
Moncuquet, Michel
Monet, David
Monin, Jean-Louis
Monnier, John
Montes, David
Montmerle, Thierry
Moodley, Kavilan
Moody, Joseph
Mookerjea, Bhaswati
Moorhead, James
Moos, Henry
Morabito, David
Moran, James

Morbidelli, Roberto
Morgan, Lawrence
Morganti, Raffaella
Morison, Ian
MORITA, Kazuhiko
MORIYAMA, Fumio
MOROKUMA, Tomoki
Morras, Ricardo
Morris, David
Morris, Mark
Morris, Patrick
Morris, Rhys
Morris, Simon
Morrison, Nancy
Morton, Donald
Moscadelli, Luca
Mosoni, Laszlo
MOTOHARA, Kentaro
Mourard, Denis
Mozurkewich, David
Muanwong, Orrarujee
Muiños Haro, José
Muller, Erik
Muller, Sebastien
Mumma, Michael
Munari, Ulisse
Mundy, Lee
Muñoz Tuñón, Casiana
MURAKAMI, Naoshi
MURAKAMI, Hiroshi
MURAOKA, Kazuyuki
MURATA, Yasuhiro
MURAYAMA, Hitoshi
Murdin, Paul
Murdoch, Hugh
Murphy, Tara
Murtagh, Fionn
Mutel, Robert
Muxlow, Thomas
Myers, Philip
Nadeau, Daniel
NAGATA, Tetsuya
Nagendra, K.
Nagnibeda, Valerij
Nahar, Sultana
NAKAJIMA, Koichi
NAKAJIMA, Junichi
NAKAMURA, Akiko
NAKAMURA, Tsuko
NAKANO, Syuichi
NAKANO, Takenori
NAKASHIMA, Jun-ichi
Nakos, Theodoros
Nammahachak, Suwit
NAN, Ren-Dong
Näränen, Jyri
Narbutis, Donatas
NARDETTO, Nicolas
NARITA, Norio
Nave, Gillian
Naylor, David
Naylor, Tim
Neeser, Mark
Nefedyev, Yury
Neff, James
Neiner, Coralie
Nelson, Burt
Nemiroff, Robert
Nesci, Roberto
Nesvadba, Nicole
Newsom, Gerald
Nguyễn, Khanh
Nguyễn, Phuong
Nguyen-Quang, Rieu
Nicastro, Luciano
Nice, David
Nicholls, Ralph
Nicolet, Bernard
Nicolson, George
Niell, Arthur
Nielsen, Krister
NIINUMA, Kotaro
Nikoghosyan, Elena
Nikoli'c, Silvana
Nilsson, Hampus
Ninkov, Zoran
NISHIKAWA, Jun
NISHIMURA, Shiro
NISHIMURA, Jun
NISHIO, Masanori
NITTA, Atsuko
NOGUCHI, Kunio
Noll, Stefan
Nollez, Gerard
Noriega-Crespo, Alberto
Norris, Raymond
Nota, Antonella
Notni, Peter
NOUMARU, Junichi
Nürnberger, Dieter
Nussbaumer, Harry
Nuza, Sebastian
Oberst, Thomas
Oblak, Edouard
O'Brian, Thomas
O'Brien, Tim
O'Byrne, John
Ochsenbein, François
O'Dea, Christopher
O'Dell, Stephen
O'Donoghue, Aileen
O'Donoghue, Darragh
Oestreicher, Roland
Oetken, L.
Ofek, Eran
Ogando, Ricardo
OGAWA, Hideo
Ohashi, Nagayoshi
OHISHI, Masatoshi
Ohnaka, Keiichi
Ojha, Roopesh
Oka, Takeshi
OKA, Tomoharu
OKUMURA, Sachiko
Olberg, Michael
Oliva, Ernesto
Oluseyi, Hakeem
Omont, Alain
ONAKA, Takashi
Önel, Hakan
ONISHI, Toshikazu
Onuora, Lesley
Oozeer, Nadeem
Orchiston, Wayne
Orellana, Rosa
Orienti, Monica
Orsatti, Ana
Orton, Glenn
Osborn, Wayne
Oscoz, Alejandro
Osmer, Patrick
Osório, José
O'Sullivan, Créidhe
O'Sullivan, John
Otmianowska-Mazur, Katarzyna
OTSUKA, Masaaki
Ott, Juergen
Owen, Frazer
OZEKI, Hiroyuki

Özel, Mehmet
Özeren, Ferhat
Oezisik, Tuncay
Pacholczyk, Andrzej
Padman, Rachael
Pagani, Laurent
Pagano, Isabella
PAK, Soojong
Pakhomov, Yury
Paletou, Frédéric
Palmer, Patrick
Palmeri, Patrick
Pamyatnykh, Alexey
Panessa, Francesca
Pankonin, Vernon
Pantoja, Carmen
Paragi, Zsolt
Paredes Poy, Josep
Paresce, Francesco
Pareschi, Giovanni
Parijskij, Yurij
Parimucha, Stefan
PARK, Yong Sun
Parkinson, William
Parma, Paola
Paron, Sergio
Parrish, Allan
Pasachoff, Jay
Pashchenko, Mikhail
Pasian, Fabio
Patat, Ferdinando
Patel, Nimesh
Pati, Ashok
Paturel, Georges
Pauls, Thomas
Paumard, Thibaut
Pauwels, Thierry
Pavani, Daniela
Pavlyuchenkov, Yaroslav
Payne, David
Peach, Gillian
Pearce, Mark
Pearson, Timothy
Peck, Alison
Pecker, Jean-Claude
Pedersen, Holger
Pedlar, Alan
Pedreros, Mario
Peel, Michael
PEI, Chunchuan
Pel, Jan
Pelt, Jaan
Pence, William
PENG, Bo
PENG, Qingyu
Penny, Alan
Penzias, Arno
Percheron, Isabelle
Percy, John
Perea-Duarte, Jaime
Perez, Fournon
Pérez-González, Pablo
Pérez Torres, Miguel
Perley, Richard
Perozzi, Ettore
Perraut, Karine
Perrin, Guy
Perrin, Marshall
Persson, Carina
Peters, William
Petit, Pascal
Petrini, Daniel
Petrova, Svetlana
Pettengill, Gordon
Pettini, Marco
Pevtsov, Alexei
Pfeiffer, Raymond
PHAM, Diep
Philip, A.G.
Philipp-May, Sabine
Phillips, Christopher
Phillips, Kenneth
Phillips, Mark
Phillips, Thomas
Piacentini, Ruben
Pick, Monique
Pickles, Andrew
Pigulski, Andrzej
Piirola, Vilppu
Pijpers, Frank
Pilachowski, Catherine
Pilkington, John
Pilling, Sergio
PING, Jinsong
Pintado, Olga
Pinte, Christophe
Pisano, Daniel
Piskunov, Anatolij
Pitkin, Matthew
Pizzichini, Graziella
Planesas, Pere
Platais, Imants
Plavchan, Peter
Plume, Rene
Pogrebenko, Sergei
Pokrzywka, Bartłomiej
Polatidis, Antonios
Polechova, Pavla
Pollas, Christian
Pompea, Stephen
Pompei, Emanuela
Ponsonby, John
Pooley, Guy
Porcas, Richard
Porceddu, Ignazio
Porras Juárez, Bertha
Portegies Zwart, Simon
Posch, Thomas
Potapov, Vladimir
Pott, Jörg-Uwe
Potter, Stephen
Pound, Marc
Pourbaix, Dimitri
Pozanenko, Alexei
Poznanski, Dovi
Prabhu, Tushar
Pradhan, Anil
Prandoni, Isabella
Pravdo, Steven
Preston, Robert
Preuss, Eugen
Price, R.
Primas, Francesca
Produit, Nicolas
Proffitt, Charles
Protsyuk, Yuri
Prša, Andrej
Pshirkov, Maxim
Pucillo, Mauro
Puetter, Richard
Pulone, Luigi
Purton, Christopher
Puschell, Jeffery
Pushkarev, Alexander
Puxley, Phil
Puzia, Thomas
Pye, John
QIAN, Shengbang
QU, Jinlu

Quanz, Sascha
Queloz, Didier
Querci, Francois
Quinet, Pascal
Quinn, Peter
Quintana, Hernan
Quirrenbach, Andreas
Rabbia, Yves
Radford, Simon
Rahimov, Ismail
Rajagopal, Jayadev
Ralchenko, Yuri
Ramírez, Jose
Rank-Lueftinger, Theresa
Ransom, Scott
Rao, A.
Rao, Pasagada
Raoult, Antoinette
Rastegaev, Denis
Rastogi, Shantanu
Rathborne, Jill
Ratnatunga, Kavan
Rauch, Thomas
Raveendran, A.
Rawlings, Mark
Ray, Alak
Ray, Paul
Ray, Tom
Readhead, Anthony
Reardon, Kevin
Redman, Matthew
Redman, Stephen
Reglero Velasco, Victor
Reich, Wolfgang
Reid, Mark
Reif, Klaus
Reitsema, Harold
Rengel, Miriam
Renson, P.
Reshetnyk, Volodymyr
Reyes, Francisco
Reyes-Ruiz, Mauricio
Reynolds, Cormac
Reynolds, John
Rhee, Jaehyon
RHEE, Myung Hyun
Ribak, Erez
Ribó, Marc
Richards, Mercedes
Richer, John
Richichi, Andrea
Rickard, Lee
Ridgway, Stephen
Ridgway, Susan
Rioja, Maria
Rivinius, Thomas
Rix, Hans-Walter
Rizzi, Luca
Rizzo, Jose
Robb, Russell
Robberto, Massimo
Roberts, David
Roberts, Morton
Robertson, Douglas
Robertson, James
Robinson, Edward
Robinson Jr, Richard
Robson, Ian
Rocha-Pinto, Hélio
Rodrigues, Claudia
Rodriguez, Luis
Roeder, Robert
Roelfsema, Peter
Roellig, Thomas
Roemer, Elizabeth
Roennaeng, Bernt
Röser, Hans-peter
Roger, Robert
Rogers, Alan
Rogers, Forrest
Rogstad, David
Rohlfs, Kristen
Roman, Nancy
Romaniello, Martino
Romanov, Andrey
Romanyuk, Yaroslav
Romero, Gustavo
Roming, Peter
Romney, Jonathan
Romoli, Marco
Rosa, Michael
Rosado, Margarita
Rosa González, Daniel
Ros Ibarra, Eduardo
Rosolowsky, Erik
Rossi, Corinne
Rostas, François
Rostopchina, Alla
Roth, Miguel
Rothberg, Barry
Rots, Arnold
Rouan, Daniel
Roueff, Evelyne
Rousset, Gérard
Rowson, Barrie
Rubio, Monica
Rubio-Herrera, Eduardo
Ruder, Hanns
Rudnick, Lawrence
Rudnitskij, Georgij
Ruffini, Remo
Ruiz, Maria Teresa
Rupprecht, Gero
Rushton, Anthony
Russell, Jane
Russo, Guido
Russo, Pedro
Rutten, Renee
Ryabchikova, Tatiana
Ryabova, Galina
Rydbeck, Gustaf
Ryder, Stuart
Ryś, Stanisław
Saar, Enn
Sabin, Laurence
Sadler, Elaine
Safi-Harb, Samar
Saha, Abhijit
Saha, Swapan
Sahai, Raghvendra
Sahal-Bréchot, Sylvie
Saikia, Dhruba
SAKAMOTO, Seiichi
SAKAMOTO, Tsuyoshi
SAKAO, Taro
Salama, Farid
Salomé, Philippe
Salter, Christopher
Samec, Ronald
Samodurov, Vladimir
Samus, Nikolay
Sánchez, Francisco
Sandell, Göran
Sanders, David
Sanders, Gary
Santillán, Alfredo
Santos Agostinho, Rui
Santos-Sanz, Pablo
Sarasso, Maria
Sargent, Annelia

Saripalli, Lakshmi
Sarma, Anuj
Sarma, N.
Sarre, Peter
Sasaki, Toshiyuki
Sastry, Ch.
SATO, Fumio
Sault, Robert
Savage, Ann
Savanevich, Vadim
Savanov, Igor
Saviane, Ivo
Savin, Daniel
Savolainen, Tuomas
Sawada, Tsuyoshi
SAWADA-SATOH, Satoko
Sawant, Hanumant
Scalise Jr, Eugenio
Schade, David
Schilbach, Elena
Schilizzi, Richard
Schilke, Peter
Schiller, Stephen
Schinckel, Antony
Schipani, Pietro
Schlesinger, Barry
Schlickeiser, Reinhard
Schmadel, Lutz
Schmidt, Edward
Schmidt, Maarten
Schmitt, Henrique
Schmitz, Marion
Schneider, Jean
Schödel, Rainer
Schöller, Markus
Schreiber, Roman
Schrijver, Johannes
Schröder, Anja
Schuch, Nelson
Schuller, Peter
Schultz, David
Schulz, R.
Schure, Klara
Schuster, William
Schwartz, Philip
Schwarz, Ulrich
Scott, John
Scott, Paul
Scuderi, Salvo
Seaman, Rob
Seaquist, Ernest
Seaton, Daniel
Sedmak, Giorgio
Sefako, Ramotholo
Segaluvitz, Alexander
šegan, Stevo
Seielstad, George
Seiradakis, John
Seitzer, Patrick
SEKIDO, Mamoru
SEKIGUCHI, Kazuhiro
SEKIMOTO, Yutaro
Semenov, Dmitry
Sen, Asoke
Sese, Rogel Mari
SETA, Masumichi
Setti, Giancarlo
Seymour, Nicholas
Shaffer, David
Shafter, Allen
SHANG, Hsien
Shankland, Paul
Shaposhnikov, Vladimir
Shara, Michael
Sharma, A.
Sharp, Christopher
Sharp, Nigel
Sharples, Ray
Shastri, Prajval
Shaver, Peter
Shaw, Richard
Shawl, Stephen
Shearer, Andrew
Shelton, Ian
SHEN, Zhiqiang
Shepherd, Debra
Shetrone, Matthew
Shevgaonkar, R.
SHI, Huoming
SHI, Jianrong
SHI, Shengcai
SHIBATA, Katsunori
SHIMOIKURA, Tomomi
Shinnaga, Hiroko
SHIRASAKI, Yuji
šmelds, Ivar
Shoemaker, David
Shone, David
Shore, Bruce
Shoyoqubov, Shoayub
SHU, Fengchun
Shulga, Valerii
Sickafoose, Amanda
Siebenmorgen, Ralf
Sieber, Wolfgang
Silva, David
Simić, Zoran
Simon, Michal
Simón-Díaz, Sergio
Simons, Douglas
Sinachopoulos, Dimitris
Singal, Ashok
Sinha, Krishnanand
Sinha, Rameshwar
Siringo, Giorgio
Sivakoff, Gregory
Skillen, Ian
Skillman, Evan
Skinner, Gerald
Skoda, Petr
Slade, Martin
Slee, O.
Smart, Richard
Smida, Radomír
Smirnov, Grigorij
Smirnova, Tatiana
Smith, Dean
Smith, Eric
Smith, Francis
Smith, Haywood
Smith, J.
Smith, Malcolm
Smith, Myron
Smith, Niall
Smith, Nigel
Smith, Peter
Smith, Randall
Smith, Robert
Smith, William
Smolentsev, Sergej
Smol'kov, Gennadij
Smyth, Michael
Snellen, Ignas
Snik, Frans
Snowden, Michael
Sobolev, Yakov
Sodin, Leonid
SOFUE, Yoshiaki
Sokolov, Konstantin
SOMA, Mitsuru

SOMANAH, Radhakhrishna
Somerville, William
SONG, In-Ok
SONG, Liming
SONG, Qian
Soonthornthum, Boonrucksar
Soria, Roberto
Sorochenko, Roman
Soszyński, Igor
Spagna, Alessandro
Spahr, Timothy
Spencer, John
Spencer, Ralph
Spielfiedel, Annie
Spite, François
Richard, Richard
Srianand, Raghunathan
Sridharan, Tirupati
Stachowski, Grzegorz
Stagg, Christopher
Stairs, Ingrid
Stancil, Philip
Stanghellini, Carlo
Stanley, G.
Stappers, Benjamin
Stark, Glenn
Stée, Philippe
Steffen, Matthias
Stehle, Chantal
Steinberg, Jean-Louis
Steinbring, Eric
Steinlin, Uli
Steinmetz, Matthias
Stencel, Robert
Sterken, Christiaan
Sterzik, Michael
Stetson, Peter
Stewart, Paul
Stewart, Ronald
Stickland, David
Stift, Martin
Stil, Jeroen
Still, Martin
Stockman Jr, Hervey
Stoehr, Felix
Stone, R.
Stonkutė, Rima
Storey, John
Storey, Michelle
Storm, Jesper
Storrie-Lombardi, Lisa
Strachan, Leonard
Straižys, Vytautas
Strassmeier, Klaus
Strauss, Michael
Strelnitski, Vladimir
Stritzinger, Maximilian
Strom, Richard
Strom, Stephen
Strukov, Igor
SU, Cheng-yue
SU, Ding-qiang
Subrahmanya, C.
Subrahmanyan, Ravi
Subramaniam, Annapurni
Sudzius, Jokubas
SUGAI, Hajime
SUGITANI, Koji
Sukumar, Sundarajan
Suleymanova, Svetlana
Sullivan, Denis
Sullivan, Mark
Sullivan, III, Woodruff
Summers, Hugh
SUNADA, Kazuyoshi
SUNG, Hyun-Il
Suntzeff, Nicholas
Surdej, Jean
Surkis, Igor
Sutherland, Ralph
Sutton, Edmund
Svalgaard, Leif
Swade, Daryl
Swarup, Govind
Swenson Jr., George
Swings, Jean-Pierre
Szalay, Alex
Szeifert, Thomas
Szkody, Paula
Szomoru, Arpad
Szymański, Michał
Szymczak, Marian
Tacconi, Linda
Tagliaferri, Gianpiero
TAKABA, Hiroshi
TAKANO, Shuro
TAKANO, Toshiaki
TAKATA, Tadafumi
TAKAYANAGI, Kazuo
Talavera, Antonio
Tallon, Michel
Tallon-Bosc, Isabelle
Tammi, Joni
TAN, Baolin
TANAKA, Ichi Makoto
TANAKA, Riichiro
Tancredi, Gonzalo
Tandon, S.
Tapping, Kenneth
Taranova, Olga
Tarenghi, Massimo
Tarter, Jill
Taş, Günay
TATEMATSU, Ken'ichi
Tatum, Jeremy
Tauber, Jan
Tayal, Swaraj
Taylor, A.
Taylor, Gregory
Tchang-Brillet, Lydia
Tedds, Jonathan
te Lintel Hekkert, Peter
Tello Bohorquez, Camilo
Templeton, Matthew
ten Brummelaar, Theo
Teng, Stacy
Tennyson, Jonathan
TERADA, Yukikatsu
TERASHITA, Yoichi
Terasranta, Harri
Terzian, Yervant
Tessema, Solomon
Testi, Leonardo
Teuben, Peter
Teyssier, Romain
Theureau, Gilles
Thevenin, Frederic
Thiébaut, Éric
Tholen, David
Thomasson, Peter
Thompson, A.
Thompson, Rodger
Thomsen, Bjarne
Thorsett, Stephen
Thorstensen, John
Thum, Clemens
Thurston, Mark
TIAN, Wenwu

Tichá, Jana
Timothy, J.
Tingay, Steven
Tiplady, Adrian
Tissera, Patricia
Titov, Vladimir
Tody, Douglas
Tofani, Gianni
Tokarev, Yurij
Tokovinin, Andrei
Tokunaga, Alan
Tolbert, Charles
Tornikoski, Merja
Torrelles, Jose Maria
Torres-Papaqui, Juan
TOSAKI, Tomoka
Toth, Laszlo
Tovmassian, Hrant
Townes, Charles
Townsend, Richard
Tozzi, Gian
Trager, Scott
Trakhtenbrot, Benny
TranMinh, Nguyet
Traub, Wesley
Trefftz, Eleonore
Tremko, Jozef
Tresse, Laurence
Trigilio, Corrado
Trimble, Virginia
Trinidad, Miguel
Tripathi, Durgesh
Tripicco, Michael
TRIPPE, Sascha
Tristram, Konrad
Tritton, Keith
Tritton, Susan
Troland, Thomas
Trujillo, Chadwick
Truong, Bach
Trushkin, Sergey
TSAI, An-Li
TSUBOI, Masato
Tsujimoto, Masahiro
TSUNETA, Saku
Tsutsumi, Takahiro
Tsvetkov, Milcho
Tuccari, Gino
Tull, Robert
Turatto, Massimo
Turcu, Vlad
Turło, Zygmunt
Turner, Jean
Turner, Kenneth
Turtle, A.
Tuthill, Peter
Tycner, Christopher
Tyson, John
Tyul'bashev, Sergei
Tzioumis, Anastasios
Ubertini, Pietro
Udaya, Shankar
UENO, Munetaka
UESUGI, Akira
Ueta, Toshiya
Ulmer, Melville
Ulrich, Marie-Helene
Ulrich, Roger
Ulvestad, James
Ulyanov, Oleg
Umana, Grazia
UMEDA, Hideyuki
UMEMOTO, Tomofumi
Unger, Stephen
Unwin, Stephen
Uomoto, Alan
Upgren, Arthur
Urama, Johnson
URATA, Yuji
Urban, Sean
Ureche, Vasile
Urov sevi'c, Dejan
Urquhart, James
Uslenghi, Michela
Uson, Juan
Usowicz, Jerzy
USUDA, Tomonori
Vaccari, Mattia
Vakili, Farrokh
Vakoch, Douglas
Valeev, Azamat
Valentijn, Edwin
Vallée, Jacques
Valls-Gabaud, David
Valtaoja, Esko
Valtonen, Mauri
Val'tts, Irina
van Belle, Gerard
van den Ancker, Mario
van den Bergh, Sidney
VandenBout, Paul
Vandenbussche, Bart
van der Bliek, Nicole
van der Hulst, Jan
van der Kruit, Pieter
van der Laan, Harry
van der Tak, Floris
van der Werf, Paul
van Dishoeck, Ewine
van Driel, Wim
van Gorkom, Jacqueline
van Haarlem, Michiel
van Kampen, Eelco
Vaňko, Martin
van Langevelde, Huib
van Leeuwen, Joeri
van Rensbergen, Walter
van Woerden, Hugo
Varela López, Jesús
Varela Perez, Antonia
Varshalovich, Dmitrij
Vasta, Magda
Vats, Hari
Vaughan, Alan
Vaughan, Arthur
Vavrek, Roland
Vega, Olga
Veillet, Christian
Velikodsky, Yuri
Velusamy, T.
Venkatakrishnan, P.
Venturi, Tiziana
Venugopal, V.
Verdes-Montenegro, Lourdes
Vereshchagin, Sergej
Verheijen, Marc
Verhoelst, Tijl
Verkhodanov, Oleg
Verma, R.
Vermeulen, Rene
Vernin, Jean
Verschuur, Gerrit
Verter, Frances
Vestergaard, Marianne
Vicente, Raimundo
Vidali, Gianfranco
Vidotto, Aline
Vilas, Faith
Vilas-Boas, José

Viotti, Roberto
Vishniac, Ethan
Vivekanand, M.
Vlahakis, Catherine
Vlemmings, Wouter
Völk, Heinrich
Vogel, Stuart
Volk, Kevin
Vollmer, Bernd
Volonte, Sergio
Volo'shina, Irina
Volvach, Alexander
von Braun, Kaspar
Voronkov, Maxim
Vrba, Frederick
Vrtilek, Jan
Vujnovic, Vladis
WADA, Takehiko
Wadadekar, Yogesh
Waddington, Ian
Wahlgren, Glenn
Wainscoat, Richard
WAJIMA, Kiyoaki
Wakelam, Valentine
Walker, Alistair
Walker, Constance
Walker, Gordon
Walker, Merle
Walker, Robert
Walker, William
Wall, Jasper
Wall, William
Wallace, James
Wallace, Patrick
Walmsley, C.
Walsh, Andrew
Walsh, Wilfred
Walton, Nicholas
Wambsganß, Joachim
WANG, Chen
WANG, Feilu
WANG, Guangli
WANG, Guomin
WANG, Hong-Guang
WANG, Jian-Min
WANG, Jingyu
WANG, Junxian
WANG, Junzhi
WANG, Min
WANG, Na
WANG, Sen
WANG, Shen
WANG, Shiang-Yu
WANG, Shouguan
WANG, Shujuan
WANG, Tinggui
WANG, Xunhao
Wannier, Peter
Wardle, John
Ward-Thompson, Derek
Warmels, Rein
Warner, Peter
Warren Jr, Wayne
Watanabe, Ken
Watson, Michael
Watson, Robert
Wayth, Randall
Webber, John
Wehrle, Ann
Wei, Mingzhi
WEI, Erhu
Weigelt, Gerd
Weilbacher, Peter
Weiler, Edward
Weiler, Kurt
Weiss, Werner
Weistrop, Donna
Welch, Douglas
Welch, William
Weller, Charles
Wellington, Kelvin
Wells, Donald
WEN, Linqing
Wendt, Harry
Wendt, Martin
Wenger, Marc
WENLEI, Shan
Werner, Klaus
Wesselius, Paul
West, Richard
Westergaard, Niels
Westmeier, Tobias
Wheatley, Peter
Whitcomb, Stanley
White, Glenn
White, Nathaniel
White, Richard
Whitelock, Patricia
Whiteoak, John
Whiting, Matthew
Wicenec, Andreas
Wickramasinghe, N.
Wiedemann, Günter
Wielebinski, Richard
Wielen, Roland
Wiese, Wolfgang
Wiik, Kaj
Wijers, Ralph
Wiklind, Tommy
Wild, Wolfgang
Wilkins, George
Wilkinson, Peter
Williams, Gareth
Williams, Robert
Willis, Anthony
Wills, Beverley
Wills, Derek
Willson, Robert
Willstrop, Roderick
Wilner, David
Wilson, Christine
Wilson, Richard
Wilson, Robert
Wilson, Thomas
Wilson, William
Windhorst, Rogier
Winiarski, Maciej
Winnberg, Anders
Wise, Michael
Wittkowski, Markus
Witzel, Arno
Woan, Graham
Woillez, Julien
Wolleben, Maik
Wolszczan, Alexander
Woltjer, Lodewijk
Womack, Maria
Wong, Tony
Wood, Douglas
Woodsworth, Andrew
Woolf, Neville
Wootten, Henry
Worrall, Diana
Woudt, Patrick
Wramdemark, Stig
Wright, Alan
Wrixon, Gerard
Wrobel, Joan
Wu, Nailong
WU, Hong

WU, Xiangping
WU, Xinji
WU, Yuefang
Wucknitz, Olaf
Wunner, Guenter
Wyrowski, Friedrich
XIAO, Dong
Yahil, Amos
YAMADA, Shimako
YAMAMOTO, Hiroaki
YAMAOKA, Hitoshi
YAMASHITA, Yasumasa
YAMAUCHI, Makoto
YAN, Yihua
Yanamandra-Fisher, Padma
YANG, Changgen
YANG, Dehua
YANG, Hong-Jin
YANG, Ji
YANG, Xiaohu
YANG, Zhigen
YANO, Hajime
YAO, Qijun
YAO, Yongqiang
YE, Binxun
YE, Shuhua
Yegorova, Irina
Yin, Qi-Feng
YONEKURA, Yoshinori
Yoshino, Kouichi
Young, Andrew
Young, Arthur
Young, John
Young, Louise
Younis, Saad
Yu, Yan
YU, Zhiyao
YUAN, Xiangyan
Yudin, Ruslan
YUJI, Ikeda
YUMIKO, Oasa
Yun, Min
Yusef-Zadeh, Farhad
YUTAKA, Shiratori
Zabolotny, Vladimir
Zacchei, Andrea
Zacharias, Norbert
Zainal Abidin, Zamri
Zaitsev, Valerij
Zamkotsian, Frederic
Zanichelli, Alessandra
Zannoni, Mario
Zarka, Philippe
Zavala, Robert
Zdanavičius, Justas
Zeiler, Michael
Zeilinger, Werner
Zeippen, Claude
Zender, Joe
ZENG, Qin
Zensus, J-Anton
Zhang, Jie
Zhang, Yuying
ZHANG, Chengmin
ZHANG, Haiyan
ZHANG, Hongbo
ZHANG, Jian
ZHANG, JiangShui
ZHANG, Jingyi
ZHANG, Jun
ZHANG, Qizhou
ZHANG, Shu
ZHANG, Shuang Nan
ZHANG, Xiuzhong
ZHANG, Xizhen
ZHANG, Yanxia
ZHANG, Yong
ZHANG, You-Hong
ZHANG, Zhibin
ZHAO, Gang
ZHAO, Jun-Hui
ZHAO, Jun Liang
ZHAO, Yongheng
Zheleznyak, Alexander
Zheleznyakov, Vladimir
ZHENG, Weimin
ZHENG, Xiaonian
ZHENG, Xinwu
ZHOU, Jianfeng
ZHOU, Jianjun
ZHOU, Xia
ZHU, LiChun
ZHU, Liying
ZHU, Ming
ZHU, Qingfeng
ZHU, Wenbai
Zięba, Stanisław
Ziegler, Bodo
Zinchenko, Igor
Zinnecker, Hans
Zirm, Andrew
Zitelli, Valentina
Zlobec, Paolo
Zlotnik, Elena
Zoccali, Manuela
Zucker, Shay
Zuckerman, Benjamin
Zwaan, Martin
Zwintz, Konstanze
Zylka, Robert

Division C Education, Outreach and Heritage

President: Mary Kay M. Hemenway
Vice-President: Hakim Luthfi Malasan

Organizing Committee Members:

Michele Gerbaldi
Xiaochun Sun
Clive Ruggles
Juan Antonio Belmonte Aviles
Lars Lindberg Christensen
Beatriz Elena Garcia
Raymond P. Norris
Pedro Russo
Mary Kay M. Hemenway
Hakim Luthfi Malasan
Jean-Pierre de Greve
Kazuhiro SEKIGUCHI

Members:

Abalakin, Viktor
Abraham, Roberto
Abrevaya, Ximena
Abt, Helmut
Accomazzi, Alberto
Acharya, Bannanje
Acker, Agnes
AGATA, Hidehiko
Aghaee, Alireza
Agüeros, Marcel
Aguilar, Maria
AHN, SANG-HYEON
AHN, Youngsook
Airapetian, Vladimir
Ajhar, Edward
AKITA, Kyo
Albanese, Lara
Alexandrov, Yuri
Ali, Sabry
ALIBERT, Yann
Allan, Alasdair
Almár, Ivan
Al-Naimiy, Hamid
Alonso-Herrero, Almudena
Alsabti, Abdul Athem
Altamore, Aldo
Alvarez, Pedro
Alvarez, Rodrigo
Alvarez del Castillo, Elizabeth
Alvarez Pomares, Oscar
Alves, João
Alves, Virgínia
Ambastha, Ashok
Anandaram, Mandayam
Andersen, Johannes
Andrews, David
Andrews, Frank
Ansari, S.M.
Antonello, Elio
Apai, Daniel
Apostolovska, Gordana
Aquilano, Roberto
ARAKIDA, Hideyoshi
Arbutina, Bojan
Arcidiacono, Carmelo
Arellano Ferro, Armando
Arenou, Frédéric
Aretxaga, Itziar
Argo, Megan
Arifyanto, Mochamad
Arion, Douglas
Asanok, Kitiyanee
Asher, David
Ashok, N.
Aslan, Zeki
Atanackovic, Olga
Aubier, Monique
Augereau, Jean-Charles
Augusto, Pedro
Auriemma, Giulio
Axelsson, Magnus
Baade, Robert
Baars, Jacob
Babayev, Elchin
Babu, G.S.D.
Babul, Arif
Badescu, Octavian
Badolati, Ennio
BAEK, Chang Hyun
Bagla, Jasjeet
Bailey, Katherine
Bailey, Mark
Baki, Paul
Balbi, Amedeo
Baliyan, Kiran
Balkowski-Mauger, Chantal
Ball, Lewis
Ballabh, Goswami
Balyshev, Marat
Bamford, Steven
Bandyopadhyay, A.
Baneke, David
Banerjee, Dipankar
Banhatti, Dilip
Barclay, Charles
Bardelli, Sandro
BARET, Bruny
Barlai, Katalin
Barlow, Nadine
Barmby, Pauline
Barnes, David
Barrantes, Marco
Barret, Didier
Barstow, Martin
Barthel, Peter
Bartkevicius, Antanas
Bartlett, Jennifer
Baruch, John
Barway, Sudhashu
Bary, Jeffrey
Basa, Stephane

Baskill, Darren
Bassett, Bruce
Battaner, Eduardo
Batten, Alan
Bauer, Amanda
Bechtold, Jill
Beck, Sara
Becker, Werner
Beckman, John
Beckmann, Volker
Beesham, Aroonkumar
Bell Burnell, Jocelyn
Belmonte Aviles, Juan Antonio
Benacchio, Leopoldo
Bendjoya, Philippe
Benedict, George
Ben-Jaffel, Lofti
Benkhaldoun, Zouhair
Bennett, Jim
Benvenuti, Piero
Berendzen, Richard
Berger, Jean-Philippe
Bernabeu, Guillermo
Berrilli, Francesco
Bertola, Francesco
Bessell, Michael
Bhardwaj, Anil
Bhatia, Vishnu
Bhatt, H.
Bhattacharjee, Pijushpani
Bhavsar, Suketu
Bianchi, Luciana
Bianchi, Simone
Bien, Reinhold
Birlan, Mirel
Bishop, Roy
Bittar, Jamal
Bjælde, Ole
Bjorkman, Karen
Blaga, Cristina
Blair, David
Blair, William
Blum, Robert
Bobrowsky, Matthew
Boccaletti, Dino
Bode, Michael
Boër, Michel
Boerngen, Freimut
Boffin, Henri
Boissier, Samuel
Bojurova, Eva
Boles, Thomas
Bon, Edi
Bond, Howard
Bongiovanni, Angel
Bònoli, Fabrizio
Booth, Roy
Borchkhadze, Tengiz
Bord, Donald
Boroson, Bram
Bosken, Sarah
Bot, Caroline
Botez, Elvira
Botti, Thierry
Bottinelli, Lucette
Bougeret, Jean-Louis
Bouton, Ellen
Bowen, David
Braes, Lucien
Brammer, Gabriel
Brandt, William
Brecher, Kenneth
Bretones, Paulo
Briand, Carine
Bridges, Terry
Brieva, Eduardo
Briot, Danielle
Bromage, Gordon
Brooks, Randall
Brosch, Noah
Brosche, Peter
Brouillet, Nathalie
Brouw, Willem
Brunet, Jean-Pierre
Brunner, Robert
Bruzual, Gustavo
Buccino, Andrea
Budding, Edwin
Burgasser, Adam
Burman, Ronald
Bzowski, Maciej
Cabanac, Remi
CAI, Kai
CAI, Michael
Calamida, Annalisa
Calvet, Nuria
Cami, Jan
Campana, Riccardo
Cannon, Russell
Cannon, Wayne
Capaccioli, Massimo
Capelato, Hugo
Caplan, James
Cappi, Alberto
Caretta, Cesar
Carignan, Claude
Carlson, John
Carter, Brian
Carter, William
Cassan, Arnaud
Cattaneo, Andrea
Cauzzi, Gianna
Celebre, Cynthia
Chaboyer, Brian
Chakrabarti, Supriya
Chamcham, Khalil
CHANG, Heon-Young
Chapman, Jessica
Charmandaris, Vassilis
Chaty, Sylvain
Chatzichristou, Eleni
CHEN, Alfred
CHEN, An-Le
CHEN, Dongni
CHEN, Li
CHEN, Lin-wen
CHEN, Wen Ping
CHEN, Xinyang
CHEN, Xuelei
CHEN, Yang
Chenevez, Jérôme
Chengalur, Jayaram
Chernyakova, Maria
CHIN, Yi-nan
Chinnici, Ileana
Chitre, Dattakumar
Choudhary, Debi Prasad
Choudhuri, Arnab
Christensen, Lars
Christian, Carol
Christie, Grant
Christlieb, Norbert
Chukwude, Augustine
Chung, Eduardo
Chyży, Krzysztof
Ciroi, Stefano
Clark, David
Clark, Thomas
Clarke, David

Clem, James
Clifton, Gloria
Cliver, Edward
Coe, Malcolm
Coffey, Deirdre
Cohen, David
Cohen, Marshall
Colafrancesco, Sergio
Colomer, Francisco
Comins, Neil
Connors, Martin
Conti, Alberto
Copin, Yannick
Cora, Alberto
Corbally, Christopher
Corbin, Brenda
Cornejo, Alejandro
Corral, Luis
COTTIN, Hervé
Cottrell, Peter
Courbin, Frederic
Courtois, Helene
Couto da Silva, Telma
Covone, Giovanni
Crabtree, Dennis
Cracco, Valentina
Crawford, David
Cristiani, Stefano
Croft, Steve
Crosta, Mariateresa
Crovisier, Jacques
Cudworth, Kyle
Cuesta Crespo, Luis
Cui, Wei
CUI, Shizhu
CUI, Zhenhua
Cunningham, Maria
Cuntz, Manfred
Cuypers, Jan
Dadhich, Naresh
Dadic, Zarko
DAI, Zhibin
DAISAKU, Nogami
DAISUKE, Iono
Dall'Ora, Massimo
Dal Ri Barbosa, Cassio
Daltabuit, Enrique
Damineli Neto, Augusto
Danezis, Emmanuel
Danford, Stephen
Daniel, Jean-Yves
Danner, Rolf
DARHMAOUI, Hassane
Darriulat, Pierre
Das, P.
Davies, Rodney
Davies, Roger
Davis, A. E. L.
Davoust, Emmanuel
Debarbat, Suzanne
De Biasi, Maria
Dechev, Momchil
De Cuyper, Jean-Pierre
de Greve, Jean-Pierre
de Grijs, Richard
de Jong, Teije
Dekker, E.
de Lange, Gert
Dell'Antonio, Ian
Delsanti, Audrey
Del Santo, Melania
Demarco, Ricardo
Demircan, Osman
Demory, Brice-Olivier
DeNisco, Kenneth
Denisse, Jean-Francois
Dennefeld, Michel
Depagne, Éric
Derman, Ethem
de Swardt, Bonita
Deustua, Susana
Devaney, Martin
DeVorkin, David
Dick, Steven
Dick, Wolfgang
Diego, Francisco
Dimitrijevic, Milan
Dintinjana, Bojan
Dipper, Nigel
Dluzhnevskaya, Olga
Dole, Herve
Dominguez, Inma
Domínguez, Mariano
Donahue, Megan
Doran, Rosa
Dorch, Søren Bertil
Doressoundiram, Alain
Dorschner, Johann
Doval, Jorge M.
Drissen, Laurent
Drury, Luke
DU, Lan
Ducati, Jorge
Ducourant, Christine
Duffard, Rene
Dukes Jr., Robert
Dumont, Simone
Duorah, Hira
Dupuy, David
Durrell, Patrick
Dutil, Yvan
Duval, Marie-France
Dworetsky, Michael
Easther, Richard
Eastwood, Kathleen
Edmunds, Michael
Edwards, Philip
Ehgamberdiev, Shuhrat
Ehle, Matthias
Ekström García Nombela, Sylvia
El Eid, Mounib
Elgarøy, Øystein
Elliott, Kenneth
Emerson, James
Emsellem, Eric
Engelbrecht, Chris
Engels, Dieter
English, Jayanne
Engvold, Oddbjørn
Esguerra, Jose Perico
Espenak, Fred
Esteban, César
Evans, Robert
Eyres, Stewart
Eze, Romanus
Fabregat, Juan
Fahey, Richard
Falceta-Goncalves, Diego
Falcon Veloz, Nelson
Falgarone, Edith
Faundez-Abans, Max
Feain, Ilana
Feinstein, Alejandro
Feinstein, Carlos
Feitzinger, Johannes
Ferlet, Roger
Fernández, Julio
Fernández, Yanga
Fernandez-Figueroa, M.

Fernie, J.
Ferrari, Attilio
Ferrusca, Daniel
Feulner, Georg
Field, J. V.
Fienberg, Richard
Fierro, Julieta
Figueiredo, Newton
Firneis, Maria
Fitzsimmons, Alan
Fleck, Robert
Flin, Piotr
Florides, Petros
Florsch, Alphonse
Floyd, David
Fluke, Christopher
Focardi, Paola
Foing, Bernard
Folgueira, Marta
Forbes, Douglas
Forero Villao, Vicente
Foukal, Peter
Francis, Paul
Franco, José
Frasca, Antonio
Frater, Robert
Freeman, Kenneth
Freitas Mourão, Ronaldo
Frew, David
Friedlander, Michael
Fruchter, Andrew
FU, Hsieh-Hai
FU, Jian-Ning
FUJISHITA, Mitsumi
FUJITA, Mitsutaka
FUJIWARA, Tomoko
FUKUE, Jun
Funes, José
Füzfa, Andre
Gábor, Pavel
Gabriel, Carlos
Galeotti, Piero
Gális, Rudolf
Gallagher, Sarah
Galli, Daniele
Gallino, Roberto
Gangui, Alejandro
Ganguly, Rajib
GAO, Jian
Garcia, Beatriz
Garcia-Barreto, José
Garcia-Lorenzo, Maria
Gargaud, Muriel
Garmany, Katy
Gary, Dale
Gasiprong, Nipon
Gastaldello, Fabio
Gavras, Panagiotis
Gavrilov, Mikhail
Gay, Pamela
Geffert, Michael
Geller, Aaron
Genova, Françoise
George, Martin
Gerbaldi, Michele
Germany, Lisa
Geyer, Edward
Ghirlanda, Giancarlo
Ghobros, Roshdy
Giacaglia, Giorgio
Gilbank, David
Gill, Peter
Gillanders, Gerard
Gillingham, Peter
Gills, Martins
Giménez, Alvaro
Gingerich, Owen
Girard, Julien
Glass, Ian
Goicoechea, Luis
Goldes, Guillermo
Golev, Valeri
Golub, Leon
Gomboc, Andreja
Gomez, Edward
Gomez, Haley
González, Gabriela
Goodman, Alyssa
Gordon, Mark
Gorenstein, Paul
Gorgas, Garcia
Goss, W. Miller
Gouguenheim, Lucienne
Govender, Kevindran
Graham, Eric
Graham, John
Gray, Richard
Green, Anne
Green, Daniel
Green, David
Greenhouse, Matthew
Gregorio, Anna
Gregorio-Hetem, Jane
Griffin, Ian
Griffin, R. Elizabeth
Grinspoon, David
Groote, Detlef
Grundstrom, Erika
Guessoum, Nidhal
Guglielmino, Salvatore
Guinan, Edward
Gull, Theodore
Gulyaev, Sergei
Gumjudpai, Burin
GUNJI, Shuichi
Günthardt, Guillermo
Gupta, Surendra
Gurshtein, Alexander
Gurzadyan, Vahagn
Guseva, Irina
Gussmann, Ernst-August
Guziy, Sergiy
Hadrava, Petr
Hafizi, Mimoza
Hainaut, Olivier
HAMABE, Masaru
Hamilton, Douglas
HAN, Wonyong
HANDA, Toshihiro
Hanlon, Lorraine
Hannah, Iain
Haque, Shirin
Harper, Graham
Hasan, Hashima
Hasan, S. Sirajul
HASEGAWA, Ichiro
HASHIMOTO, Osamu
Hau, George
Haubold, Hans
Haupt, Hermann
Haverkorn, Marijke
Havlen, Robert
Hawkes, Robert
Hayli, Abraham
Haynes, Raymond
Haynes, Roslynn
Haywood, J.
He, Han
Hearnshaw, John
Heck, Andre

Heddle, Douglas
Hegedues, Tibor
Hegmann, Michael
Heiser, Arnold
Helfand, David
Helou, George
Hemenway, Mary
Hempel, Marc
Henry, Richard
Herpin, Fabrice
Herrmann, Dieter
Hesser, James
Heudier, Jean-Louis
Heydari-Malayeri, Mohammad
Hicks, Amalia
Hidayat, Bambang
Hillwig, Todd
HIRAI, Masanori
HIRAMATSU, Masaaki
Hjalmarsdotter, Linnea
Hobbs, George
Hockey, Thomas
Høg, Erik
Hönig, Sebastian
Hollow, Robert
Holmberg, Gustav
Homeier, Derek
Hopkins, Andrew
Horn, Martin
Horton, Anthony
Hotan, Aidan
HOU, Jinliang
Houziaux, Leo
Howard, Sethanne
Howard III, William
HSU, Rue-Ron
HU, Tiezhu
Huan, Nguyen
Hudec, Rene
Huertas-Company, Marc
Huettemeister, Susanne
Hughes, Stephen
Hunstead, Richard
HURUKAWA, Kiichirō
HWANG, Chorng-Yuan
Hysom, Edmund
Hyung, Siek
Ibadinov, Khursand
Ibrahim, Alaa
Iliev, Ilian
Ilyas, Mohammad
Ilyasov, Sabit
Impey, Christopher
Inglis, Michael
Ioannou, Zacharias
Ishida, Toshihito
ISHIZAKA, Chiharu
ITOH, Yoichi
Iwaniszewska, Cecylia
Izzard, Robert
Jafelice, Luiz
Jahnke, Knud
Jahreiss, Hartmut
Jakobsen, Peter
Jankovics, Istvan
Jasniewicz, Gerard
Jatenco-Pereira, Vera
Jauncey, David
Jayawardhana, Ray
JEON, Young Beom
JEONG, Jang-Hae
JIANG, Biwei
JIANG, Xiaoyuan
Jiménez-Vicente, Jorge
Johns, Bethany
Johnston, Helen
Johnston, Kenneth
Johnston-Hollitt, Melanie
Jones, David
Jones, Derek
Jones, Paul
Jopek, Tadeusz
Jordi, Carme
Joseph, Robert
Jubier, Xavier
Kablak, Nataliya
Kafka, Styliani (Stella)
Kahane, Claudine
Kahil, Magd
Kalemci, Emrah
Kalirai, Jason
Kammeyer, Peter
Kandel, Robert
Kantharia, Nimisha
Kapoor, Ramesh
Karas, Vladimir
Karetnikov, Valentin
Karouzos, Marios
Karttunen, Hannu
Kascheev, Rafael
Katgert-Merkelijn, J.
Kay, Laura
Kazantseva, Liliya
Keay, Colin
Keeney, Brian
Keller, Hans-Ulrich
Kellermann, Kenneth
Kembhavi, Ajit
KENTARO, Matsuda
Kepler, S.
Kerschbaum, Franz
Kervella, Pierre
Khan, J
Khanna, Ramon
Khare, Pushpa
Khodachenko, Maxim
Khosroshahi, Habib
Kiasatpour, Ahmad
Kikwaya Eluo, Jean-Baptiste
Kilambi, G.
Kim, Yoo Jea
KIM, Chun-Hwey
KIM, Kwang tae
KIM, Sang Chul
KIM, SANG HYUK
KIM, Yong Cheol
KIM, Yonggi
KIM, Young-Soo
Kinemuchi, Karen
King, David
KINUGASA, Kenzo
Kippenhahn, Rudolf
Kirilova, Daniela
Kitaeff, Vyacheslav
Kjaergaard, Per
Klapp, Jaime
Klinglesmith III, Daniel
Kneib, Jean-Paul
Knight, Matthew
Kocer, Dursun
Kochhar, Rajesh
Koeberl, Christian
Koechlin, Laurent
Kolb, Ulrich
Kolenberg, Katrien
Kolka, Indrek
Kolláth, Zoltan
Kollerstrom, Nicholas

Kołomański, Sylwester
Kolomiyets, Svitlana
Komonjinda, Siramas
KONG, Xu
Kononovich, Edvard
Korakitis, Romylos
Koribalski, Bärbel
KOSAI, Hiroki
Koschny, Detlef
Kosovichev, Alexander
Kostama, Veli-Petri
Kotulla, Ralf
Kouwenhoven, M.B.N. (Thijs)
Kovács, József
KOZAI, Yoshihide
Krajnović, Davor
Kramer, Busaba
Kreiner, Jerzy
Krisciunas, Kevin
Krishna, Gopal
Krishnan, Thiruvenkata
Kristiansen, Jostein
Kroll, Peter
Krupp, Edwin
Kryvodubskyj, Valery
KUAN, Yi-Jehng
Kuiper, Rolf
Kuiper, Thomas
Kunjaya, Chatief
Kunth, Daniel
Kutner, Marc
Kuznetsov, Eduard
Kwok, Sun
Lago, Maria
Lähteenmäki, Anne
Lai, Sebastiana
Lamontagne, Robert
Lanciano, Nicoletta
Landolt, Arlo
Lanfranchi, Gustavo
Lang, Kenneth
Lang, Mark
Langston, Glen
LAS VERGNAS, Olivier
Laufer, Diana
Launay, Françoise
Lazauskaite, Romualda
Lebofsky, Larry
Lebreton, Yveline
Lebrón, Mayra
Lebzelter, Thomas
Lecavelier des Etangs, Alain
Le Contel, Jean-Michel
Lee, Hyun-chul
Lee, William
LEE, Eun Hee
LEE, Jun
LEE, Kang Hwan
LEE, Ki-Won
LEE, Woo baik
LEE, Yong Bok
LEE, Yong Sam
Le Guet Tully, Françoise
Leibacher, John
Lequeux, James
Lerner, Michel-Pierre
Letarte, Bruno
Leto, Giuseppe
Leung, Chun
Leung, Kam
Levasseur-Regourd, Anny-Chantal
Levato, Orlando
Levenson, Nancy
Levine, Randolph
Levine, Robyn
Levy, Eugene
LI, Ji
LI, Jinzeng
LI, Min
LI, Qingkang
LI, Yong
LI, Zongwei
Liller, William
Lima Neto, Gastao
LIN, Chuang-Jia
LIN, Qing
LIN, Weipeng
Linden-Vørnle, Michael
Ling, Josefina
Lintott, Chris
Liritzis, Ioannis
Little-Marenin, Irene
LIU, Ciyuan
LIU, Sheng-Yuan
LIU, Xiaoqun
LIU, Yu
Loaring, Nicola
Locher, Kurt
Loiseau, Nora
łokas, Ewa
Lomb, Nicholas
Longair, Malcolm
Longo, Giuseppe
Lopes, Paulo
Lopes, Rosaly
Lopes de Oliveira, Raimundo
López, Alejandro
Lopez, Carlos
Lopez, Sebastian
López-Sánchez, Angel
Loubser, Ilani
Loukitcheva, Maria
Lowenthal, James
Lubowich, Donald
Luminet, Jean-Pierre
Luri, Xavier
Lutz, Barry
Ma, Chunyu
MA, Xingyuan
Maccacaro, Tommaso
MacConnell, Darrell
Machado, Marcos
Machado Folha, Daniel
Maciel, Walter
Mackie, Glen
Maddison, Ronald
Maddison, Sarah
Madjarska, Maria
Madsen, Claus
Magalhaes, Antonio Mario
Maggio, Antonio
Mahoney, Terence
Maíz Apellániz, Jesús
Majumdar, Subhabrata
Malasan, Hakim
Malbet, Fabien
Malin, David
Malkan, Matthew
Mallamaci, Claudio
Mallik, D.
Malville, J.
Mamadazimov, Mamadmuso
Mampaso, Antonio
Manchanda, R.
Manchester, Richard

Mancìni, Dario
Mandrini, Cristina
Manimanis, Vassilios
Manoharan, P.
Manset, Nadine
Marchi, Simone
Marco, Olivier
Marcy, Geoffrey
Maričić, Darije
Marleau, Francine
Marschall, Laurence
Marshalov, Dmitriy
Martin, Christopher
Martín, Eduardo
Martinet, Louis
Martinez, Peter
Martínez Bravo, Oscar
Martínez-Delgado, David
Martinez-Frias, Jesus
Martins, Carlos
Martins, Donald
Masiero, Joseph
Mason, Brian
Mason, Helen
Massey, Robert
Masters, Karen
Mathewson, Donald
Mathieu, Robert
MATSUMURA, Masafumi
Matthews, Jaymie
Mattox, John
Mauersberger, Rainer
Maury, Alain
Maza, José
Mazumdar, Anwesh
McAdam, Bruce
McAteer, R. T. James
McCammon, Dan
McConnell, David
McGimsey Jr, Ben
McKenna-Lawlor, Susan
McKinnon, David
McLean, Donald
McNally, Derek
McNeil, Stephen
Meadows, A.
Medina, Etelvina
Meech, Karen
Meidav, Meir
Melbourne, Jason
Melita, Mario
Melnick, Jorge
Melo, Claudio
Méndez, Mariano
Mendillo, Michael
Menon, T.
Merín Martín, Bruno
Merlo, David
Merrifield, Michael
Messerotti, Mauro
Metaxa, Margarita
Meyer, Michael
Mickaelian, Areg
Mickelson, Michael
Middleton, Christopher
Miesch, Mark
MIKAMI, Takao
Mikhail, Joseph
Miley, George
Milkey, Robert
Miller, Eric
Miller, Hugh
Miller, Neal
Milne, Douglas
Milone, Eugene
Minier, Vincent
Minikulov, Nasridin
Miralles, Joan-Marc
Mitton, Simon
Miyawaki, Ryosuke
MIZUNO, Takao
Modali, Sarma
Moesgaard, Kristian
Mohd, Zambri
Molenda-żakowicz, Joanna
Molinari, Emilio
Molnar, Michael
Monin, Jean-Louis
Monteiro, Mario Joao
Montes, David
Montgomery, Michele
Montmerle, Thierry
MOON, Hong-Kyu
Moore-Weiss, John
Moorhead, James
Moreels, Guy
Morrell, Nidia
Morris, Patrick
Morton, Donald
Mota, David
Mothé-Diniz, Thais
Mourão, Ana Maria
Moussas, Xenophon
Muanwong, Orrarujee
Mueller, Beatrice
Müller, Andreas
Mújica, Raul
Mukai, Koji
Mukherjee, Krishna
Mumford, George
MURAKAMI, Hiroshi
MURAYAMA, Hitoshi
Murphy, John
Nadal, Robert
Nahar, Sultana
Nair, Sunita
Najid, Nour-Eddine
NAKAMURA, Tsuko
NAKAMURA, Yasuhisa
NAKANO, Makoto
NAKASHIMA, Jun-ichi
Nakos, Theodoros
Nammahachak, Suwit
Nandi, Dibyendu
Naoz, Smadar
NARITA, Norio
Narlikar, Jayant
Navone, Hugo
Nayar, S.R.Prabhakaran
Naze, Yael
Nefedyev, Yury
Nelson, George
Nemiroff, Robert
Nesci, Roberto
Nesvadba, Nicole
Netzer, Hagai
Netzer, Nathan
Nguyễn, Lan
Nguyễn, Khanh
Nguyễn, Phuong
Nguyen-Quang, Rieu
NHA, Il Seong
Niarchos, Panagiotis
Nicolaidis, Efthymios
Nicolson, Iain
Niemczura, Ewa
NING, Xiaoyu
Ninković, Slobodan
Nitschelm, Christian
NITTA, Atsuko

Noels, Arlette
Noll, Keith
Norci, Laura
Nordström, Birgitta
Norris, Raymond
Norton, Andrew
Nota, Antonella
Nussbaumer, Harry
Nymark, Tanja
Oberst, Thomas
O'Brien, Tim
O'Byrne, John
Ocvirk, Pierre
Odell, Andrew
Ödman, Carolina
O'Donoghue, Aileen
Ogando, Ricardo
Ogłoza, Waldemar
OH, Kyu-Dong
OH, Suyeon
O'Handley, Douglas
Ohashi, Nagayoshi
Ôhashi, Yukio
OHNISHI, Kouji
Oja, Heikki
Okeke, Pius
Okpala, Kingsley
Olive, Don
Olivier, Enrico
Olowin, Ronald
Olsen, Hans
Olthof, Henk
Oluseyi, Hakeem
Önel, Hakan
Onuora, Lesley
Oozeer, Nadeem
Oproiu, Tiberiu
Orchiston, Wayne
Orellana, Rosa
Ortiz Gil, Amelia
Osborn, Wayne
Osmer, Patrick
Osório, José
Ossendrijver, Mathieu
Oswalt, Terry
O'Toole, Simon
OTSUKA, Masaaki
OWAKI, Naoaki
Özeren, Ferhat
Oezisik, Tuncay
Padmanabhan, Janardhan
Pagani, Laurent
PALACIOS, Ana
Pandey, Uma
Pang, Kevin
Pantoja, Carmen
Papathanasoglou, Dimitrios
Parenti, Susanna
Parisot, Jean-Paul
Partridge, Robert
Pasachoff, Jay
Pat-El, Igal
Pati, Ashok
Pavani, Daniela
Pavlovski, Kresimir
Pecker, Jean-Claude
Pedrosa, António
Peel, Michael
Pẽa Saint-Martin, José
Penston, Margaret
Percy, John
Pérez, Enrique
Pérez-González, Pablo
Pérez Torres, Miguel
Perozzi, Ettore
Perrin, Marshall
Peterson, Charles
Petit, Jean-Marc
Pettersen, Bjørn
Pfleiderer, Jorg
PHAM, Diep
PHAN, Dong
Picazzio, Enos
Pigatto, Luisa
Pilachowski, Catherine
Pilbratt, Göran
Pineault, Serge
Pineda de Carias, Maria
Pinigin, Gennadiy
Pintado, Olga
Pisano, Daniel
Planesas, Pere
Polcaro, V.
Politi, Romolo
Pollas, Christian
Polyakhova, Elena
Pompea, Stephen
Ponce, Gustavo
Popov, Sergey
Popović, Luka
Porceddu, Ignazio
Porras Juárez, Bertha
Postnov, Konstantin
Poulle, Emmanuel
Pović, Mirjana
Pozhalova, Zhanna
Preston, Robert
Price, Charles
Price, Daniel
Prokakis, Theodore
Proverbio, Edoardo
Puschell, Jeffery
Pustil'nik, Lev
Quamar, Jawaid
Querci, Francois
Quirós, Israel
Radeva, Veselka
Radiman, Iratio
Rafferty, Theodore
Ramadurai, Souriraja
Ramella, Massimo
Rassat, Anais
Rastegaev, Denis
Ratnatunga, Kavan
Rauch, Thomas
Rauscher, Thomas
Ravindranath, Swara
Ray, Tom
Reardon, Kevin
Reboul, Henri
Rebull, Luisa
Rector, Travis
Reid, Michael
Rekola, Rami
Rengel, Miriam
Reshetnikov, Vladimir
Reyes, Reinabelle
Reylé, Céline
Reynolds, Cormac
RHEE, Myung Hyun
Rice, John
Richer, John
Ridgway, Susan
Rijsdijk, Case
Rizzi, Luca
Roark, Terry
Robb, Russell
Roberts, Douglas
Roberts, Mallory
Roberts, Morton

Robertson, James
Robinson, Garry
Robson, Ian
Roca Cortés, Teodoro
Rocca-Volmerange, Brigitte
Rodriguez Espinosa, Jose
Rodríguez Hidalgo, Inés
Rojas, Gustavo
Rojo, Patricio
Román-Zúñiga, Carlos
Ros, Rosa
Roša, Dragan
Rosa, Michael
Rosado, Margarita
Rosa González, Daniel
Rosenzweig-Levy, Patrica
Ros Ibarra, Eduardo
Rosolowsky, Erik
Rosvick, Joanne
Roth, Miguel
Rothberg, Barry
Rovithis-Livaniou, Helen
Różańska, Agata
Rubio-Herrera, Eduardo
Rudnick, Lawrence
Ruffini, Remo
Ruggles, Clive
Ruiz, Maria Teresa
Rupprecht, Gero
Russo, Pedro
Ryde, Nils
Ryder, Stuart
Saad, Somaya
Sabra, Bassem
Sadat, Rachida
Saenz, Eduardo
Saffari, Reza
Safi-Harb, Samar
Safko, John
Sage, Leslie
Saggion, Antonio
Sahai, Raghvendra
Sahni, Varun
SAKAMOTO, Tsuyoshi
Salomé, Philippe
Samodurov, Vladimir
Sampson, Russell
Sánchez-Blázquez, Patricia
Sánchez Doreste, Néstor
Sandqvist, Aage
Sandrelli, Stefano
Sankrit, Ravi
Santillán, Alfredo
Santos-Lleó, Maria
Saraiva, Maria de Fatima
Saripalli, Lakshmi
SATO, Fumio
Sattarov, Isroil
Saucedo Morales, Julio
Sawicki, Marcin
Saxena, P.
Saygac, Ahmet
Schaefer, Bradley
Schechner, Sara
Schinckel, Antony
Schleicher, David
Schlosser, Wolfhard
Schmadel, Lutz
Schmidt, Maarten
Schmitter, Edward
Schnell, Anneliese
Schneps, Matthew
Schönherr, Gabriele
Schreiber, Roman
Schroeder, Daniel
Schuler, Simon
Schulz, R.
Scorza, Cecilia
Seaman, Rob
Seaton, Daniel
Seck, Friedrich
Sedmak, Giorgio
Seeds, Michael
Segaluvitz, Alexander
šegan, Stevo
Seggewiss, Wilhelm
Seiradakis, John
SEKIGUCHI, Kazuhiro
Serber, Alexander
Serjeant, Stephen
Sese, Rogel Mari
Shakht, Natalia
Shank, Michael
Shankland, Paul
Sharp, Nigel
Shastri, Prajval
Shaver, Peter
Shaw, James
Shearer, Andrew
Shelton, Ian
SHEN, Chun-Shan
SHIMOIKURA, Tomomi
Shingareva, Kira
Shinnaga, Hiroko
Shipman, Harry
Shore, Steven
Short, Christopher
Shostak, G.
Shukre, C.
Sigismondi, Costantino
Signore, Monique
SIHER, El Arbi
Sima, Zdislav
Simkin, Susan
Simonia, Irakli
Simons, Douglas
Simpson, Allen
Sinachopoulos, Dimitris
Singh, Harinder
Singh, Jagdev
Sinha, Krishnanand
Sivakoff, Gregory
Slater, Timothy
šlechta, Miroslav
Slee, O.
Smail, Ian
Smith, Francis
Smith, Horace
Smith, Howard
Smith, Malcolm
Smith, Michael
Smith, Niall
Smits, Derck
Snik, Frans
Soares, Domingos Savio
Sobouti, Yousef
Sobreira, Paulo
Soderblom, David
Solc, Martin
Solheim, Jan
SOMA, Mitsuru
Soonthornthum, Boonrucksar
Soriano, Bernardo
Sosa, Andrea
Souchay, Jean
Sowell, James
Squires, Gordon
Srivastava, Nandita
Stachowski, Grzegorz

Stam, Daphne
Stanga, Ruggero
Stanghellini, Carlo
Stathopoulou, Maria
Stavinschi, Magda
Steel, Duncan
Steele, John
štefl, Vladimir
Steinle, Helmut
Stencel, Robert
Stenholm, Björn
Stephenson, F.
Sterken, Christiaan
Stinebring, Daniel
Stoev, Alexey
Stonkutė, Rima
Storey, John
Storey, Michelle
Storrs, Alexander
Strafella, Francesco
Strom, Richard
Strubbe, Linda
Subramanian, K.
Sudzius, Jokubas
Sukartadiredja, Darsa
Sullivan, III, Woodruff
SUMI, Takahiro
Sun, Xiaochun
Sundin, Maria
SUNG, Hyun-Il
Surdej, Jean
Surdin, Vladimir
Svalgaard, Leif
Svestka, Jiri
Svolopoulos, Sotirios
Svoren, Jan
Swarup, Govind
Swerdlow, Noel
Szabados, Laszlo
Szabo, Robert
Szostak, Roland
Tadross, Ashraf
TAKAGI, Toshinobu
Tamazian, Vakhtang
Tammann, Gustav
Tammi, Joni
TAMURA, Shin'ichi
TANAKA, Ichi Makoto
Tancredi, Gonzalo
Tanga, Paolo
TANIKAWA, Kiyotaka
Tanzella-Nitti, Giuseppe
Taş, Günay
Taub, Liba
Tautvaisiene, Gražina
Teixeira, Paula
Teixeira, Ramachrisna
Tektunali, H.
Templeton, Matthew
ten Brummelaar, Theo
Teng, Stacy
Tepper Garcia, Thorsten
Terzian, Yervant
Tessema, Solomon
Theis, Christian
Theodossiou, Efstratios
Thomsen, Bjarne
Thorsett, Stephen
Thorstensen, John
Tichá, Jana
Tignalli, Horacio
Tiplady, Adrian
Tobin, William
Toffolatti, Luigi
Tolbert, Charles
Tomasella, Lina
Tomita, Akihiko
José Miguel Torrejón, Jose Miguel
Torres, Jesus Rodrigo
Torres-Peimbert, Silvia
TOSAKI, Tomoka
Toth, Laszlo
Touma, Jihad
Townes, Charles
Tozzi, Gian
Trakhtenbrot, Benny
Tr`ân, Ha
Tresse, Laurence
Trimble, Virginia
Trinidad, Miguel
Tripathy, Sushanta
Tritton, Keith
Trujillo, Chadwick
Trullols, I.
TSAI, An-Li
Tsvetkov, Milcho
Türler, Marc
Tugay, Anatoliy
Turcu, Vlad
Turner, Kenneth
Tzioumis, Anastasios
Úbeda, Leonardo
Ubertini, Pietro
Udaya, Shankar
UEDA, Haruhiko
Ueta, Toshiya
Ugolnikov, Oleg
Ulla Miguel, Ana
UMEMOTO, Tomofumi
Unda-Sanzana, Eduardo
Urama, Johnson
Urban, Sean
Urquhart, James
Usher, Peter
Vahia, Mayank
Väisänen, Petri
Vakoch, Douglas
Valdes Parra, Jose
Valdivielso, Luisa
Valentijn, Edwin
Valls-Gabaud, David
Valluri, Monica
Valsecchi, Giovanni
van den Heuvel, Edward
van der Kruit, Pieter
van der Werf, Paul
van Eymeren, Janine
van Gent, Robert
van Groningen, Ernst
van Loon, Jacco
van Santvoort, Jacques
van Woerden, Hugo
Varvoglis, Harry
Vass, Gheorghe
Vats, Hari
Vauclair, Sylvie
Vaughan, Alan
Vavilova, Iryna
Vázquez, Roberto
Vazquez, Ruben
Venkatakrishnan, P.
Venters, Tonia
Verdes-Montenegro, Lourdes
Verdet, Jean-Pierre
Verdoes Kleijn, Gijsbert
Verdun, Andreas
Vereshchagin, Sergej
Vergani, Daniela

Verma, Aprajita
Vicente, Raimundo
Vidal-Madjar, Alfred
Videira, Antonio
Viegas, Sueli
Vienne, Alain
Vierdayanti, Kiki
Vilinga, Jaime
Vilks, Ilgonis
Villar Martin, Montserrat
Viñuales Gavín, Ederlinda
Viollier, Raoul
Vlahakis, Nektarios
Voelzke, Marcos
Voges, Wolfgang
Voigt, Hans
Voit, Gerard
Volyanska, Margaryta
Vujnovic, Vladis
Wadadekar, Yogesh
Wagner, Robert
Wainscoat, Richard
Walker, Constance
Walsh, Robert
Walsh, Wilfred
Walter, Frederick
Wambsganß, Joachim
WANG, Chen
WANG, Guangchao
WANG, Na
WANG, Rongbin
WANG, Shouguan
Ward, Richard
Warner, Brian
Watanabe, Ken
WATANABE, Junichi
WATARAI, Kenya
Watson, Frederick
Webster, Rachel
Webster, Zodiac
WEI, Erhu
Weisberg, Joel
Weiss, Werner
Welch, Douglas
Welsh, William
Wendt, Harry
West, Michael
West, Richard
Westera, Pieter
Whelan, Emma
Whitaker, Ewen
White, Graeme
White, Richard
Whitelock, Patricia
Whiteoak, John
Whiting, Alan
Wielen, Roland
Wilkins, George
Williamon, Richard
Williams, Robert
Williams, Thomas
Willmore, A.
Willstrop, Roderick
Wilson, Curtis
Wilson, Gillian
Winkler, Christoph
Winkler, Paul
Winter, Lisa
Wittenmyer, Robert
Wnuk, Edwin
Wolfschmidt, Gudrun
Wolter, Anna
Woudt, Patrick
Wozniak, Hervé
Wright, Alan
Wright, Nicholas
Wrixon, Gerard
WU, Jiun-Huei
WU, Zhen-Yu
Wulandari, Hesti
XIE, Xianchun
XIONG, Jianning
Yahil, Amos
Yair, Yoav
YAMAOKA, Hitoshi
YAN, Yihua
Yanamandra-Fisher, Padma
YANG, Hong-Jin
YANO, Hajime
Yatskiv, Yaroslav
Yau, Kevin
YE, Shuhua
Yegorova, Irina
Yeomans, Donald
YIM, Hong Suh
YOSHIDA, Hiroshi
YOSHIOKA, Kazuo
YUMIKO, Oasa
YUMIN, Wang
Yun, Joao
YUTAKA, Shiratori
Zadnik, Marjan
Zagars, Juris
Zaggia, Simone
Zakirov, Mamnum
Zamorano, Jaime
Zanini, Valeria
Zarnecki, John
Zavagno, Annie
Zealey, William
Zeiler, Michael
Zeilik, Michael
Zeilinger, Werner
Zhang, Jie
ZHANG, Huawei
ZHANG, JiangShui
ZHANG, Shouzhong
ZHANG, Yang
ZHANG, Yong
ZHANG, You-Hong
ZHAO, Jun Liang
ZHENG, Xiaonian
ZHOU, Yonghong
ZHU, Jin
Zirm, Andrew
Zitelli, Valentina
Zsoldos, Endre

Division D High Energy Phenomena and Fundamental Physics

President: Diana Mary Worrall
Vice-President: Felix Albert Aharonian

Steering Committee Members:

Christine Jones
Chryssa Kouveliotou
Tadayuki TAKAHASHI
Diana Mary Worrall
Felix Albert Aharonian
Xavier Barcons
John G. Kirk
Anna Wolter
Elena Pian

Members:

Abramowicz, Marek
Acharya, Bannanje
Acton, Loren
Adams, Thomas
Aghaee, Alireza
Agrawal, P.
Aguiar, Odylio
Aharonian, Felix
Ahluwalia, Harjit
Ahmad, Imad
Alcaniz, Jailson
Alexander, Joseph
Allington-Smith, Jeremy
Almleaky, Yasseen
Amati, Lorenzo
AN, TAO
Andersen, Bo Nyborg
Anderson, Kurt
Andruchow, Ileana
Anton, Sonia
Antonelli, Lucio Angelo
Antoniou, Vallia
Antonuccio-Delogu, Vincenzo
Apparao, K.
ARAFUNE, Jiro
Arbutina, Bojan
Arefiev, Vadim
Arnaud, Monique
Arnett, W.
Arnould, Marcel
Arons, Jonathan
ASANO, Katsuaki
Aschenbach, Bernd
Asseo, Estelle
Asvarov, Abdul
Audard, Marc
Audley, Michael
Audouze, Jean
Augereau, Jean-Charles
Auriemma, Giulio
AWAKI, Hisamitsu
Axelsson, Magnus
Ayres, Thomas
Azcarate, Diana
Baan, Willem
BAEK, Chang Hyun
BAI, Jinming
Bailyn, Charles
Bajkova, Anisa
Baki, Paul
Balikhin, Michael
Baliunas, Sallie
Balman, Solen
Bamba, Aya
Barcons, Xavier
Bardeen, James
Baring, Matthew
Barkhouse, Wayne
Barrantes, Marco
Barret, Didier
Barrow, John
Barstow, Martin
Basa, Stephane
Basart, John
Baskill, Darren
Baym, Gordon
Bazzano, Angela
Becker, Robert
Becker, Werner
Beckmann, Volker
Begelman, Mitchell
Behar, Ehud
Beiersdorfer, Peter
Bekenstein, Jacob
Beklen, Elif
Bell Burnell, Jocelyn
Belloni, Tomaso
Bender, Peter
Benedict, George
Benford, Gregory
Bennett, Charles
Bennett, Kevin
Benvenuto, Omar
Bergeron, Jacqueline
Bergström, Lars
Berkhuijsen, Elly
Bernardini, Federico
Berta, Stefano
Beskin, Gregory
Beskin, Vasily
Bhattacharjee, Pijushpani
Bhattacharya, Dipankar
Bhattacharyya, Sudip
Bhavsar, Suketu
Bianchi, Luciana
Bianchi, Stefano
Bicak, Jiri
Bicknell, Geoffrey
Biermann, Peter
Bigdeli, Mohsen
Bignall, Hayley
Bignami, Giovanni
Bingham, Robert
Birkinshaw, Mark

Birkmann, Stephan
Bisikalo, Dmitrij
Bjælde, Ole
Björnsson, Gunnlaugur
Black, John
Blair, David
Blamont, Jacques-Emile
Blandford, Roger
Bleeker, Johan
Bless, Robert
Blinnikov, Sergey
Bloemen, Hans
Blondin, John
Bludman, Sidney
Bocchino, Fabrizio
Bodaghee, Arash
Bodo, Gianluigi
Boër, Michel
Boggess, Albert
Boggess, Nancy
Bohlin, Ralph
Boksenberg, Alec
Boller, Thomas
Bombaci, Ignazio
Bonaccini, Domenico
Bonanno, Alfio
Bonazzola, Silvano
Bonnet, Roger
Bonnet-Bidaud, Jean-Marc
Bonometto, Silvio
Borka Jovanović, Vesna
Boroson, Bram
Borozdin, Konstantin
Bosch-Ramon, Valenti
Bouchet, François
Bougeret, Jean-Louis
Bower, Geoffrey
Bowyer, C.
Bradley, Arthur
Bradt, Hale
Braga, João
Braithwaite, Jonathan
Branchesi, Marica
Brandt, John
Brandt, Soeren
Brandt, William
Branduardi-Raymont, Graziella
Brecher, Kenneth
Bregman, Joel
Brenneman, Laura
Breslin, Ann
Breton, Rene
Brinkman, Bert
Bromberg, Omer
Brosch, Noah
Brown, Alexander
Browne, Ian
Bruhweiler, Frederick
Bruner, Marilyn
Brunetti, Gianfranco
Bulik, Tomasz
Bumba, Vaclav
Bunner, Alan
Buote, David
Burderi, Luciano
Burenin, Rodion
Burger, Marijke
Burikham, Piyabut
Burke, Bernard
Burrows, Adam
Burrows, David
Bursa, Michal
Butler, Christopher
Butterworth, Paul
Bykov, Andrei
Caccianiga, Alessandro
CAI, Michael
CAI, Mingsheng
Camenzind, Max
Campana, Riccardo
Campbell, Murray
Campusano, Luis
Canizares, Claude
Cannon, Kipp
CAO, Li
Capelato, Hugo
Cappi, Massimo
Caputi, Karina
Caraveo, Patrizia
Cardini, Daniela
Carlson, Per
Carpenter, Kenneth
Carramiñana, Alberto
Carrasco, Bertha
Carroll, P.
Casandjian, Jean-Marc
Cash Jr, Webster
Cassano, Rossella
Cassé, Michel
Cassinelli, Joseph
Castro-Tirado, Alberto
Catura, Richard
Cavaliere, Alfonso
Celotti, Anna Lisa
Cenko, Stephen
Cesarsky, Catherine
Chakrabarti, Sandip
Chakraborty, Deo
Chamel, Nicolas
Champion, David
CHANG, Heon-Young
CHANG, Hsiang-Kuang
Channok, Chanruangrit
Chapman, Robert
Chapman, Sandra
Charles, Philip
Chartas, George
Chaty, Sylvain
Chechetkin, Valerij
Chelliah Subramonian, Stalin
Chelouche, Doron
CHEN, Lin-wen
CHEN, Pisin
CHEN, Yang
Chenevez, Jérôme
CHENG, Kwongsang
Chernyakova, Maria
Cheung, Cynthia
Chevalier, Roger
Chian, Abraham
Chiappetti, Lucio
CHIBA, Takeshi
CHIKAWA, Michiyuki
Chitre, Shashikumar
Chochol, Drahomir
CHOE, Gwangson
CHOI, Chul-Sung
CHOU, Yi
Choudhury, Tirthankar
Christensen, Per
Chu, You-Hua
Chupp, Edward
Churazov, Eugene
Ciotti, Luca
Clark, David
Clark, George
Clark, Thomas
Clay, Roger

Clowe, Douglas
Coe, Malcolm
Cohen, David
Collin, Suzy
Comastri, Andrea
Combes, Françoise
Comins, Neil
Condon, James
Contopoulos, Ioannis
Corbel, Stéphane
Corbet, Robin
Corbett, Ian
Corcoran, Michael
Cordova, France
Cornelisse, Remon
Córsico, Alejandro
Costa, Enrico
Costantini, Elisa
Courtes, Georges
Courvoisier, Thierry
Cowie, Lennox
Cowsik, Ramanath
Crannell, Carol
Crawford, Fronefield
Crenshaw, Daniel
Cristiani, Stefano
Crocker, Roland
Cropper, Mark
Croston, Judith
Croton, Darren
Cruise, Adrian
Cuadra, Jorge
Cui, Wei
Culhane, John
Cunniffe, John
Curir, Anna
Cusumano, Giancarlo
Czerny, Bożena
da Costa, António Armando
da Costa, Jose
Dadhich, Naresh
Dagkesamanskii, Rustam
DAI, Zigao
Daigne, Frederic
Dalla Bontà, Elena
D'Amico, Flavio
D'Ammando, Filippo
Danner, Rolf
Darriulat, Pierre
da Silveira, Enio
Davidson, William
Davis, Michael
Davis, Robert
Dawson, Bruce
De Becker, Michaël
Decourchelle, Anne
de Felice, Fernando
de Gouveia Dal Pino, Elisabete
de Graauw, Mattheus
Deiss, Bruno
de Jager, Cornelis
Della Ceca, Roberto
Del Olmo, Ascension
Del Santo, Melania
Del Zanna, Luca
de Martino, Domitilla
Dempsey, Robert
den Herder, Jan-Willem
DeNisco, Kenneth
Dennerl, Konrad
Dennis, Brian
Dermer, Charles
de Rop, Yves
De Rossi, María
Deshpande, Avinash
de Ugarte Postigo, Antonio
Diaz Trigo, Maria
Dickman, Robert
Di Cocco, Guido
Diehl, Roland
Dietrich, Jörg
Digel, Seth
Disney, Michael
Dokuchaev, Vyacheslav
Dolan, Joseph
Domingo, Vicente
Dominguez, Inma
Dominis Prester, Dijana
Donea, Alina
Dong, Xiao-Bo
DOTANI, Tadayasu
DOU, Jiangpei
Dovciak, Michal
Downes, Turlough
Downs, George
Drake, Frank
Drake, Jeremy
Drake, Stephen
Drury, Luke
Dryer, Murray
Dubner, Gloria
Dubus, Guillaume
Duorah, Hira
Dupree, Andrea
Durouchoux, Philippe
Duthie, Joseph
Easther, Richard
Edelson, Rick
Edwards, Paul
Edwards, Philip
Ehle, Matthias
Eichler, David
Eilek, Jean
Elgarøy, Øystein
Elizalde, Emilio
El Raey, Mohamed
Elvis, Martin
Elyiv, Andrii
Emanuele, Alessandro
Engelbrecht, Chris
Enßlin, Torsten
ESAMDIN, Ali
ESIMBEK, Jarken
Ettori, Stefano
Eungwanichayapant, Anant
Evans, Daniel
Evans, W.
Eze, Romanus
Fabian, Andrew
Fabiani, Sergio
Fabricant, Daniel
Fahey, Richard
Falceta-Goncalves, Diego
Falize, Emeric
FAN, Junhui
Faraggiana, Rosanna
Farrell, Sean
Fatkhullin, Timur
Faundez-Abans, Max
Faure, Cécile
Fazio, Giovanni
Fedorova, Elena
Feldman, Paul
Fender, Robert
Fendt, Christian
Ferland, Gary
Ferrari, Attilio
Fichtel, Carl

Field, George
Filipov, Lachezar
Filipovic, Miroslav
Finn, Lee
Fisher, Philip
Fishman, Gerald
Florido, Estrella
Foing, Bernard
Fomin, Valery
Fonseca Gonzalez, Maria
Forman, William
Foschini, Luigi
Frail, Dale
Franceschini, Alberto
Frandsen, Soeren
Frank, Juhan
Fransson, Claes
Fraschetti, Federico
Fredga, Kerstin
Friedlander, Michael
Fruchter, Andrew
Fruscione, Antonella
FUJIMOTO, Shin-ichiro
FUJITA, Mitsutaka
FUJITA, Yutaka
FUKUE, Jun
Furniss, Ian
Füzfa, Andre
Fyfe, Duncan
Gabriel, Alan
Gaensler, Bryan
Gai, Mario
Gaisser, Thomas
Galeotti, Piero
Gallagher, Sarah
Galloway, Duncan
Gal-Yam, Avishay
Gammie, Charles
GAN, Weiqun
Gangadhara, R.T.
GAO, Yu
Garcia, Beatriz
Garmire, Gordon
Gaskell, C.
Gastaldello, Fabio
Gathier, Roel
Gedalin, Michael
Gehrels, Neil
Gendre, Bruce
Genzel, Reinhard
Georgantopoulos, Ioannis
Geppert, Ulrich
Gezari, Daniel
Gezari, Suvi
Ghia, Piera Luisa
Ghirlanda, Giancarlo
Ghisellini, Gabriele
Giacconi, Riccardo
Giampapa, Mark
Gilfanov, Marat
Gillanders, Gerard
Gilra, Daya
Gioia, Isabella
Giroletti, Marcello
Gitti, Myriam
Glaser, Harold
Glasner, Shimon
Goicoechea, Luis
Goldsmith, Donald
Goldwurm, Andrea
Golub, Leon
Gomboc, Andreja
Gomez, Haley
Gómez de Castro, Ana
Gómez Fernández, Jose
GONG, Biping
Gonzales'a, Walter
González, Gabriela
González de Buitrago, Jesús
Gordon, Chris
Gorenstein, Paul
Gotthelf, Eric
Götz, Diego
Graffagnino, Vito
Grebenev, Sergei
Greenhill, John
Greenhill, Lincoln
Gregorio, Anna
Grenier, Isabelle
Grewing, Michael
Greyber, Howard
Griffiths, Richard
Grindlay, Jonathan
Grosso, Nicolas
Grygar, Jiri
GU, Minfeng
Gudehus, Donald
Guessoum, Nidhal
Gull, Theodore
Gumjudpai, Burin
Gün, Gulnur
GUNJI, Shuichi
Gunn, James
Gupta, Yashwant
Gurvits, Leonid
Gutiérrez, Carlos
Guziy, Sergiy
Gwinn, Carl
Hadrava, Petr
Haehnelt, Martin
Haerendel, Gerhard
Hakkila, Jon
Halevin, Alexandros
Hallam, Kenneth
Hambaryan, Valeri
Hameury, Jean-Marie
Hanasz, Michał
Hanlon, Lorraine
Hanna, David
Hannah, Iain
Hannikainen, Diana
HARA, Tetsuya
Hardcastle, Martin
Harms, Richard
Harris, Daniel
Hartmann, Dieter
Harvey, Christopher
Harvey, Paul
Harwit, Martin
Hasan, Hashima
Hasinger, Günther
HATSUKADE, Isamu
Hau, George
Haubold, Hans
Haugbølle, Troels
Hauser, Michael
Hawkes, Robert
Hawking, Stephen
Hawley, John
HAYAMA, Kazuhiro
Haymes, Robert
HAZUMI, Masashi
Heckathorn, Harry
Heger, Alexander
Hein, Righini
Heinz, Sebastian
Heise, John
Helfand, David
Helmken, Henry

Helou, George
Hempel, Marc
Heng, Kevin
Henoux, Jean-Claude
Henriksen, Richard
Henry, Richard
Hermsen, Willem
Hernández, Xavier
Hernanz, Margarita
Hertz, Paul
Hervik, Sigbjorn
Heske, Astrid
Hessels, Jason
Heyl, Jeremy
Hicks, Amalia
Hill, Adam
Hillebrandt, Wolfgang
Hiriart, David
Hjalmarsdotter, Linnea
Hnatyk, Bohdan
Ho, Wynn
Hoffman, Jeffrey
Hogan, Craig
Holberg, Jay
Holloway, Nigel
Holt, Stephen
Holz, Daniel
Hora, Joseph
Hörandel, Jörg
Horiuchi, Shinji
Horns, Dieter
Hornschemeier, Ann
Hornstrup, Allan
Houde, Martin
Houziaux, Leo
Howarth, Ian
Hoyng, Peter
HSU, Rue-Ron
HU, Wenrui
Huang, Jiasheng
HUANG, YongFeng
Huber, Martin
Hudec, Rene
Huenemoerder, David
Hughes, John
Hulth, Per
Humphries, Colin
Hurley, Kevin
Hutchings, John
Hütsi, Gert
HWANG, Chorng-Yuan
Ibadov, Subhon
Ibrahim, Alaa
ICHIMARU, Setsuo
Ikhsanov, Nazar
Illarionov, Andrei
Imamura, James
Imhoff, Catherine
INOUE, Hajime
INOUE, Makoto
in't Zand, Johannes
IOKA, Kunihito
Ipser, James
ISHIDA, Manabu
Isliker, Heinz
Israel, Werner
ITO, Kensai
ITOH, Naoki
ITOH, Masayuki
Ivanova, Natalia
Jackson, John
Jacob, Uri
Jaffe, Walter
Jakobsson, Pall
Jamar, Claude
Jamrozy, Marek
Janka, Hans
Jaranowski, Piotr
Jenkins, Edward
Jetzer, Philippe
JI, Li
Jin, Liping
Johns, Bethany
Jokipii, Jack
Jones, Thomas
Jonker, Peter
Jordan, Carole
Jordan, Stuart
Joss, Paul
Kaastra, Jelle
Kafatos, Menas
Kahil, Magd
Kalemci, Emrah
Kalomeni, Belinda
KAMENO, Seiji
Kaminker, Alexander
Kanbach, Gottfried
KANEDA, Hidehiro
KANG, Hyesung
Kaper, Lex
Kapoor, Ramesh
Karakas, Amanda
Karami, Kayoomars
Karas, Vladimir
Karpov, Sergey
Kaspi, Victoria
Kassim, Namir
Kasturirangan, K.
Katarzyński, Krzysztof
KATO, Yoshiaki
KATO, Tsunehiko
Katsova, Maria
Katz, Jonathan
KAWABATA, Koji
KAWAI, Nobuyuki
KAWAKATU, Nozomu
Kelemen, Janos
Kellermann, Kenneth
Kellogg, Edwin
Kelly, Brandon
Kembhavi, Ajit
KENJI, Nakamura
Keppens, Rony
Kessler, Martin
Khanna, Ramon
Khumlemlert, Thiranee
Killeen, Neil
KIM, Kwang tae
KIM, Minsun
KIM, Yonggi
Kimble, Randy
KINUGASA, Kenzo
Kirilova, Daniela
Kirk, John
Kitaeff, Vyacheslav
KITAMOTO, Shunji
KIYOSHI, Hayashida
Klinkhamer, Frans
Klioner, Sergei
Klose, Sylvio
Knapp, Johannes
Kneib, Jean-Paul
KO, Chung-Ming
Kobayashi, Shiho
Kocharovskij, Vladimir
Kocharovsky, Vitaly
Koch-Miramond, Lydie
KODAMA, Hideo
Kohmura, Takayoshi
KOIDE, Shinji

KOJIMA, Yasufumi
Kokubun, Motohide
Kolb, Edward
Kolb, Ulrich
Konar, Sushan
KONDO, Yoji
KONDO, Masaaki
Kong, Albert
Konigl, Arieh
Kontar, Eduard
Koopmans, Leon
Kopeikin, Sergei
KOSHIBA, Masatoshi
KOSUGI, George
Koupelis, Theo
Kouveliotou, Chryssa
Kovačević, Jelena
Kovalev, Yuri
Kovář, Jiří
KOYAMA, Katsuji
Kozłowski, Maciej
Kozma, Cecilia
Kraft, Ralph
Kramer, Michael
Kretschmar, Peter
Kreykenbohm, Ingo
Kriss, Gerard
Kristiansson, Krister
Krittinatham, Watcharawuth
Kryvdyk, Volodymyr
Kubát, Jiri
Kudryavtseva, Nadezhda (Nadia)
Kuin, Paul
Kuiper, Lucien
Kulhánek, Petr
Kulsrud, Russell
KUMAGAI, Shiomi
Kuncic, Zdenka
Kundt, Wolfgang
KUNIEDA, Hideyo
Kunz, Martin
Kurt, Vladimir
Kurtanidze, Omar
KUSUNOSE, Masaaki
Labeyrie, Antoine
La Franca, Fabio
Lagache, Guilaine
Lähteenmäki, Anne
Lal, Dharam
Lamb, Frederick
Lamb, Susan
Lamb Jr, Donald
Lamers, Henny
Lampton, Michael
La Mura, Giovanni
Landolt, Arlo
Lang, Kenneth
Lang, Mark
Langston, Glen
Lapenta, Giovanni
Lapi, Andrea
Lapington, Jon
Larsson, Stefan
Lasher, Gordon
Lasota-Hirszowicz, Jean-Pierre
Lattimer, James
Laurin, Denis
Lea, Susan
Leahy, Denis
Leckrone, David
Lee, William
LEE, Sang-Sung
LEE, Wo-Lung
Leighly, Karen
Lemaire, Philippe
Le Poncin-Lafitte, Christophe
Levenson, Nancy
Levin, Yuri
Levine, Robyn
Levinson, Amir
Lewin, Walter
Li, Bo
LI, Li-Xin
LI, Min
LI, Qingkang
LI, Tipei
LI, Xiangdong
LI, Yuanjie
LI, Zhongyuan
LI, Zongwei
Liang, Edison
LIN, Xuan-bin
Lingenfelter, Richard
Linsky, Jeffrey
LipunOv, Vladimir
Lister, Matthew
Litvak, Marvin
Liu, Bifang
LIU, Fukun
LIU, Guoqing
LIU, Jifeng
LIU, Siming
LIU, Xiang
Loaring, Nicola
Lobanov, Andrei
Lobel, Alex
Lochner, James
Long, Knox
Longair, Malcolm
Lopes, Paulo
Lopes de Oliveira, Raimundo
Lopez, Sebastian
LOU, Yu-Qing
Loubser, Ilani
Lovelace, Richard
LU, Fangjun
LU, Jufu
LU, Ye
LU, Youjun
Lucchesi, David
Lüst, Reimar
Luminet, Jean-Pierre
Lundquist, Charles
Luo, Qinghuan
Luque-Escamilla, Pedro
Lutovinov, Alexander
Lynden-Bell, Donald
Lyubarsky, Yury
MA, YuQian
Maccacaro, Tommaso
Maccarone, Maria Concetta
Maccarone, Thomas
Macchetto, Ferdinando
MACHIDA, Mami
Maciesiak, Krzysztof
Mackie, Glen
Macquart, Jean-Pierre
Madsen, Jes
MAEDA, Keiichi
MAEDA, Kei-ichi
Maggio, Antonio
Maguire, Kate
Mainieri, Vincenzo
Majumdar, Subhabrata
Makarov, Valeri

MAKINO, Fumiyoshi
MAKISHIMA, Kazuo
Malesani, Daniele
Malitson, Harriet
Malkan, Matthew
Malov, Igor
Manara, Alessandro
Manchanda, R.
Manchester, Richard
Mandolesi, Nazzareno
Mangalam, Arun
Mangano, Vanessa
Maran, Stephen
Marar, T.
Margon, Bruce
Maričić, Darije
Marino, Antonietta
Markoff, Sera
Marov, Mikhail
Marranghello, Guilherme
Marscher, Alan
Martin, Inácio
Martínez Bravo, Oscar
Martinis, Mladen
Martins, Carlos
Marziani, Paola
MASAI, Kuniaki
Masnou, Jean-Louis
Mason, Glenn
Mason, Keith
Massaro, Francesco
Mather, John
Mathews, William
MATSUMOTO, Hironori
MATSUMOTO, Ryoji
MATSUOKA, Masaru
MATSUSHITA, Kyoko
Matt, Giorgio
Mattox, John
Matz, Steven
Mazurek, Thaddeus
McBreen, Brian
McBride, Vanessa
McCammon, Dan
McClintock, Jeffrey
McCluskey Jr, George
McCray, Richard
McMahon, Richard
McWhirter, R.
Medina, Jose
Medina Tanco, Gustavo
Meier, David
Meiksin, A.
Meisenheimer, Klaus
Melatos, Andrew
Melia, Fulvio
Melikidze, Giorgi
Mellier, Yannick
Melnick, Gary
Melnyk, Olga
Melrose, Donald
Méndez, Mariano
Mereghetti, Sandro
Merlo, David
Mestel, Leon
Meszaros, Peter
Métris, Gilles
Meyer, Friedrich
Meyer, Jean-Paul
Micela, Giuseppina
Michel, F.
Miller, Eric
Miller, Guy
Miller, Hugh
Miller, John
Miller, Michael
Mineo, Teresa
Minikulov, Nasridin
Miranda, Oswaldo
Miroshnichenko, Alla
Mishra, Kameshwar
Mitra, Abhas
MITSUDA, Kazuhisa
Miyaji, Takamitsu
MIYAJI, Shigeki
Miyata, Emi
MIZUMOTO, Yoshihiko
MIZUNO, Yosuke
MIZUTANI, Kohei
Moderski, Rafał
Modisette, Jerry
Moffat, Anthony
Molla, Mercedes
Monet, David
Montmerle, Thierry
Moodley, Kavilan
MOON, Shin-Haeng
Moorhead, James
Moos, Henry
Morgan, Thomas
MORI, Koji
MORI, Masaki
MORI, Masao
MOROKUMA, Tomoki
Morsony, Brian
Morton, Donald
Mota, David
Motch, Christian
MOTIZUKI, Yuko
Mouchet, Martine
Mourão, Ana Maria
Moussas, Xenophon
Mucciarelli, Paola
Mukhopadhyay, Banibrata
Mulchaey, John
MURAKAMI, Hiroshi
MURAKAMI, Hiroshi
MURAKAMI, Toshio
MURAYAMA, Hitoshi
Murdock, Thomas
Murthy, Jayant
Mushotzky, Richard
NAGATAKI, Shigehiro
Nagendra, K.
NAITO, Tsuguya
Nakar, Ehud
NAKASHIMA, Jun-ichi
NAKAYAMA, Kunji
Naoz, Smadar
Naselsky, Pavel
Neff, Susan
Nelemans, Gijs
Nemiroff, Robert
Neslusan, Lubos
Ness, Norman
Netzer, Hagai
Netzer, Nathan
Neuhaeuser, Ralph
Neupert, Werner
Neustroev, Vitaly
Nicastro, Luciano
Nichols, Joy
Nicollier, Claude
Nielsen, Krister
Nikołajuk, Marek
Nilsson, Kari
Nishikawa, Ken-Ichi
NISHIMURA, Osamu
NISHIMURA, Jun
NITTA, Atsuko

NITTA, Shin-ya
Nityananda, Rajaram
NOMOTO, Ken'ichi
Norci, Laura
Nordh, Lennart
Norman, Colin
Noyes, Robert
Nulsen, Paul
Nymark, Tanja
Obregón Díaz, Octavio
O'Brien, Paul
O'Brien, Tim
O'Connell, Robert
O'Dell, Stephen
OGAWARA, Yoshiaki
OKAMOTO, Isao
Okeke, Pius
OKUDA, Haruyuki
OKUDA, Toru
Olthof, Henk
O'Mongain, Eon
Onken, Christopher
OOHARA, Ken-ichi
Oozeer, Nadeem
Orellana, Mariana
Orford, Keith
Orio, Marina
Orlandini, Mauro
Orlando, Salvatore
Osborne, Julian
Oskinova, Lidia
Osten, Rachel
Ostorero, Luisa
Ostriker, Jeremiah
Ostrowski, Michał
O'Sullivan, Denis
Ott, Juergen
Owen, Tobias
Owers, Matthew
Owocki, Stanley
OZAKI, Masanobu
Özel, Mehmet
Pacholczyk, Andrzej
Paciesas, William
Padmanabhan, Thanu
Padovani, Paolo
Pagano, Isabella
Page, Clive
Page, Mathew
PAK, Soojong
Paltani, Stéphane
Palumbo, Giorgio
Panagia, Nino
Pandey, Uma
Panessa, Francesca
Papadakis, Iossif
Papaelias, Philip
Papitto, Alessandro
Paragi, Zsolt
Paredes Poy, Josep
Pareschi, Giovanni
PARK, Myeong Gu
Parker, Eugene
Parkinson, William
Patten, Brian
Paul, Biswajit
Paul, Jacques
Pavlidou, Vasiliki
Pavlov, George
Peacock, Anthony
Pearce, Mark
Pearson, Kevin
Pekünlü, E.
Pellegrini, Silvia
Pellizza, Leonardo
PENG, Qingyu
PENG, Qiuhe
Perets, Hagai
Perez, Mario
Pérez-González, Pablo
Perola, Giuseppe
Perry, Peter
Peters, Geraldine
Peterson, Bradley
Peterson, Bruce
Peterson, Laurence
Pethick, Christopher
Petitjean, Patrick
Petkaki, Panagiota
Petre, Robert
Petro, Larry
Petrosian, Vahe
Petrova, Svetlana
PHAM, Diep
Phillips, Kenneth
Pian, Elena
Pierre, Marguerite
Pineault, Serge
Pinkau, K.
Pinto, Philip
Pipher, Judith
Piran, Tsvi
Piro, Luigi
Pirronello, Valerio
Pitkin, Matthew
Polidan, Ronald
Polletta, Maria del Carmen
Ponman, Trevor
Pooley, David
Popov, Sergey
Porquet, Delphine
Postnov, Konstantin
Pottschmidt, Katja
Pounds, Kenneth
Poutanen, Juri
Pozanenko, Alexei
Prasanna, A.
Pravdo, Steven
Preuss, Eugen
Produit, Nicolas
Protheroe, Raymond
Prouza, Michael
Prusti, Timo
Pshirkov, Maxim
Pustil'nik, Lev
Puy, Denis
Pye, John
QIU, Yulei
QU, Jinlu
QU, Qinyue
Quintana, Hernan
Quirós, Israel
Raiteri, Claudia
Ramadurai, Souriraja
Ramírez, Jose
Ranalli, Piero
Rankin, Joanna
Rao, Arikkala
Rao, Ramachandra
Rasio, Frederic
Rasmussen, Ib
Rasmussen, Jesper
Rastegaev, Denis
Rauscher, Thomas
Rauw, Gregor
Ray, Alak
Ray, Paul
Raychaudhury, Somak
Raymond, John
Rea, Nanda

Reale, Fabio
Rees, Martin
Reeves, Hubert
Regev, Oded
Reichert, Gail
Reig, Pablo
Reimer, Olaf
Reiprich, Thomas
Reisenegger, Andreas
Reitze, David
Rengarajan, Thinniam
Rephaeli, Yoel
Revnivtsev, Mikhail
Rhoads, James
Riazi, Nematollah
Ribó, Marc
Ricker, Paul
Rijsdijk, Case
Riles, Keith
Rioja, Maria
Risaliti, Guido
Robba, Natale
Roberts, Mallory
Roberts, Timothy
Robinson, Andrew
Robinson, Edward
Rödiger, Elke
Roman, Nancy
Romano, Patrizia
Romero, Gustavo
Roming, Peter
Rosado, Margarita
Rosendhal, Jeffrey
Rosner, Robert
Rosquist, Kjell
Rossi, Elena
Rovero, Adrián
Rowell, Gavin
Roxburgh, Ian
Różańska, Agata
Rubenstein, Eric
Rubino-Martin, Jose Alberto
Rubio-Herrera, Eduardo
Ruder, Hanns
Rudnick, Lawrence
Ruffert, Maximilian
Ruffini, Remo
Ruffolo, David
Rushton, Anthony
Russell, Alexander
Ruszkowski, Mateusz
Rutledge, Robert
Saar, Enn
Sabau-Graziati, Lola
Sabra, Bassem
Saffari, Reza
Safi-Harb, Samar
Sagdeev, Roald
Saggion, Antonio
Sahai, Raghvendra
Sahlén, Martin
Sahni, Varun
Saiz, Alejandro
Sakano, Masaaki
SAKAO, Taro
Sakelliou, Irini
Salvati, Marco
Samimi, Jalal
Sanchez, Norma
Sanders, Gary
Sanders III, Wilton
Santos, Nilton
Santos-Lleó, Maria
Sarazin, Craig
Sari, Re'em
Saripalli, Lakshmi
Sartori, Leo
SASAKI, Misao
Saslaw, William
SATO, Katsuhiko
Savage, Blair
Savedoff, Malcolm
Savin, Daniel
Savolainen, Tuomas
Sazonov, Sergey
Sbarufatti, Boris
Scargle, Jeffrey
Schaefer, Gerhard
Schaffner-Bielich, Jurgen
Schartel, Norbert
Schatten, Kenneth
Schilizzi, Richard
Schmidt, Robert
Schmitt, Juergen
Schnopper, Herbert
Schönherr, Gabriele
Schramm, Thomas
Schreier, Ethan
Schulz, Norbert
Schure, Klara
Schwartz, Daniel
Schwartz, Steven
Schwehm, Gerhard
Schwope, Axel
Sciortino, Salvatore
Scott, John
Segaluvitz, Alexander
Seielstad, George
Selvelli, Pierluigi
Semerák, Oldrich
SEON, Kwang il
Sequeiros, Juan
Serber, Alexander
Setti, Giancarlo
Severgnini, Paola
Seward, Frederick
Shaham, Jacob
Shahbaz, Tariq
Shakhov, Boris
Shakura, Nikolaj
Shalchi, Andreas
Sharma, Prateek
Shaver, Peter
Shaviv, Giora
Shearer, Andrew
Sheikh, Suneel
SHEN, Zhiqiang
SHIBAI, Hiroshi
Shibanov, Yuri
SHIBATA, Masaru
SHIBAZAKI, Noriaki
Shields, Gregory
SHIGEYAMA, Toshikazu
SHIMURA, Toshiya
SHIN, Watanabe
SHIRASAKI, Yuji
Shivanandan, Kandiah
Shoemaker, David
Shukre, C.
Shustov, Boris
Signore, Monique
Sikora, Marek
Silvestro, Giovanni
Simic, Sasa
Simon, Paul
Simon, Vojtech
Sims, Mark
Simunac, Kristin
Singh, Kulinder Pal

Sivakoff, Gregory
Skilling, John
Skinner, Gerald
Skinner, Stephen
Skjæraasen, Olaf
Slane, Patrick
Slaný, Petr
Sleath, John
Smale, Alan
Smida, Radomír
Smith, Bradford
Smith, Linda
Smith, Nigel
Smith, Peter
Smith, Randall
Snow, Theodore
Sodnomsambuu, Demberel
Sofia, Sabatino
Soker, Noam
Sokolov, Vladimir
Sol, Helene
Soloviev, Alexandr
Soltynski, Maciej
Somasundaram, Seetha
SONG, Qian
Sonneborn, George
Sood, Ravi
Soria, Roberto
Spaans, Marco
Spallicci di Filottrano, Alessandro
Spencer, Ralph
Spyromilio, Jason
Sreekumar, Parameswaran
Srinivasan, Ganesan
Srivastava, Dhruwa
Stanga, Ruggero
Starrfield, Sumner
Staubert, Rüdiger
Stebbins, Robin
Stecher, Theodore
Stecker, Floyd
Steenbrugge, Katrien
Steigman, Gary
Steinberg, Jean-Louis
Steiner, Joao
Stencel, Robert
Stephens, S.
Stergioulas, Nikolaos
Stern, Robert
Stevens, Ian
Stier, Mark
Still, Martin
Stinebring, Daniel
Stockman Jr, Hervey
Stoehr, Felix
Stone, R.
Straumann, Norbert
Stringfellow, Guy
Strohmayer, Tod
Strong, Andrew
Strong, Ian
Struminsky, Alexei
Stuchlík, Zdenek
Sturrock, Peter
SU, Cheng-yue
šubr, Ladislav
Subramanian, Kandaswamy
Subramanian, Prasad
Suleimanov, Valery
SUMIYOSHI, Kosuke
SUN, Wei-Hsin
Sunyaev, Rashid
Sutherland, Ralph
SUZUKI, Hideyuki
SUZUKI, Tomoharu
Swank, Jean
Swings, Jean-Pierre
Tagliaferri, Gianpiero
TAGOSHI, Hideyuki
TAKAHARA, Fumio
TAKAHARA, Mariko
TAKAHASHI, Masaaki
TAKAHASHI, Rohta
TAKEI, Yoh
Tammi, Joni
Tanaka, Yasuo
TASHIRO, Makoto
TATEHIRO, Mihara
Tauris, Thomas
Tavecchio, Fabrizio
Tempel, Elmo
Templeton, Matthew
Teng, Stacy
TERADA, Yukikatsu
TERASHIMA, Yuichi
Tessema, Solomon
Theureau, Gilles
Thöne, Christina
Thomas, Roger
Thomsen, Bjarne
Thorne, Kip
Thorsett, Stephen
Thorstensen, John
Thronson Jr, Harley
TIAN, Wenwu
Toffolatti, Luigi
Tokarev, Yurij
TOMIMATSU, Akira
TOMINAGA, Nozomu
Torkelsson, Ulf
Török, Gabriel
José Miguel Torrejón, Jose Miguel
Torres, Carlos Alberto
Torres, Diego
TOTANI, Tomonori
Tovmassian, Gagik
Tovmassian, Hrant
Trakhtenbrot, Benny
Traub, Wesley
Tresse, Laurence
Trevese, Dario
Trimble, Virginia
Trinchieri, Ginevra
Truemper, Joachim
Truran, James
Trussoni, Edoardo
TSAI, An-Li
Tsujimoto, Masahiro
TSUNEMI, Hiroshi
TSUNETA, Saku
TSURU, Takeshi
Tsuruta, Sachiko
Tsutsumi, Takahiro
Tsygan, Anatolij
Tsygankov, Sergey
Türler, Marc
Turolla, Roberto
Tylka, Allan
Ubertini, Pietro
UEDA, Yoshihiro
Ulmer, Melville
Ulyanov, Oleg
Underwood, James
Upson, Walter
Urama, Johnson
URATA, Yuji
Urry, Claudia

Uslenghi, Michela
Usov, Vladimir
Uttley, Philip
Vaccari, Mattia
Vagnetti, Fausto
Vahia, Mayank
Väliviita, Jussi-Pekka
Valluri, Monica
Valtaoja, Esko
Valtonen, Mauri
van den Berg, Maureen
van den Heuvel, Edward
van der Hucht, Karel
van der Klis, Michiel
van der Walt, Diederick
van Duinen, R.
van Kerkwijk, Marten
van Putten, Maurice
van Riper, Kenneth
Vaughan, Simon
Venter, Christo
Venters, Tonia
Vercellone, Stefano
Vestergaard, Marianne
Vial, Jean-Claude
Vidal, Nissim
Vidal-Madjar, Alfred
Vierdayanti, Kiki
Vignali, Cristian
Vikhlinin, Alexey
Vilhu, Osmi
Mónica, Monica
Villata, Massimo
Vink, Jacco
Viollier, Raoul
Viotti, Roberto
Vlahakis, Catherine
Vlahakis, Nektarios
Völk, Heinrich
Volonteri, Marta
Vrtilek, Jan
Vrtilek, Saeqa
Vucetich, Héctor
WADA, Takehiko
Wagner, Alexander
Wagner, Robert
Wagoner, Robert
Walder, Rolf
Walker, Helen
Walker, Mark
Walter, Frederick
Wanas, Mamdouh
Wandel, Amri
Wang, Q. Daniel
Wang, Yi-ming
WANG, Chen
WANG, Ding-Xiong
WANG, Feilu
WANG, Hong-Guang
WANG, jiancheng
WANG, Sen
WANG, Shiang-Yu
WANG, Shouguan
WANG, Shui
WANG, Shujuan
WANG, Tinggui
WANG, Zhenru
Wardle, John
Warner, John
Watanabe, Ken
WATARAI, Kenya
Watson, Darach
Watson, Michael
Watts, Anna
Waxman, Eli
Weaver, Kimberly
Weaver, Thomas
Webster, Adrian
Wehrle, Ann
WEI, Daming
WEI, Jianyan
Weiler, Edward
Weiler, Kurt
Weinberg, Jerry
Weisheit, Jon
Weisskopf, Martin
Wells, Donald
WEN, Linqing
Werner, Klaus
Wesselius, Paul
Westergaard, Niels
Wheatley, Peter
Wheeler, J.
Whitcomb, Stanley
White, Nicholas
Whiting, Matthew
Wijers, Ralph
Wijnands, Rudy
Wilkes, Belinda
Will, Clifford
Willis, Allan
Willis, Anthony
Willner, Steven
Wilms, Jörn
Wilson, Gillian
Winkler, Christoph
Winter, Lisa
Wise, Michael
Woan, Graham
Wolfendale FRS, Sir Arnold
Wolstencroft, Ramon
Wolter, Anna
Woltjer, Lodewijk
WOO, Jong-Hak
Woosley, Stanford
Worrall, Diana
Wray, James
Wrobel, Joan
WU, Jiun-Huei
WU, Shaoping
WU, Xue-bing
WU, Xue-Feng
WU, Xuejun
Wulandari, Hesti
Wunner, Guenter
XIANGYU, Wang
XU, Dawei
XU, Renxin
Yadav, Jagdish
Yahil, Amos
Yakovlev, Dmitrij
Yamada, Shoichi
YAMAMOTO, Hiroaki
YAMAMOTO, Yoshiaki
Yamasaki, Tatsuya
YAMASAKI, Noriko
YAMASHITA, Kojun
YAMAUCHI, Makoto
YAMAUCHI, Shigeo
Yock, Philip
YOICHIRO, Suzuki
YOKOSAWA, Masayoshi
YOKOYAMA, Takaaki
YONETOKU, Daisuke
YOSHIDA, Atsumasa
YOSHIDA, Shin'ichirou
YOU, Junhan
YU, Cong
YU, Qingjuan

YU, Wenfei
Yuan, Weimin
YUAN, Feng
YUAN, Ye-fei
Zacchei, Andrea
Zamorani, Giovanni
Zampieri, Luca
Zane, Silvia
Zannoni, Mario
Zarnecki, John
Zdziarski, Andrzej
Zezas, Andreas
Zhang, Jie
Zhang, Yuying
ZHANG, Chengmin
ZHANG, Jialu
ZHANG, JiangShui
ZHANG, Jingyi
ZHANG, Li
ZHANG, Shu
ZHANG, Shuang Nan
ZHANG, William
ZHANG, Yanxia
ZHANG, You-Hong
ZHANG, Zhibin
Zharkova, Valentina
Zhekov, Svetozar
Zheleznyakov, Vladimir
ZHENG, Wei
ZHENG, Xiaoping
ZHOU, Jianfeng
ZHOU, Xia
Zhuk, Alexander
Zombeck, Martin
Zwintz, Konstanze

Division E Sun and Heliosphere

President: Lidia van Driel-Gesztelyi
Vice-President: Yihua YAN

Steering Committee Members:

Sarah Gibson
Marc L DeRosa
Peng-Fei CHEN
Arnab Rai Choudhuri
Lyndsay Fletcher
Ingrid Mann
Karel J. Schrijver
Rudolf von Steiger
Lidia van Driel-Gesztelyi
Yihua YAN

Members:

Abbett, William
Abdelatif, Toufik
Aboudarham, Jean
Ábrahám, Péter
Abramenko, Valentina
Acton, Loren
Afram, Nadine
Ahluwalia, Harjit
AI, Guoxiang
Aime, Claude
Airapetian, Vladimir
AKITA, Kyo
Alexander, Joseph
Alissandrakis, Costas
Almleaky, Yasseen
Altrock, Richard
Altschuler, Martin
Altyntsev, Alexandre
Aly, Jean-Jacques
Ambastha, Ashok
Ambroz, Pavel
Amer, Morsi
Ananthakrishnan, Subramaniam
Anastasiadis, Anastasios
Andersen, Bo Nyborg
Anderson, Jay
ANDO, Hiroyasu
Andrei, Alexandre
Andretta, Vincenzo
Andries, Jesse
Ansari, S.M.
Antia, H.
Antiochos, Spiro
Antonucci, Ester
Anzer, Ulrich
Arnett, W.
Arons, Jonathan
Arregui, Inigo
Artzner, Guy
ASAI, Ayumi
Asareh, Habibolah
Aschwanden, Markus
Asplund, Martin
Atac, Tamer
Aurass, Henry
Avrett, Eugene
Ayres, Thomas
Aznar Cuadrado, Regina
Babayev, Elchin
Babin, Arthur
Bagala, Liria
Bagare, S.
Baki, Paul
Balasubramaniam, K.
Balikhin, Michael
Baliunas, Sallie
Ballester, Jose
Balthasar, Horst
Banerjee, Dipankar
BAO, Shudong
Baranovsky, Edward
Barklem, Paul

Barnes, Aaron
Barnes, Graham
Barrantes, Marco
Barrow, Colin
Barta, Miroslav
Basu, Sarbani
Batchelor, David
Batmunkh, Damdin
Baturin, Vladimir
Beckers, Jacques
Beckman, John
Bedding, Timothy
Beebe, Herbert
Beiersdorfer, Peter
Bell, Barbara
Bell, Steven
Bellot Rubio, Luis
Belvedere, Gaetano
Bemporad, Alessandro
Benevolenskaya, Elena
Benford, Gregory
Ben-Jaffel, Lofti
Benkhaldoun, Zouhair
Bentley, Robert
Benz, Arnold
Berger, Mitchell
Berghmans, David
Berrilli, Francesco
Bertaux, Jean-Loup
Bertello, Luca
Bewsher, Danielle
Bhardwaj, Anil
BI, Shaolan
Bianda, Michele
Bingham, Robert
Björnsson, Gunnlaugur
Blamont, Jacques-Emile
Blandford, Roger
Bobylev, Vadim
Bocchia, Romeo
Bochsler, Peter
Bogdan, Thomas
Bogod, Vladimir
Bohm-Vitense, Erika
Bommier, Veronique
Bonaccini, Domenico
Bonanno, Alfio
Bondal, Krishna
Bonnet, Roger
Book, David
Bornmann, Patricia
Borovik, Valerya
Botha, Gert
Bothmer, Volker
Bouchard, Antoine
Bougeret, Jean-Louis
Boyer, René
Braithwaite, Jonathan
Brajša, Roman
Brandenburg, Axel
Brandt, John
Brandt, Peter
Branduardi-Raymont, Graziella
Braun, Douglas
Bray, Robert
Breckinridge, James
Brekke, Pål
Briand, Carine
Bromage, Barbara
Brooke, John
Brosius, Jeffrey
Brown, John
Browning, Matthew
Browning, Philippa
Bruls, Jo
Brun, Allan
Bruner, Marilyn
Bruning, David
Bruno, Roberto
Bruns, Andrey
Brynildsen, Nils
Buccino, Andrea
Buchlin, Eric
Buecher, Alain
Buechner, Joerg
Bumba, Vaclav
Burlaga, Leonard
Busa', Innocenza
Bushby, Paul
Buti, Bimla
Butler, Christopher
Bzowski, Maciej
čadež, Vladimir
CAI, Mingsheng
Cairns, Iver
Cally, Paul
Cameron, Robert
Cane, Hilary
Canfield, Richard
Carbonell, Marc
Cargill, Peter
Carlsson, Mats
Carpenter, Kenneth
Cassinelli, Joseph
Cauzzi, Gianna
Cavallini, Fabio
Cecconi, Baptiste
Ceppatelli, Guido
CHAE, Jongchul
Chakrabarti, Supriya
Chambe, Gilbert
CHAN, Kwing
Chandra, Suresh
CHANG, Heon-Young
Channok, Chanruangrit
Chapanov, Yavor
Chaplin, William
Chapman, Gary
Chapman, Sandra
Charbonneau, Paul
Chashei, Igor
Chassefiere, Eric
CHEN, Peng-Fei
CHEN, Zhiyuan
Chernov, Gennadij
Chertok, Ilya
Chertoprud, Vadim
Chitre, Shashikumar
Chiuderi-Drago, Franca
CHIUEH, Tzihong
CHO, Kyung Suk
CHOE, Gwangson
CHOU, Chih-Kang
CHOU, Dean-Yi
Choudhary, Debi Prasad
Choudhuri, Arnab
Christensen-Dalsgaard, Jørgen
Chupp, Edward
Cionco, Rodolfo
Clark, Thomas
Clette, Frederic
Cliver, Edward
Coffey, Helen
Collados, Manuel
Collet, Remo
Conway, Andrew
Cora, Alberto
Correia, Emilia

Córsico, Alejandro
Costa, Joaquim
Couturier, Pierre
Couvidat, Sebastien
Craig, Ian
Cramer, Neil
Crannell, Carol
Cremades Fernández, María
Culhane, John
Cuperman, Sami
Curdt, Werner
Daglis, Ioannis
Dalla, Silvia
Damé, Luc
Dara, Helen
Dasso, Sergio
Datlowe, Dayton
Davila, Joseph
Dechev, Momchil
Degenhardt, Detlev
De Groof, Anik
de Jager, Cornelis
de Kertanguy, Amaury
De Keyser, Johan
Deliyannis, John
Delmas, Christian
Del Toro Iniesta, Jose
Del Zanna, Luca
Demarque, Pierre
Deming, Leo
Démoulin, Pascal
DENG, YuanYong
DeNisco, Kenneth
Dennerl, Konrad
Dennis, Brian
Dere, Kenneth
DeRosa, Marc
de Toma, Giuliana
Deubner, Franz-Ludwig
Dialetis, Dimitris
Didkovsky, Leonid
Di Mauro, Maria Pia
DING, Mingde
Dinulescu, Simona
Diver, Declan
Dobler, Wolfgang
Dobrzycka, Danuta
do Nascimento, José
Donea, Alina
Dorch, Søren Bertil
Dorotovic, Ivan
Doval, Jorge M.
Doyle MRIA, John
Drake, Jeremy
Dravins, Dainis
Druckmüller, Miloslav
Dryer, Murray
Dubau, Jacques
Dubois, Marc
Duchlev, Peter
Duldig, Marcus
Dumitrache, Cristiana
Dumont, Simone
Durney, Bernard
Duvall Jr, Thomas
Dwivedi, Bhola
Dzifcáková, Elena
Efimenko, Volodymyr
Ehgamberdiev, Shuhrat
Einaudi, Giorgio
Elliott, Ian
Elmhamdi, Abouazza
Elste, Günther
Emslie, Gordon
Engvold, Oddbjørn
Epstein, Gabriel
Erdelyi, Robertus
Ermolli, Ilaria
Esenoğlu, Hasan
Eshleman, Von
Esser, Ruth
Eviatar, Aharon
Fabiani, Sergio
Fahr, Hans
Falciani, Roberto
Falewicz, Robert
FAN, Yuhong
FANG, Cheng
Fárník, Frantisek
Feldman, Uri
Fernandes, Francisco
Fernández, Laura
Ferreira, João
Ferriz Mas, Antonio
Feulner, Georg
Feynman, Joan
Fichtner, Horst
Field, George
Filippov, Boris
Finsterle, Wolfgang
Fisher, George
Fisher, Richard
Fleck, Bernhard
Fletcher, Lyndsay
Fludra, Andrzej
Fluri, Dominique
Fofi, Massimo
Foing, Bernard
Fomichev, Valerij
Fontenla, Juan
Forbes, Terry
Forgács-Dajka, Emese
Fossat, Eric
Foukal, Peter
Foullon, Claire
Fraenz, Markus
Frasca, Antonio
Fraschetti, Federico
Fraser, Brian
Frazier, Edward
Frisch, Helene
Froehlich, Claus
Frutos-Alfaro, Francisco
FU, Hsieh-Hai
Gabriel, Alan
Gaizauskas, Victor
Galal, A.
Galloway, David
Galsgaard, Klaus
Galvin, Antoinette
GAN, Weiqun
Gangadhara, R.T.
Garcia, Rafael
García-Berro, Enrique
García de la Rosa, Ignacio
Gary, Dale
Gary, Gilmer
Gedalin, Michael
Georgoulis, Manolis
Gergely, Tomas
Getling, Alexander
Ghizaru, Mihai
Giampapa, Mark
Gibson, David
Gibson, Sarah
Gill, Michael
Gill, Peter
Gilliland, Ronald
Gilman, Peter

Giménez de Castro, Carlos
Gizon, Laurent
Glasner, Shimon
Glatzmaier, Gary
Gleisner, Hans
Goedbloed, Johan
Gokhale, Moreshwar
Golden, Aaron
Goldman, Martin
Golub, Leon
Gómez, Daniel
Gomez, Maria
Gömöry, Peter
Gontikakis, Constantin
Goode, Philip
Goossens, Marcel
Gopalswamy, Natchimuthuk
Gopasyuk, Olga
Gosling, John
Graffagnino, Vito
Grandpierre, Attila
Grant, Athay
Gray, Richard
Grec, Gérard
Grechnev, Victor
Gregorio, Anna
Grevesse, Nicolas
Grib, Sergei
Grzędzielski, Stanisław
Gudiksen, Boris
Güdel, Manuel
Guglielmino, Salvatore
Guhathakurta, Madhulika
Guinan, Edward
Gulyaev, Rudolf
Gunár, Stanislav
GUNJI, Shuichi
Gupta, Surendra
Gurman, Joseph
Guzik, Joyce
Györi, Lajos
Habbal, Shadia
Haberreiter, Margit
Haerendel, Gerhard
Hagyard, Mona
Hamedivafa, Hashem
Hammer, Reiner
Hanaoka, Yoichiro
Hanasz, Jan
Hannah, Iain
Hansen, Richard
Hanslmeier, Arnold
HARA, Hirohisa
Harra, Louise
Harrison, Richard
Harvey, Christopher
Harvey, John
Hasan, S. Sirajul
Hathaway, David
Haubold, Hans
Haugan, Stein Vidar
HAYASHI, Keiji
Hayward, John
He, Han
Hein, Righini
Heinzel, Petr
Hejna, Ladislav
Hempel, Marc
Henoux, Jean-Claude
Heras, Ana
Herdiwijaya, Dhani
Heyl, Jeremy
Heynderickx, Daniel
HIEI, Eijiro
Hilaris, Alexander
Hildebrandt, Joachim
Hildner, Ernest
Hill, Frank
Hoang, Binh
Hochedez, Jean-François
Hoeksema, Jon
Hohenkerk, Catherine
Hollweg, Joseph
Holman, Gordon
Holzer, Thomas
Hood, Alan
Houdebine, Eric
House, Lewis
Howard, Robert
Hoyng, Peter
HUANG, Guangli
Huber, Martin
Hudson, Hugh
Hughes, David
Humble, John
Humphries, Colin
Hurford, Gordon
HWANG, Junga
Ibadov, Subhon
Illing, Rainer
INAGAKI, Shogo
Irbah, Abdanour
ISHII, Takako
Ishitsuka, Mutsumi
Isliker, Heinz
Ivanchuk, Victor
Ivanov, Evgenij
Ivchenko, Vasily
Jabbar, Sabeh
Jackson, Bernard
Jacobs, Carla
Jain, Rajmal
Jakimiec, Jerzy
Janches, Diego
Janssen, Katja
Jardine, Moira
Jatenco-Pereira, Vera
Jefferies, Stuart
JI, Haisheng
JIANG, Aimin
JIANG, Yunchun
Jimenez, Mancebo
JING, Hairong
Jokipii, Jack
Jones, Harrison
Jordan, Carole
Jordan, Stuart
Joselyn, Jo
Jubier, Xavier
Jurčák, Jan
KABURAKI, Osamu
Kahler, Stephen
KAKINUMA, Takakiyo
Kalkofen, Wolfgang
Kallenbach, Reinald
Kálmán, Bela
Kaltman, Tatyana
Kamal, Fouad
Kane, Sharad
KANG, Jin Sok
Käpylä, Petri
Karami, Kayoomars
Karlický, Márian
Karoff, Christoffer
Karpen, Judith
Karpov, Nikolai
Kasiviswanathan, Sankarasubramanian
Kašparová, Jana

KATO, Yoshiaki
Katsova, Maria
Kaufmann, Pierre
Keil, Stephen
Keller, Horst
Kentischer, Thomas
Keppens, Rony
Khan, J
Khodachenko, Maxim
Khomenko, Elena
Khumlemlert, Thiranee
KILCIK, Ali
Kim, Iraida
KIM, Kap sung
KIM, Yong Cheol
Kiplinger, Alan
KITAI, Reizaburo
Kitchatinov, Leonid
Kitiashvili, Irina
Kjeldseth-Moe, Olav
Klein, Karl
Kliem, Bernhard
Klimchuk, James
Klvana, Miroslav
Kneer, Franz
Knoelker, Michael
KO, Chung-Ming
KOJIMA, Masayoshi
Kolobov, Dmitri
Kołomański, Sylwester
Kondrashova, Nina
Kononovich, Edvard
Kontar, Eduard
Kopp, Roger
Kopylova, Yulia
Korhonen, Heidi
Kosovichev, Alexander
Kostik, Roman
Kotov, Valery
Kotrc, Pavel
Koutchmy, Serge
Kovacs, Agnes
Koza, Julius
Kozak, Lyudmyla
Kozlovsky, Ben
Kramer, William
Kretzschmar, Matthieu
Krimigis, Stamatios
Krittinatham, Watcharawuth
Krivova, Natalie
Krucker, Sam
Kryshtal, Alexander
Kryvodubskyj, Valery
Kubičela, Aleksandar
KUBOTA, Jun
Kučera, Aleš
Kučinskas, Arunas
Kulhánek, Petr
Kuperus, Max
Kupka, Friedrich
Kurochka, Evgenia
KUROKAWA, Hiroki
Kurt, Vladimir
Kurtz, Stanley
KUSANO, Kanya
Kuznetsov, Vladimir
Labrosse, Nicolas
Lafon, Jean-Pierre
Lai, Sebastiana
Lallement, Rosine
Lamy, Philippe
Landecker, Peter
Landi, Simone
Landi Degl'Innocenti, Egidio
Landman, Donald
Landolfi, Marco
Landolt, Arlo
Lang, Kenneth
Lanza, Antonino
Lanzafame, Alessandro
Lapenta, Giovanni
Lario, David
Lawrence, John
Lazrek, Mohamed
Leenaarts, Jorrit
Leer, Egil
Leibacher, John
Leiko, Uliana
Leka, K.D.
Leroy, Bernard
Leroy, Jean-Louis
Levine, Randolph
Levy, Eugene
Li, Bo
LI, Hui
LI, Kejun
LI, Qingkang
LI, Wei
LI, Zhi
Lie-Svendsen, Oystein
Lima, Joao
Lin, Yong
LIN, Jun
Lingenfelter, Richard
Linsky, Jeffrey
Liritzis, Ioannis
Liu, Zhong
LIU, Siming
LIU, Xinping
LIU, Yang
LIU, Yu
Livadiotis, George
Livingston, William
Livshits, Moisey
Locke, Jack
Löfdahl, Mats
Lopez Arroyo, M.
Lopez Fuentes, Marcelo
Lotova, Natalja
LOU, Yu-Qing
Loukitcheva, Maria
Low, Boon
Lozitskij, Vsevolod
Ludmány, Andras
Lüst, Reimar
Lundstedt, Henrik
LUO, Xianhan
MA, Guanyi
Machado, Marcos
Mackay, Duncan
MacKinnon, Alexander
MacQueen, Robert
Madjarska, Maria
MAKITA, Mitsugu
Malandraki, Olga
Malara, Francesco
Malherbe, Jean-Marie
Malitson, Harriet
Malville, J.
MANABE, Seiji
Mandel, Holger
Mandrini, Cristina
Mangalam, Arun
Mangeney, André
Mann, Gottfried
Mann, Ingrid
Manoharan, P.
Marcu, Alexandru

Maričić, Darije
Marilena, Mierla
Marilli, Ettore
Maris, Georgeta
Mariska, John
Marková, Eva
Marmolino, Ciro
Marsch, Eckart
Marsden, Stephen
Marshall, Herman
Martel, Hugo
Martens, Petrus
Martínez Pillet, Valentin
Mason, Glenn
MASUDA, Satoshi
Mathioudakis, Mihalis
Matsuura, Oscar
Matthews, Sarah
Mattig, W.
Mauas, Pablo
Mavromichalaki, Helen
Maxwell, Alan
McAteer, R. T. James
McCabe, Marie
McIntosh, Patrick
McKenna-Lawlor, Susan
McLean, Donald
McMullin, Joseph
Mein, Pierre
Meisel, David
Meister, Claudia
Mel'nik, Valentin
Melrose, Donald
Mendes, Da
Mendes, Luiz
Messerotti, Mauro
Messmer, Peter
Mestel, Leon
Mészárosová, Hana
Meyer, Friedrich
Michałek, Grzegorz
Michard, Raymond
Michel, F.
Miesch, Mark
Miletsky, Eugeny
Milkey, Robert
Milone, Eugene
Miralles, Mari Paz
MITRA, Dhrubaditya
Mohan, Anita
Mohd, Zambri
Moncuquet, Michel
Monteiro, Mario Joao
Montes, David
Moore, Ronald
Morabito, David
Moravec, Zdeněk
Moreno-Insertis, Fernando
Morita, Satoshi
MORIYAMA, Fumio
Moss, David
Motta, Santo
Mouradian, Zadig
Moussas, Xenophon
Mukherjee, Krishna
Muller, Richard
MUNETOSHI, Tokumaru
Munro, Richard
Musielak, Zdzislaw
Nagendra, K.
Nahar, Sultana
NAITO, Tsuguya
NAKAJIMA, Hiroshi
Nakariakov, Valery
Namba, Osamu
Nandi, Dibyendu
Naoz, Smadar
Narain, Udit
Nasiri, Sadollah
Neidig, Donald
Nesis, Anastasios
Ness, Norman
Netzer, Nathan
Neukirch, Thomas
Neupert, Werner
Nguyễn, Phuong
Nickeler, Dieter
Nicolas, Kenneth
NING, Zongjun
Nocera, Luigi
Noens, Jacques-Clair
Nordlund, Aake
Noyes, Robert
Nozawa, Satoshi
Nussbaumer, Harry
Obridko, Vladimir
Ofman, Leon
OH, Suyeon
OHKI, Kenichiro
OKAMOTO, Takenori Joten
Okeke, Francisca
Okpala, Kingsley
Oláh, Katalin
Oliver, Ramón
Oluseyi, Hakeem
Önel, Hakan
Opara, Fidelix
Orlando, Salvatore
Ortiz Carbonell, Ada
Ossendrijver, Mathieu
Ostrowski, Michał
Owocki, Stanley
Ozguc, Atila
Oezisik, Tuncay
Padmanabhan, Janardhan
Paletou, Frédéric
Pallé, Pere
Pallé Bagó, Enric
Pa'uš, Pavel
PAN, Liande
Pandey, Birendra
Pantoja, Carmen
Pap, Judit
Papathanasoglou, Dimitrios
Parenti, Susanna
Paresce, Francesco
Parfinenko, Leonid
Parhi, Shyamsundar
Pariat, Etienne
PARK, Young Deuk
Parker, Eugene
Parkinson, William
Parnell, Clare
Pasachoff, Jay
Paternò, Lucio
Pecker, Jean-Claude
Pelt, Jaan
Peres, Giovanni
Pérez Hernández, Fernando
Perkins, Francis
Peter, Hardi
Petit, Pascal
Petkaki, Panagiota
Petrie, Gordon
Petrosian, Vahe
Petrov, Nikola
Petrovay, Kristof
Pevtsov, Alexei

Pflug, Klaus
PHAM, Diep
PHAN, Dong
Phillips, Kenneth
Picazzio, Enos
Pick, Monique
PING, Jinsong
Pipin, Valery
Plainaki, Christina
Podesta, John
Poedts, Stefaan
Pötzi, Werner
Pohjolainen, Silja
Poland, Arthur
Poletto, Giannina
Poquerusse, Michel
Preka-Papadema, Panagiota
Preś, Paweł
Priest, Eric
Proctor, Michael
Prokakis, Theodore
Pustil'nik, Lev
QU, Zhongquan
Quémerais, Eric
Raadu, Michael
Radick, Richard
Radiman, Iratio
Ramelli, Renzo
Rao, A.
Raoult, Antoinette
Raulin, Jean-Pierre
Raymond, John
Rayrole, Jean
Readhead, Anthony
Reale, Fabio
Reames, Donald
Reardon, Kevin
Reeves, Hubert
Regnier, Stephane
Régulo, Clara
Reinard, Alysha
Rendtel, Juergen
Rengel, Miriam
Reshetnyk, Volodymyr
RI, Son Jae
Richards, Mercedes
Riddle, Anthony
Riehokainen, Aleksandr
Rieutord, Michel
Riley, Pete
Ripken, Hartmut
Robinson Jr, Richard
Roca Cortés, Teodoro
Roddier, Francois
Rodríguez Hidalgo, Inés
Roemer, Max
Romano, Paolo
Romoli, Marco
Rompolt, Bogdan
Roša, Dragan
Rosa, Reinaldo
Rosner, Robert
Roth, Ilan
Roth, Markus
Roth, Michel
Roudier, Thierry
Rouppe van der Voort, Luc
Rovira, Marta
Roxburgh, Ian
Rozelot, Jean-Pierre
Rubenstein, Eric
Rudawy, Paweł
Ruderman, Michael
Ruediger, Guenther
Ruffolo, David
Ruiz Cobo, Basilio
Rusin, Vojtech
Russell, Alexander
Russell, Christopher
Rust, David
Rutten, Robert
Ruždjak, Domagoj
Ruzickova-Topolova, B.
Ryabov, Boris
Rybák, Jan
Rybansky, Milan
Saemundsson, Thorsteinn
Safari, Hossein
Sagdeev, Roald
Sahai, Raghvendra
Sahal-Bréchot, Sylvie
SAITO, Mamoru
Saiz, Alejandro
SAKAI, Junichi
SAKAO, Taro
SAKURAI, Takashi
SAKURAI, Kunitomo
Samain, Denys
Sanahuja Parera, Blai
Sánchez Almeida, Jorge
Saniga, Metod
Santillán, Alfredo
Sarris, Emmanuel
Sasso, Clementina
Sastri, Hanumath
Sattarov, Isroil
Sauval, A.
Savanov, Igor
Savin, Daniel
Sawyer, Constance
Scharmer, Goeran
Scherb, Frank
Scherrer, Philip
Schindler, Karl
Schleicher, Helmold
Schlichenmaier, Rolf
Schmahl, Edward
Schmelz, Joan
Schmidt, Wolfgang
Schmieder, Brigitte
Schmitt, Dieter
Schmutz, Werner
Schober, Hans
Schou, Jesper
Schreiber, Roman
Schrijver, Karel
Schuessler, Manfred
Schwartz, Pavol
Schwartz, Steven
Seaton, Daniel
Sefako, Ramotholo
Segaluvitz, Alexander
Seiradakis, John
Setti, Giancarlo
Severino, Giuseppe
Shalchi, Andreas
Sharma, A.
Shchukina, Nataliia
Shea, Margaret
Sheeley, Neil
Sheminova, Valentina
SHIBAHASHI, Hiromoto
SHIBASAKI, Kiyoto
SHIBATA, Kazunari
Shimizu, Toshifumi
Shimojo, Masumi
Shine, Richard
SHIN'ICHI, Nagata
Shore, Steven

Short, Christopher
Sigalotti, Leonardo
Sigismondi, Costantino
Sigwarth, Michael
Simnett, George
Simon, George
Simon, Guy
Simunac, Kristin
Singh, Jagdev
Sinha, Krishnanand
Sivaraman, Koduvayur
Skumanich, Andrew
Smaldone, Luigi
Smith, Dean
Smith, Peter
Smol'kov, Gennadij
Snegirev, Sergey
Snik, Frans
Sobolev, Andrey
Sobotka, Michal
Socas-Navarro, Hector
Soderblom, David
Sofia, Sabatino
Solanki, Sami
Soliman, Mohamed
Soloviev, Alexandr
Somov, Boris
Sonett, Charles
Soonthornthum, Boonrucksar
Spadaro, Daniele
Spicer, Daniel
Spiegel, Edward
Spruit, Hendrik
Srivastava, Nandita
Stathopoulou, Maria
Staude, Juergen
Stebbins, Robin
Steffen, Matthias
Stein, Robert
Steiner, Oskar
Stellmacher, Götz
Stencel, Robert
Stenflo, Jan
štěpán, Jiří
Stepanian, Natali
Stepanov, Alexander
Steshenko, N.
Stewart, Ronald
Stix, Michael
Stodilka, Myroslav
Stone, R.
Strachan, Leonard
Straus, Thomas
Strong, Andrew
Strong, Keith
Struminsky, Alexei
Sturrock, Peter
Subramanian, K.
Subramanian, Kandaswamy
Subramanian, Prasad
Sudar, Davor
SUEMATSU, Yoshinori
Suess, Steven
Sukartadiredja, Darsa
SUZUKI, Takeru
Svalgaard, Leif
švanda, Michal
Sylwester, Barbara
Sylwester, Janusz
Szalay, Alex
TAKAHASHI, Kunio
TAKANO, Toshiaki
Talon, Raoul
TAMENAGA, Tatsuo
TAN, Baolin
TANG, Yuhua
Tapping, Kenneth
Tarashchuk, Vera
Tarter, C.
Teplitskaya, Raisa
TERADA, Yukikatsu
Teriaca, Luca
Ternullo, Maurizio
Teske, Richard
Thomas, John
Thomas, Roger
Thomsen, Bjarne
Tikhomolov, Evgeniy
Timothy, J.
Tlatov, Andrej
Tobias, Steven
Tokarev, Yurij
Tomczak, Michał
Torelli, M.
tr`ân, Ha
Tripathi, Durgesh
Tripathy, Sushanta
Tritakis, Basil
Trottet, Gerard
Trujillo Bueno, Javier
Tsap, Yuriy
Tsiklauri, David
Tsinganos, Kanaris
Tsiropoula, Georgia
TSUNETA, Saku
Turck-Chièze, Sylvaine
Tyul'bashev, Sergei
Tziotziou, Konstantinos
Uddin, Wahab
Ueta, Toshiya
Ulrich, Roger
Underwood, James
Usoskin, Ilya
Usowicz, Jerzy
Vahia, Mayank
Vainshtein, Leonid
Válio, Adriana
Vandas, Marek
van den Oord, Bert
van der Heyden, Kurt
van der Linden, Ronald
Van Doorsselaere, Tom
van Driel-Gesztelyi, Lidia
Van Hoven, Gerard
van't Veer, Frans
Vaughan, Arthur
Veck, Nicholas
Vekstein, Gregory
Velli, Marco
Venkatakrishnan, P.
Ventura, Paolo
Ventura, Rita
Vergez, Madeleine
Verheest, Frank
Verma, V.
Veronig, Astrid
Verwichte, Erwin
Vial, Jean-Claude
Viall, Nicholeen
Vidal-Madjar, Alfred
Vidotto, Aline
Vieytes, Mariela
Vilinga, Jaime
Vilkoviskij, Emmanuil
Vilmer, Nicole
Vinod, S.
Vlahakis, Nektarios
Voitenko, Yuriy

Voitsekhovska, Anna
Volkmer, Reiner
Volonte, Sergio
von der Lühe, Oskar
von Steiger, Rudolf
Vršnak, Bojan
Vucetich, Héctor
Vukicevic-Karabin, Mirjana
Wachlin, Felipe
WADA, Takehiko
Walsh, Robert
Walter, Frederick
Wang, Yi-ming
WANG, Dongguang
WANG, Haimin
WANG, Huaning
WANG, Jingxiu
WANG, Jingyu
WANG, Min
WANG, Shujuan
WANG, Xiaoya
Warren Jr, Wayne
WATANABE, Takashi
WATANABE, Tetsuya
WATARI, Shinichi
Webb, David
Weiss, Nigel
Weller, Charles
White, Stephen
Wiegelmann, Thomas
Wiehr, Eberhard
Wiik Toutain, Jun
Wikstol, Oivind
Winebarger, Amy
Winter, Lisa
Wittmann, Axel
Woehl, Hubertus
Wolfson, C.
Wolfson, Richard
Woltjer, Lodewijk
Wood, Brian
Worden, Simon
Wrixon, Gerard
Wu, Chin-Chun
Wu, Shi
WU, De Jin
WU, Hsin-Heng
Wülser, Jean-Pierre
XIE, Xianchun
XU, Aoao
XU, Jun
XU, Zhi
Yahil, Amos
Yair, Yoav
YAN, Yihua
YANG, Hong-Jin
YANG, Jing
YANG, Lei
YANG, Zhiliang
YE, Binxun
Yeh, Tyan
Yeşilyurt, Ibrahim
YI, Yu
YOICHIRO, Suzuki
YOKOYAMA, Takaaki
YOSHIMURA, Hirokazu
Youssef, Nahed
YU, Cong
YU, Dai
Yun, Hong-Sik
Zachariadis, Theodosios
Zagars, Juris
Zaitsev, Valerij
Zampieri, Luca
Zappalà, Rosario
Zarro, Dominic
Zeiler, Michael
Zelenka, Antoine
Zender, Joe
Zhang, Jie
Zhang, Tielong
ZHANG, Hongqi
ZHANG, Jun
ZHANG, Mei
ZHAO, Junwei
Zharkova, Valentina
Zheleznyakov, Vladimir
Zhitnik, Igor
ZHOU, Daoqi
ZHOU, Guiping
Zhugzhda, Yuzef
Zirker, Jack
Zlobec, Paolo
Zlotnik, Elena
Zuccarello, Francesca

Division F Planetary Systems and Bioastronomy

President: Giovanni B. Valsecchi
Vice-President: Nader Haghighipour

Steering Committee Members:

Pamela Kilmartin
Alberto Cellino
Dominique Bockelee-Morvan
Mark T Lemmon
Didier Queloz
William M. Irvine
Steven R. Chesley
Petrus Matheus Marie Jenniskens
Pascale Ehrenfreund
Alain Lecavelier des Etangs
Jiří Borovička
Paul Winchester Chodas
Giovanni B. Valsecchi
Nader Haghighipour

Members:

Abalakin, Viktor
ABE, Shinsuke
Abrevaya, Ximena
Absil, Olivier
Adams, Fred
Aerts, Conny
Afram, Nadine
AGATA, Hidehiko
A'Hearn, Michael
Aigrain, Suzanne
Aikman, G.
Akimov, Leonid
Aksnes, Kaare
Alexandrov, Alexander
Alexandrov, Yuri
ALIBERT, Yann
Allan, Alasdair
Allard, France
Allegre, Claude
Allison, Michael
Almár, Ivan
Al-Naimiy, Hamid
Alonso Sobrino, Roi
Alsabti, Abdul Athem
Altwegg, Kathrin
Alvarez del Castillo, Elizabeth
Alves, João
Amado, Pedro
Ammons, Stephen
Anderson, Kurt
ANDO, Hiroyasu
Andrade, Manuel
Andreic, Zeljko
Angeli, Claudia
Apai, Daniel
Apostolovska, Gordana
Appleby, John
Archinal, Brent
Ardila, David
Arlot, Jean-Eudes
Arpigny, Claude
Arthur, David
Artymowicz, Pawel
Asanok, Kitiyanee
Asher, David
Atkinson, David
Atreya, Sushil
Audley, Michael
Augereau, Jean-Charles
BABA, Naoshi
Babadjanov, Pulat
Baggaley, William
Bagrov, Alexander
Bailey, Mark
Baines, Ellyn
Baki, Paul
Balázs, Bela
Balbi, Amedeo
Balikhin, Michael
Baliyan, Kiran
Ball, John
Bania, Thomas
Barabanov, Sergey
Baran, Andrzej
Baransky, Olexander
Barber, Robert
Barbieri, Cesare
Barge, Pierre
Barker, Edwin
Barkin, Yuri
Barklem, Paul
Barlow, Nadine
Bar-Nun, Akiva
Barrado Navascués, David
Barriot, Jean-Pierre
Barrow, Colin
Barsony, Mary
Bartczak, Przemysław
Barucci, Maria
Baruch, John
Baryshev, Andrey
Basu, Baidyanath
Basu, Kaustuv
Battaner, Eduardo
Bear, Ealeal
Beaudet, Gilles
Beaugé, Cristian
Beaulieu, Jean-Philippe
Beckman, John
Beckwith, Steven
Beebe, Reta
Beer, Reinhard
Behrend, Raoul
Belkovich, Oleg
Bell, Jeffrey
Bell III, James
Bellot Rubio, Luis
Belmonte Aviles, Juan Antonio
Belskaya, Irina
Belton, Michael
Bemporad, Alessandro
Bender, Peter
Bendjoya, Philippe
Benedict, George
Benest, Daniel
Benisty, Myriam
Ben-Jaffel, Lofti
Benkhoff, Johannes
Bennett, David
Benz, Willy
Berendzen, Richard
Berge, Glenn
Berger, Jean-Philippe
Bergin, Edwin
Bergstralh, Jay
Bernardi, Fabrizio
Bertaux, Jean-Loup
Berthier, Jerôme
Bezard, Bruno
Bhandari, N.
Bhardwaj, Anil
Bhavsar, Suketu
Biazzo, Katia
Bien, Reinhold
Billebaud, Francoise
Billingham, John
Bingham, Robert
Binzel, Richard
Biraud, François
Birch, Peter
Birlan, Mirel
Bisikalo, Dmitrij
Biver, Nicolas
Bjorkman, Karen
Black, John
Blamont, Jacques-Emile
Blanco, Armando
Blanco, Carlo
Bless, Robert

Bodaghee, Arash
Bodenheimer, Peter
Bodewits, Dennis
Boechat-Roberty, Heloisa
Boerngen, Freimut
Boffin, Henri
Bohlender, David
Boice, Daniel
Bond, Ian
Bondarenko, Lyudmila
Bonev, Tanyu
Bonfils, Xavier
Bordé, Pascal
Borisov, Borislav
Borisov, Galin
Borovička, Jiří
Borysenko, Serhii
Bosma, Pieter
Boss, Alan
Bouchet, Patrice
Bowell, Edward
Bowyer, C.
Boyajian, Tabetha
Boyce, Peter
Brahic, André
Brandeker, Alexis
Brandner, Wolfgang
Brandt, John
Branduardi-Raymont, Graziella
Branham, Richard
Brecher, Aviva
Bretones, Paulo
Briot, Danielle
Britt, Daniel
Broadfoot, A.
Broderick, John
Brosch, Noah
Brouwer, D.
Brown, Peter
Brown, Robert
Brownlee, Donald
Brož, Miroslav
Brozovic, Marina
Brun, Allan
Brunini, Adrian
Brunk, William
Buccino, Andrea
Buchhave, Lars
Budaj, Jan
Bueno de Camargo, Julio
Buie, Marc
Buratti, Bonnie
Burba, George
Burke, Bernard
Burlaga, Leonard
Burns, Joseph
Burrows, Adam
Burton, Michael
Busarev, Vladimir
Butkovskaya, Varvara
Butler, Bryan
Butler, Paul
CAI, Kai
Caldwell, John
Calvin, William
Cameron, Andrew
Campbell, Donald
Campbell-Brown, Margaret
Campins, Humberto
Campo Bagatin, Adriano
Campusano, Luis
Cannon, John
Canto Martins, Bruno
Capaccioni, Fabrizio
čapek, David
Capria, Maria
Cardoso Santos, Nuno
Caretta, Cesar
Carigi, Leticia
Carlson, John
Carpenter, Kenneth
Carpino, Mario
Carruba, Valerio
Carruthers, George
Carsenty, Uri
Carsmaru, Maria
Carusi, Andrea
Carvano, Jorge
casassus, simon
Cash Jr, Webster
Casoli, Fabienne
Cassan, Arnaud
Ceccarelli, Cecilia
Cecconi, Baptiste
Cellino, Alberto
Cernicharo, Jose
Cerroni, Priscilla
Cevolani, Giordano
Chabrier, Gilles
Chaisson, Eric
Chakrabarti, Sandip
Chakrabarti, Supriya
Chakraborty, Deo
CHAN, Kwing
Chandrasekhar, Thyagarajan
Chapanov, Yavor
Chapman, Clark
Chapman, Robert
Chapront-Touze, Michelle
Chaty, Sylvain
Chauvin, Gael
CHEN, Daohan
CHEN, Xinyang
Chernetenko, Yulia
Chesley, Steven
Chevrel, Serge
CHIN, Yi-nan
Chodas, Paul
Christensen, Lars
Christie, Grant
Christodoulou, Dimitris
Chubko, Larysa
Chung, Eduardo
Churyumov, Klim
Ciardi, David
Cieza, Lucas
Cionco, Rodolfo
Cirkovic, Milan
Clairemidi, Jacques
Clarke, John
Claudi, Riccardo
Clayton, Donald
Clayton, Geoffrey
Clifton, Kenneth
Clube, S.
Cochran, Anita
Cochran, William
Cohen, David
Collet, Remo
Colom, Pierre
Combi, Michael
Connes, Janine
Connes, Pierre
Connolly, Harold
Connors, Martin
Conrad, Albert
Consolmagno, Guy
Cook, Kem

Cooper, Nicholas
Cooper, Timothy
Cora, Alberto
Corbet, Robin
Cosmovici, Cristiano
COTTIN, Hervé
Cottini, Valeria
Coudé du Foresto, Vincent
Couper, Heather
Courtin, Régis
Coustenis, Athena
Covino, Elvira
Cowley, Charles
Crawford, Ian
Creech-Eakman, Michelle
Cremonese, Gabriele
Crovisier, Jacques
Cruikshank, Dale
Cuadra, Jorge
Cuesta Crespo, Luis
Cunningham, Maria
Cuntz, Manfred
Cuypers, Jan
DAI, Zhibin
Daigne, Gerard
Dall, Thomas
Danchi, William
Danks, Anthony
Danner, Rolf
DARHMAOUI, Hassane
da Silveira, Enio
Davidsson, Björn
Davies, Ashley
Davies, John
Davis, Gary
Davis, Michael
Day-Jones, Avril
de Almeida, Amaury
De Becker, Michaël
de Bergh, Catherine
De Cuyper, Jean-Pierre
Deeg, Hans
de Jager, Cornelis
de Jonge, J.
De Keyser, Johan
Delbo, Marco
DELEUIL, Magali
Dell' Oro, Aldo
de Loore, Camiel
Delplancke, Francoise
Del Popolo, Antonino
Delsanti, Audrey
Delsemme, Armand
de Medeiros, Jose
Demory, Brice-Olivier
DeNisco, Kenneth
Dent, William
de Pater, Imke
Derekas, Aliz
Dermott, Stanley
de Sanctis, Giovanni
de Sanctis, Maria
Despois, Didier
Deutschman, William
de Val-Borro, Miguel
de Vincenzi, Donald
Dick, Steven
Dickel, John
Dickey, Jean
Dickman, Steven
Dieleman, Pieter
Di Martino, Mario
Di Sisto, Romina
Dixon, Robert
Djorgovski, Stanislav
Dlugach, Zhanna
Domiciano de Souza, Armando
Dominis Prester, Dijana
do Nascimento, José
Donnison, John
Doressoundiram, Alain
Dorschner, Johann
Dotto, Elisabetta
DOU, Jiangpei
Doubinskij, Boris
Dourneau, Gerard
Doval, Jorge M.
Downes Wallace, Juan
Downs, George
Doyle, Laurance
Doyon, Rene
Drake, Frank
Drake, Jeremy
Dreher, John
Dreizler, Stefan
Drossart, Pierre
Dryer, Murray
Dubin, Maurice
Duffard, Rene
Duncan, Martin
Dunham, David
(Dunkin) Beardsley, Sarah
Duorah, Hira
Durech, Josef
Durisen, Richard
Durrance, Samuel
Dutil, Yvan
Dvorak, Rudolf
Dybczyński, Piotr
Dyson, Freeman
Dzhapiashvili, Victor
Edwards, Suzan
Ehrenfreund, Pascale
Ehrenreich, David
Eisner, Josh
El- Baz, Farouk
Elford, William
Elias II, Nicholas
Elipe, Antonio
Ellis, George
Elst, Eric
Elston, Wolfgang
Emelianov, Nikolaj
Emel'yanenko, Vacheslav
Encrenaz, Therese
Epishev, Vitali
Epstein, Eugene
Erard, Stéphane
Ershkovich, Alexander
Escapa, Alberto
Esenoğlu, Hasan
Eshleman, Von
Esposito, Larry
Eubanks, Thomas
Evans, Michael
Evans, Neal
Eviatar, Aharon
Eyres, Stewart
FAN, Yufeng
Farnham, Tony
Fazio, Giovanni
Feigelson, Eric
Feldman, Paul
Feldman, Paul
Ferlet, Roger
Fernández, Julio
Fernández, Yanga
Fernández Lajús, Eduardo
Ferrari, Cécile

Ferraz-Mello, Sylvio
Ferreri, Walter
Ferrin, Ignacio
Ferrusca, Daniel
Feulner, Georg
Field, George
Filacchione, Gianrico
Filipov, Lachezar
Fink, Uwe
Firneis, Maria
Fisher, Philip
Fitzsimmons, Alan
Fleck, Robert
Fletcher, Leigh
Foing, Bernard
Folkner, William
Ford, Eric
Fornasier, Sonia
Forrest, William
Fors, Octavi
Forveille, Thierry
Foryta, Dietmar
Foschini, Luigi
Fouchard, Marc
Fouqué, Pascal
Fraenz, Markus
Franklin, Fred
Fraser, Brian
Fraser, Helen
Fredrick, Laurence
Freire Ferrero, Rubens
Freitas Mourão, Ronaldo
Fridlund, Malcolm
Froeschle, Christiane
Froeschle, Claude
Fromang, Sebastien
FUJIMOTO, Masa-Katsu
FUJISHITA, Mitsumi
FUJIWARA, Akira
FUKAGAWA, Misato
Fulchignoni, Marcello
FURUSHO, Reiko
FUSE, Tetsuharu
Gábor, Pavel
Gänsicke, Boris
Gajdoš, Stefan
Galád, Adrián
Gallardo Castro, Carlos
Galletta, Giuseppe
Gammelgaard, Peter
Gandolfi, Davide
GAO, Jian
García de María, Juan
Garcia-Hernandez, Domingo
Gargaud, Muriel
Gatewood, George
Gautier, Daniel
Gawroński, Marcin
Geiss, Johannes
Geller, Aaron
Gerakines, Perry
Gerard, Eric
Gerard, Jean-Claude
Ghigo, Francis
Giani, Elisabetta
Gibson, James
Gierasch, Peter
Gil-Hutton, Ricardo
Gillon, Michaël
Gilmore, Alan
Giorgini, Jon
Giovane, Frank
Giovannelli, Franco
Girard, Julien
Glass, Billy
Glatzmaier, Gary
Gledhill, Timothy
Godwin, Jon
Golden, Aaron
Goldreich, Peter
Goldsmith, Donald
Goldstein, Richard
Gomez, Edward
Gomez, Mercedes
Gómez de Castro, Ana
Gonzalez, Jean-Francois
Goody, Richard
Gorbanev, Jury
Gorenstein, Paul
Gor'kavyi, Nikolai
Gorshanov, Denis
Goswami, J.
Gott, J.
Goudas, Constantine
Goudis, Christos
Gounelle, Matthieu
Goździewski, Krzysztof
Gradie, Jonathan
Grady, Monica
Granvik, Mikael
Gratton, Raffaele
Grav, Tommy
Green, Daniel
Green, Jack
Green, Simon
Greenberg, Richard
Greenhouse, Matthew
Gregory, Philip
Grieger, Bjoern
Griffin, Ian
Grillmair, Carl
Grinspoon, David
Gronchi, Giovanni
Gronkowski, Piotr
Grossman, Lawrence
Grueff, Gavril
Grün, Eberhard
Grundy, William
Guenther, Eike
Guinan, Edward
Gulkis, Samuel
GUNJI, Shuichi
Gunn, James
GUO, Jianheng
Gurshtein, Alexander
Gustafson, Bo
Guzik, Joyce
Hadamcik, Edith
Haghighipour, Nader
Hahn, Gerhard
Hainaut, Olivier
Haisch, Bernard
Hajdukova, Maria
Hajduková, Jr., Maria
Hale, Alan
Halliday, Ian
Hallinan, Gregg
Hamilton, Douglas
Hammel, Heidi
HAN, Inwoo
HANAWA, Tomoyuki
Hanner, Martha
Hänninen, Jyrki
Hapke, Bruce
Haque, Shirin
HARA, Tetsuya
Harper, David
Harris, Alan
Harris, Alan

Hart, Michael
Hartmann, William
Hartogh, Paul
Harvey, Gale
Harwit, Martin
HASEGAWA, Ichiro
HASEGAWA, Sunao
Hatzes, Artie
Haugbølle, Troels
Haupt, Hermann
Hauschildt, Peter
Hawkes, Robert
HE, Miao-fu
Heck, Andre
Helled, Ravit
Hellier, Coel
Helmich, Frank
Helou, George
Hemenway, Paul
Hempel, Marc
Heng, Kevin
Herbst, Eric
Hernández, Jesús
HERSANT, Franck
Hershey, John
Hestroffer, Daniel
Heudier, Jean-Louis
Heyl, Jeremy
Hidayat, Taufiq
Hinners, Noel
HIRABAYASHI, Hisashi
HIRAO, Takanori
HIROSE, Shigenobu
Hoang, Binh
Hoard, Donald
Hobbs, David
Hodapp, Klaus
Hodge, Paul
Hönig, Sebastian
Högbom, Jan
Hohenkerk, Catherine
Hol, Pedro
Holberg, Jay
Hollis, Jan
Holm, Nils
Homeier, Derek
HONG, Seung-Soo
Horedt, Georg
Horiuchi, Shinji
Horowitz, Paul
HOU, Xiyun
Hovenier, J.
Howard, Andrew
Howell, Ellen
Hsieh, Henry
HU, Zhong wen
Hubbard, William
Hudkova, Ludmila
Huebner, Walter
Hunt, Garry
Hunter, James
Huntress, Wesley
Hurnik, Hieronim
HURUKAWA, Kiichirō
Husárik, Marek
HWANG, Junga
Hysom, Edmund
Ianna, Philip
Ibadinov, Khursand
Ibadov, Subhon
Inglis, Michael
INOUE, Akio
IP, Wing-Huen
Ireland, Michael
Iro, Nicolas
Irvine, William
Irwin, Patrick
Isaak, Kate
ISHIGURO, Masateru
Israel, Frank
Israelevich, Peter
ITO, Takashi
ITOH, Yoichi
Ivanov, Pavel
Ivanova, Oleksandra
Ivanova, Violeta
Ivantsov, Anatoliy
Ivezic, Zeljko
IWASAKI, Kyosuke
Izmailov, Igor
Jackson, Bernard
Jackson, William
Jacobson, Robert
Jakobsson, Pall
Jakubík, Marian
Jalali, Mir Abbas
Janches, Diego
Janes, Kenneth
Janssen, Michael
Jardine, Moira
Jastrow, Robert
Jayawardhana, Ray
Jedicke, Robert
Jeffers, Sandra
Jehin, Emmanuel
JEON, Young Beom
JI, Jianghui
JIANG, Ing-Guey
Jin, Liping
Jockers, Klaus
Johansen, Anders
Johnson, Torrence
Johnston, Kenneth
Jones, Derek
Jones, Eric
Jones, James
Jopek, Tadeusz
Jorda, Laurent
Jordán, Andrés
Jura, Michael
Jurgens, Raymond
Kaasalainen, Mikko
Kabath, Petr
Kablak, Nataliya
Käufl, Hans Ulrich
Kafatos, Menas
Kafka, Styliani (Stella)
KAIFU, Norio
Kalenichenko, Valentin
Kaltenegger, Lisa
Kamal, Fouad
Kandel, Robert
Kane, Stephen
Kaňuchová, Zuzana
Karakas, Amanda
Karatekin, Özgür
Kardashev, Nicolay
Karoff, Christoffer
Karttunen, Hannu
Kascheev, Rafael
Kasuga, Toshihiro
Kaufmann, Pierre
Kavelaars, JJ.
Kawada, Mitsunobu
KAWAKITA, Hideyo
Kawaler, Steven
Kaydash, Vadym
Kazantsev, Anatolii
Keay, Colin
Kedziora-Chudczer, Lucyna

Keheyan, Yeghis
Keil, Klaus
Kelemen, Janos
Keller, Hans-Ulrich
Keller, Horst
Kellermann, Kenneth
Kenworthy, Matthew
Kern, Pierre
Kervella, Pierre
Keskin, Varol
Khanna, Ramon
Khodachenko, Maxim
Khosroshahi, Habib
Khovritchev, Maxim
Khrutskaya, Evgenia
Kibblewhite, Edward
Kidger, Mark
Kikwaya Eluo, Jean-Baptiste
Killen, Rosemary
Kilston, Steven
Kim, Joo Hyeon
Kim, Yoo Jea
KIM, Kwang tae
KIM, Sang Joon
KIM, Seung-Lee
KIM, Yongha
KIMURA, Hiroshi
KING, Sun-Kun
KINOSHITA, Hiroshi
Kiselev, Nikolai
Kiss, Csaba
Kisseleva, Tamara
Kitiashvili, Irina
Kjaergaard, Per
Kjeldsen, Hans
Klacka, Jozef
Klahr, Hubert
Klemola, Arnold
Kley, Wilhelm
Klioner, Sergei
Knacke, Roger
Knežević, Zoran
Knight, Matthew
Knoelker, Michael
Knowles, Stephen
KO, Chung-Ming
Kobayashi, Shiho
KOBAYASHI, Yukiyasu
Kocer, Dursun
Koeberl, Christian
Kohoutek, Lubos
Kokhirova, Gulchehra
KOKUBO, Eiichiro
Kolb, Ulrich
Kolokolova, Ludmilla
Kolomiyets, Svitlana
Komonjinda, Siramas
Konacki, Maciej
Konigl, Arieh
Korhonen, Heidi
Kornoš, Leonard
Korokhin, Viktor
Korpela, Eric
Korsun, Pavlo
KOSAI, Hiroki
Koschny, Detlef
Koshkin, Nikolay
Kostama, Veli-Petri
Koten, Pavel
Kouwenhoven, M.B.N. (Thijs)
Kovačević, Andjelka
Kovacs, Geza
Kovács, József
KOZAI, Yoshihide
Kozak, Lyudmyla
Kozak, Pavlo
KOZASA, Takashi
Kozłowski, Szymon
Kraft, Ralph
Kramer, Kh
Kraus, Stefan
Krimigis, Stamatios
Krishna, Swamy
Kristensen, Leif
Krivov, Alexander
Królikowska-Sołtan, Małgorzata
Kroll, Peter
Kruchinenko, Vitaliy
Krüger, Harald
Krugly, Yurij
Kryszczynska, Agnieszka
Kryvodubskyj, Valery
Ksanfomality, Leonid
KUAN, Yi-Jehng
Küppers, Michael
Kürster, Martin
Kuiper, Thomas
Kulikova, Nelly
Kumar, Shiv
Kupka, Friedrich
Kurt, Vladimir
Kurtev, Radostin
Kuzmanoski, Mike
Kwiatkowski, Tomasz
Kwok, Sun
Labeyrie, Antoine
Lacour, Sylvestre
Lacy, John
Lafon, Jean-Pierre
Lagage, Pierre-Olivier
Lagerkvist, Claes-Ingvar
Lahulla, J.
Lala, Petr
Lammer, Helmut
Lamontagne, Robert
Lamy, Philippe
Landolt, Arlo
Lane, Arthur
Lang, Kenneth
Lanz, Thierry
Lanza, Antonino
Lanzafame, Alessandro
Laques, Pierre
Lara, Luisa
Larsen, Jeffrey
Larson, Harold
Larson, Stephen
Laskar, Jacques
La Spina, Alessandra
Latham, David
Lattanzio, John
Laufer, Diana
Laurin, Denis
Lawson, Peter
Lazio, Joseph
Lazzarin, Monica
Lazzaro, Daniela
Leach, Sydney
Lebofsky, Larry
Lebreton, Yveline
Lecavelier des Etangs, Alain
LEE, Chung-Uk
LEE, Jae Woo
LEE, Jun
LEE, Ki-Won
LEE, Myung Gyoon

LEE, Sang-Gak
LEE, Thyphoon
Lefloch, Bertrand
Leger, Alain
Leinert, Christoph
Lellouch, Emmanuel
Lemaire, Joseph
Lemaître, Anne
Lemmon, Mark
Lester, Daniel
Lestrade, Jean-François
Leto, Giuseppe
Levasseur-Regourd, Anny-Chantal
Levy, Eugene
Lewis, John
Li, Jian-Yang
LI, Guangyu
LI, Lifang
LI, Yong
LIAO, Xinhao
Libert, Anne-Sophie
Licandro, Javier
Lichtenegger, Herbert
Lieske, Jay
Liller, William
Lilley, Edward
Lillie, Charles
Lin, Douglas
Lindqvist, Michael
Lindsey, Charles
Lineweaver, Charles
Lintott, Chris
Lippincott Zimmerman, Sarah
Lipschutz, Michael
Lis, Dariusz
Lissauer, Jack
Lisse, Carey
LIU, Chengzhi
LIU, Fukun
LIU, Michael
LIU, Sheng-Yuan
LIU, Xiaoqun
LIU, Yujuan
Livio, Mario
Lo, Wing-Chi Nadia
Lockwood, G.
Lo Curto, Gaspare
Lodders, Katharina
Lodieu, Nicolas
Lomb, Nicholas
Longmore, Andrew
Lopes, Rosaly
Lopez Moreno, Jose
Lopez Puertas, Manuel
Lopez Valverde, M.
LOU, Yu-Qing
Lovas, Miklos
Lovis, Christophe
Lowe, Robert
Lucchesi, David
Lugaro, Maria
Lukyanyk, Igor
LUO, Ali
Lutz, Barry
Luu, Jane
Luz, David
Lyon, Ian
MA, Guanyi
MA, Yuehua
Mackay, Duncan
Maddison, Sarah
Magee-Sauer, Karen
Maggio, Antonio
Magnusson, Per
Mainzer, Amy
Makalkin, Andrei
Malbet, Fabien
Malhotra, Renu
Mamajek, Eric
Manara, Alessandro
Mancìni, Dario
Manimanis, Vassilios
Mann, Ingrid
Manset, Nadine
Maran, Stephen
Marchi, Simone
Marcialis, Robert
Marciniak, Anna
Marcy, Geoffrey
Mardling, Rosemary
Marengo, Massimo
Margot, Jean-Luc
Margrave Jr, Thomas
Maris, Michele
Marley, Mark
Marois, Christian
Marov, Mikhail
Marsh, Thomas
Martín, Eduardo
Martin, Maria
Martinez Fiorenzano, Aldo
Martinez-Frias, Jesus
Martin-Pintado, Jesus
Martins, Roberto
Marvin, Ursula
Marzari, Francesco
Marzo, Giuseppe
Masciadri, Elena
Masiero, Joseph
Mason, John
Matese, John
Matsakis, Demetrios
Matson, Dennis
MATSUDA, Takuya
MATSUI, Takafumi
Matsuura, Oscar
Matthews, Brenda
Matthews, Clifford
Matthews, Jaymie
Mauas, Pablo
Maury, Alain
Mawet, Dimitri
Mayor, Michel
Mazeh, Tsevi
Mazzotta Epifani, Elena
McAlister, Harold
McCord, Thomas
McCullough, Peter
McDonough, Thomas
McElroy, Michael
McFadden, Lucy
McGimsey Jr, Ben
McGrath, Melissa
McIntosh, Bruce
McKenna-Lawlor, Susan
McKinnon, William
McMillan, Robert
McNaught, Robert
Meadows, A.
Medvedev, Yurij
Meech, Karen
Meeus, Gwendolyn
Megeath, S.
Meibom, Soren
Meisel, David
Melita, Mario
Melo, Claudio
Melott, Adrian

Méndez Bussard, Rene
Mendillo, Michael
Mendoza, V.
Menzies, John
Merín Martín, Bruno
Merline, William
Messerotti, Mauro
Messina, Sergio
Meyer, Michael
Michałowski, Tadeusz
Michel, Patrick
Mickelson, Michael
Migaszewski, Cezary
Mignard, François
MIKAMI, Takao
Mikhail, Joseph
Milani Comparetti, Andrea
Miles, Howard
Millan-Gabet, Rafael
Millis, Robert
Mills, Franklin
Milone, Eugene
MINH, Young-Chol
Minier, Vincent
Minikulov, Nasridin
Mink, Jessica
MINN, Young-Ki
Minniti, Dante
Mintz Blanco, Betty
Mirabel, Igor
Misconi, Nebil
MITRA, Dhrubaditya
Mkrtichian, David
Modali, Sarma
Moehlmann, Diedrich
Moerchen, Margaret
Mohd, Zambri
Mokhele, Khotso
Molina, Antonio
Molinari, Emilio
MOMOSE, Munetake
Moncuquet, Michel
Monet, Alice
Monin, Jean-Louis
Monteiro, Mario Joao
Montes, David
Montmerle, Thierry
Montmessin, Franck
Mookerjea, Bhaswati
MOON, Hong-Kyu
Moór, Attila
Moore, Elliott
Moore, Marla
Moore-Weiss, John
Moravec, Zdeněk
Morbidelli, Alessandro
Mordasini, Christoph
Moreira Morais, Maria
Moreno-Insertis, Fernando
Morozhenko, A.
Morris, Charles
Morris, Mark
Morrison, David
Mosser, Benoît
Mothé-Diniz, Thais
Mousis, Olivier
Moussas, Xenophon
Mueller, Beatrice
Mueller, Thomas
Muinonen, Karri
MUKAI, Tadashi
Mukherjee, Krishna
Muller, Richard
Mumma, Michael
MURAKAMI, Naoshi
MURATA, Yasuhiro
Murphy, Robert
Murray, Carl
Murray, John
Musielak, Zdzislaw
Nacozy, Paul
Nadeau, Daniel
Naef, Dominique
NAGAHARA, Hiroko
NAKAGAWA, Yoshitsugu
NAKAJIMA, Tadashi
NAKAMOTO, Taishi
NAKAMURA, Akiko
NAKAMURA, Tsuko
NAKAMURA, Takuji
NAKANO, Syuichi
NAKASHIMA, Jun-ichi
NAKAZAWA, Kiyoshi
Namouni, Fathi
Nandi, Dibyendu
Naoz, Smadar
Napier, William
NARITA, Norio
Nedelcu, Dan
Nelson, Richard
Nelson, Robert
Neslusan, Lubos
Ness, Norman
Neuhaeuser, Ralph
Newburn Jr, Ray
Nguyễn, Phuong
Niarchos, Panagiotis
Niedner, Malcolm
Niedzielski, Andrzej
Niemczura, Ewa
Nikoli'c, Silvana
Ninkov, Zoran
Nitschelm, Christian
Nixon, Conor
Nobili, Anna
Noerdlinger, Peter
Nolan, Michael
Noll, Keith
Nordh, Lennart
Norris, Raymond
Novaković, Bojan
Noyelles, Benoît
Nuth, Joseph
Oberst, Thomas
O'Brien, Tim
O'Dell, Charles
Ofir, Aviv
O'Handley, Douglas
OHISHI, Masatoshi
OHNISHI, Kouji
OHTSUKI, Keiji
OKUMURA, Shin-ichiro
Oliva, Ernesto
Olive, Don
Ollongren, A.
ONAKA, Takashi
Opara, Fidelix
Orchiston, Wayne
Orellana, Rosa
Ortiz, Jose
Orton, Glenn
Ostriker, Jeremiah
O'Toole, Simon
Owen, Tobias
Owen Jr, William
Oezisik, Tuncay
Padmanabhan, Janardhan
Pagano, Isabella
Page, Gary
Pál, András

Pallé Bagó, Enric
Palmer, Patrick
Palumbo, Maria Elisabetta
Pandey, A.
Pang, Kevin
Pantoja, Carmen
Paolicchi, Paolo
Papaelias, Philip
Papaloizou, John
Parijskij, Yurij
Parisot, Jean-Paul
PARK, Byeong-Gon
PARK, Yong Sun
Pascu, Dan
Pascucci, Ilaria
Pat-El, Igal
Patten, Brian
Pauwels, Thierry
Peale, Stanton
Pecina, Petr
Peixinho, Nuno
Pellas, Paul
Pellinen-Wannberg, Asta
Pendleton, Yvonne
Penny, Alan
Penny, Matthew
Pepe, Francesco
Perek, Luboš
Perets, Hagai
Perez, Mario
Perez de Tejada, Hector
Pérez-Garrido, Antonio
Perozzi, Ettore
Perrin, Marshall
Persson, Carina
Petit, Jean-Marc
Petr-Gotzens, Monika
Pettengill, Gordon
Picazzio, Enos
Pierce, David
Piironen, Jukka
Pilat-Lohinger, Elke
Pilbratt, Göran
Pilcher, Carl
Pilling, Sergio
PING, Jinsong
Pinte, Christophe
Piotto, Giampaolo
Pirronello, Valerio
Piskunov, Nikolai
Pitjeva, Elena
Pittich, Eduard
Pittichova, Jana
Plavchan, Peter
Plavec, Zdenka
Podolak, Morris
Polidan, Ronald
Politi, Romolo
Pollacco, Don
Polyakhova, Elena
Ponsonby, John
Pontoppidan, Klaus
Poole, Graham
Poretti, Ennio
Porubcan, Vladimir
Potter, Andrew
Pozhalova, Zhanna
Pravec, Petr
Press, William
Prialnik, Dina
Proisy, Paul
Prša, Andrej
Psaryov, Volodymyr
Puxley, Phil
Puzia, Thomas
QIAO, Rongchuan
QIU, Yaohui
Quanz, Sascha
Queloz, Didier
Quintana, Hernan
Quintana, José
Quirrenbach, Andreas
Radiman, Iratio
Rajamohan, R.
Raju, Vasundhara
Rank-Lueftinger, Theresa
Rao, M.
Rapaport, Michel
Rasio, Frederic
Rasmussen, Ib
Rastegaev, Denis
Rauer, Heike
Rawlings, Mark
Reach, William
Rebolo, Rafael
Rebull, Luisa
Reddy, Bacham
Redman, Stephen
Rees, Martin
Reffert, Sabine
Reid, Iain
Reiners, Ansgar
Reitsema, Harold
Rekola, Rami
Rendtel, Juergen
Rengel, Miriam
Reyes-Ruiz, Mauricio
RHEE, Myung Hyun
Ribak, Erez
Ribas, Ignasi
Ribeiro, Marcelo
Richardson, Derek
Rickman, Hans
Ridgway, Stephen
Ripken, Hartmut
Robb, Russell
Roberts Jr, Lewis
Rodionova, Zhanna
Rodrigo, Rafael
Rodriguez, Luis
Roemer, Elizabeth
Roeser, Siegfried
Roig, Fernando
Rojo, Patricio
Romero Pérez, Maria
Ron, Cyril
Roos-Serote, Maarten
Roques, Françoise
Rosenbush, Vera
Rossi, Alessandro
Rouan, Daniel
Rousselot, Philippe
Rovithis-Livaniou, Helen
Rowan-Robinson, Michael
Roxburgh, Ian
Rubenstein, Eric
Rudnitskij, Georgij
RUI, Qi
Ruiz, Maria Teresa
Ruskol, Evgeniya
Russel, Sara
Russell, Christopher
Russell, Jane
Russell, Kenneth
Russo, Pedro
Ryabova, Galina
Saad, Somaya
Saffari, Reza
Saffe, Carlos
Sagdeev, Roald

Sahai, Raghvendra
Sahu, Kailash
Saissac, Joseph
SAKURAI, Kunitomo
Salama, Farid
Salitis, Antonijs
Salo, Heikki
Samarasinha, Nalin
Sampson, Russell
Sánchez Béjar, Victor
Sanchez-Lavega, Agustin
Sancisi, Renzo
Santos-Sanz, Pablo
Sari, Re'em
Sarre, Peter
Sasaki, Sho
SATO, Bunei
SATO, Isao
Sault, Robert
Savanevich, Vadim
Savanov, Igor
Savin, Daniel
Scargle, Jeffrey
Schaller, Emily
Scheeres, Daniel
Scheirich, Peter
Schild, Rudolph
Schleicher, David
Schloerb, F.
Schmadel, Lutz
Schmidt, Maarten
Schneider, Jean
Schneider, Nicholas
Schober, Hans
Scholl, Hans
Schubart, Joachim
Schuch, Nelson
Schuler, Simon
Schulz, Rita
Schuster, William
Segaluvitz, Alexander
Seidelmann, P.
Seielstad, George
Seiradakis, John
Sekanina, Zdenek
SEKIGUCHI, Tomohiko
Selam, Selim
Semenov, Dmitry
Sen, Asoke
Sengupta, Sujan
SEO, Haingja
Sergis, Nick
Serra Ricart, Miquel
Sese, Rogel Mari
SHANG, Hsien
Shankland, Paul
Shanklin, Jonathan
Shapiro, Irwin
Sharma, A.
Sharp, Christopher
Shelus, Peter
SHEN, Chun-Shan
SHEN, Kaixian
Shevchenko, Ivan
Shevchenko, Vasilij
Shevchenko, Vladislav
SHI, Huli
Shkuratov, Yurii
Shor, Viktor
Shostak, G.
Shoyoqubov, Shoayub
Shrbený, Lukáš
Shulga, Oleksandr
Shustov, Boris
Sicardy, Bruno
Sickafoose, Amanda
Sidore'nkov, Nikolay
Siebenmorgen, Ralf
Silva, Laura
Silvotti, Roberto
Simek, Milos
Simon, Amy
Simonia, Irakli
Sims, Mark
Sinachopoulos, Dimitris
Singh, Harinder
Sitarski, Grzegorz
Sitko, Michael
Sivaram, C.
Sivaraman, Koduvayur
Sizonenko, Yuri
Sjogren, William
Skillen, Ian
Slade, Martin
Smalley, Barry
Smith, Bradford
Smith, Michael
Smith, Verne
Snellen, Ignas
Snik, Frans
Snyder, Lewis
Soberman, Robert
Soderblom, David
Soderblom, Larry
Soderhjelm, Staffan
SOFIA LYKAWKA, Patryk
SOFUE, Yoshiaki
Solc, Martin
Solovaya, Nina
SOMA, Mitsuru
Sonett, Charles
SONG, In-Ok
Soonthornthum, Boonrucksar
Sosa, Andrea
Southworth, John
Sozzetti, Alessandro
Spaans, Marco
Spahr, Timothy
Sparks, William
Spencer, John
Spinrad, Hyron
Sprague, Ann
Spurny, Pavel
Stalio, Roberto
Stallard, Thomas
Stam, Daphne
Standish, E.
Stapelfeldt, Karl
Steel, Duncan
Stein, John
Stellmacher, Irène
Stencel, Robert
Stern, S.
Sterzik, Michael
Stewart-Mukhopadhyay, Sarah
Stoev, Alexey
Stokes, Grant
Stone, Edward
Stone, Remington
Storey, John
Storrs, Alexander
Strafella, Francesco
Straižys, Vytautas
Strazzulla, Giovanni
Strelnitski, Vladimir
Stringfellow, Guy
Strobel, Darrell

Strom, Robert
Sturrock, Peter
Süli, Áron
Sullivan, III, Woodruff
Sultanov, G.
SUMI, Takahiro
SUNADA, Kazuyoshi
Surdej, Jean
SUTO, Yasushi
Svestka, Jiri
Svoren, Jan
Swade, Daryl
Sybiryakova, Yegeniya
Sykes, Mark
Synnott, Stephen
Szabó, Gyula
Szabo, Robert
Szego, Karoly
Székely, Péter
Szutowicz, Sławomira
Tacconi-Garman, Lowell
Tafalla, Mario
TAKABA, Hiroshi
TAKADA-HIDAI, Masahide
TAKAHARA, Mariko
TAKEDA, Hidenori
TAMURA, Motohide
TANABE, Hiroyoshi
Tancredi, Gonzalo
Tanga, Paolo
TAO, Jun
Tarashchuk, Vera
Tarter, Jill
TARUYA, Atsushi
Taş, Günay
Tatum, Jeremy
Tavakol, Reza
Taylor, Andrew
Taylor, Donald
Taylor, Fredric
Tchouikova, Nadezhda
Tedesco, Edward
Teixeira, Paula
Teixeira, Ramachrisna
Tejfel, Victor
Tennyson, Jonathan
Tepper Garcia, Thorsten
Terentjeva, Alexandra
Terrile, Richard
Terzian, Yervant
Testi, Leonardo
Thaddeus, Patrick
Tholen, David
Thomas, Nicolas
Thomsen, Bjarne
Thuillot, William
TIAN, Feng
Tichá, Jana
Tinney, Christopher
Tiscareno, Matthew
Tisserand, Patrick
Titov, Vladimir
Tolbert, Charles
Torres, Guillermo
Tosi, Federico
Toth, Imre
Toth, Juraj
Tovmassian, Hrant
Townes, Charles
Tozzi, Gian
Trafton, Laurence
Traub, Wesley
Tremaine, Scott
Trigo-Rodríguez, Josep
Trimble, Virginia
Tritton, Keith
Trujillo, Chadwick
TSAI, An-Li
Tsiganis, Kleomenis
Tsuchida, Masayoshi
TSUNETA, Saku
Tsvetanov, Zlatan
Tuccari, Gino
Tull, Robert
Tungalag, Namkhai
Turner, Edwin
Turner, Kenneth
Tyler Jr, G.
Udry, Stephane
UENO, Munetaka
Ugolnikov, Oleg
UMEMOTO, Tomofumi
Unda-Sanzana, Eduardo
Unwin, Stephen
Usowicz, Jerzy
Vakoch, Douglas
Valdés-Sada, Pedro
Valdivielso, Luisa
Válio, Adriana
Vallée, Jacques
Valsecchi, Giovanni
Valtaoja, Esko
van Belle, Gerard
Vandenbussche, Bart
van der Hucht, Karel
van Dishoeck, Ewine
van Hamme, Walter
Van Hoolst, Tim
van Houten-Groeneveld, Ingrid
Vaňko, Martin
Varshalovich, Dmitrij
Varvoglis, Harry
Vashkov'yak, Mikhail
Vaubaillon, Jérémie
Vauclair, Gérard
Vauclair, Sylvie
Vavrek, Roland
Vazan, Allona
Vázquez, Manuel
Vázquez, Roberto
Veeder, Glenn
Veiga, Carlos
Veillet, Christian
Velikodsky, Yuri
Venugopal, V.
Verschuur, Gerrit
Veverka, Joseph
Vidal-Madjar, Alfred
Vidmachenko, Anatoliy
Vidotto, Aline
Vieira Neto, Ernesto
Vienne, Alain
Vilas, Faith
Villaver, Eva
Virtanen, Jenni
Voelzke, Marcos
Vogt, Nikolaus
Voloshchuk, Yuri
von Braun, Kaspar
von Hippel, Theodore
Vorobyov, Eduard
Vukotic, Branislav
WADA, Takehiko
Wainscoat, Richard
Walker, Alta
Walker, Gordon
Wallace, James
Wallace, Lloyd

Wallis, Max
Walsh, Andrew
Walsh, Wilfred
Walton, Nicholas
Wambsganß, Joachim
Wandel, Amri
WANG, Hongchi
WANG, Na
WANG, Shiang-Yu
WANG, Xiao-bin
WANG, Xiaoya
Wannawichian, Suwicha
Wardle, Mark
Wasserman, Lawrence
Wasson, John
WATANABE, Junichi
Watson, Frederick
Wdowiak, Thomas
Weaver, Harold
Webster, Alan
Wegner, Gary
Wehinger, Peter
Weidenschilling, S.
Weinberg, Jerry
Weiss, Werner
Weissman, Paul
Welch, William
Wellington, Kelvin
Wells, Eddie
Welsh, William
Wesselius, Paul
Wesson, Paul
West, Richard
Wheatley, Peter
Whelan, Emma
Whipple, Arthur
Whitaker, Ewen
White, Glenn
Whittet, Douglas
Wiebe, Dmitri
Wiedemann, Günter
Wiegert, Paul
Wielebinski, Richard
Wilkening, Laurel
Williams, Gareth
Williams, Iwan
Williams, James
Williams, Jonathan
Willson, Robert
Wilner, David
Wilson, Lionel
Wilson, Thomas
Winn, Joshua
Wittenmyer, Robert
Wnuk, Edwin
Wolstencroft, Ramon
Wolszczan, Alexander
Womack, Maria
Wooden, Diane
Woolfson, Michael
Wootten, Henry
Wright, Alan
Wright, Ian
Wu, Yanqin
WU, Lianda
WU, Zhen-Yu
Wurz, Peter
Wyckoff, Susan
XIAO, Dong
XIONG, Jianning
XU, Weibiao
YABUSHITA, Shin
Yahil, Amos
Yair, Yoav
YAMAMOTO, Masayuki
YAMASHITA, Takuya
YANAGISAWA, Masahisa
Yanamandra-Fisher, Padma
YANG, Jongmann
YANG, Xiaohu
YANG, Zhigen
YANO, Hajime
YE, Shuhua
Yeomans, Donald
Yeşilyurt, Ibrahim
YI, Yu
YIM, Hong Suh
Yoder, Charles
YOSHIDA, Fumi
YOSHIKAWA, Makoto
YOSHIOKA, Kazuo
Young, Andrew
Young, Louise
Yseboodt, Marie
YU, Cong
YU, Qingjuan
YUASA, Manabu
Yüce, Kutluay
YUMIKO, Oasa
Yun, Joao
YUTAKA, Shiratori
Zadnik, Marjan
Zagretdinov, Renat
Zahn, Jean-Paul
Zapatero-Osorio, Maria Rosa
Zappalà, Vincenzo
Zarka, Philippe
Zarnecki, John
Zellner, Benjamin
Zender, Joe
Zhang, Tielong
ZHANG, Jiaxiang
ZHANG, Jun
ZHANG, Mian
ZHANG, Qiang
ZHANG, Wei
ZHANG, Xiaoxiang
ZHANG, Yang
ZHANG, You-Hong
ZHAO, Gang
ZHAO, Haibin
Zharkov, Vladimir
ZHOU, Li-Yong
ZHU, Jin
Zinnecker, Hans
Ziołkowski, Krzysztof
Zucker, Shay
Zuckerman, Benjamin

Division G Stars and Stellar Physics

President: Ignasi Ribas
Vice-President: Corinne Charbonnel

Steering Committee Members:

Conny Aerts
Karen Pollard
Richard O. Gray
Marco Limongi
Katia Cunha
Beatriz Barbuy
Mercedes T. Richards
David R. Soderblom
Virginia Trimble
Corinne Charbonnel
Birgitta Nordström
Francesca D'Antona
Brian D. Mason
Joachim Puls
Ignasi Ribas

Members:

Abbott, David
Abia, Carlos
Ables, Harold
Abraham, Zulema
Abrevaya, Ximena
Absil, Olivier
Abt, Helmut
Adams, Mark
Adelman, Saul
Aerts, Conny
Afram, Nadine
Agüeros, Marcel
Ahumada, Javier
Aigrain, Suzanne
AIKAWA, Toshiki
Aikman, G.
Airapetian, Vladimir
Aizenman, Morris
AK, Serap
AK, Tansel
Akashi, Muhammad
Ake III, Thomas
AKITAYA, Hiroshi
Albinson, James
Albrecht, Rudolf
Albrow, Michael
Alcalá, Juan Manuel
Alcolea, Javier
Alecian, Evelyne
Alecian, Georges
Alencar, Silvia
Alfaro, Emilio
ALIBERT, Yann
Alksnis, Andrejs
Allan, David
Allard, France
Allen, Christine
Allende Prieto, Carlos
Al-Naimiy, Hamid
Alpar, Mehmet
Althaus, Leandro
Altrock, Richard
Alves, João
Amado, Pedro
Amer, Morsi
Anandaram, Mandayam
Andersen, Johannes
Anderson, Joseph
Anderson, Kurt
ANDO, Hiroyasu
Andrade, Manuel
Andretta, Vincenzo
Andreuzzi, Gloria
Andrievsky, Sergei
Andrillat, Yvette
Andronov, Ivan
Angel, J.
Angelov, Trajko
Angione, Ronald
Annuk, Kalju
Anosova, Joanna
Antia, H.
Antipin, Sergei
Antipova, Lyudmila
Antokhin, Igor
Antokhina, Eleonora
Antonello, Elio
Antoniou, Vallia
Antonopoulou, Evgenia
Antonova, Antoaneta
Antonyuk, Kirill
Antov, Alexandar
Anupama, G.
AOKI, Wako
Appenzeller, Immo
Aquilano, Roberto
ARAI, Kenzo
Arbutina, Bojan
Ardeberg, Arne
Ardila, David
Arefiev, Vadim
Arellano Ferro, Armando
Arenou, Frédéric
Arentoft, Torben
Aret, Anna
Argast, Dominik
Argyle, Robert
Arias, Maria
ARIMOTO, Nobuo
Arion, Douglas
Arkharov, Arkadij
Arkhipova, Vera
Armstrong, John
Arnaboldi, Magda
Arnett, W.
Arnould, Marcel
Arpigny, Claude
Arribas Mocoroa, Santiago
Arsenijevic, Jelisaveta
Asanok, Kitiyanee
Ashley, Michael
Ashok, N.

Aspin, Colin
Asplund, Martin
Asteriadis, Georgios
Atac, Tamer
Atanackovic, Olga
Audard, Marc
Audouze, Jean
Auer, Lawrence
Aufdenberg, Jason
Auman, Jason
Aungwerojwit, Amornrat
Auriere, Michel
Avgoloupis, Stavros
Avrett, Eugene
Awadalla, Nabil
Ayres, Thomas
Azzopardi, Marc
Baade, Dietrich
Baade, Robert
BABA, Hajime
BABA, Naoshi
Babkovskaia, Natalia
Babu, G.S.D.
Baglin, Annie
Bagnulo, Stefano
Bagnuolo Jr., William
Bailer-Jones, Coryn
Bailes, Matthew
Bailyn, Charles
Baird, Scott
Bakker, Eric
Baldinelli, Luigi
Balega, Yurij
Baliunas, Sallie
Baliyan, Kiran
Ballereau, Dominique
Balman, Solen
Balona, Luis
Banerjee, Dipankar
Baptista, Raymundo
Baran, Andrzej
Baransky, Olexander
Baratta, Giovanni
Barbá, Rodolfo
Barban, Caroline
Barber, Robert
Barbuy, Beatriz
Barcza, Szabolcs
Barkin, Yuri
Barklem, Paul
Barlow, Michael
Barnes, Graham
Barnes, Sydney
Barnes III, Thomas
Baron, Edward
Barone, Fabrizio
Barrado Navascués, David
Barrett, Paul
Barrientos, Luis
Barsony, Mary
Barstow, Martin
Bartašiute, Stanislava
Bartaya, R.
Bartkevicius, Antanas
Bartkiewicz, Anna
Bartolini, Corrado
Barway, Sudhashu
Barwig, Heinz
Bary, Jeffrey
Baschek, Bodo
Baskill, Darren
Basri, Gibor
Bastien, Pierre
Basu, Sarbani
Batalha, Celso
Batten, Alan
Baudry, Alain
Bauer, Wendy
Baume, Gustavo
Baym, Gordon
Bazot, Michael
Bear, Ealeal
Beasley, Anthony
Beaudet, Gilles
Beaulieu, Jean-Philippe
Beavers, Willet
Becker, Stephen
Beckman, John
Bedding, Timothy
Bedogni, Roberto
Beers, Timothy
Beiersdorfer, Peter
Beklen, Elif
Belkacem, Kevin
Bell, Steven
Bellas-Velidis, Ioannis
Bellazzini, Michele
Belmonte Aviles, Juan Antonio
Belserene, Emilia
Belvedere, Gaetano
Benacquista, Matthew
Benaglia, Paula
Bendjoya, Philippe
Benedict, George
Benetti, Stefano
Benkő, Jozsef
Bennett, Philip
Bensby, Thomas
Benz, Willy
Berdnikov, Leonid
Berdyugin, Andrei
Berdyugina, Svetlana
Berg, Richard
Bergeat, Jacques
Bergeron, Pierre
Bernabeu, Guillermo
Bernat, Andrew
Bersier, David
Bertelli, Gianpaolo
Berthomieu, Gabrielle
Bertone, Emanuele
Bertout, Claude
Bessell, Michael
Bianchi, Luciana
Bianchini, Antonio
Biazzo, Katia
Bikmaev, Ilfan
Bingham, Robert
Birkmann, Stephan
Bisikalo, Dmitrij
Bisnovatyi-Kogan, Gennadij
Bjorkman, Jon
Bjorkman, Karen
Black, John
Blaga, Cristina
Blair, William
Blanco, Carlo
Bless, Robert
Blommaert, Joris
Blomme, Ronny
Bludman, Sidney
Blundell, Katherine
Boboltz, David
Bocchia, Romeo
Bochkarev, Nikolai
Bodaghee, Arash
Bode, Michael
Boden, Andrew

Bodenheimer, Peter
Bodo, Gianluigi
Boehm, Torsten
Boesgaard, Ann
Boffin, Henri
Boggess, Albert
Bohlender, David
Bohm-Vitense, Erika
Bolton, Charles
Bombaci, Ignazio
Bon, Natasa
Bonanno, Alfio
Bonazzola, Silvano
Bond, Howard
Bonifacio, Piercarlo
Bonneau, Daniel
Bonnet-Bidaud, Jean-Marc
Bono, Giuseppe
Bopp, Bernard
Borczyk, Wojciech
Bord, Donald
Borgman, Jan
Borisov, Nikolay
Borisova, Ana
Borissova, Jordanka
Boroson, Bram
Borra, Ermanno
Bortoletto, Alexandre
Boss, Alan
Bouchet, Patrice
Boumis, Panayotis
Bouvier, Jerôme
Boyajian, Tabetha
Boyd, David
Boyle, Richard
Boyle, Stephen
Bozic, Hrvoje
Bradley, Paul
Bradstreet, David
Bradt, Hale
Bragaglia, Angela
Braithwaite, Jonathan
Brandi, Elisande
Brandner, Wolfgang
Brassard, Pierre
Bravo, Eduardo
Breger, Michel
Bressan, Alessandro
Breysacher, Jacques
Bricẽo, Cesar
Brickhouse, Nancy
Brinchmann, Jarle
Briot, Danielle
Briquet, Maryline
Brocato, Enzo
Broglia, Pietro
Bromage, Gordon
Bromberg, Omer
Brosche, Peter
Brown, Alexander
Brown, David
Brown, Douglas
Brown, John
Brown, Thomas
Brown, Warren
Browning, Matthew
Brownlee, Robert
Bruch, Albert
Bruenn, Stephen
Bruhweiler, Frederick
Brun, Allan
Bruning, David
Bruntt, Hans
Bruzual, Gustavo
Buccino, Andrea
Buchhave, Lars
Buchlin, Eric
Budaj, Jan
Budding, Edwin
Bues, Irmela
Bulik, Tomasz
Bunner, Alan
Burderi, Luciano
Burgasser, Adam
Burikham, Piyabut
Burkhart, Claude
Burki, Gilbert
Burwitz, Vadim
Busa', Innocenza
Buser, Roland
Busko, Ivo
Busso, Maurizio
Butkovskaya, Varvara
Butler, Christopher
Butler, Dennis
Butler, Keith
Butler, Ray
Buzasi, Derek
Buzzoni, Alberto
Cacciari, Carla
CAI, Mingsheng
Calamida, Annalisa
Caldwell, John
Callanan, Paul
Callebaut, Dirk
Caloi, Vittoria
Calura, Francesco
Cameron, Andrew
Campbell, Simon
Canal, Ramon
Canalle, Joao
Cantiello, Michele
Canto Martins, Bruno
CAO, Huilai
Capelato, Hugo
Caputo, Filippina
Carbon, Duane
Carciofi, Alex
Carlin, Jeffrey
Carlsson, Mats
Carney, Bruce
Carpenter, Kenneth
Carraro, Giovanni
Carretta, Eugenio
Carrier, Fabien
Carson, T.
Carter, Bradley
Carter, Brian
Casares, Jorge
casassus, simon
Cassinelli, Joseph
Castelaz, Micheal
Castelli, Fiorella
Castor, John
Catala, Claude
Catalano, Franco
Catanzaro, Giovanni
Catchpole, Robin
Catelan, Márcio
Cauzzi, Gianna
Cayrel, Roger
Cenko, Stephen
Cernicharo, Jose
Cesetti, Mary
Cester, Bruno
Chaboyer, Brian
Chabrier, Gilles
Chadid, Merieme
Chambliss, Carlson
Chamel, Nicolas

Chan, Roberto
CHAN, Kwing
Chandler, Claire
Chaplin, William
Chapman, Jessica
Chapman, Robert
Charbonnel, Corinne
Charles, Philip
Charpinet, Stéphane
Chaty, Sylvain
Chaubey, Uma
Chavez-Dagostino, Miguel
Chechetkin, Valerij
CHEN, Alfred
CHEN, An-Le
CHEN, Peisheng
CHEN, Wen Ping
CHEN, Xuefei
CHEN, Yuqin
Cherchneff, Isabelle
Cherepashchuk, Anatolij
Chevalier, Roger
Chini, Rolf
Chiosi, Cesare
Chitre, Shashikumar
Chkhikvadze, Iakob
CHO, Se Hyung
Chochol, Drahomir
CHOI, Chul-Sung
CHOI, Kyu Hong
CHOU, Mei-Yin
CHOU, Yi
Choudhary, Debi Prasad
Christensen-Dalsgaard, Jørgen
Christie, Grant
Christlieb, Norbert
Christodoulou, Dimitris
Christopoulou, Panagiota-Eleftheria
Christy, James
Chugai, Nikolaj
Chukwude, Augustine
Ciardi, David
Cidale, Lydia
Cioni, Maria-Rosa
Claria, Juan
Clark, David
Claudi, Riccardo
Clem, James
Clement, Christine
Clementini, Gisella
Climenhaga, John
Clocchiatti, Alejandro
Coe, Malcolm
Cohen, David
Cohen, Judith
Cohen, Martin
Collet, Remo
Colomer, Francisco
Connolly, Leo
Contadakis, Michael
Conti, Peter
Cook, Kem
Copin, Yannick
Corbally, Christopher
Cornelisse, Remon
Cornide, Manuel
Corradi, Romano
Corradi, Wagner
Corral, Luis
Córsico, Alejandro
Cosentino, Rosario
Costa, Vitor
Côte, Patrick
Cottrell, Peter
Coulson, Iain
Covino, Elvira
Cowan, John
Cowley, Anne
Cowley, Charles
Coyne, S.J, George
Cram, Lawrence
Crause, Lisa
Crawford, David
Creech-Eakman, Michelle
Cristallo, Sergio
Cropper, Mark
Crowther, Paul
Cruzado, Alicia
Cugier, Henryk
CUI, Wenyuan
Culver, Roger
Cunha, Katia
Cunha, Margarida
Cuntz, Manfred
Cuny, Yvette
Curé, Michel
Cutispoto, Giuseppe
Cuypers, Jan
Cvetkovi'c, Zorica
Dachs, Joachim
Dav ci'c, Miodrag
Da Costa, Gary
Daflon, Simone
Dahn, Conard
DAI, Zhibin
DAISAKU, Nogami
Dall, Thomas
Dall'Ora, Massimo
Dal Ri Barbosa, Cassio
D'Amico, Nicolo
Damineli Neto, Augusto
Danchi, William
Danford, Stephen
D'Antona, Francesca
Danziger, I.
Das, Mrinal
da Silva, Licio
Daszynska-Daszkiewicz, Jadwiga
David, Marc
Davidge, Timothy
Davies, Benjamin
Davis, Christopher
Davis Jr, Cecil
Day-Jones, Avril
Dearborn, David
De Becker, Michaël
De Biasi, Maria
de Bruijne, Jos
de Castro, Elisa
De Cat, Peter
Decin, Leen
De Cuyper, Jean-Pierre
de Greve, Jean-Pierre
Deinzer, W.
de Jager, Cornelis
de Koter, Alex
de Laverny, Patrick
DELEUIL, Magali
Delgado, Antonio
Deliyannis, Constantine
de Loore, Camiel
del Peloso, Eduardo
Delplancke, Francoise
Del Santo, Melania
De Marchi, Guido
Demarque, Pierre
de Martino, Domitilla

de Medeiros, Jose
Demers, Serge
de Mink, Selma
Demircan, Osman
Demory, Brice-Olivier
DENG, LiCai
DeNisco, Kenneth
Denisenkov, Pavel
Depagne, Éric
Derekas, Aliz
de Ridder, Joris
Derman, Ethem
De Rossi, María
Deshpande, M.
de Silva, Gayandhi
Deupree, Robert
Deustua, Susana
de Val-Borro, Miguel
Dhillon, Vikram
Diaz, Marcos
Diehl, Roland
Di Mauro, Maria Pia
Dimitrijevic, Milan
Dionatos, Odysseas
Divan, Lucienne
Djurasevic, Gojko
Dluzhnevskaya, Olga
Dobrotka, Andrej
Dobrzycka, Danuta
Docobo, Jose-Angel
Dolan, Joseph
Dolidze, Madona
Domiciano de Souza, Armando
Dominguez, Inma
Dominis Prester, Dijana
Donahue, Robert
do Nascimento, José
Donati, Jean-Francois
Donnison, John
Doppmann, Gregory
Dorch, Søren Bertil
Dorfi, Ernst
Dotter, Aaron
DOU, Jiangpei
Dougados, Catherine
Dougherty, Sean
Downes, Ronald
Downs, George
Doyle MRIA, John
Dragunova, Alina
Drake, Jeremy
Drake, Natalia
Drake, Stephen
Dravins, Dainis
Drechsel, Horst
Dreizler, Stefan
Drew, Janet
Drilling, John
Duari, Debiprosad
Dubout, Renee
Dubus, Guillaume
Ducati, Jorge
Ducourant, Christine
Dudorov, Aleksandr
Dufour, Patrick
Dufton, Philip
Dukes Jr., Robert
Duncan, Douglas
Dunham, David
Dunlop, Storm
Dupree, Andrea
Dupuis, Jean
Dupuy, David
Durisen, Richard
Duschl, Wolfgang
Duvert, Gilles
Dwek, Eli
Dworetsky, Michael
Eaton, Joel
Echevarria, Juan
Edalati Sharbaf, Mohammad
Ederoclite, Alessandro
Edvardsson, Bengt
Edwards, Alan
Edwards, Paul
Edwards, Suzan
Eenens, Philippe
Efremov, Yurij
Eggenberger, Patrick
Eggleton, Peter
Eglitis, Ilgmars
Egret, Daniel
Eisner, Josh
Ekström García Nombela, Sylvia
El Basuny, Ahmed
Eldridge, John
El Eid, Mounib
Elias II, Nicholas
Elkin, Vladimir
Elmhamdi, Abouazza
Elste, Günther
Emerson, James
Eminzade, T.
Endal, Andrew
Engelbrecht, Chris
Engels, Dieter
Engvold, Oddbjørn
Ercan, Enise
ERIGUCHI, Yoshiharu
Eriksson, Kjell
ESAMDIN, Ali
Esenoğlu, Hasan
Eskioğlu, A.
Esser, Ruth
Etzel, Paul
Evangelidis, E.
Evans, Aneurin
Evans, Christopher
Evans, Dafydd
Evans, Nancy
Evren, Serdar
Eyer, Laurent
Eyres, Stewart
Eze, Romanus
Fabiani, Sergio
Fabregat, Juan
Fabrika, Sergei
Fadeyev, Yurij
Fahlman, Gregory
Falceta-Goncalves, Diego
Falcon Veloz, Nelson
Falize, Emeric
Faraggiana, Rosanna
Farrell, Sean
Faulkner, John
Faurobert, Marianne
Feast, Michael
Federici, Luciana
Feigelson, Eric
Feinstein, Alejandro
Fekel, Francis
Felenbok, Paul
Feltzing, Sofia
Ferland, Gary
Ferluga, Steno
Fernandes, Joao
Fernandez-Figueroa, M.

Fernández Lajús, Eduardo
Fernie, J.
Ferrario, Lilia
Ferreira, João
Ferrer, Osvaldo
Feynman, Joan
Fichtner, Horst
Figer, Donald
Filipov, Lachezar
Filipovic, Miroslav
Finch, Charlie
Fisher, George
Fisher, Richard
Fitzpatrick, Edward
Flannery, Brian
Fletcher, J.
Floquet, Michele
Fluri, Dominique
Foing, Bernard
Fokin, Andrei
Fontaine, Gilles
Fontenla, Juan
Forbes, J.
Forgács-Dajka, Emese
Formiggini, Lilliana
Forrest, William
Fors, Octavi
Forte, Juan
Forveille, Thierry
Fossat, Eric
Foukal, Peter
Fox-Machado, Lester
Foy, Renaud
Franchini, Mariagrazia
Franco, Gabriel Armando
Francois, Patrick
Frandsen, Soeren
Frank, Juhan
Fransson, Claes
Franz, Otto
Frasca, Antonio
Fredrick, Laurence
Freimanis, Juris
Freire Ferrero, Rubens
Freitas Mourão, Ronaldo
Frémat, Yves
Frew, David
Freytag, Bernd
Fridlund, Malcolm
Friel, Eileen
Frisch, Helene
Frisch, Uriel
Froeschle, Christiane
Fromang, Sebastien
Fruchter, Andrew
FU, Hsieh-Hai
FU, Jian-Ning
FUJIMOTO, Masayuki
FUJIWARA, Tomoko
FUKUDA, Ichiro
Fulbright, Jon
Fullerton, Alexander
Gabriel, Maurice
Gänsicke, Boris
Gahm, Goesta
Gail, Hans-Peter
Galadi-Enriquez, David
Gális, Rudolf
Gallagher III, John
Gallino, Roberto
Gal-Yam, Avishay
Gameiro, Jorge
Gamen, Roberto
Gandolfi, Davide
Gangadhara, R.T.
Garcia, Domingo
Garcia, Lia
García García, Miriam
Garcia, Rafael
García-Berro, Enrique
García de María, Juan
Garcia-Hernandez, Domingo
García López, Ramón
Garcia-Lorenzo, Maria
Garmany, Katy
Garrido, Rafael
Garrison, Robert
Gary, Dale
Gasiprong, Nipon
Gatewood, George
Gaudenzi, Silvia
Gautier, Daniel
Gautschy, Alfred
Gavras, Panagiotis
Gay, Pamela
Gebbie, Katharine
Geffert, Michael
Gehren, Thomas
Gehrz, Robert
Geisler, Douglas
Geldzahler, Barry
Geller, Aaron
Genet, Russell
Georgy, Cyril
Gerbaldi, Michele
Geroyannis, Vassilis
Gershberg, R.
Gęsicki, Krzysztof
Geyer, Edward
Ghez, Andrea
Ghosh, Kajal
Ghosh, Swarna
Giampapa, Mark
Giannone, Pietro
Gieren, Wolfgang
Gies, Douglas
Gigas, Detlef
Gillet, Denis
Gilliland, Ronald
Gillon, Michaël
Gilra, Daya
Giménez, Alvaro
Giorgi, Edgard
Giovannelli, Franco
Girardi, Leo
Giridhar, Sunetra
Gizis, John
Gizon, Laurent
Glagolevskij, Yurij
Glasner, Shimon
Glass, Ian
Glatzmaier, Gary
Glazunova, Ljudmila
Gledhill, Timothy
Goebel, John
Goedhart, Sharmila
Golay, Marcel
Golden, Aaron
Goldman, Itzhak
Gomboc, Andreja
Gomez, Haley
Gomez, Mercedes
Gómez Rivero, José
Gonçalves, Denise
Gondoin, Philippe
GONG, Biping
González, Gabriela
Gonzalez, Guillermo
Gonzalez, Jean-Francois

González Martínez-Pais, Ignacio
Goode, Philip
Goodwin, Simon
Goossens, Marcel
Gopka, Vera
Gorbaneva, Tatyana
Gordon, Charlotte
Goriely, Stephane
Gosset, Eric
Goswami, Aruna
Gottlieb, Carl
Gough, Douglas
Gouliermis, Dimitrios
Goupil, Marie-Jose
Grabowski, Bolesław
Grady, Carol
Graffagnino, Vito
Graham, Eric
Graham, John
Grant, Athay
Grant, Ian
Gratton, Raffaele
Grauer, Albert
Gray, David
Gray, Richard
Grebel, Eva
Grec, Gérard
Green, Daniel
Green, Elizabeth
Greenhill, Lincoln
Greenhouse, Matthew
Greggio, Laura
Gregorio-Hetem, Jane
Grenon, Michel
Grevesse, Nicolas
Grewing, Michael
Griffin, R. Elizabeth
Griffin, Roger
Grinin, Vladimir
Groenewegen, Martin
Groh, Jose
Groot, Paul
Groote, Detlef
Grosso, Monica
Grundahl, Frank
Grundstrom, Erika
Grygar, Jiri
GU, Shenghong
GU, Wei-Min
Gudehus, Donald
Güdel, Manuel
Guenther, David
Guerrero, Gianantonio
Guetter, Harry
Guglielmino, Salvatore
Guinan, Edward
Gull, Theodore
Gulliver, Austin
Gülmen, Ömür
Gün, Gulnur
GUNJI, Shuichi
Gunn, Alastair
Günthardt, Guillermo
GUO, Jianheng
Gupta, Ranjan
Gussmann, Ernst-August
Gustafsson, Bengt
Guthrie, Bruce
Guzik, Joyce
Haas, Martin
Haberreiter, Margit
HACHISU, Izumi
Hackman, Thomas
Hackwell, John
Hadrava, Petr
Haefner, Reinhold
Haisch, Bernard
Hakala, Pasi
Hakkila, Jon
Halbwachs, Jean-Louis
Hall, Douglas
Hallam, Kenneth
Hallinan, Gregg
Hamann, Wolf-Rainer
Hambaryan, Valeri
Hamdy, M.
Hameury, Jean-Marie
Hammond, Gordon
Hamuy, Mario
HAN, Inwoo
HAN, Zhanwen
HANAWA, Tomoyuki
Handler, Gerald
Hanson, Margaret
Hantzios, Panayiotis
Hanuschik, Reinhard
HAO, Jinxin
Harmanec, Petr
Harmer, Charles
Harmer, Dianne
Harper, Graham
Harris, William
Hartigan, Patrick
Hartkopf, William
Hartman, Henrik
Hartmann, Dieter
Hartmann, Lee
Hartwick, F.
Harutyunian, Haik
HASHIMOTO, Masa-aki
HASHIMOTO, Osamu
Hassall, Barbara
Haswell, Carole
Hauck, Bernard
Hauschildt, Peter
Hawley, Suzanne
HAYASAKI, Kimitake
Hayes, Donald
Hazlehurst, John
HE, Jinhua
Heacox, William
Heap, Sara
Hearnshaw, John
Heasley, James
Heber, Ulrich
Heck, Andre
Hegedues, Tibor
Heger, Alexander
Heiser, Arnold
Heiter, Ulrike
Hejna, Ladislav
Hekker, Saskia
Hellier, Coel
Helt, Bodil
Hempel, Marc
Hempelmann, Alexander
Henden, Arne
Henrichs, Hubertus
Henry, Richard
Hensberge, Herman
Hensler, Gerhard
Herbst, Eric
Herbst, William
Hernanz, Margarita
Herrero Davó, Artemio
Hershey, John
Heske, Astrid
Hessels, Jason
Hesser, James

Hessman, Frederic
Heydari-Malayeri, Mohammad
Heyl, Jeremy
Hidayat, Bambang
Hilditch, Ronald
Hill, Graham
Hill, Grant
Hill, Henry
Hill, Vanessa
Hillebrandt, Wolfgang
Hillenbrand, Lynne
Hillier, John
Hills, Jack
Hillwig, Todd
Hindsley, Robert
Hinkle, Kenneth
Hintz, Eric
HIRAI, Masanori
HIRATA, Ryuko
HIROSE, Shigenobu
Hirschi, Raphael
Ho, Wynn
Hoard, Donald
Hoare, Melvin
Hobbs, Lewis
Hodapp, Klaus
Hoeflich, Peter
Höfner, Susanne
Hojaev, Alisher
Hollowell, David
Holmberg, Johan
Holmgren, David
Holt, Stephen
Holzer, Thomas
Homeier, Derek
Honda, Satoshi
Honeycutt, R.
Hopp, Ulrich
HORAGUCHI, Toshihiro
Horch, Elliott
HORIUCHI, Ritoku
Horner, Scott
Houdek, Gunter
Houk, Nancy
House, Lewis
Houziaux, Leo
Howarth, Ian
Howell, Steve
Hric, Ladislav
Hrivnak, Bruce
Hron, Josef
HU, Zhong wen
HUANG, He
Hube, Douglas
Hubeny, Ivan
Hubert-Delplace, Anne-Marie
Hubrig, Swetlana
Huenemoerder, David
Huggins, Patrick
Hughes, David
Hughes, John
Hui bon Hoa, Alain
Hummel, Christian
Hummel, Wolfgang
Humphreys, Elizabeth
Humphreys, Roberta
Huovelin, Juhani
Hutchings, John
Hutter, Donald
Hyland, Harry
Ianna, Philip
Ibadov, Subhon
Ibanoglu, Cafer
Iben Jr, Icko
Ignace, Richard
Ignjatovi'c, Ljubinko
Iijima, Takashi
Ikhsanov, Nazar
Ikonnikova, Natalia
Iliev, Ilian
Ilyin, Ilya
IMAI, Hiroshi
Imamura, James
Imbert, Maurice
Imbroane, Alexandru
Imshennik, Vladimir
Ioannou, Zacharias
Ireland, Michael
Irwin, Alan
Irwin, Michael
Isern, Jordi
Ishida, Toshihito
ISHIDA, Manabu
ISHIHARA, Daisuke
ISHIZUKA, Toshihisa
Ismayilov, Nariman
Israelian, Garik
ITA, Yoshifusa
ITO, Yutaka
ITOH, Naoki
ITOH, Yoichi
Ivanov, Vsevolod
Ivanova, Natalia
Ivans, Inese
Ivezic, Zeljko
IWAMOTO, Nobuyuki
Izmailov, Igor
IZUMIURA, Hideyuki
Izzard, Robert
Jabbar, Sabeh
Jablonski, Francisco
Jackson, Bernard
Jahn, Krzysztof
Jahreiss, Hartmut
James, Richard
Janík, Jan
Janka, Hans
Jankov, Slobodan
Janot Pacheco, Eduardo
Jardine, Moira
Jasniewicz, Gerard
Jassur, Davoud
Jatenco-Pereira, Vera
Jayawardhana, Ray
Jeffers, Sandra
Jeffery, Simon
Jehin, Emmanuel
JEON, Young Beom
JEONG, Jang-Hae
Jerzykiewicz, Mikołaj
Jetsu, Lauri
Jevremovic, Darko
Jewell, Philip
Jha, Saurabh
JIANG, Biwei
JIANG, Zhibo
Jin, Liping
JIN, Zhenyu
Johnson, Christian
Johnson, Hollis
Johnson, Jennifer
Johnston, Helen
Joncas, Gilles
Joner, Michael
Jones, Albert
Jones, Carol
Jones, David
Jones, Derek

Jonker, Peter
Jordan, Carole
Jordan, Stefan
Jordi, Carme
Jørgensen, Jes
José, Jordi
Joshi, Umesh
Joss, Paul
Josselin, Eric
Judge, Philip
Jurcsik, Johanna
Jurdana-šepić, Rajka
Justtanont, Kay
Kadouri, Talib
Kaehler, Helmuth
Käufl, Hans Ulrich
Kafka, Styliani (Stella)
Kaitchuck, Ronald
Kaler, James
Kalirai, Jason
Kalkofen, Wolfgang
Kalomeni, Belinda
Kałużny, Janusz
Kamal, Fouad
KAMBE, Eiji
Kaminker, Alexander
Kamiński, Krzysztof
Kamp, Inga
Kamp, Lucas
KANAMITSU, Osamu
Kanbach, Gottfried
Kanbur, Shashi
Kandel, Robert
KANG, Young Woon
Kantharia, Nimisha
Kanyo, Sandor
Kaper, Lex
Käpylä, Petri
Karakas, Amanda
Karami, Kayoomars
Karetnikov, Valentin
Karitskaya, Evgeniya
Karoff, Christoffer
Karovska, Margarita
Karp, Alan
Kasiviswanathan, Sankarasubramanian
Kašparová, Jana
KATO, Ken-ichi
KATO, Mariko
KATO, Yoshiaki
KATO, Taichi
Katsova, Maria
Kaufer, Andreas
KAWABATA, Koji
KAWABATA, Shusaku
Kawaler, Steven
KAWARA, Kimiaki
Kawka, Adela
Kaye, Anthony
Kazantseva, Liliya
Kazarovets, Elena
Kazlauskas, Algirdas
Kebede, Legesse
Keller, Stefan
KENJI, Nakamura
Kenny, Harold
KENTARO, Matsuda
Kenyon, Scott
Kepler, S.
Kerber, Florian
Kerschbaum, Franz
Kervella, Pierre
Keskin, Varol
Khalesseh, Bahram
Khodachenko, Maxim
KholtYgin, Alexander
Khomenko, Elena
KIGUCHI, Masayoshi
Kilkenny, David
Kilpio, Elena
KIM, Chulhee
KIM, Chun-Hwey
KIM, Ho-il
KIM, Sang Chul
KIM, Seung-Lee
KIM, Tu Whan
KIM, Woong-Tae
KIM, Yong Cheol
KIM, Young-Soo
Kinemuchi, Karen
King, Andrew
King, David
King, Ivan
Kiplinger, Alan
Kippenhahn, Rudolf
Kipper, Tonu
Kiraga, Marcin
Kirkpatrick, Joseph
Kiselman, Dan
Kiss, László
Kisseleva-Eggleton, Ludmila
Kitchatinov, Leonid
Kitiashvili, Irina
Kitsionas, Spyridon
Kiziloglu, Nilgun
Kjeldsen, Hans
Kjurkchieva, Diana
Klapp, Jaime
Klein, Richard
Kley, Wilhelm
Klochkova, Valentina
Knoelker, Michael
KOBAYASHI, Masakazu
Koch, Andreas
Kocharovskij, Vladimir
Kocharovsky, Vitaly
Kochhar, Rajesh
Kochukhov, Oleg
KODAIRA, Keiichi
Köhler, Rainer
Koen, Marthinus
Koenigsberger, Gloria
Koester, Detlev
KOGURE, Tomokazu
Kolb, Ulrich
Kolenberg, Katrien
Kolesnikov, Sergey
Kolesov, Aleksandr
Kolka, Indrek
Kolláth, Zoltan
Komonjinda, Siramas
Komžík, Richard
Konacki, Maciej
Konar, Sushan
KONDO, Yoji
Konigl, Arieh
Konstantinova-Antova, Renada
Kontizas, Evangelos
KOO, Bon-Chul
Kopacki, Grzegorz
Kopatskaya, Evgenia
Kopp, Roger
Kopylova, Yulia
Korcakova, Daniela
Kordi, Ayman
Korhonen, Heidi
Korn, Andreas

Kornilov, Victor
Korotin, Sergey
Kosovichev, Alexander
Kóspál, Ágnes
Kotnik-Karuza, Dubravka
Koubsky, Pavel
Kouwenhoven, M.B.N. (Thijs)
Kovačević, Andjelka
Kovachev, Bogomil
Kovacs, Geza
Kovács, József
Kovaleva, Dana
Kővári, Zsolt
Kovetz, Attay
Kovtyukh, Valery
KOZASA, Takashi
Kozłowski, Maciej
Kraft, Robert
Kraicheva, Zdravka
Kramer, Busaba
Kraus, Michaela
Kraus, Stefan
Krautter, Joachim
Kreiner, Jerzy
Krempeć-Krygier, Janina
Kretzschmar, Matthieu
Kreykenbohm, Ingo
Krikorian, Ralph
Krisciunas, Kevin
Krishna, Swamy
Kriwattanawong, Wichean
Kroll, Peter
Kroupa, Pavel
Krticka, Jiri
Kruchinenko, Vitaliy
Krumholz, Mark
Kruszewski, Andrzej
Kryvdyk, Volodymyr
Kryvodubskyj, Valery
Krzesiński, Jerzy
Kubát, Jiri
Kubiak, Marcin
Kučinskas, Arunas
Kudashkina, Larisa
Kudritzki, Rolf-Peter
Kudryavtsev, Dmitry
Kudryavtseva, Nadezhda (Nadia)
Kuhi, Leonard
Kuiper, Rolf
Kulkarni, Prabhakar
Kumar, Shiv
Kumsiashvily, Mzia
Kunjaya, Chatief
Kupka, Friedrich
KURAYAMA, Tomoharu
Kurtanidze, Omar
Kurtz, Donald
Kurtz, Michael
Kurucz, Robert
Kwee, K.
Kwok, Sun
Labay, Javier
Labeyrie, Antoine
Lacour, Sylvestre
Lacy, Claud
Lago, Maria
Lagrange, Anne-Marie
Laird, John
Lamb, Susan
Lambert, David
Lamb Jr, Donald
Lamers, Henny
Lamontagne, Robert
Lampens, Patricia
Lamzin, Sergei
Lançon, Ariane
Landolt, Arlo
Landstreet, John
Laney, Clifton
Lang, Kenneth
Langer, Norbert
Lanz, Thierry
Lanza, Antonino
Lanzafame, Alessandro
Lapasset, Emilio
Larionov, Valeri
Larson, Richard
Larsson, Stefan
Larsson-Leander, Gunnar
Lasala Jr., Gerald
Laskarides, Paul
Lasota-Hirszowicz, Jean-Pierre
LAS VERGNAS, Olivier
Latham, David
Lattanzi, Mario
Lattanzio, John
Lattimer, James
Laugalys, Vygandas
Lavrov, Mikhail
Lawlor, Timothy
Lawson, Warrick
Lázaro Hernando, Carlos
Lazauskaite, Romualda
Leahy, Denis
Lẽo, Joao Rodrigo
Le Bertre, Thibaut
Le Borgne, Jean-Francois
Lebovitz, Norman
Lèbre, Agnes
Lebreton, Yveline
Lebzelter, Thomas
Leckrone, David
Le Contel, Jean-Michel
Lee, Hyun-chul
Lee, William
LEE, Chung-Uk
LEE, Jae Woo
LEE, Jae Woo
LEE, Jun
LEE, Myung Gyoon
LEE, Sang-Gak
LEE, Thyphoon
LEE, Woo baik
LEE, Yong Sam
Leedjarv, Laurits
Leenaarts, Jorrit
Leggett, Sandy
Lehman, Holger
Leibacher, John
Leibowitz, Elia
Leinert, Christoph
Leister, Nelson
Leitherer, Claus
Leloudas, Georgios
Lemke, Michael
Lenzen, Rainer
Leone, Francesco
Lépine, Jacques
Lepine, Sebastien
Le Poole, Rudolf
Leroy, Jean-Louis
Lester, Daniel
Lester, John
Letarte, Bruno
Leto, Giuseppe
Leung, Kam
Leushin, Valerij

Levato, Orlando
LI, Ji
LI, Jinzeng
LI, Lifang
LI, Min
LI, Qingkang
LI, Xiangdong
LI, Yan
LI, Zhi
LI, Zhiping
LI, Zhongyuan
LI, Zongwei
LIANG, Guiyun
LIANG, Yanchun
Liebendörfer, Matthias
Liebert, James
Liermann, Adriane
Lignieres, François
Ligori, Sebastiano
LIM, Jeremy
Limongi, Marco
LIN, Yi-qing
Linde, Peter
Lindqvist, Michael
Ling, Josefina
Lingenfelter, Richard
Linnell, Albert
Linsky, Jeffrey
Lippincott Zimmerman, Sarah
Little-Marenin, Irene
Littleton, John
LIU, Caipin
LIU, Fukun
LIU, Guoqing
LIU, Jifeng
LIU, Michael
LIU, Qingzhong
LIU, Yujuan
Livio, Mario
Lloyd, Christopher
Lobel, Alex
Lockwood, G.
Lodders, Katharina
Loktin, Alexhander
Lomb, Nicholas
Longmore, Andrew
Lopes, Dalton
Lopez, De
Lorenz-Martins, Silvia
LOU, Yu-Qing
Louys, Mireille
Lu, Phillip
Lub, Jan
Lubowich, Donald
Lucatello, Sara
Luck, R.
Lucy, Leon
Ludwig, Hans
Lugaro, Maria
Luna, Homero
Lundstrom, Ingemar
Luo, Qinghuan
LUO, Ali
Luque-Escamilla, Pedro
Luri, Xavier
Luttermoser, Donald
Lutz, Julie
Lyubarsky, Yury
Lyubchik, Yuri
Lyubimkov, Leonid
Maccarone, Thomas
MacConnell, Darrell
MacDonald, James
Maceroni, Carla
Machado, Maria
Machado Folha, Daniel
Mackey, Alasdair
Macri, Lucas
Maddison, Sarah
Madej, Jerzy
Madore, Barry
MAEDA, Keiichi
Maeder, Andre
MAEHARA, Hideo
Magain, Pierre
Magalhaes, Antonio Mario
Magazzu, Antonio
Maggio, Antonio
Magnan, Christian
Magnier, Eugene
Magrini, Laura
Maguire, Kate
Maharaj, Sunil
Maheswaran, Murugesapillai
Mahmoud, Farouk
Maigurova, Nadiia
Maillard, Jean-Pierre
Mainzer, Amy
Maitzen, Hans
Maíz Apellániz, Jesús
Makarov, Valeri
Malagnini, Maria
Malaroda, Stella
Malasan, Hakim
Malbet, Fabien
Malkov, Oleg
Mallik, D.
Malogolovets, Evgeny
Malov, Igor
Mamajek, Eric
Mandel, Ilya
Manfroid, Jean
Manimanis, Vassilios
Manset, Nadine
Mantegazza, Luciano
Manteiga Outeiro, Minia
Mapelli, Michela
Marchev, Dragomir
Marco, Amparo
Marconi, Marcella
Mardirossian, Fabio
Mardling, Rosemary
Marengo, Massimo
Margon, Bruce
Margrave Jr, Thomas
Marilli, Ettore
Marín-Franch, Antonio
Mariska, John
Markkanen, Tapio
Markoff, Sera
Markworth, Norman
Marlborough, Michael
Marley, Mark
Marraco, Hugo
Marsakova, Vladyslava
Marsden, Stephen
Marsh, Thomas
Marston, Anthony
Martayan, Christophe
Martí, Josep
Martic, Milena
Martín, Eduardo
Martinez, Peter
Martinez Fiorenzano, Aldo
Martinez Roger, Carlos
Martins, Fabrice
Mashonkina, Lyudmila
Maslennikov, Kirill

Mason, Brian
Mason, Paul
Massaglia, Silvano
Massey, Philip
Mateu, Cecilia
Mathias, Philippe
Mathieu, Robert
Mathioudakis, Mihalis
Mathis, Stephane
Mathys, Gautier
MATSUMOTO, Katsura
MATSUNAGA, Noriyuki
Matsuura, Mikako
Matteucci, Francesca
Matthews, Jaymie
Matthews, Lynn
Mauas, Pablo
Mauche, Christopher
Mauerhan, Jon
Mawet, Dimitri
Mayer, Pavel
Mazeh, Tsevi
Mazumdar, Anwesh
Mazurek, Thaddeus
Mazzali, Paolo
Mazzitelli, Italo
McAlister, Harold
McBride, Vanessa
McCluskey Jr, George
McDavid, David
McGraw, John
McGregor, Peter
McLean, Brian
McLean, Ian
McSaveney, Jennifer
McSwain, Mary
Medupe, Rodney
Megeath, S.
Mégessier, Claude
Meibom, Soren
Meintjes, Petrus
Meištas, Edmundas
Melbourne, Jason
Melendez, Jorge
Melia, Fulvio
Meliani, Mara
Melikian, Norair
Melo, Claudio
Mendes, Luiz
Mendoza, V.
Mennickent, Ronald
Mereghetti, Sandro
Merlo, David
Messina, Sergio
Mestel, Leon
Metcalfe, Travis
Meyer, Michael
Meyer-Hofmeister, Emmi
Meynet, meynet
Mezzetti, Marino
Mianes, Pierre
Michaud, Georges
Michel, Eric
Michel, Raul
Mickaelian, Areg
Middleton, Christopher
Miesch, Mark
Migaszewski, Cezary
Migenes, Victor
Miglio, Andrea
Mihajlov, Anatolij
MIKAMI, Takao
Mikkola, Seppo
Mikołajewska, Joanna
Mikołajewski, Maciej
Mikulášek, Zdeněk
Milano, Leopoldo
Miller, Joseph
Millour, Florentin
Milone, Eugene
Milone, Luis
MINESHIGE, Shin
Minier, Vincent
Minikulov, Nasridin
Mintz Blanco, Betty
Mironov, Aleksey
Mishenina, Tamara
Miszalski, Brent
Mitalas, Romas
Mitra, Abhas
MIYAJI, Shigeki
Mkrtichian, David
Mochnacki, Stefan
Moehler, Sabine
Möllenhoff, Claus
Moffat, Anthony
Moffett, Thomas
Mohan, Chander
Mohan, Vijay
Mohd, Zambri
Moiseenko, Sergey
Moitinho, André
Molaro, Paolo
Molenda-żakowicz, Joanna
Monaco, Lorenzo
Monaghan, Joseph
Monard, Libert
Monier, Richard
Monin, Dmitry
Montalb'an, Josefina
Monteiro, Mario Joao
Montes, David
Montgomery, Michele
Moore, Daniel
Moos, Henry
Morales Rueda, Luisa
Morbey, Christopher
Morbidelli, Roberto
Morel, Pierre-Jacques
Morel, Thierry
Morgan, John
Morgan, Thomas
Morossi, Carlo
Morrell, Nidia
Morris, Patrick
Morris, Rhys
Morrison, Nancy
Moskalik, Paweł
Mosoni, Laszlo
Moss, David
Mouchet, Martine
Mould, Jeremy
Mourard, Denis
Mowlavi, Nami
Mozurkewich, David
Mueller, Ewald
Muench, Guido
Mukai, Koji
Mukherjee, Krishna
Mukhopadhyay, Banibrata
Mumford, George
Munari, Ulisse
Murdin, Paul
Murray, James
Musielak, Zdzislaw
Mutel, Robert
Mutschlecner, Joseph
Nadyozhin, Dmitrij
NAGATAKI, Shigehiro
Nagendra, K.

Nagirner, Dmitrij
Nahar, Sultana
NAITO, Tsuguya
Najarro de la Parra, Francisco
NAKAMURA, Takashi
NAKAMURA, Yasuhisa
NAKANO, Takenori
NAKAO, Yasushi
NAKASHIMA, Jun-ichi
NAKAZAWA, Kiyoshi
Nandi, Dibyendu
Naoz, Smadar
Napiwotzki, Ralf
Näränen, Jyri
Narasimha, Delampady
Narbutis, Donatas
NARDETTO, Nicolas
NARIAI, Kyoji
NARITA, Shinji
Nasiri, Sadollah
Naylor, Tim
Nazarenko, Victor
Naze, Yael
Neff, James
Negueruela, Ignacio
Neiner, Coralie
Nelemans, Gijs
Nelson, Burt
Netzer, Nathan
Neuhaeuser, Ralph
Neustroev, Vitaly
Newman, Michael
Newsom, Gerald
NGEOW, Chow Choong
Nguyễn, Lan
NHA, Il Seong
Niarchos, Panagiotis
Nicastro, Luciano
Nicholls, Ralph
Nichols, Joy
Nicolet, Bernard
Niedzielski, Andrzej
Nielsen, Krister
Niemczura, Ewa
Nieuwenhuijzen, Hans
Nieva, Maria
NIINUMA, Kotaro
Nikoghossian, Arthur
Nikoghosyan, Elena
Nikolov, Andrej
Nilsson, Hampus
Ninkov, Zoran
NISHIDA, Minoru
NISHIMURA, Masayoshi
NISHIMURA, Shiro
Nissen, Poul
Nitschelm, Christian
NITTA, Atsuko
Noels, Arlette
NOGUCHI, Kunio
NOMOTO, Ken'ichi
Norci, Laura
Nordlund, Aake
Nordström, Birgitta
Norris, John
North, Pierre
Norton, Andrew
Nota, Antonella
Notni, Peter
Nowotny-Schipper, Walter
Nürnberger, Dieter
Nugis, Tiit
Nuth, Joseph
Oblak, Edouard
O'Brien, Tim
Odell, Andrew
O'Donoghue, Darragh
Oestreicher, Roland
Oey, Sally
Ofir, Aviv
Ogłoza, Waldemar
OH, Kyu-Dong
OHYAMA, Noboru
Oja, Tarmo
OKAMOTO, Isao
OKAZAKI, Akira
OKAZAKI, Atsuo
OKUMURA, Shin-ichiro
Oláh, Katalin
Oliva, Ernesto
Oliveira, Alexandre
Oliveira, Joana
Olivier, Enrico
Olsen, Erik
Olson, Edward
Oluseyi, Hakeem
Omizzolo, Alessandro
O'Neal, Douglas
Orellana, Rosa
Orlandini, Mauro
Orlov, Viktor
Orsatti, Ana
Ortolani, Sergio
OSAKI, Yoji
Osborn, Wayne
Oskinova, Lidia
Østensen, Roy
Ostriker, Jeremiah
Oswalt, Terry
O'Toole, Simon
OTSUKA, Masaaki
OTSUKI, Kaori
Oudmaijer, Rene
Owocki, Stanley
Özeren, Ferhat
Ozkan, Mustafa
Pacharin-Tanakun, P.
Padalia, T.
Pagano, Isabella
PAK, Soojong
Pakhomov, Yury
Pál, András
PALACIOS, Ana
Paletou, Frédéric
Palouš, Jan
Pamyatnykh, Alexey
Panagia, Nino
Pande, Girish
Pandey, Birendra
Pandey, Uma
Panei, Jorge
Panek, Robert
Pannunzio, Renato
Panov, Kiril
Pantoja, Carmen
Papaloizou, John
Paparo, Margit
Parimucha, Stefan
Parisi, Maria
PARK, Byeong-Gon
PARK, Hong-Seo
Paron, Sergio
Parravano, Antonio
Parsamyan, Elma
Parsons, Sidney
Parthasarathy, Mudumba
Pastori, Livio
Patat, Ferdinando
Pat-El, Igal

釣

Paternò, Lucio
Pati, Ashok
Patkos, Laszlo
Patten, Brian
Pauls, Thomas
Paunzen, Ernst
Pavani, Daniela
Pavlenko, Elena
Pavlenko, Yakov
Pavlov, George
Pavlovski, Kresimir
Pazhouhesh, Reza
Pearce, Gillian
Pearson, Kevin
Pecker, Jean-Claude
Pedoussaut, André
Pedreros, Mario
Pel, Jan
Pelt, Jaan
Pẽa Saint-Martin, José
Penny, Alan
Penny, Matthew
Peraiah, Annamaneni
Percy, John
Perdang, Jean
Pereira, Claudio
Perets, Hagai
Perez, Mario
Pérez Hernández, Fernando
Perrin, Guy
Perrin, Marshall
Peters, Geraldine
Petersen, J.
Peterson, Deane
Peterson, Ruth
Petit, Pascal
Petr-Gotzens, Monika
Petrov, Peter
Pettersen, Bjørn
Pfeiffer, Raymond
Philip, A.G.
Phillips, Mark
Piccioni, Adalberto
Pickles, Andrew
Pietrukowicz, Paweł
Pigulski, Andrzej
Piirola, Vilppu
Pijpers, Frank
Pikichian, Hovhannes
Pilachowski, Catherine
Pines, David
Pinotsis, Antonis
Pinsonneault, Marc
Pintado, Olga
Pinto, Philip
Piotto, Giampaolo
Pipin, Valery
Pires Martins, Lucimara
Piskunov, Nikolai
Pizzichini, Graziella
Plachinda, Sergei
Platais, Imants
Plavchan, Peter
Plez, Bertrand
Podsiadlowski, Philipp
Pogodin, Mikhail
Pojmanski, Grzegorz
Pokrzywka, Bartłomiej
Polcaro, V.
Polidan, Ronald
Pollacco, Don
Polosukhina-Chuvaeva, Nina
Pompeia, Luciana
Pongracic, Helen
Pont, Frédéric
Pontoppidan, Klaus
Pooley, David
Pooley, Guy
Pop, Alexandru
Pop, Vasile
Popov, Sergey
Popovic, Georgije
Poretti, Ennio
Porfir'ev, Vladimir
Portinari, Laura
Porto de Mello, Gustavo
Postnov, Konstantin
Pott, Jörg-Uwe
Pottasch, Stuart
Potter, Stephen
Pourbaix, Dimitri
Poveda, Arcadio
Pravdo, Steven
Prentice, Andrew
Preston, George
Pretorius, Magaretha
Prialnik, Dina
Pribulla, Theodor
Price, Charles
Pricopi, Dumitru
Prieto, Cristina
Prieur, Jean-Louis
Primas, Francesca
Pringle, James
Prinja, Raman
Pritchet, Christopher
Pritzl, Barton
Proffitt, Charles
Prokhorov, Mikhail
Provost, Janine
Prša, Andrej
Prugniel, Philippe
Prusti, Timo
Przybilla, Norbert
Pugach, Alexander
Pulone, Luigi
Puls, Joachim
Pustõnski, Vladislav-Veniamin
Puzia, Thomas
Pye, John
QIAN, Shengbang
QIAO, Guojun
QU, Qinyue
Querci, Francois
Querci, Monique
Quirrenbach, Andreas
Raassen, Ion
Radiman, Iratio
Raedler, K.
Rafert, James
Rahunen, Timo
Ramadurai, Souriraja
Ramsey, Lawrence
Randich, Sofia
Rangarajan, K.
Rank-Lueftinger, Theresa
Ransom, Scott
Rao, N.
Rao, Pasagada
Rasio, Frederic
Rastegaev, Denis
Rastogi, Shantanu
Ratcliff, Stephen
Rauch, Thomas
Rauscher, Thomas
Rautela, B.
Rauw, Gregor
Raveendran, A.

Rawlings, Mark
Ray, Alak
Rayet, Marc
Reale, Fabio
Rebolo, Rafael
Rebull, Luisa
Recio-Blanco, Alejandra
Reddy, Bacham
Reeves, Hubert
Reglero Velasco, Victor
Regnier, Stephane
Rego, Fernandez
Reid, Mark
Reid, Warren
Reimers, Dieter
Reiners, Ansgar
Reinsch, Klaus
Reipurth, Bo
Reisenegger, Andreas
Renson, P.
Renzini, Alvio
Reshetnyk, Volodymyr
Rettig, Terrence
Rey, Soo-Chang
Reylé, Céline
Reyniers, Maarten
Rhee, Jaehyon
Ribas, Ignasi
Ribó, Marc
Rice, John
Rich, Robert
Richard, Olivier
Richards, Mercedes
Richer, Harvey
Richichi, Andrea
Ricker, Paul
Rieutord, Michel
Riles, Keith
Riley, Pete
Ringuelet, Adela
Ringwald, Frederick
Ritter, Hans
Rivinius, Thomas
Rix, Hans-Walter
Rizzi, Luca
Roark, Terry
Robb, Russell
Robberto, Massimo
Roberts Jr, Lewis
Robertson, John
Robin, Annie
Robinson, Edward
Robinson, Garry
Roca Cortés, Teodoro
Rodrigues, Claudia
Rodrigues de Oliveira Filho, Irapuan
Rodriguez, Eloy
Rohrmann, Rene
Rojas, Gustavo
Roman, Nancy
Romaniello, Martino
Romanov, Yuri
Romanyuk, Iosif
Romanyuk, Yaroslav
Romero-Colmenero, Encarnacion
Rose, James
Rosenbush, Alexander
Rossi, Corinne
Rossi, Lucio
Rossi, Silvia
Rostas, François
Rostopchina, Alla
Roth, Markus
Rountree, Janet
Rovira, Marta
Rovithis, Peter
Rovithis-Livaniou, Helen
Roxburgh, Ian
Royer, Frédéric
Różańska, Agata
Różyczka, Michal
Rucinski, Slavek
Rudnitskij, Georgij
Ruffert, Maximilian
Ruiz, Maria Teresa
Ruiz-Lapuente, María
Russell, Jane
Russev, Ruscho
Russo, Guido
Rutten, Robert
Ruždjak, Domagoj
Ryabchikova, Tatiana
Ryan, Sean
Rybicki, George
Ryde, Nils
Saad, Somaya
Sabin, Laurence
Sachkov, Mikhail
Sackmann, Inge
SADAKANE, Kozo
Sadik, Aziz
Safari, Hossein
Saffari, Reza
Saffe, Carlos
Sagar, Ram
Saha, Abhijit
Sahai, Raghvendra
Sahal-Bréchot, Sylvie
SAIJO, Keiichi
SAIO, Hideyuki
SAKAMOTO, Tsuyoshi
SAKASHITA, Shiro
Sakhibullin, Nail
SAKON, Itsuki
Salaris, Maurizio
Salukvadze, G.
Samec, Ronald
Samus, Nikolay
Sánchez Almeida, Jorge
Sandmann, William
Sansom, Anne
Santos, Filipe
Santos Agostinho, Rui
Santos-Sanz, Pablo
Sanwal, Basant
Sanyal, Ashit
Sapar, Arved
Sapar, Lili
Sarajedini, Ata
Sareyan, Jean-Pierre
Sarna, Marek
Sarre, Peter
Sarty, Gordon
Sasselov, Dimitar
Sasso, Clementina
SATO, Bunei
SATO, Katsuhiko
Sauty, Christophe
Savanov, Igor
Savedoff, Malcolm
Saviane, Ivo
Savin, Daniel
Savonije, Gerrit
Saygac, Ahmet
Sbordone, Luca
Scalo, John
Scardia, Marco
Scarfe, Colin

Schaefer, Bradley
Schaerer, Daniel
Scharmer, Goeran
Schartel, Norbert
Schatten, Kenneth
Schild, Rudolph
Schiller, Stephen
Schlegel, Eric
Schmalberger, Donald
Schmid, Hans
Schmid-Burgk, J.
Schmidt, Edward
Schmidtke, Paul
Schmidtobreick, Linda
Schmitt, Henrique
Schmutz, Werner
Schnell, Anneliese
Schober, Hans
Schöller, Markus
Schoenberner, Detlef
Scholz, Alexander
Scholz, M.
Scholz, Ralf-Dieter
Schrijver, Karel
Schroeder, Klaus
Schröder, Rolf
Schuessler, Manfred
Schuh, Sonja
Schuler, Simon
Schulte-Ladbeck, Regina
Schuster, William
Schutz, Bernard
Schwartz, Philip
Schwarzenberg-Czerny, Alex
Schwope, Axel
Sciortino, Salvatore
Scuderi, Salvo
Scuflaire, Richard
Sedlmayer, Erwin
Seeds, Michael
Segaluvitz, Alexander
Seggewiss, Wilhelm
Seidov, Zakir
SEKIGUCHI, Kazuhiro
Selam, Selim
Selvelli, Pierluigi
Semeniuk, Irena
Semkov, Evgeni
Sen, Asoke
Sengbusch, Kurt
Sengupta, Sujan
Serber, Alexander
Sese, Rogel Mari
Shafter, Allen
Shahbaz, Tariq
Shahul Hameed, Mohin
Shakhovskaya, Nadejda
Shakht, Natalia
Shakura, Nikolaj
SHANG, Hsien
Shankland, Paul
Shara, Michael
Sharma, Dharma
Sharpless, Stewart
Shatsky, Nicolai
Shaviv, Giora
Shaw, James
Shawl, Stephen
Shchukina, Nataliia
Shelton, Ian
Shenavrin, Victor
Sherwood, William
Shetrone, Matthew
SHI, Huoming
SHI, Jianrong
SHIBAHASHI, Hiromoto
SHIBATA, Masaru
SHIBATA, Yukio
SHIGEYAMA, Toshikazu
Shimansky, Vladislav
SHIMOIKURA, Tomomi
Shine, Richard
Shinnaga, Hiroko
Shipman, Harry
Sholukhova, Olga
Shore, Steven
Short, Christopher
Shoyoqubov, Shoayub
SHU, Frank
Shustov, Boris
Shvelidze, Teimuraz
Sickafoose, Amanda
Siess, Lionel
Sigalotti, Leonardo
Signore, Monique
Sigut, T. A. Aaron
Sills, Alison
Silvestro, Giovanni
Silvotti, Roberto
Sima, Zdislav
Simić, Zoran
Simmons, John
Simon, Klaus
Simon, Michal
Simon, Theodore
Simón-Díaz, Sergio
Simonneau, Eduardo
Sinachopoulos, Dimitris
Singh, Harinder
Singh, Mahendra
Sinnerstad, Ulf
Sion, Edward
Siopis, Christos
Sistero, Roberto
Sivakoff, Gregory
Skillen, Ian
Skinner, Stephen
Skoda, Petr
Skokos, Charalampos
Skopal, Augustin
Skumanich, Andrew
Slane, Patrick
šlechta, Miroslav
Slovak, Mark
Smak, Józef
Smalley, Barry
Smeyers, Paul
Smirnova, Olesja
Smith, Graeme
Smith, Horace
Smith, Howard
Smith, J.
Smith, Myron
Smith, Robert
Smith, Verne
Smits, Derck
Smolec, Radosław
Smyth, Michael
Sneden, Chris
Snik, Frans
Snow, Theodore
Snowden, Michael
Sobieski, Stanley
Sobolev, Andrey
Sobouti, Yousef
Socas-Navarro, Hector
Soderblom, David
Soderhjelm, Staffan
Sódor, Ádám

Sofia, Sabatino
Soker, Noam
Solanki, Sami
Solf, Josef
Solheim, Jan
Soliman, Mohamed
Soltynski, Maciej
Somasundaram, Seetha
SONG, Liming
Sonneborn, George
Sonti, Sreedhar
Soonthornthum, Boonrucksar
Soszyński, Igor
Soubiran, Caroline
Southworth, John
Sowell, James
Sparks, Warren
Spiegel, Edward
Spite, François
Spite, Monique
Spruit, Hendrik
Sreenivasan, S.
Srivastava, J.
Srivastava, Ram
Stachowski, Grzegorz
Stagg, Christopher
Stahl, Otmar
Stalio, Roberto
Stancliffe, Richard
Stanishev, Vallery
Starrfield, Sumner
Stateva, Ivanka
Stathakis, Raylee
Stauffer, John
Stawikowski, Antoni
Stecher, Theodore
Stée, Philippe
Steffen, Matthias
Steiman-Cameron, Thomas
Stein, John
Stein, Robert
Steiner, Joao
Steiner, Oskar
Steinlin, Uli
Stellingwerf, Robert
Stello, Dennis
Stencel, Robert
Stępień, Kazimierz
Stergioulas, Nikolaos
Sterken, Christiaan
Stern, Robert
Sterzik, Michael
Stetson, Peter
Stift, Martin
Stil, Jeroen
Still, Martin
St-Louis, Nicole
Stockman Jr, Hervey
Stonkutė, Rima
Storey, John
Storm, Jesper
Stoyanov, Kiril
Straižys, Vytautas
Strassmeier, Klaus
Strigachev, Anton
Stringfellow, Guy
Strittmatter, Peter
Stritzinger, Maximilian
Strobel, Andrzej
Strom, Karen
Strom, Stephen
Stuik, Remko
Subramaniam, Annapurni
Suda, Takuma
Sudar, Davor
Sudzius, Jokubas
SUGIMOTO, Daiichiro
Sullivan, Denis
Sullivan, Mark
Sundqvist, Jon
SUNG, Hwankyung
Suntzeff, Nicholas
Suran, Marian
SUZUKI, Tomoharu
Svolopoulos, Sotirios
Sweigart, Allen
Swings, Jean-Pierre
Sylwester, Janusz
Szabados, Laszlo
Szabo, Robert
Szatmary, Karoly
Szeifert, Thomas
Székely, Péter
Szkody, Paula
Szymański, Michał
Szymczak, Marian
Taam, Ronald
TAKADA-HIDAI, Masahide
TAKAHARA, Mariko
TAKAHASHI, Hidenori
TAKAHASHI, Rohta
TAKASHI, Hasegawa
TAKATA, Masao
TAKEDA, Yoichi
TAKEUTI, Mine
Talavera, Antonio
Tallon, Michel
Tamazian, Vakhtang
Tammann, Gustav
TAMURA, Motohide
TAMURA, Shin'ichi
TAN, Huisong
TANABE, Kenji
TANABE, Toshihiko
TANAKA, Masuo
Tandon, S.
Tango, William
Tantalo, Rosaria
Taranova, Olga
Tarasov, Anatolii
Tarasova, Taya
Taş, Günay
Tassoul, Jean-Louis
Tauris, Thomas
Tautvaisiene, Gražina
Teays, Terry
Teixeira, Paula
Teixeira, Ramachrisna
Tektunali, H.
Templeton, Matthew
ten Brummelaar, Theo
Tennyson, Jonathan
TERADA, Yukikatsu
Terquem, Caroline
Terrell, Dirk
Terzan, Agop
Tessema, Solomon
Thejll, Peter
Thevenin, Frederic
Thielemann, Friedrich-Karl
Thomas, Hans
Thompson, Ian
Thompson, Rodger
Thomsen, Bjarne
Thorsett, Stephen
Thorstensen, John
Thoul, Anne
Thurston, Mark

Timothy, J.
Tinney, Christopher
Tiplady, Adrian
Tisserand, Patrick
Titov, Vladimir
Tjin-a-Djie, Herman
Todt, Helge
Tohline, Joel
Tokovinin, Andrei
Tokunaga, Alan
Tolbert, Charles
Tomasella, Lina
TOMINAGA, Nozomu
Tomov, Nikolai
Tomov, Toma
Toomre, Juri
Tornambe, Amedeo
José Miguel Torrejón, Jose Miguel
Torres, Carlos Alberto
Torres, Guillermo
Torres-Papaqui, Juan
Tout, Christopher
Tovmassian, Gagik
Townsend, Richard
Trager, Scott
Traulsen, Iris
Tremko, Jozef
Trimble, Virginia
Tripathi, Durgesh
Tripicco, Michael
Trullols, I.
Truran, James
Trushkin, Sergey
TSAY, Wean-Shun
Tscharnuter, Werner
Tsinganos, Kanaris
TSUJI, Takashi
TSUNETA, Saku
Tsvetkov, Milcho
Tsvetkova, Katja
Tsygankov, Sergey
Tull, Robert
Turatto, Massimo
Turck-Chièze, Sylvaine
Turcu, Vlad
Turner, David
Turner, Nils
Turolla, Roberto
Tutukov, Aleksandr
Twarog, Bruce
Tycner, Christopher
Tylenda, Romuald
Úbeda, Leonardo
Ubertini, Pietro
UCHIDA, Juichi
ud-Doula, Asif
Udovichenko, Sergei
Udry, Stephane
UEMURA, Makoto
UESUGI, Akira
Ueta, Toshiya
UKITA, Nobuharu
Ulla Miguel, Ana
Ulmschneider, Peter
Ulrich, Roger
Ulyanov, Oleg
UMEDA, Hideyuki
Unda-Sanzana, Eduardo
UNNO, Wasaburo
Upgren, Arthur
Ureche, Vasile
Urquhart, James
Usenko, Igor
Usher, Peter
Uslenghi, Michela
Usov, Vladimir
Utrobin, Victor
UTSUMI, Kazuhiko
Uytterhoeven, Katrien
Vaccaro, Todd
Vakili, Farrokh
Valcheva, Antoniya
Valdivielso, Luisa
Valeev, Azamat
Valenti, Jeff
Válio, Adriana
Vallenari, Antonella
Valls-Gabaud, David
Valtier, Jean-Claude
Valtonen, Mauri
Valyavin, Gennady
van Altena, William
van Belle, Gerard
van den Berg, Maureen
VandenBerg, Don
Vandenbussche, Bart
van den Heuvel, Edward
van der Bliek, Nicole
van der Borght, Rene
van der Hucht, Karel
van der Kruit, Pieter
Vandervoort, Peter
van Dessel, Edwin
van de Steene, Griet
Van Doorsselaere, Tom
Van Dyk, Schuyler
van Eck, Sophie
van Genderen, Arnoud
van Hamme, Walter
Van Hoolst, Tim
van Horn, Hugh
van Kerkwijk, Marten
Vaňko, Martin
van Leeuwen, Floor
van Loon, Jacco
van Riper, Kenneth
van't Veer, Frans
van't Veer-Menneret, Claude
Van Winckel, Hans
Vardavas, Ilias
Vasta, Magda
Vasu-Mallik, Sushma
Vauclair, Gérard
Vauclair, Sylvie
Vaughan, Arthur
Vaz, Luiz Paulo
Velusamy, T.
Venn, Kim
Vennes, Stephane
Ventura, Paolo
Ventura, Rita
Verdugo, Eva
Vereshchagin, Sergej
Verheest, Frank
Verheijen, Marc
Verhoelst, Tijl
Verma, R.
Vesperini, Enrico
Viala, Yves
Vidotto, Aline
Vierdayanti, Kiki
Vieytes, Mariela
Viik, Tõnu
Vila, Samuel
Vilhu, Osmi
Vilkoviskij, Emmanuil
Mónica, Monica
Villaver, Eva

Vink, Jorick
Viotti, Roberto
Vishniac, Ethan
Vivas, Anna
Vladilo, Giovanni
Vlahakis, Nektarios
Vlemmings, Wouter
Vogt, Nikolaus
Vogt, Steven
Volk, Kevin
Volo'shina, Irina
von Braun, Kaspar
von Hippel, Theodore
Votruba, Viktor
Vrba, Frederick
Vreux, Jean
Wachlin, Felipe
Wachter, Stefanie
WADA, Takehiko
Wade, Gregg
Wade, Richard
Waelkens, Christoffel
Wahlgren, Glenn
Walborn, Nolan
Walder, Rolf
Walker, Alistair
Walker, Edward
Walker, Gordon
Walker, Merle
Walker, William
Wallerstein, George
Walter, Frederick
Walton, Nicholas
WANG, Bo
WANG, Ding-Xiong
WANG, Feilu
WANG, Hongchi
WANG, Jia-Ji
WANG, Min
WANG, Xunhao
WANG, Zhenru
Ward, Martin
Ward, Richard
Warner, Brian
Warren Jr, Wayne
WATANABE, Tetsuya
Waters, Laurens
Waterworth, Michael
Watson, Robert
Waxman, Eli
Weaver, Thomas
Weaver, William
Webbink, Ronald
Weber, Stephen
Wegner, Gary
Wehinger, Peter
Wehlau, Amelia
Weigelt, Gerd
Weiler, Edward
Weis, Edward
Weis, Kerstin
Weisberg, Joel
Weiss, Achim
Weiss, Nigel
Weiss, Werner
Weistrop, Donna
Welch, Douglas
Welsh, William
WEN, Linqing
Werner, Klaus
Wesselius, Paul
Wesson, Roger
Wheatley, Peter
Wheeler, J.
Whelan, Emma
White, Nathaniel
White, Richard
White, Stephen
White II, James
Whitelock, Patricia
Wickramasinghe, N.
Wiedemann, Günter
Wielebinski, Richard
Wijers, Ralph
Williamon, Richard
Williams, Glen
Williams, John
Williams, Peredur
Williams, Robert
Willson, Lee Anne
Willstrop, Roderick
Wilson, Robert
Wing, Robert
Winiarski, Maciej
Winkler, Christoph
Winkler, Karl-Heinz
Winkler, Paul
Winnberg, Anders
Wittenmyer, Robert
Wittkowski, Markus
Woehl, Hubertus
Wolf, Marek
Wolff, Sidney
Womack, Maria
Wood, Brian
Wood, H.
Wood, Matthew
Wood, Peter
Woodsworth, Andrew
Woosley, Stanford
Worters, Hannah
Woudt, Patrick
Wramdemark, Stig
Wright, Nicholas
Wrixon, Gerard
WU, Hsin-Heng
Wyckoff, Susan
XIONG, Da Run
XUE, Li
Yahil, Amos
Yakovlev, Dmitrij
Yamada, Shoichi
YAMADA, Shimako
YAMAOKA, Hitoshi
YAMASAKI, Atsuma
YAMASHITA, Yasumasa
YAMASHITA, Takuya
Yanovitskij, Edgard
YAO, Yongqiang
Yegorova, Irina
Yengibarian, Norair
YI, Sukyoung
Yong, David
Yoon, Suk-Jin
YOON, Tae-Seog
Yorke, Harold
YOSHIDA, Shin'ichirou
YOSHIDA, Takashi
YOSHIOKA, Kazuo
Young, Andrew
Young, John
YU, Cong
Yüce, Kutluay
Yudin, Ruslan
YUJI, Ikeda
YUMIKO, Oasa
Yun, Joao
Yungelson, Lev
Yushkin, Maxim
YUTAKA, Shiratori

Začs, Laimons
Zaggia, Simone
Zahn, Jean-Paul
Zakirov, Mamnum
Zamanov, Radoslav
Zampieri, Luca
Zapatero-Osorio, Maria Rosa
Zaritsky, Dennis
Zasche, Petr
Zavagno, Annie
Zavala, Robert
Zdanavičius, Kazimeras
Zdanavičius, Justas
Zeilik, Michael
Zejda, Miloslav
Zhang, Jie
Zhang, Yuying
ZHANG, Bo
ZHANG, Chengmin
ZHANG, Er-Ho
ZHANG, Fenghui
ZHANG, Haotong
ZHANG, Huawei
ZHANG, Shu
ZHANG, Xiaobin
ZHANG, Yanxia
ZHAO, Gang
ZHAO, Junwei
Zharikov, Sergey
Zhekov, Svetozar
Zheleznyak, Alexander
Zheleznyakov, Vladimir
Zhilkin, Andrey
ZHOU, Daoqi
ZHOU, Hongnan
ZHOU, Xia
ZHU, Liying
ZHU, Zhenxi
Zickgraf, Franz
Zijlstra, Albert
Zinnecker, Hans
Ziółkowski, Janusz
Ziurys, Lucy
Zoccali, Manuela
Zoła, Stanisław
Zorec, Juan
Zsoldos, Endre
Zucker, Shay
Zuckerman, Benjamin
Zverko, Juraj
Zwintz, Konstanze
Zwitter, Tomaž

Division H Interstellar Matter and Local Universe

President: Ewine F. van Dishoeck
Vice-President: Jonathan Bland-Hawthorn

Steering Committee Members:

Thomas Henning
Sun Kwok
Giovanni Carraro
Eileen D. Friel
Annie C. Robin
Holger Baumgardt
Bruce G. Elmegreen
Diego Mardones
Michael R. Meyer
Birgitta Nordström
Ewine F. van Dishoeck

Members:

Aalto, Susanne
Aannestad, Per
Aarseth, Sverre
Abgrall, Herve
Abou'el-ella, Mohamed
Abraham, Zulema
Acker, Agnes
Acosta Pulido, Jose
Adams, Fred
Afanas'ev, Viktor
Aguilar, Luis
Ahumada, Andrea
Ahumada, Javier
Aigrain, Suzanne
AIKAWA, Yuri
Aitken, David
Ajhar, Edward
AK, Serap
AK, Tansel
AKABANE, Kenji
Akashi, Muhammad
Akeson, Rachel
Alcobé, Santiago
Alcolea, Javier
Alfaro, Emilio
Alksnis, Andrejs
Allen, Christine
Allen, Lori
Allen, Ronald
Allende Prieto, Carlos
Al-Mostafa, Zaki
Al-Naimiy, Hamid
Altenhoff, Wilhelm
Alves, João
Alves, Virgínia
Ambastha, Ashok
Ambrocio-Cruz, Silvia
Andersen, Anja
Andersen, Johannes
Andersen, Morten
Anderson, Joseph
Anderson, Kurt
Andersson, B-G
Andreuzzi, Gloria
Andrillat, Yvette

Andronov, Ivan
Anglada, Guillem
Anthony-Twarog, Barbara
Antoniou, Vallia
Aparicio, Antonio
Arbutina, Bojan
Ardeberg, Arne
Ardi, Eliani
Ardila, David
Arenou, Frédéric
Aret, Anna
Arifyanto, Mochamad
Arkhipova, Vera
Armandroff, Taft
Arnaboldi, Magda
Arnold, Richard
Arny, Thomas
Arthur, Jane
ASANO, Katsuaki
Asanok, Kitiyanee
Asplund, Martin
Asteriadis, Georgios
Asvarov, Abdul
Athanassoula, Evangelie (Lia)
Audard, Marc
Auriere, Michel
Avila-Reese, Vladimir
Azcarate, Diana
Baade, Dietrich
Baan, Willem
Baars, Jacob
BABA, Junichi
Babkovskaia, Natalia
Babusiaux, Carine
Bachiller, Rafael
BAEK, Chang Hyun
Baier, Frank
Bailin, Jeremy
Bailyn, Charles
Baker, Andrew
Balasubramanyam, Ramesh
Balázs, Bela
Balazs, Lajos
Balbus, Steven
Balcells, Marc
Balick, Bruce
Ballesteros-Paredes, Javier
Balman, Solen
Balser, Dana
Baluteau, Jean-Paul
Banhatti, Dilip
Bania, Thomas
Barbá, Rodolfo
Barberis, Bruno
Barge, Pierre
Barlow, Michael
Barmby, Pauline
Barnes, Aaron
Bar-Nun, Akiva
Barrado Navascués, David
Barsony, Mary
Barstow, Martin
Bartašiute, Stanislava
Bary, Jeffrey
Baryshev, Andrey
Bash, Frank
Bassino, Lilia
Bastian, Nathan
Basu, Baidyanath
Basu, Shantanu
Baud, Boudewijn
Baudry, Alain
Bauer, Amanda
Baume, Gustavo
Baumgardt, Holger
Bautista, Manuel
Bayet, Estelle
Beck, Sara
Becklin, Eric
Beckman, John
Beckwith, Steven
Bedogni, Roberto
Bekki, Kenji
Bellazzini, Michele
Benacquista, Matthew
Benaglia, Paula
Benaydoun, Jean-Jacques
Benisty, Myriam
Benjamin, Robert
Bensby, Thomas
Berdyugin, Andrei
Bergeron, Jacqueline
Bergin, Edwin
Bergman, Per
Bergström, Lars
Berkhuijsen, Elly
Bernat, Andrew
Bersier, David
Bertout, Claude
Bettoni, Daniela
Bhat, Ramesh
Bhatt, H.
Bianchi, Luciana
Bianchi, Simone
Biazzo, Katia
Bieging, John
Bienaymé, Olivier
Bignall, Hayley
Bignell, R.
Bijaoui, Albert
Binette, Luc
Binney, James
Black, John
Blades, John
Blair, Guy
Blair, William
Bland-Hawthorn, Jonathan
Bless, Robert
Blitz, Leo
Bloemen, Hans
Blommaert, Joris
Blum, Robert
Bobrowsky, Matthew
Bobylev, Vadim
Bocchino, Fabrizio
Bochkarev, Nikolai
Bodaghee, Arash
Bode, Michael
Bodenheimer, Peter
Boechat-Roberty, Heloisa
Böker, Torsten
Boeshaar, Gregory
Boggess, Albert
Bohlin, Ralph
Boily, Christian
Boisse, Patrick
Boland, Wilfried
Bolatto, Alberto
Bon, Edi
Bonatto, Charles
Bond, Howard
Bonifacio, Piercarlo
Bontemps, Sylvain
Booth, Roy
Boquien, Méderic
Bordbar, Gholam
Borgman, Jan
Borissova, Jordanka
Borka Jovanović, Vesna

Borkowski, Kazimierz
Boroson, Bram
Bosch, Guillermo
Bosma, Albert
Bot, Caroline
Boulanger, Francois
Boumis, Panayotis
Bourke, Tyler
Bouvier, Jerôme
Bowen, David
Bragaglia, Angela
Braine, Jonathan
Braithwaite, Jonathan
Brand, Jan
Brand, Peter
Brandl, Bernhard
Brandner, Wolfgang
Braun, Robert
Bregman, Joel
Bresolin, Fabio
Bressan, Alessandro
Bricẽo, Cesar
Brinchmann, Jarle
Brocato, Enzo
Bromage, Gordon
Bronfman, Leonardo
Brooks, Kate
Brosch, Noah
Brouillet, Nathalie
Brown, Anthony
Brown, Warren
Bruhweiler, Frederick
Brunthaler, Andreas
Bruzual, Gustavo
Bujarrabal, Valentin
Buonanno, Roberto
Burderi, Luciano
Bureau, Martin
Burikham, Piyabut
Burke, Bernard
Burkhead, Martin
Burton, Michael
Burton, W.
Butler, Dennis
Butler, Ray
Buzzoni, Alberto
Bychkov, Konstantin
Bykov, Andrei
Byrd, Gene
Bzowski, Maciej
Cacciari, Carla
CAI, Kai
Calamida, Annalisa
Caldwell, John
Callebaut, Dirk
Caloi, Vittoria
Calura, Francesco
Calzetti, Daniela
Cambrésy, Laurent
Cami, Jan
Campbell, Simon
Cane, Hilary
Canizares, Claude
Cannon, John
Cannon, Russell
Cantiello, Michele
Canto, Jorge
Canto Martins, Bruno
CAO, Zhen
Capelato, Hugo
Caplan, James
Cappa de Nicolau, Cristina
Capriotti, Eugene
Caputo, Filippina
Capuzzo Dolcetta, Roberto
Carciofi, Alex
Caretta, Cesar
Carigi, Leticia
Carignan, Claude
Carlin, Jeffrey
Carney, Bruce
Carollo, Daniela
Carpintero, Daniel
Carrasco, Luis
Carretta, Eugenio
Carretti, Ettore
Carruthers, George
Casandjian, Jean-Marc
Casasola, Viviana
casassus, simon
Caselli, Paola
Casetti, Dana
Castañeda, Héctor
Castelletti, Gabriela
Caswell, James
Cattaneo, Andrea
cazaux, Stephanie
Ceccarelli, Cecilia
Cecchi-Pestellini, Cesare
Cenko, Stephen
Centurión Martin, Miriam
Cernicharo, Jose
Cerruti Sola, Monica
Cersosimo, Juan
Cesarsky, Catherine
Cesarsky, Diego
Cesetti, Mary
CHA, Seung-Hoon
Chaboyer, Brian
Chaisson, Eric
Chamcham, Khalil
Chandler, Claire
Chandra, Suresh
Chapman, Jessica
Charbonnel, Corinne
Chatzichristou, Eleni
Chavarria-K, Carlos
Chelouche, Doron
Chemin, Laurent
CHEN, Huei-Ru
CHEN, Li
CHEN, Li
CHEN, Wen Ping
CHEN, Xuefei
CHEN, Yafeng
CHEN, Yang
CHEN, Yang
CHEN, Yuqin
CHENG, Kwang
Chengalur, Jayaram
Cherchneff, Isabelle
Chevalier, Roger
CHIBA, Masashi
CHIHARA, Hiroki
CHIN, Yi-nan
Chini, Rolf
Chiosi, Cesare
Chopinet, Marguerite
CHOU, Mei-Yin
Christensen-Dalsgaard, Jørgen
Christian, Carol
Christodoulou, Dimitris
Christopoulou, Panagiota-Eleftheria
Chryssovergis, Michael
Chu, You-Hua
CHUN, Mun-suk
Churchwell, Edward
Chyży, Krzysztof

Ciardullo, Robin
Cichowolski, Silvina
Cincotta, Pablo
Cioni, Maria-Rosa
Ciroi, Stefano
Claria, Juan
Clark, David
Clark, Frank
Clarke, David
Clegg, Robin
Clem, James
Clemens, Dan
Clementini, Gisella
Clube, S.
Codella, Claudio
Coffey, Deirdre
Cohen, Marshall
Cohen, Richard
Colangeli, Luigi
Colin, Jacques
Collin, Suzy
Combes, Françoise
Comerón, Sébastien
Comins, Neil
Contopoulos, George
Corbelli, Edvige
Corradi, Romano
Corradi, Wagner
Corral, Luis
Costa, Edgardo
Costantini, Elisa
Costero, Rafael
Courtes, Georges
Covino, Elvira
Cowie, Lennox
Cox, Donald
Cox, Pierre
Coyne, S.J, George
Cracco, Valentina
Crampton, David
Crane, Philippe
Crause, Lisa
Crawford, David
Crawford, Ian
Crézé, Michel
Crocker, Roland
Cropper, Mark
Croton, Darren
Crovisier, Jacques
Crowther, Paul
Cruvellier, Paul
Cubarsi, Rafael
Cudworth, Kyle
Cuesta Crespo, Luis
Cunningham, Maria
Cuperman, Sami
Da Costa, Gary
Dahn, Conard
DAISUKE, Iono
Dale, James
Dalgarno, Alexander
Dalla Bontà, Elena
Dambis, Andrei
D'Amico, Nicolo
Danford, Stephen
Danks, Anthony
Danly, Laura
Dannerbauer, Helmut
D'Antona, Francesca
Danziger, I.
Dapergolas, Anastasios
Daube-Kurzemniece, Ilga
Dauphole, Bertrand
David, Marc
Davies, Benjamin
Davies, Melvyn
Davies, Rodney
Davis, Christopher
Davis, Timothy
Dawson, Peter
de Almeida, Amaury
de Avillez, Miguel
De Becker, Michaël
De Bernardis, Paolo
de Boer, Klaas
de Bruijne, Jos
De Buizer, James
Decourchelle, Anne
de Gouveia Dal Pino, Elisabete
de Gregorio-Monsalvo, Itziar
de Grijs, Richard
DEGUCHI, Shuji
Deharveng, Lise
Dehghani, Mohammad
Deiss, Bruno
de Jager, Gerhard
de Jong, Roelof
de Jong, Teije
Dejonghe, Herwig
Dekel, Avishai
de La Noë, Jerome
de Laverny, Patrick
De Marchi, Guido
De Marco, Orsola
Demarque, Pierre
Demers, Serge
de Mink, Selma
DENG, LiCai
DeNisco, Kenneth
Dennefeld, Michel
De Rossi, María
Derriere, Sebastien
Désert, François-Xavier
Deshpande, Avinash
de Silva, Gayandhi
Despois, Didier
Dettmar, Ralf-Juergen
Dewdney, Peter
d'Hendecourt, Louis
Diaferio, Antonaldo
Dias da Costa, Roberto
Diaz, Ruben
Díaz-Santos, Tanio
Dib, Sami
Dickel, Helene
Dickel, John
Dickey, John
Dickman, Robert
Diehl, Roland
Dieleman, Pieter
Dieter Conklin, Nannielou
Di Fazio, Alberto
Di Francesco, James
Dinerstein, Harriet
Dinh, Trung
Dionatos, Odysseas
Disney, Michael
Djamaluddin, Thomas
Djorgovski, Stanislav
Djupvik, Anlaug Amanda
Dluzhnevskaya, Olga
Docenko, Dmitrijs
d'Odorico, Sandro
Dokuchaev, Vyacheslav
Dokuchaeva, Olga
Dominik, Carsten
Donahue, Megan
do Nascimento, José

Dopita, Michael
Dorschner, Johann
Dottori, Horacio
Dougados, Catherine
Downes, Dennis
Downes, Turlough
Downes Wallace, Juan
Draine, Bruce
Drake, Jeremy
Dreher, John
Drew, Janet
Drilling, John
Drimmel, Ronald
Drissen, Laurent
Drury, Luke
Dubner, Gloria
Dubout, Renee
Duc, Pierre-Alain
Ducati, Jorge
Ducourant, Christine
Dudorov, Aleksandr
Dufour, Reginald
Dufton, Philip
Duley, Walter
Dunne, Loretta
Duorah, Hira
Dupree, Andrea
Durrell, Patrick
Dutrey, Anne
Duvert, Gilles
Dwarakanath, K.
Dwarkadas, Vikram
Dwek, Eli
Dzigvashvili, R.
Eastwood, Kathleen
Edwards, Suzan
Efremov, Yurij
Egan, Michael
Egret, Daniel
Ehlerová, Soňa
Einasto, Jaan
Eisloeffel, Jochen
El Basuny, Ahmed
Eldridge, John
Elia, Davide
Elitzur, Moshe
Ellingsen, Simon
Elliott, Kenneth
Ellison, Sara
Elmegreen, Debra
Emerson, James
Encrenaz, Pierre
Enoch, Melissa
Enßlin, Torsten
ESAMDIN, Ali
Escalante, Vladimir
Esguerra, Jose Perico
ESIMBEK, Jarken
Esipov, Valentin
Esteban, César
Evangelidis, E.
Evans, Aneurin
Evans, Dafydd
Evans, Ian
Evans, Neal
Faber, Sandra
Falceta-Goncalves, Diego
Falcón Barroso, Jesus
Falgarone, Edith
Falize, Emeric
Falk Jr, Sydney
Fall, S.
Falle, Samuel
Fathi, Kambiz
Feast, Michael
Feautrier, Nicole
Federici, Luciana
Federman, Steven
Feinstein, Alejandro
Feitzinger, Johannes
Felli, Marcello
Feltzing, Sofia
Fendt, Christian
Ferguson, Annette
Ferland, Gary
Ferlet, Roger
Fernandes, Amadeu
Fernández, David
Ferrarese, Laura
Ferrari, Fabricio
Ferriere, Katia
Ferrini, Federico
Fesen, Robert
Fich, Michel
Fichtner, Horst
Fiebig, Dirk
Field, David
Field, George
Fierro, Julieta
Figer, Donald
Figueras, Francesca
Filipovic, Miroslav
Fischer, Jacqueline
Flannery, Brian
Fleck, Robert
Fletcher, Andrew
Florido, Estrella
Flower, David
Flynn, Chris
Foing, Bernard
Folini, Doris
Forbes, Douglas
Ford, Holland
Forrest, William
Forster, James
Forte, Juan
Forveille, Thierry
Foster, Tyler
Fox, Andrew
Fox-Machado, Lester
Franco, Gabriel Armando
Franco, José
Fraschetti, Federico
Fraser, Helen
Freeman, Kenneth
Freimanis, Juris
Frew, David
Friberg, Per
Fridlund, Malcolm
Friedman, Scott
Frisch, Priscilla
Fromang, Sebastien
Fuchs, Burkhard
Fuente, Asuncion
FUJIMOTO, Masa-Katsu
FUJIWARA, Takao
Fukuda, Naoya
FUKUI, Yasuo
FUKUSHIGE, Toshiyuki
Fuller, Gary
Furniss, Ian
FURUYA, Ray
Fusi-Pecci, Flavio
Gaensler, Bryan
Gahm, Goesta
Gai, Mario
Gail, Hans-Peter
Galli, Daniele
Gammelgaard, Peter
Gandolfi, Davide

Gangadhara, R.T.
Ganguly, Rajib
GAO, Jian
GAO, Yu
Garay, Guido
Garcia, Beatriz
Garcia, Paulo
Garcia-Burillo, Santiago
Garcia-Hernandez, Domingo
Garcia-Lario, Pedro
Garcia-Segura, Guillermo
Garnett, Donald
Garzón, Francisco
Gathier, Roel
Gaume, Ralph
Gaustad, John
Gay, Jean
Geballe, Thomas
Geffert, Michael
Geisler, Douglas
Geller, Aaron
Gemmo, Alessandra
Genkin, Igor
Genzel, Reinhard
Georgelin, Yvon
Gerard, Eric
Gerhard, Ortwin
Gerin, Maryvonne
Gerola, Humberto
Gezari, Daniel
Ghanbari, Jamshid
Giacani, Elsa
Gibson, Steven
Giersz, Mirosław
Gilmore, Gerard
Gilra, Daya
Giorgi, Edgard
Giovanelli, Riccardo
Giridhar, Sunetra
Glasner, Shimon
Gledhill, Timothy
Glinski, Robert
Glover, Simon
Glushkova, Elena
Goddi, Ciriaco
Goebel, John
Golay, Marcel
Goldes, Guillermo
Goldreich, Peter
Goldsmith, Donald
Goldsmith, Paul
Golev, Valeri
Golovatyj, Volodymyr
Gomez, Ana
Gomez, Gonzalez
Gomez, Haley
Gómez Rivero, José
Gonçalves, Denise
Gontcharov, George
Gonzales-Alfonso, Eduardo
Goodman, Alyssa
Goodwin, Simon
Gordon, Courtney
Gordon, Karl
Gordon, Mark
Gorenstein, Paul
Gosachinskij, Igor
Goss, W. Miller
Goswami, Aruna
Gottesman, Stephen
Gottlieb, Carl
Gouliermis, Dimitrios
Graham, David
Granato, Gian Luigi
Gratton, Raffaele
Grayzeck, Edwin
Grebel, Eva
Gredel, Roland
Green, Anne
Green, Anne
Green, David
Green, Elizabeth
Green, James
Greenhill, Lincoln
Greenhouse, Matthew
Gregorio-Hetem, Jane
Greisen, Eric
Grenier, Isabelle
Grenon, Michel
Grewing, Michael
Griffiths, William
Grillmair, Carl
Grindlay, Jonathan
Groenewegen, Martin
Grosso, Nicolas
Grundahl, Frank
Gry, Cecile
GU, Minfeng
Guelin, Michel
Guertler, Joachim
Guesten, Rolf
Guetter, Harry
GUILLOTEAU, Stéphane
Guinan, Edward
Gulkis, Samuel
Gull, Theodore
Gulyaev, Sergei
GUNJI, Shuichi
Günthardt, Guillermo
GUO, Jianheng
Gupta, Ranjan
Gupta, Sunil
Gupta, Yashwant
Gwinn, Carl
HABE, Asao
Habing, Harm
Hackwell, John
Haghi, Hosein
Haisch Jr, Karl
Hajduková, Jr., Maria
Hakkila, Jon
Hakopian, Susanna
HAMAJIMA, Kiyotoshi
Hammer, François
HAN, JinLin
HANAMI, Hitoshi
HANDA, Toshihiro
Hanes, David
Hanson, Margaret
HAO, Lei
Harju, Jorma
Harrington, J.
Harris, Alan
Harris, Gretchen
Harris, Hugh
Harris, William
(Harris) Law, Stella
Hartigan, Patrick
Hartkopf, William
Hartl, Herbert
Hartmann, Dieter
Hartquist, Thomas
Harvel, Christopher
Harvey, Paul
HASEGAWA, Tatsuhiko
Hatchell, Jennifer
Hatzidimitriou, Despina
Haugbølle, Troels
Haverkorn, Marijke

Hawkins, Michael
Hayashi, Saeko
Hayes, Matthew
Hayli, Abraham
Haynes, Martha
Haynes, Raymond
Haywood, Misha
HE, Jinhua
Heald, George
Heap, Sara
Hebrard, Guillaume
Hecht, James
Heggie, Douglas
Hegmann, Michael
Heikkilä, Arto
Heiles, Carl
Hein, Righini
Heinz, Sebastian
Helfer, H.
Helmi, Amina
Helmich, Frank
Helou, George
Heng, Kevin
Henkel, Christian
Henney, William
Henry, Richard
Hensler, Gerhard
Herbst, Eric
Herbst, William
Hernández, Jesús
Hernández-Pajares, Manuel
Herpin, Fabrice
Herrero Davó, Artemio
HERSANT, Franck
Hess, Kelley
Hessels, Jason
Hesser, James
Hetem Jr., Annibal
Heudier, Jean-Louis
Heydari-Malayeri, Mohammad
Heyer, Mark
Heyl, Jeremy
Hidayat, Bambang
Higgs, Lloyd
Hildebrand, Roger
Hilker, Michael
Hillenbrand, Lynne
Hills, Jack
Hinkle, Kenneth
Hippelein, Hans
HIRANO, Naomi
HIRAO, Takanori
Hiriart, David
HIROMOTO, Norihisa
HIROSE, Shigenobu
Hjalmarson, Ake
Hobbs, Lewis
Hodapp, Klaus
Hollenbach, David
Hollis, Jan
Holmberg, Johan
HONDA, Mitsuhiko
HONG, Seung-Soo
HONMA, Mareki
Hora, Joseph
Horácek, Jiri
HORI, Genichiro
Horiuchi, Shinji
Horns, Dieter
HOU, Jinliang
Houde, Martin
Houziaux, Leo
Howard III, William
HOZUMI, Shunsuke
Hron, Josef
Hua, Chon Trung
Hudson, Reggie
Hünsch, Matthias
Huggins, Patrick
Hughes, John
Humphreys, Elizabeth
Humphreys, Roberta
Humphries, Colin
Hunt, Leslie
Hunter, Todd
Hut, Piet
Hutchings, John
Hutsemékers, Damien
Hyung, Siek
Ibadov, Subhon
Ibanez, S.
Ibata, Rodrigo
Iben Jr, Icko
IGUCHI, Osamu
IKEDA, Norio
IKEUCHI, Satoru
Ikonnikova, Natalia
Il'in, Vladimir
Illingworth, Garth
IMAI, Hiroshi
IMANISHI, Masatoshi
INOUE, Akio
INUTSUKA, Shu-ichiro
Irvine, William
ISHIHARA, Daisuke
ISHII, Miki
Israel, Frank
Issa, Issa
ITA, Yoshifusa
ITOH, Hiroshi
ITOH, Yoichi
Ivanova, Natalia
Ivezic, Zeljko
Iwaniszewska, Cecylia
IYE, Masanori
Izzard, Robert
Jablonka, Pascale
Jáchym, Pavel
Jackson, James
Jackson, Peter
Jackson, William
Jacoby, George
Jacq, Thierry
Jaffe, Daniel
Jahnke, Knud
Jahreiss, Hartmut
Jalali, Mir Abbas
Janes, Kenneth
Jasniewicz, Gerard
Jenkins, Edward
Jenner, David
JEON, Young Beom
JEONG, Woong-Seob
Jerjen, Helmut
JIANG, Biwei
JIANG, Dongrong
JIANG, Zhibo
JIANG, Ing-Guey
Jiménez-Vicente, Jorge
JIN, Zhenyu
JING, Yipeng
Joblin, Christine
Jog, Chanda
Johansson, Peter
Johnson, Christian
Johnson, Fred
Johnston, Kenneth
Johnstone, Douglas
Jonas, Justin

Joncas, Gilles
Jones, Christine
Jones, David
Jones, Derek
Jones, Paul
Jordi, Carme
Jørgensen, Jes
Jørgensen, Uffe
Joshi, Umesh
Jourdain de Muizon, Marie
Jura, Michael
Just, Andreas
Justtanont, Kay
Juvela, Mika
Kafatos, Menas
Kaftan, May
Kahane, Claudine
Kahil, Magd
KAIFU, Norio
Kaisin, Serafim
Kalenskii, Sergei
Kaler, James
Kalirai, Jason
Kalnajs, Agris
Kałużny, Janusz
Kamal, Fouad
KAMAYA, Hideyuki
KAMAZAKI, Takeshi
KAMEGAI, Kazuhisa
Kamp, Inga
Kamp, Lucas
Kanekar, Nissim
KANG, Xi
KANG, Yong-Hee
Kantharia, Nimisha
Kaper, Lex
Käpylä, Maarit
Karakas, Amanda
Kassim, Namir
Kasumov, Fikret
KATO, Shoji
Kaviraj, Sugata
Kawada, Mitsunobu
KAWAGUCHI, Kentarou
KAWARA, Kimiaki
Kazlauskas, Algirdas
Kedziora-Chudczer, Lucyna
Keene, Jocelyn
Kegel, Wilhelm
Keheyan, Yeghis
Keller, Horst
Kemper, Francisca
Kennicutt, Robert
Kerber, Florian
Kervella, Pierre
Keshet, Uri
Kessler, Martin
Kharchenko, Nina
Khare, Pushpa
Khesali, Ali
KholtYgin, Alexander
Khomenko, Elena
Khosroshahi, Habib
Khovritchev, Maxim
Khrutskaya, Evgenia
Kilambi, G.
KIM, Ji Hoon
KIM, Jongsoo
KIM, Kwang tae
KIM, Sang Chul
KIM, Seung-Lee
KIM, Sungsoo
KIM, Woong-Tae
KIMURA, Toshiya
King, David
King, Ivan
Kinman, Thomas
Kirkpatrick, Ronald
Kirshner, Robert
Kiss, Csaba
Kitaeff, Vyacheslav
Kitsionas, Spyridon
Klapp, Jaime
Klare, Gerhard
Klessen, Ralf
Knacke, Roger
Knapp, Gillian
Knee, Lewis
Knezek, Patricia
Knoelker, Michael
Knude, Jens
KO, Chung-Ming
KOBAYASHI, Naoto
Koch, Andreas
Kocifaj, Miroslav
KOHNO, Kotaro
Kohoutek, Lubos
KOIKE, Chiyoe
KONDO, Yoji
KONG, Xu
Konigl, Arieh
Kontizas, Evangelos
Kontizas, Mary
KOO, Bon-Chul
Koornneef, Jan
Korchagin, Vladimir
Koribalski, Bärbel
Kormendy, John
Korpela, Eric
Kóspál, Ágnes
Kotulla, Ralf
Kouwenhoven, M.B.N. (Thijs)
KOZASA, Takashi
Kozłowski, Szymon
Krabbe, Angela
Kraft, Robert
Krajnović, Davor
Kramer, Busaba
Krause, Marita
Krautter, Joachim
Kravchuk, Sergei
Kreysa, Ernst
Krishna, Swamy
Kroupa, Pavel
Krüger, Harald
Krumholz, Mark
KUAN, Yi-Jehng
Kučinskas, Arunas
KUDOH, Takahiro
Kuiper, Rolf
Kuiper, Thomas
Kulhánek, Petr
Kulsrud, Russell
Kumar, C.
Kumar, Shiv
Kun, Maria
Kundu, Arunav
Kunth, Daniel
Kurt, Vladimir
Kurtev, Radostin
Kurtz, Stanley
Kutner, Marc
Kutuzov, Sergei
Kwitter, Karen
Kwok, Sun
Kylafis, Nikolaos
Lacey, Cedric
Lacy, John
Lada, Charles

Lafon, Jean-Pierre
Lagache, Guilaine
LAI, Shih-Ping
Laloum, Maurice
Landolt, Arlo
Lang, Kenneth
Langer, William
Langston, Glen
Lanz, Thierry
Lapasset, Emilio
Larsen, Søren
Larson, Richard
Larsson-Leander, Gunnar
Latham, David
Latter, William
Laugalys, Vygandas
Laureijs, Rene
Laurent, Claudine
Lauroesch, James
Laval, Annie
Lazarian, Alexandre
Lazio, Joseph
Leach, Sydney
Leahy, Denis
Lẽo, Joao Rodrigo
Le Bertre, Thibaut
Lebrón, Mayra
Lee, Hyun-chul
Lee, Young-Wook
LEE, Dae Hee
LEE, Hee Won
LEE, Hyung-Mok
LEE, Jae Woo
LEE, Jeong-Eun
LEE, Jung-Won
LEE, Kang Hwan
LEE, Myung Gyoon
LEE, Sang-Gak
Lefloch, Bertrand
Leger, Alain
Lehtinen, Kimmo
Leinert, Christoph
Leisawitz, David
Lemaire, Jean-louis
Leonard, Peter
Lépine, Jacques
Lepine, Sebastien
Le Poole, Rudolf
Lequeux, James
Le Squeren, Anne-Marie
Lester, Daniel
Leto, Giuseppe
Leung, Chun
LI, Jinzeng
LI, Qingkang
LIANG, Yanchun
Liebert, James
Ligori, Sebastiano
Likkel, Lauren
Liller, William
Limongi, Marco
LIN, Qing
LIN, Weipeng
Lindblad, Per
Lingenfelter, Richard
Linnartz, Harold
Lis, Dariusz
Liseau, René
Liszt, Harvey
Litvak, Marvin
LIU, Fukun
LIU, Michael
LIU, Sheng-Yuan
LIU, Siming
LIU, Xiang
LIU, Xiaowei
Lizano, Susana
Lloyd, Myfanwy
Lo, Fred K. Y.
Lo, Wing-Chi Nadia
Lobel, Alex
Lockman, Felix
Lodders, Katharina
Lodieu, Nicolas
Loinard, Laurent
łokas, Ewa
Loktin, Alexhander
Long, Knox
Lopez-Corredoira, Martin
López Garcia, Jose
Lopez Hermoso, Maria
Lorenz-Martins, Silvia
LOU, Yu-Qing
Louise, Raymond
Louys, Mireille
Lovas, Francis
Lozinskaya, Tatjana
Lu, Phillip
LU, Youjun
Lucas, Robert
Lucatello, Sara
Lucero, Danielle
LUO, Ali
LUO, Shaoguang
Lutz, Barry
Lynden-Bell, Donald
Lynds, Beverly
Lyon, Ian
MA, Jun
Maccarone, Thomas
MacConnell, Darrell
Maciejewski, Witold
Maciel, Walter
Mackey, Alasdair
MacLeod, John
Mac Low, Mordecai-Mark
Macquart, Jean-Pierre
Madden, Suzanne
Maddison, Sarah
Madsen, Gregory
MAEDA, Keiichi
Maeder, Andre
Magalhaes, Antonio Mario
Magnani, Loris
Magrini, Laura
Maier, Christian
MAIHARA, Toshinori
Maiolino, Roberto
Maíz Apellániz, Jesús
Majumdar, Subhabrata
Makalkin, Andrei
Makarov, Dmitry
MAKINO, Junichiro
MAKIUTI, Sin'itirou
Malbet, Fabien
Mallik, D.
Mamajek, Eric
Mampaso, Antonio
Manchado, Arturo
Manchester, Richard
Mandel, Ilya
Manfroid, Jean
Mangalam, Arun
Mangum, Jeffrey
Mann, Ingrid
MAO, Rui-Qing
Marco, Amparo
Mardling, Rosemary
Mardones, Diego
Maret, Sébastien

Marín-Franch, Antonio
Marino, Antonietta
Markkanen, Tapio
Markov, Haralambi
Marleau, Francine
Marmolino, Ciro
Marochnik, Leonid
Marraco, Hugo
Marsden, Stephen
Marston, Anthony
Martin, Christopher
Martin, Peter
Martin, Robert
Martín, Sergio
Martínez-Delgado, David
Martinez Roger, Carlos
Martini, Paul
Martin-Pintado, Jesus
Martins, Donald
Martos, Marco
Masheder, Michael
Masson, Colin
Mateu, Cecilia
Mather, John
Mathews, William
Mathewson, Donald
Mathis, John
MATSUHARA, Hideo
MATSUMOTO, Tomoaki
MATSUMURA, Tomotake
MATSUMURA, Masafumi
MATSUNAGA, Noriyuki
Matsuura, Mikako
Matteucci, Francesca
Matthews, Brenda
Mattila, Kalevi
Mauersberger, Rainer
Mayor, Michel
McBride, Vanessa
McCall, Benjamin
McCall, Marshall
McCammon, Dan
McClure-Griffiths, Naomi
Mccombie, June
McCray, Richard
McCutcheon, William
McDavid, David
McGee, Richard
McGehee, Peregrine
McGregor, Peter
McKee, Christopher
McLean, Ian
McMillan, Paul
McNally, Derek
Meaburn, John
Mebold, Ulrich
Meeus, Gwendolyn
Megeath, S.
Meibom, Soren
Meier, Robert
Meisel, David
Meixner, Margaret
Mellema, Garrelt
Melnick, Gary
Melnick, Jorge
Mel'nik, Anna
Mendez, Roberto
Méndez Bussard, Rene
Mennella, Vito
Menon, T.
Menten, Karl
Menzies, John
Merrifield, Michael
Meszaros, Peter
Meyer, David
Meyer, Martin
Meyer, Michael
Meylan, Georges
Migaszewski, Cezary
Migenes, Victor
Miglio, Andrea
Mikkola, Seppo
Millar, Thomas
Miller, Eric
Miller, Joseph
Miller, Richard
Milne, Douglas
Milone, Eugene
MINH, Young-Chol
Minier, Vincent
Minikulov, Nasridin
MINN, Young-Ki
Minniti, Dante
Minter, Anthony
Mirabel, Igor
Mironov, Aleksey
Misawa, Toru
Mishurov, Yury
Miszalski, Brent
Mitchell, George
MIYAMA, Syoken
Miyawaki, Ryosuke
MIYAZAKI, Atsushi
MIZUNO, Shun
Mo, Jinger
Moehler, Sabine
Møller, Palle
Moffat, Anthony
Mohammed, Ali
Mohan, Vijay
Moiseev, Alexei
Moitinho, André
Molenda-żakowicz, Joanna
Momjian, Emmanuel
MOMOSE, Munetake
Monaco, Lorenzo
Monet, David
Monin, Jean-Louis
Monnet, Guy
Montalb'an, Josefina
Montmerle, Thierry
Mookerjea, Bhaswati
Moór, Attila
Moore, Marla
Moos, Henry
Morales Rueda, Luisa
Moreno-Corral, Marco
Moreno Lupiañez, Manuel
MORI, Masao
Morris, Mark
Morris, Patrick
Morris, Rhys
Morton, Donald
Mosoni, Laszlo
Mould, Jeremy
Mouschovias, Telemachos
Muench, Guido
Mufson, Stuart
Mukherjee, Krishna
Mukhopadhyay, Banibrata
Mulas, Giacomo
Muller, Erik
Muller, Sebastien
Muminov, Muydinjon
Munari, Ulisse
MURAOKA, Kazuyuki
MURATA, Yasuhiro
Murray, Stephen
Murthy, Jayant
Muzzio, Juan

Myers, Philip
NAGAHARA, Hiroko
NAGATA, Tetsuya
Nahar, Sultana
NAITO, Tsuguya
NAKADA, Yoshikazu
NAKAGAWA, Takao
NAKAMOTO, Taishi
NAKAMURA, Fumitaka
NAKANO, Makoto
NAKANO, Takenori
NAKASATO, Naohito
NAKASHIMA, Jun-ichi
Namboodiri, P.
Nammahachak, Suwit
Naoz, Smadar
Napolitano, Nicola
Narbutis, Donatas
Natta, Antonella
Navone, Hugo
Naylor, David
Naylor, Tim
Naze, Yael
Neckel, Th.
Nedialkov, Petko
Negueruela, Ignacio
Nelemans, Gijs
Nelson, Alistair
Nemec, James
Nesci, Roberto
Nesvadba, Nicole
Netzer, Nathan
Neuhaeuser, Ralph
Newberg, Heidi
Newell, Edward
Nguyen-Quang, Rieu
Nieva, Maria
Nikiforov, Igor
Nikoghosyan, Elena
Nikoli'c, Silvana
Ninkov, Zoran
Ninković, Slobodan
NISHI, Ryoichi
NISHIDA, Minoru
NISHIDA, Mitsugu
NISHIMURA, Jun
NOMURA, Hideko
Nordh, Lennart
Noriega-Crespo, Alberto
Norman, Colin
Nota, Antonella
Noterdaeme, Pasquier
Nürnberger, Dieter
Nulsen, Paul
Nummelin, Albert
Nuritdinov, Salakhutdin
Nussbaumer, Harry
Nuth, Joseph
Oberst, Thomas
Oblak, Edouard
O'Brien, Tim
O'Connell, Robert
Ocvirk, Pierre
O'Dell, Charles
O'Dell, Stephen
Oey, Sally
OGURA, Katsuo
OH, Kap Soo
OHISHI, Masatoshi
OHTANI, Hiroshi
Oja, Tarmo
Ojha, Devendra
OKA, Tomoharu
OKAMOTO, Takashi
OKAMURA, Sadanori
OKUDA, Haruyuki
OKUMURA, Shin-ichiro
Olano, Carlos
Oliveira, Joana
Ollongren, A.
Olmi, Luca
Olofsson, Hans
Oluseyi, Hakeem
Omont, Alain
OMUKAI, Kazuyuki
ONAKA, Takashi
Onello, Joseph
ONISHI, Toshikazu
Origlia, Livia
Orlando, Salvatore
Orlov, Viktor
Ortiz, Roberto
Ortolani, Sergio
Osborne, John
Oskinova, Lidia
Osman, Anas
Ostorero, Luisa
Ostriker, Eve
Ostriker, Jeremiah
OTSUKA, Masaaki
OTSUKI, Kaori
Ott, Juergen
Oudmaijer, Rene
Pagani, Laurent
Pagano, Isabella
PAK, Soojong
PALACIOS, Ana
Palla, Francesco
Palmer, Patrick
Palouš, Jan
Palumbo, Maria Elisabetta
Panagia, Nino
Pandey, A.
Pandey, Birendra
Panei, Jorge
Pankonin, Vernon
Pantoja, Carmen
Papayannopoulos, Theodoros
PARK, Byeong-Gon
PARK, Yong Sun
Parker, Eugene
Parker, Quentin
Parmentier, Geneviève
Paron, Sergio
Parravano, Antonio
Parsamyan, Elma
Parthasarathy, Mudumba
Patat, Ferdinando
Patsis, Panos
Patten, Brian
Patzer, A. Beate C.
Pauls, Thomas
Paunzen, Ernst
Pavani, Daniela
Pavlyuchenkov, Yaroslav
Pearson, Timothy
Pecker, Jean-Claude
Pedreros, Mario
Peel, Michael
Peeters, Els
Peimbert, Manuel
Pellegrini, Silvia
Pena, Miriam
Pendleton, Yvonne
Peng, Eric
PENG, Qingyu
Penny, Alan
Penzias, Arno
Pequignot, Daniel

Perault, Michel
Perek, Luboš
Perets, Hagai
Pérez, Enrique
Perez, Mario
Perryman, Michael
Persi, Paolo
Persson, Carina
Pesch, Peter
Peters, William
Peterson, Charles
Petre, Robert
Petrosian, Vahe
Petrov, Georgi
Petrovskaya, Margarita
Petuchowski, Samuel
Peykov, Zvezdelin
Pfenniger, Daniel
Phelps, Randy
Philip, A.G.
Philipp-May, Sabine
Phillips, Thomas
Piatti, Andrés
Pier, Jeffrey
Pietrukowicz, Paweł
Pigulski, Andrzej
Pihlström, Ylva
Pikichian, Hovhannes
Pilachowski, Catherine
Pilbratt, Göran
Pilling, Sergio
Pineau des Forets, Guillaume
Pineault, Serge
Pinte, Christophe
Pirronello, Valerio
Pirzkal, Norbert
Pisano, Daniel
Piskunov, Anatolij
Pitjev, Nikolaj
Planesas, Pere
Platais, Imants
Plume, Rene
Podio, Linda
Pöppel, Wolfgang
Pogge, Richard
Polyachenko, Evgeny
Pongracic, Helen
Pontoppidan, Klaus
Pooley, David
Porceddu, Ignazio
Porquet, Delphine
Porras Juárez, Bertha
Portegies Zwart, Simon
Portinari, Laura
Pottasch, Stuart
Pound, Marc
Pouquet, Annick
Poveda, Arcadio
Poznanski, Dovi
Prasad, Sheo
Preite Martinez, Andrea
Price, Daniel
Price, R.
Pritchet, Christopher
Prochaska, Jason
Prusti, Timo
Przybilla, Norbert
Puetter, Richard
Puget, Jean-Loup
Pulone, Luigi
Puxley, Phil
Puzia, Thomas
QIN, Zhihai
Quanz, Sascha
Rabolli, Monica
Radford, Simon
Raharto, Moedji
Raimondo, Gabriella
Ramírez, Jose
Ramos-Larios, Gerardo
Ranalli, Piero
Randich, Sofia
Rastegaev, Denis
Rastogi, Shantanu
Ratag, Mezak
Rathborne, Jill
Ratnatunga, Kavan
Rauch, Thomas
Rauw, Gregor
Ravindranath, Swara
Rawlings, Jonathan
Rawlings, Mark
Ray, Tom
Raymond, John
Reach, William
Read, Justin
Rebull, Luisa
Recchi, Simone
Recio-Blanco, Alejandra
Redman, Matthew
Reid, Iain
Reid, Mark
Reid, Michael
Reif, Klaus
Reipurth, Bo
Rejkuba, Marina
Rengarajan, Thinniam
Rengel, Miriam
Renzini, Alvio
Reshetnyk, Volodymyr
Revaz, Yves
Rey, Soo-Chang
Reyes, Rafael
Reylé, Céline
Reynolds, Cormac
Reynolds, Ronald
Reynolds, Stephen
Reynoso, Estela
Rhee, Jaehyon
Rho, Jeonghee
Rich, Robert
Richard, Olivier
Richer, Harvey
Richer, John
Richer, Michael
Richter, Philipp
Richtler, Tom
Rickard, Lee
Riegel, Kurt
Rix, Hans-Walter
Rizzo, Jose
Roark, Terry
Robberto, Massimo
Roberge, Wayne
Roberts, Douglas
Roberts, Mallory
Roberts, Morton
Roberts Jr, William
Robinson, Garry
Rocha-Pinto, Hélio
Roche, Patrick
Rodrigues, Claudia
Rodrigues de Oliveira Filho, Irapuan
Rodriguez, Luis
Rodríguez, Monica
Rodríguez-Franco, Arturo
Rödiger, Elke
Roelfsema, Peter

Roellig, Thomas
Röser, Hans-peter
Roger, Robert
Rogers, Alan
Rohlfs, Kristen
Romaniello, Martino
Román-Zúñiga, Carlos
RONG, Jianxiang
Rosa, Michael
Rosado, Margarita
Rosolowsky, Erik
Rossi, Silvia
Rothberg, Barry
Rouan, Daniel
Roueff, Evelyne
Rountree, Janet
Rowell, Gavin
Roxburgh, Ian
Royer, Frédéric
Royer, Pierre
Różyczka, Michal
Rubin, Vera
Rudnitskij, Georgij
Ruelas-Mayorga, R.
Ruiz, Maria Teresa
Russeva, Tatjana
Růžička, Adam
Ryabov, Michael
Rybicki, George
Saar, Enn
Sabbadin, Franco
Sabin, Laurence
Safi-Harb, Samar
Sagar, Ram
Sage, Leslie
Sahai, Raghvendra
Sahu, Kailash
SAIGO, Kazuya
SAKAMOTO, Tsuyoshi
Sakano, Masaaki
SAKON, Itsuki
Sala, Ferran
Salama, Farid
Salinari, Piero
Salomé, Philippe
Salter, Christopher
Salukvadze, G.
Samodurov, Vladimir
Samus, Nikolay
Sánchez Béjar, Victor
Sánchez Doreste, Néstor
Sanchez-Saavedra, M.
Sancisi, Renzo
Sandell, Göran
Sanders, David
Sanders, Walter
Sanders III, Wilton
Sandqvist, Aage
Sankrit, Ravi
Santiago, Basilio
Santillán, Alfredo
Santos Jr., Joao
Sanz, Jaume
Sarajedini, Ata
Sarazin, Craig
Sargent, Annelia
Sarma, N.
Sarre, Peter
SATO, Fumio
SATO, Shuji
Savage, Blair
Savedoff, Malcolm
Savin, Daniel
Scalo, John
Scappini, Flavio
Schechter, Paul
Scherb, Frank
Schilke, Peter
Schinnerer, Eva
Schlemmer, Stephan
Schmid-Burgk, J.
Schmidt, Maarten
Schneider, Raffaella
Schödel, Rainer
Scholz, Alexander
Scholz, Ralf-Dieter
Schröder, Anja
Schuler, Simon
Schulte-Ladbeck, Regina
Schulz, R.
Schure, Klara
Schuster, William
Schwartz, Philip
Schwarz, Ulrich
Schweizer, François
Scott, Eugene
Scoville, Nicholas
Segaluvitz, Alexander
Seggewiss, Wilhelm
Seigar, Marc
Seimenis, John
Seitzer, Patrick
SEKI, Munezo
Sellgren, Kristen
Sellwood, Jerry
Sembach, Kenneth
Semenov, Dmitry
Semkov, Evgeni
Sen, Asoke
SEON, Kwang il
Serabyn, Eugene
Serjeant, Stephen
SETA, Masumichi
Shadmehri, Mohsen
Shalchi, Andreas
SHAN, Hongguang
Shane, William
Shapiro, Stuart
Sharina, Margarita
Sharma, Prateek
Sharpless, Stewart
Shaver, Peter
Shaw, Richard
Shawl, Stephen
Shchekinov, Yuri
Shelton, Ian
Shematovich, Valerij
SHEN, Juntai
Sher, David
Sherwood, William
SHI, Huoming
Shields, Gregory
SHIMOIKURA, Tomomi
SHINN, Jong-Ho
Shinnaga, Hiroko
Shipman, Russell
šmelds, Ivar
Shore, Steven
SHU, Chenggang
SHU, Frank
Shull, John
Shull, Peter
Shustov, Boris
Siebenmorgen, Ralf
Siebert, Arnaud
Sigalotti, Leonardo
Sil'chenko, Olga
Silich, Sergey
Silk, Joseph
Sills, Alison

Silva, Laura
Silvestro, Giovanni
SIMODA, Mahiro
Simón-Díaz, Sergio
Simonia, Irakli
Simonson, S.
Siopis, Christos
Sitko, Michael
Sivan, Jean-Pierre
Skilling, John
Skinner, Stephen
Skulskyj, Mychajlo
Slane, Patrick
Sloan, Gregory
Smirnov, Grigorij
Smirnova, Tatiana
Smith, Craig
Smith, Graeme
Smith, Howard
Smith, J.
Smith, Michael
Smith, Peter
Smith, Randall
Smith, Robert
Smith, Tracy
Smits, Derck
Smolinski, Jason
Snell, Ronald
Snow, Theodore
Sobolev, Andrey
Sobouti, Yousef
Soderblom, David
Sofia, Sabatino
Sofia, Ulysses
SOFUE, Yoshiaki
Solc, Martin
Solf, Josef
Soltynski, Maciej
Somerville, William
Song, Inseok
SONG, In-Ok
SONG, Liming
SONG, Qian
Sotnikova, Natalia
Soubiran, Caroline
Southworth, John
Spaans, Marco
Spagna, Alessandro
Sparke, Linda
Spergel, David
Spiegel, Edward
Spurzem, Rainer
Stahler, Steven
Stanga, Ruggero
Stanghellini, Letizia
Stanimirovic, Snezana
Stapelfeldt, Karl
Stark, Ronald
Stasinska, Grazyna
Stauffer, John
Stecher, Theodore
Stecker, Floyd
Stecklum, Bringfried
Steiman-Cameron, Thomas
Steinlin, Uli
Steinmetz, Matthias
Stenholm, Björn
Stetson, Peter
Stinebring, Daniel
St-Louis, Nicole
Stoehr, Felix
Stone, James
Stone, Jennifer
Storey, John
Strafella, Francesco
Strazzulla, Giovanni
Strelnitski, Vladimir
Stringfellow, Guy
Strobel, Andrzej
Strom, Richard
Strong, Andrew
Strubbe, Linda
Struck, Curtis
Stuik, Remko
SU, Cheng-yue
Subramaniam, Annapurni
Subramanian, Kandaswamy
Sudzius, Jokubas
SUGIMOTO, Daiichiro
SUGITANI, Koji
Suh, Kyung-Won
Suleymanova, Svetlana
SUMI, Takahiro
SUN, Jin
SUNADA, Kazuyoshi
SUNG, Hwankyung
SUNG, Hyun-Il
Suntzeff, Nicholas
Surdin, Vladimir
SUSA, Hajime
Sutherland, Ralph
Sutton, Edmund
SUZUKI, Tomoharu
Svolopoulos, Sotirios
Swade, Daryl
Sygnet, Jean-Francois
Sylvester, Roger
Szczerba, Ryszard
Székely, Péter
Szymczak, Marian
TACHIHARA, Kengo
Tadross, Ashraf
Tafalla, Mario
TAKADA-HIDAI, Masahide
TAKAGI, Toshinobu
TAKAHASHI, Hidenori
TAKAHASHI, Junko
TAKAHASHI, Koji
TAKANO, Toshiaki
TAKASHI, Hasegawa
Tamm, Antti
Tammann, Gustav
TAMURA, Motohide
TAMURA, Shin'ichi
TANAKA, Ichi Makoto
TANAKA, Masuo
Tantalo, Rosaria
Tarter, C.
Taş, Günay
Tauber, Jan
Tautvaisiene, Gražina
Taylor, A.
Taylor, Kenneth
Teixeira, Paula
Tempel, Elmo
Tennyson, Jonathan
Tenorio-Tagle, Guillermo
Tepper Garcia, Thorsten
TERADA, Yukikatsu
Terranegra, Luciano
Terzan, Agop
Terzian, Yervant
Terzides, Charalambos
Testi, Leonardo
Teyssier, Romain
Thaddeus, Patrick
Thé, Pik-Sin
Thielheim, Klaus

Thöne, Christina
Thomas, Claudine
Thompson, A.
Thomsen, Bjarne
Thonnard, Norbert
Thoul, Anne
Thronson Jr, Harley
Thuan, Trinh
Thum, Clemens
Thurston, Mark
TIAN, Wenwu
Tikhonov, Nikolai
Tilanus, Remo
Tinney, Christopher
Tissera, Patricia
Tobin, William
Tokarev, Yurij
TOMISAKA, Kohji
Toomre, Alar
Toomre, Juri
Tornambe, Amedeo
Torra, Jordi
Torrelles, Jose Maria
Torres-Papaqui, Juan
Torres-Peimbert, Silvia
TOSA, Makoto
TOSAKI, Tomoka
Tosi, Monica
Toth, Laszlo
Tothill, Nicholas
Tovmassian, Gagik
Townes, Charles
Trammell, Susan
Treffers, Richard
Tremonti, Christy
Trenti, Michele
Trinchieri, Ginevra
Trinidad, Miguel
Tripicco, Michael
Trullols, I.
Trushkin, Sergey
TSAI, An-Li
TSENG, Yao-Huan
Tsinganos, Kanaris
TSUBOI, Masato
TSUJIMOTO, Takuji
Tsvetkov, Milcho
Tsvetkova, Katja
Tully, Richard
Turatto, Massimo
Turner, David
Turner, Jean
Turner, Kenneth
Turon, Catherine
Twarog, Bruce
Tyul'bashev, Sergei
Úbeda, Leonardo
Ueta, Toshiya
Ulrich, Marie-Helene
Ulyanov, Oleg
UMEMOTO, Tomofumi
Upgren, Arthur
Urov sevi'c, Dejan
Urquhart, James
USUDA, Tomonori
Valcheva, Antoniya
Vallenari, Antonella
Valls-Gabaud, David
Valluri, Monica
Valtonen, Mauri
Val'tts, Irina
van Altena, William
van den Ancker, Mario
van den Berg, Maureen
VandenBerg, Don
van den Bergh, Sidney
VandenBout, Paul
van der Bliek, Nicole
van der Hulst, Jan
van der Kruit, Pieter
van der Laan, Harry
van der Tak, Floris
Vandervoort, Peter
van der Werf, Paul
van de Steene, Griet
van Dishoeck, Ewine
van Gorkom, Jacqueline
van Langevelde, Huib
van Loon, Jacco
van Woerden, Hugo
Varela Perez, Antonia
Varshalovich, Dmitrij
Vasta, Magda
Vavrek, Roland
Vázquez, Roberto
Vazquez, Ruben
Vega, E.
Vega, Olga
Velázquez, Pablo
Veltchev, Todor
Ventura, Paolo
Venugopal, V.
Verdoes Kleijn, Gijsbert
Vereshchagin, Sergej
Vergne, María
Verheijen, Marc
Verma, Aprajita
Verner, Ekaterina
Verschueren, Werner
Verschuur, Gerrit
Vesperini, Enrico
Viala, Yves
Viallefond, Francois
Vidal-Madjar, Alfred
Viegas, Sueli
Viironen, Kerttu
Vijh, Uma
Vilchez, José
Villas da Rocha, Jaime
Villaver, Eva
Vink, Jacco
Viti, Serena
Vivas, Anna
Vlahakis, Catherine
Vlahakis, Nektarios
Vlemmings, Wouter
Voit, Gerard
Volk, Kevin
Volkov, Evgeni
Volonteri, Marta
von Hippel, Theodore
Vorobyov, Eduard
Voronkov, Maxim
Voshchinnikov, Nikolai
Vrba, Frederick
Wachlin, Felipe
WADA, Takehiko
Wagner, Alexander
Wakelam, Valentine
Wakker, Bastiaan
Walborn, Nolan
Walder, Rolf
Walker, Gordon
Walker, Mark
Walker, Merle
Wallerstein, George
Walmsley, C.
Walsh, Andrew
Walsh, Wilfred
Walterbos, Rene

Walton, Nicholas
Wang, Q. Daniel
WANG, Chen
WANG, Hongchi
WANG, Hong-Guang
WANG, Jun-Jie
WANG, Junzhi
WANG, Min
Wannier, Peter
Wardle, Mark
Ward-Thompson, Derek
Warren Jr, Wayne
Watt, Graeme
Weaver, Harold
Webster, Adrian
Wehlau, Amelia
Weiler, Kurt
Weinberger, Ronald
Weis, Kerstin
Weisberg, Joel
Weisheit, Jon
Weistrop, Donna
Welch, Douglas
Weller, Charles
Wendt, Martin
Wesselius, Paul
Wesson, Roger
Weymann, Ray
Whelan, Emma
White, Glenn
White, Richard
White, Richard
Whitelock, Patricia
Whiteoak, John
Whittet, Douglas
Whitworth, Anthony
Wickramasinghe, N.
Wiebe, Dmitri
Wielebinski, Richard
Wielen, Roland
Wild, Wolfgang
Wilkin, Francis
Williams, David
Williams, Jonathan
Williams, Robert
Williams, Robin
Willis, Allan
Willner, Steven
Wilson, Christine
Wilson, Robert
Wilson, Thomas
Winkler, Paul
Winnberg, Anders
Witt, Adolf
Wofford, Aida
Wolff, Michael
Wolfire, Mark
Wolstencroft, Ramon
Wolszczan, Alexander
Woltjer, Lodewijk
Wong, Tony
Wood, Brian
Wood, Douglas
Woodward, Paul
Woolf, Neville
Wootten, Henry
Worters, Hannah
Wouterloot, Jan
Wramdemark, Stig
Wright, Edward
Wright, Nicholas
WU, Hong
WU, Hsin-Heng
WU, Zhen-Yu
Wulandari, Hesti
Wünsch, Richard
Wurz, Peter
Wynn-Williams, Gareth
Wyrowski, Friedrich
Wyse, Rosemary
Xilouris, Emmanouel
Xiradaki, Evangelia
XU, Ye
YABUSHITA, Shin
Yahil, Amos
YAMADA, Masako
YAMADA, Shimako
YAMAGATA, Tomohiko
YAMAMOTO, Hiroaki
YAMAMOTO, Satoshi
YAMAMURA, Issei
YAMASHITA, Takuya
YAN, Jun
YANG, Xiaohu
YANO, Hajime
Yegorova, Irina
YI, Sukyoung
YIM, Hong Suh
YONEKURA, Yoshinori
Yong, David
Yoon, Suk-Jin
York, Donald
Yorke, Harold
YOSHIDA, Shigeomi
YOSHII, Yuzuru
Young, Lisa
Younis, Saad
YU, Qingjuan
YUI, Yukari
YUMIKO, Oasa
Yun, Joao
Yun, Min
YUTAKA, Shiratori
Zachilas, Loukas
Zaggia, Simone
Zakharova, Polina
Zapatero-Osorio, Maria Rosa
Zasov, Anatoly
Zavagno, Annie
Zdanavičius, Justas
Zealey, William
Zeilik, Michael
Zeilinger, Werner
ZENG, Qin
Zezas, Andreas
Zhang, Yuying
ZHANG, Cheng-Yue
ZHANG, Fenghui
ZHANG, Haotong
ZHANG, JiangShui
ZHANG, Jingyi
ZHANG, Qizhou
ZHANG, Shuang Nan
ZHAO, Gang
ZHAO, Jun Liang
Zhekov, Svetozar
ZHOU, Jianjun
ZHU, Ming
ZHU, Qingfeng
ZHU, Wenbai
Zibetti, Stefano
Ziegler, Bodo
Zijlstra, Albert
Zinchenko, Igor
Zinn, Robert
Ziurys, Lucy
Zoccali, Manuela
Zuckerman, Benjamin
Zwintz, Konstanze

Division J Galaxies and Cosmology

President: Françoise Combes
Vice-President: Thanu Padmanabhan

Steering Committee Members:

Ofer Lahav
Brian P. Schmidt
John S. Gallagher III
Marijn Franx
Jayant Murthy
Elaine M. Sadler
Thaisa Storchi-Bergmann
Françoise Combes
Thanu Padmanabhan
Andrew J. Bunker
Stéphane J. Courteau
Monica Rubio

Members:

Aalto, Susanne
Abadi, Mario
Abbas, Ummi
Ables, Harold
Abraham, Roberto
Abraham, Zulema
Abrahamian, Hamlet
Abu Kassim, Hasan
Adami, Christophe
Adams, Jenni
Adler, David
Afanas'ev, Viktor
Aghaee, Alireza
Aguero, Estela
Aguilar, Luis
Aharonian, Felix
Ahmad, Farooq
AHN, SANG-HYEON
Aizenman, Morris
Ajhar, Edward
Akashi, Muhammad
AKIYAMA, Masayuki
Alard, Christophe
Alcaniz, Jailson
Aldaya, Victor
Alexander, Tal
Ali, Sabry
Alimi, Jean-Michel
Alladin, Saleh
Allan, Peter
Allen, Ronald
Allington-Smith, Jeremy
Alloin, Danielle
Almaini, Omar
Aloisi, Alessandra
Alonso, Maria
Alonso, Maria
Alonso-Herrero, Almudena
Amati, Lorenzo
Amendola, Luca
Ammons, Stephen
Amram, Philippe
AN, TAO
Andernach, Heinz
Andersen, Michael
Anderson, Joseph
Anderson, Kurt
Andreani, Paola
Andreon, Stefano
Andrillat, Yvette
Andruchow, Ileana
Angione, Ronald
ANN, Hong-Bae
Anosova, Joanna
Ansari, S.M.
Anton, Sonia
Antonelli, Lucio Angelo
Antoniou, Vallia
Antonuccio-Delogu, Vincenzo
Aoki, Kentaro
Aparicio, Antonio
Appenzeller, Immo
Appleton, Philip
Aragón-Salamanca, Alfonso
ARAKIDA, Hideyoshi
Ardeberg, Arne
Aretxaga, Itziar
Argo, Megan
Argüeso, Francisco
Arkhipova, Vera
Arnaboldi, Magda
Arnaud, Monique
Arribas Mocoroa, Santiago
Artamonov, Boris
ASANO, Katsuaki
Athanassoula, Evangelie (Lia)
Atrio Barandela, Fernando
Audouze, Jean
Augusto, Pedro
Auriemma, Giulio
Aussel, Herve
Avelino, Pedro
Avila-Reese, Vladimir
AYANI, Kazuya
AZUMA, Takahiro
Azzopardi, Marc
Baade, Robert
Baan, Willem
BABA, Junichi
Babul, Arif
Bachev, Rumen
Baddiley, Christopher
Baes, Maarten
Baggaley, William
Bagla, Jasjeet
Bahcall, Neta
BAI, Jinming
Bailey, Mark
Bailin, Jeremy
Bajaja, Esteban
Bajtlik, Stanisław
Baker, Andrew
Baki, Paul

Balasubramanyam, Ramesh
Balazs, Lajos
Balbi, Amedeo
Baldry, Ivan
Baldwin, Jack
Baliyan, Kiran
Balkowski-Mauger, Chantal
Ballabh, Goswami
Balland, Christophe
Balman, Solen
Balogh, Michael
Bamford, Steven
Banday, Anthony
Banerji, Sriranjan
Banhatti, Dilip
Bannikova, Elena
Barberis, Bruno
Barbon, Roberto
Barbuy, Beatriz
Barcons, Xavier
Bardeen, James
Bardelli, Sandro
Barden, Marco
Barger, Amy
Barkana, Rennan
Barkhouse, Wayne
Barmby, Pauline
Barnes, David
Barret, Didier
Barrientos, Luis
Barrow, John
Bartelmann, Matthias
Barth, Aaron
Barthel, Peter
Barton, Elizabeth
Barway, Sudhashu
Baryshev, Andrey
Basa, Stephane
Basart, John
Bassett, Bruce
Bassino, Lilia
Basu, Baidyanath
Basu, Kaustuv
Battaner, Eduardo
Battinelli, Paolo
Battye, Richard
Bauer, Amanda
Baum, Stefi
Bautista, Manuel
Bayet, Estelle
Beaulieu, Sylvie
Bechtold, Jill
Beck, Rainer
Beck, Sara
Becker, Robert
Beckman, John
Beckmann, Volker
Beesham, Aroonkumar
Begelman, Mitchell
Begeman, Kor
Behar, Ehud
Bekenstein, Jacob
Bekki, Kenji
Belinski, Vladimir
Belkovich, Oleg
Bellas-Velidis, Ioannis
Bender, Ralf
Benedict, George
Benetti, Stefano
Benítez, Erika
Bennett, Charles
Bennett, David
Bensby, Thomas
Bentz, Misty
Berczik, Peter
Bergeron, Jacqueline
Bergström, Lars
Bergvall, Nils
Berkhuijsen, Elly
Berman, Marcelo
Berman, Vladimir
Berta, Stefano
Bertin, Giuseppe
Bertola, Francesco
Bertschinger, Edmund
Betancor Rijo, Juan
Bettoni, Daniela
Bharadwaj, Somnath
Bhavsar, Suketu
BIAN, Yulin
Bianchi, Simone
Bicak, Jiri
Bicknell, Geoffrey
Biermann, Peter
Bignall, Hayley
Bignami, Giovanni
Bijaoui, Albert
Bijleveld, Willem
Binetruy, Pierre
Binette, Luc
Binggeli, Bruno
Binney, James
Biretta, John
Birkinshaw, Mark
Biviano, Andrea
Bjælde, Ole
Björnsson, Gunnlaugur
Björnsson, Claes-Ingvar
Black, John
Blakeslee, John
Blamont, Jacques-Emile
Blanchard, Alain
Bleyer, Ulrich
Blitz, Leo
Block, David
Bludman, Sidney
Blumenthal, George
Blundell, Katherine
Bochkarev, Nikolai
Bodaghee, Arash
Böhringer, Hans
Böker, Torsten
Böhm, Asmus
Boissier, Samuel
Boisson, Catherine
Boksenberg, Alec
Bolatto, Alberto
Boles, Thomas
Bolzonella, Micol
Bomans, Dominik
Bon, Natasa
Bonaccini, Domenico
Bonatto, Charles
Bond, John
Bondi, Marco
Bongiovanni, Angel
Bonnor, W.
Booth, Roy
Boquien, Médéric
Borchkhadze, Tengiz
Borgani, Stefano
Borne, Kirk
Boroson, Bram
Boschin, Walter
Bosma, Albert
Bot, Caroline
Bottinelli, Lucette
Bouchet, François
Boulanger, Francois
Bouwens, Rychard

Bowen, David
Bower, Gary
Bowyer, C.
Boyle, Brian
Braine, Jonathan
Braithwaite, Jonathan
Brammer, Gabriel
Branchesi, Marica
Brandt, William
Branduardi-Raymont, Graziella
Braun, Robert
Bravo-Alfaro, Hector
Brecher, Kenneth
Bregman, Joel
Brentjens, Michiel
Brescia, Massimo
Bresolin, Fabio
Bressan, Alessandro
Bridges, Terry
Bridle, Sarah
Briggs, Franklin
Brinchmann, Jarle
Brinks, Elias
Broadfoot, A.
Brodie, Jean
Bromberg, Omer
Brosch, Noah
Brotherton, Michael
Brough, Sarah
Brouillet, Nathalie
Brown, Michael
Brown, Thomas
Browne, Ian
Brunner, Robert
Bruzual, Gustavo
Bryant, Julia
Buat, Véronique
Bunker, Andrew
Buote, David
Burbidge, Eleanor
Burdyuzha, Vladimir
Bureau, Martin
Burgarella, Denis
Burikham, Piyabut
Burkert, Andreas
Burns, Jack
Busarello, Giovanni
Buta, Ronald
Butcher, Harvey
Buzzoni, Alberto
Byrd, Gene
BYUN, Yong Ik
Cabanac, Remi
CAI, Michael
CAI, Mingsheng
Calderón, Jesús
Calura, Francesco
Calvani, Massimo
Calzetti, Daniela
Campusano, Luis
Canizares, Claude
Cannon, John
Cannon, Russell
Cantiello, Michele
Canzian, Blaise
CAO, Li
CAO, Xinwu
Caon, Nicola
Capaccioli, Massimo
Capelato, Hugo
Cappellari, Michele
Cappi, Alberto
Cappi, Massimo
Caproni, Anderson
Caputi, Karina
Caretta, Cesar
Carigi, Leticia
Carignan, Claude
Carlberg, Raymond
Carlin, Jeffrey
Carollo, Marcella
Carr, Bernard
Carraro, Giovanni
Carretta, Eugenio
Carretti, Ettore
Carrillo, Rene
Carswell, Robert
Carter, David
Casasola, Viviana
Casoli, Fabienne
Cassano, Rossella
Castagnino, Mario
Cattaneo, Andrea
Cavaliere, Alfonso
Cayatte, Veronique
Cellone, Sergio
Cenko, Stephen
Centurión Martin, Miriam
Cepa, Jordi
Cesarsky, Catherine
Cesarsky, Diego
Cesetti, Mary
CHA, Seung-Hoon
CHAE, Kyu Hyun
Chaisson, Eric
Chakrabarti, Sandip
Chamaraux, Pierre
Chamcham, Khalil
CHANG, Heon-Young
CHANG, Kyongae
CHANG, Ruixiag
Charbonnel, Corinne
Charlot, Stephane
Charmandaris, Vassilis
Chatterjee, Tapan
Chatzichristou, Eleni
Chauvineau, Bertrand
Chavushyan, Vahram
Chelliah Subramonian, Stalin
Chelouche, Doron
Chemin, Laurent
Chen, Yanmei
CHEN, DaMing
CHEN, Dongni
CHEN, Hsiao-Wen
CHEN, Jiansheng
CHEN, Lin-wen
CHEN, Pisin
CHEN, Xuelei
CHEN, Yang
CHEN, Yang
CHENG, Fuzhen
Chengalur, Jayaram
Chernyakova, Maria
Chiang, Hsin
Chiappini, Cristina
CHIBA, Masashi
CHIBA, Takeshi
Chincarini, Guido
Chiosi, Cesare
Chitre, Dattakumar
Chodorowski, Michał
CHOU, Chih-Kang
CHOU, Mei-Yin
Choudhury, Tirthankar
Christensen, Lise
Christensen, Per
Christodoulou, Dimitris

Christy, James
CHU, Yaoquan
Chugai, Nikolaj
Chukwude, Augustine
CHUN, Sun
Chyży, Krzysztof
Ciardullo, Robin
Cid Fernandes, Roberto
Ciliegi, Paolo
Cinzano, Pierantonio
Cioni, Maria-Rosa
Ciotti, Luca
Cirkovic, Milan
Ciroi, Stefano
Claeskens, Jean-François
Clairemidi, Jacques
Claria, Juan
Clark, David
Clarke, David
Clarke, Tracy
Clarkson, Chris
Clavel, Jean
Clementini, Gisella
Clocchiatti, Alejandro
Clowe, Douglas
Clowes, Roger
Cocke, William
Cohen, Marshall
Cohen, Ross
Colafrancesco, Sergio
Colbert, Edward
Cole, Shaun
Coles, Peter
Colina, Luis
Colless, Matthew
Colombi, Stephane
Comastri, Andrea
Combes, Françoise
Comerón, Sébastien
Comins, Neil
Comte, Georges
Condon, James
Conselice, Christopher
Conti, Alberto
Contopoulos, George
Cook, Kem
Cooray, Asantha
Copin, Yannick
Cora, Sofia
Corbelli, Edvige
Corbett, Ian
Corbin, Michael
Corsini, Enrico
Corwin Jr, Harold
Côte, Patrick
Côté, Stéphanie
Couch, Warrick
Couchman, Hugh
Courbin, Frederic
Courteau, Stéphane
Courtes, Georges
Courtois, Helene
Courvoisier, Thierry
Couto da Silva, Telma
Covone, Giovanni
Cowsik, Ramanath
Coziol, Roger
Cracco, Valentina
Crane, Patrick
Crane, Philippe
Crawford, Carolin
Crawford, Steven
Crenshaw, Daniel
Cress, Catherine
Cristiani, Stefano
Cristóbal, David
Croft, Steve
Croom, Scott
Crosta, Mariateresa
Croston, Judith
Croton, Darren
Csabai, Istvan
Cubarsi, Rafael
Cuby, Jean-Gabriel
Cui, Wei
CUI, Wenyuan
Cunniffe, John
Curir, Anna
Curran, Stephen
Cypriano, Eduardo
Czerny, Bożena
da Costa, Luiz
Da Costa, Gary
Dadhich, Naresh
Dagkesamanskii, Rustam
Dahle, Hakon
Daigne, Frederic
DAISUKE, Iono
Dalla Bontà, Elena
Dallacasa, Daniele
Dalton, Gavin
D'Ammando, Filippo
Danese, Luigi
Danks, Anthony
Dannerbauer, Helmut
Dantas, Christine
Da Rocha, Cristiano
Darriulat, Pierre
Das, P.
Dasyra, Kalliopi
Davidge, Timothy
Davidson, William
Davies, Benjamin
Davies, Paul
Davies, Rodney
Davies, Roger
Davis, Marc
Davis, Michael
Davis, Timothy
De Bernardis, Paolo
De Blok, Erwin
de Boer, Klaas
de Bruyn, A.
de Carvalho, Reinaldo
de Diego Onsurbe, Jose
de Grijs, Richard
Deharveng, Jean-Michel
Deiss, Bruno
de Jong, Roelof
Dejonghe, Herwig
Dekel, Avishai
de Lapparent, Valérie
de Lima, José
Dell'Antonio, Ian
Del Olmo, Ascension
Del Popolo, Antonino
De Lucia, Gabriella
Demarco, Ricardo
de Mello, Duilia
Demers, Serge
Demiański, Marek
DENG, Zugan
DeNisco, Kenneth
Dennefeld, Michel
de Petris, Marco
de Propris, Roberto
de Rijcke, Sven
Dermott, Stanley
De Rossi, María
de Ruiter, Hans

Désert, François-Xavier
de Silva, Gayandhi
de Silva, Lindamulage
Dessauges-Zavadsky, Miroslava
de Swardt, Bonita
Dettmar, Ralf-Juergen
de Ugarte Postigo, Antonio
Deustua, Susana
Devost, Daniel
de Zeeuw, Pieter
de Zotti, Gianfranco
d'Hendecourt, Louis
Dhurandhar, Sanjeev
Diaferio, Antonaldo
Diaz, Angeles
Diaz, Ruben
Díaz-Santos, Tanio
Dib, Sami
Dickey, John
Dickman, Steven
Diehl, Roland
Dietrich, Matthias
Dietrich, Jörg
Dionysiou, Demetrios
di Serego Alighieri, Sperello
Djorgovski, Stanislav
Dobbs, Matt
Dobrzycki, Adam
Dodonov, Sergej
d'Odorico, Sandro
D' Odorico, Valentina
DOI, Mamoru
Dokuchaev, Vyacheslav
Dole, Herve
Domínguez, Mariano
Dominis Prester, Dijana
Donahue, Megan
Donas, Jose
Donea, Alina
Dong, Xiao-Bo
Donner, Karl
D'Onofrio, Mauro
Donzelli, Carlos
Dopita, Michael
Doroshenko, Valentina
Dottori, Horacio
Dovciak, Michal
Doyon, Rene
Dressel, Linda
Dressler, Alan
Drinkwater, Michael
Driver, Simon
Dubin, Maurice
Duc, Pierre-Alain
Dufay, Maurice
Dufour, Reginald
Dultzin-Hacyan, Deborah
Dumke, Michael
Dumont, Anne-Marie
Dumont, Rene
Dunlop, James
Dunne, Loretta
Dunsby, Peter
Duorah, Hira
Durrell, Patrick
Durret, Florence
Duval, Marie-France
Dwarakanath, K.
Dwek, Eli
Dyer, Charles
Eales, Stephen
Easther, Richard
Edelson, Rick
Edmunds, Michael
Edsjö, Joakim
Edwards, Louise
Edwards, Philip
Efstathiou, George
Egret, Daniel
Ehle, Matthias
Einasto, Jaan
Einasto, Maret
Ekers, Ronald
Elgarøy, Øystein
Elizalde, Emilio
Ellingsen, Simon
Ellis, George
Ellis, Richard
Ellis, Simon
Ellison, Sara
Elmegreen, Bruce
Elmegreen, Debra
Elvis, Martin
Elyiv, Andrii
Emsellem, Eric
Enginol, Turan
English, Jayanne
ENOKI, Motohiro
Enßlin, Torsten
Ercan, Enise
Espey, Brian
Ettori, Stefano
Eungwanichayapant, Anant
Evans, Robert
EZAWA, Hajime
Fabbiano, Giuseppina
Faber, Sandra
Fabricant, Daniel
Fahey, Richard
Fahlman, Gregory
Falceta-Goncalves, Diego
Falco, Emilio
Falcón Barroso, Jesus
Falcon Veloz, Nelson
Falgarone, Edith
Falk Jr, Sydney
Fall, S.
Famaey, Benoit
FAN, Junhui
FAN, Zuhui
Fanaroff, Bernard
Farrell, Sean
Fasano, Giovanni
Fassnacht, Christopher
Fathi, Kambiz
Fatkhullin, Timur
Faundez-Abans, Max
Faure, Cécile
Fazio, Giovanni
Feain, Ilana
Feast, Michael
Fedeli, Cosimo
Fedorova, Elena
Feinstein, Carlos
Feitzinger, Johannes
Feldman, Paul
FENG, Long Long
Ferguson, Annette
Ferland, Gary
Ferrarese, Laura
Ferrari, Attilio
Ferrari, Fabricio
Ferreira, Pedro
Ferreras, Ignacio
Ferrini, Federico
Fich, Michel
Field, George
Figueiredo, Newton
Filipov, Lachezar

Filipovic, Miroslav
Filippenko, Alexei
Finn, Lee
Fletcher, Andrew
Flin, Piotr
Florides, Petros
Florido, Estrella
Florsch, Alphonse
Floyd, David
Fluke, Christopher
Focardi, Paola
Foltz, Craig
Fong, Richard
Forbes, Duncan
Ford, Holland
Ford Jr, W.
Forman, William
Foschini, Luigi
FOUCAUD, Sebastien
Fouqué, Pascal
Fox, Andrew
Fraix-Burnet, Didier
Franceschini, Alberto
Francis, Paul
Franx, Marijn
Freedman, Wendy
Freeman, Kenneth
Frenk, Carlos
Freudling, Wolfram
Friaca, Amancio
Fricke, Klaus
Fried, Josef
Fritze, Klaus
Frogel, Jay
Fruchter, Andrew
Frutos-Alfaro, Francisco
Fuchs, Burkhard
FUJITA, Yutaka
FUJIWARA, Akira
FUKUGITA, Masataka
FUKUI, Takao
FUNATO, Yoko
Funes, José
Furlanetto, Steven
FURUYA, Ray
Füzfa, Andre
Fynbo, Johan
Gabanyi, Krisztina
Gadotti, Dimitri
Gaensler, Bryan
Gallagher, Sarah
Gallagher III, John
Gallart, Carme
Gallazzi, Anna
Gallego, Jesús
Galletta, Giuseppe
Gallimore, Jack
Gamaleldin, Abdulla
Gammelgaard, Peter
Gangui, Alejandro
Ganguly, Rajib
GAO, Jian
GAO, Yu
Garcia-Barreto, José
Garcia-Burillo, Santiago
Garcia-Lorenzo, Maria
Gardner, Jonathan
Garilli, Bianca
Garrido, César
Garrison, Robert
Gaskell, C.
Gastaldello, Fabio
Gavazzi, Giuseppe
Gavignaud, Isabelle
Gay, Pamela
Gehrels, Neil
Gelderman, Richard
Geller, Margaret
GENG, Lihong
Gentile, Gianfranco
Genzel, Reinhard
Georgiev, Tsvetan
Gerhard, Ortwin
Germany, Lisa
Ghigo, Francis
Ghirlanda, Giancarlo
Ghosh, P.
Giacani, Elsa
Giallongo, Emanuele
Giani, Elisabetta
Gibson, Brad
Gieren, Wolfgang
Gigoyan, Kamo
Gilbank, David
Gilfanov, Marat
Gioia, Isabella
Giovanardi, Carlo
Giovane, Frank
Giovanelli, Riccardo
Girardi, Marisa
Giroletti, Marcello
Gitti, Myriam
Glass, Ian
Glazebrook, Karl
Glover, Simon
Godłowski, Włodzimierz
Godwin, Jon
Goicoechea, Luis
Goldsmith, Donald
Gomez, Haley
Gómez Fernández, Jose
Gonçalves, Denise
GONG, Biping
Gonzalez Delgado, Rosa
González Sánchez, Alejandro
Gonzalez-Serrano, Jose Ignacio
Goobar, Ariel
Goodrich, Robert
Gordon, Chris
Gordon, Karl
Gorenstein, Paul
Goret, Philippe
Gorgas, Garcia
Goss, W. Miller
Gosset, Eric
GOTO, Tomotsugu
Gottesman, Stephen
Gottlöber, Stefan
GOUDA, Naoteru
Gouguenheim, Lucienne
Gouliermis, Dimitrios
Gounelle, Matthieu
Govinder, Keshlan
Goyal, Ashok
Graham, Alister
Graham, John
Gramann, Mirt
Granato, Gian Luigi
Gray, Meghan
Gray, Richard
Grebel, Eva
Green, Anne
Green, Richard
Greenhill, Lincoln
Greenhouse, Matthew
Gregg, Michael
Gregorini, Loretta
Gregorio, Anna

Gregory, Stephen
Greve, Thomas
Greyber, Howard
Grieger, Bjoern
Griest, Kim
Griffiths, Richard
Grillmair, Carl
Grishchuk, Leonid
Griv, Evgeny
Gronwall, Caryl
Grosbøl, Preben
Grove, Lisbeth
Grueff, Gavril
Grün, Eberhard
Grupe, Dirk
GU, Minfeng
GU, Qiusheng
Gudehus, Donald
Gudmundsson, Einar
Gulkis, Samuel
Gumjudpai, Burin
GUNJI, Shuichi
Gunn, James
Günthardt, Guillermo
Gurzadyan, Vahagn
Guseva, Natalia
Gustafson, Bo
Gutiérrez, Carlos
Guzzo, Luigi
Gyulbudaghian, Armen
Haas, Martin
Haehnelt, Martin
Hagen, Hans-Juergen
Hagen-Thorn, Vladimir
Haghi, Hosein
Hakopian, Susanna
Hall, Patrick
HAMABE, Masaru
Hambaryan, Valeri
Hamilton, Andrew
Hammer, François
Hamuy, Mario
HAN, Cheongho
HANAMI, Hitoshi
Hanasz, Michał
HANDA, Toshihiro
Hanes, David
Hanisch, Robert
Hanna, David
Hanner, Martha
Hannestad, Steen
Hansen, Frode
HAO, Lei
Haque, Shirin
HARA, Tetsuya
Hardcastle, Martin
Hardy, Eduardo
Harms, Richard
Harnett, Julienne
Harris, Daniel
Harris, William
Hartmann, Dieter
Hartwick, F.
Harwit, Martin
Hasan, Hashima
Hashimoto, Yasuhiro
HATTORI, Takashi
HATTORI, Makoto
Hatziminaoglou, Evanthia
Hau, George
Haugbølle, Troels
Hauser, Michael
Hawking, Stephen
Hayes, Matthew
Haynes, Martha
HAZUMI, Masashi
HE, XiangTao
Heacox, William
Heald, George
Heap, Sara
Heavens, Alan
Hecht, James
Heckman, Timothy
Heidt, Jochen
Heiles, Carl
Heinamaki, Pekka
Heinz, Sebastian
Held, Enrico
Helfand, David
Hellaby, Charles
Heller, Michał
Helmi, Amina
Helou, George
Hendry, Martin
Henning, Patricia
Henriksen, Mark
Henriksen, Richard
Henry, Richard
Henry, Richard
Hensler, Gerhard
Héraudeau, Philippe
Herbst, Eric
Hernández, Xavier
Hernández-Monteagudo, Carlos
Hervik, Sigbjorn
Hess, Kelley
Hesser, James
Hewett, Paul
Hewitt, Adelaide
Hewitt, Anthony
Heyl, Jeremy
Heyrovský, David
Hicks, Amalia
Hickson, Paul
HIDEKI, Asada
Hilker, Michael
Hill, Vanessa
Hillebrandt, Wolfgang
Hintzen, Paul
HIRASHITA, Hiroyuki
Hiriart, David
Hirv, Anti
Hjalmarson, Ake
Hjorth, Jens
Hnatyk, Bohdan
Ho, Luis
Hodge, Paul
Hoekstra, Hendrik
Hönig, Sebastian
Hofmann, Wilfried
Hogan, Craig
Hogg, David
Holz, Daniel
HONG, Seung-Soo
Hopkins, Andrew
Hopp, Ulrich
Horellou, Cathy
Horiuchi, Shinji
Horns, Dieter
Hornschemeier, Ann
Hornstrup, Allan
HOU, Jinliang
Houdashelt, Mark
Hough, James
HOZUMI, Shunsuke
Hu, Esther
HU, Fuxing
HU, Hongbo
Hua, Chon Trung

Huang, Jiasheng
HUANG, Keliang
Huchtmeier, Walter
Hudson, Michael
Huertas-Company, Marc
Huettemeister, Susanne
Hughes, David
Hughes, John
Humphreys, Elizabeth
Humphreys, Roberta
Hunstead, Richard
Hunt, Leslie
Hunter, James
Hurwitz, Mark
Hutchings, John
Hutsemékers, Damien
Hütsi, Gert
Huynh, Minh
HWANG, Chorng-Yuan
HWANG, Jai-chan
Ibata, Rodrigo
ICHIKAWA, Shin-ichi
ICHIKAWA, Takashi
Icke, Vincent
Idiart, Thais
IKEUCHI, Satoru
Ilić, Dragana
Iliev, Ilian
Illingworth, Garth
IM, Myungshin
IMANISHI, Masatoshi
Impey, Christopher
INADA, Naohisa
Infante, Leopoldo
INOUE, Akio
INOUE, Makoto
IOKA, Kunihito
Iovino, Angela
Irwin, Judith
ISHIHARA, Hideki
ISHIMARU, Yuhri
Israel, Frank
Issa, Issa
ITOH, Naoki
Ivanov, Pavel
Ivanov-Kholodny, Gor
Ivezic, Zeljko
Ivison, Robert
IWAMURO, Fumihide
IWATA, Ikuru
IYE, Masanori
Iyer, Balasubramanian
Izotov, Yuri
Izotova, Iryna
Jablonka, Pascale
Jáchym, Pavel
Jackson, Bernard
Jaffe, Walter
Jahnke, Knud
Jakobsen, Peter
Jakobsson, Pall
Jalali, Mir Abbas
James, John
Jamrozy, Marek
JANG, Minwhan
Jannuzi, Buell
Jaroszyński, Michał
Jarrett, Thomas
Jassur, Davoud
Jauncey, David
Jaunsen, Andreas
Jedamzik, Karsten
Jelić, Vibor
Jenner, David
Jensen, Brian
JEONG, Woong-Seob
Jerjen, Helmut
Jetzer, Philippe
Jha, Saurabh
JIANG, Ing-Guey
Jiménez-Vicente, Jorge
JING, Yipeng
Jog, Chanda
Johansson, Peter
Johns, Bethany
Johnston, Helen
Johnston-Hollitt, Melanie
Joly, Monique
Jonas, Justin
Jones, Bernard
Jones, Christine
Jones, Heath
Jones, Paul
Jones, Thomas
Jordán, Andrés
Jorgensen, Inger
Joseph, Robert
Joshi, Umesh
Joubert, Martine
Jovanović, Predrag
Joy, Marshall
Jungwiert, Bruno
Junkes, Norbert
Junkkarinen, Vesa
Junor, William
Just, Andreas
Kaastra, Jelle
Kaisin, Serafim
KAJINO, Toshitaka
KAJISAWA, Masaru
Kalloglian, Arsen
Kalomeni, Belinda
Kamal, Fouad
KAMENO, Seiji
Kaminker, Alexander
Kandalyan, Rafik
Kanekar, Nissim
KANEKO, Noboru
KANG, Hyesung
KANG, Xi
Kantharia, Nimisha
Kapoor, Ramesh
Karachentsev, Igor
Karachentseva, Valentina
Karami, Kayoomars
KAROJI, Hiroshi
Karouzos, Marios
Karygina, Zoya
KASHIKAWA, Nobunari
Kaspi, Shai
Kassim, Namir
Katgert, Peter
Katgert-Merkelijn, J.
KATO, Shoji
Katsiyannis, Athanassios
Kauffmann, Guinevere
Kaufman, Michele
Kaul, Chaman
Kausch, Wolfgang
Kaviraj, Sugata
KAWABATA, Kiyoshi
Kawada, Mitsunobu
KAWAKATU, Nozomu
KAWARA, Kimiaki
KAWASAKI, Masahiro
KAYO, ISSHA
Keel, William
Keenan, Ryan
Keeney, Brian
Kellermann, Kenneth

Kelly, Brandon
Kembhavi, Ajit
Kemp, Simon
Kennicutt, Robert
Kent, Brian
Keshet, Uri
Khachikian, Edward
Khanna, Ramon
Khare, Pushpa
Khmil, Sergiy
KholtYgin, Alexander
Khosroshahi, Habib
Kilborn, Virginia
KIM, Dong Woo
KIM, Ji Hoon
KIM, Jik
KIM, Kwang tae
KIM, Minsun
KIM, Sang Chul
KIM, Sungsoo
KIM, Woong-Tae
Kimball, Amy
Kimble, Randy
King, Ivan
King, Lindsay
Kinman, Thomas
Kirilova, Daniela
Kirshner, Robert
Kissler-Patig, Markus
Kitaeff, Vyacheslav
KIYOTOMO, Ichiki
Kjaergaard, Per
Klapp, Jaime
Klein, Ulrich
Knapen, Johan
Knapp, Gillian
Kneib, Jean-Paul
Knezek, Patricia
Kniazev, Alexei
Knudsen, Kirsten
KO, Chung-Ming
Kobayashi, Chiaki
KOBAYASHI, Masakazu
KOBAYASHI, Yukiyasu
Koch, Andreas
Kocharovskij, Vladimir
Kocharovsky, Vitaly
Kochhar, Rajesh
KODAIRA, Keiichi
KODAMA, Hideo
KODAMA, Tadayuki
Koekemoer, Anton
Kogoshvili, Natela
KOHNO, Kotaro
Koivisto, Tomi
Kokkotas, Konstantinos
Kolb, Edward
Kollatschny, Wolfram
KOMIYAMA, Yutaka
Kompaneets, Dmitrij
KONG, Xu
Kontizas, Evangelos
Kontizas, Mary
Kontorovich, Victor
Koo, David
Koopmans, Leon
Kopeikin, Sergei
Kopp, Roger
Kopylov, Alexander
Koratkar, Anuradha
Koribalski, Bärbel
Kormendy, John
Korpela, Eric
KOSUGI, George
Kotilainen, Jari
Kotulla, Ralf
Koutchmy, Serge
Kouveliotou, Chryssa
Kovačević, Andjelka
Kovačević, Jelena
Kovalev, Yuri
Kovalev, Yuri
Kovetz, Attay
KOZAI, Yoshihide
Kozak, Lyudmyla
KOZASA, Takashi
Kozlovsky, Ben
Kozłowski, Szymon
Krabbe, Angela
Kraft, Ralph
Krajnović, Davor
Kramer, Busaba
Krasiński, Andrzej
Krause, Marita
Kravchuk, Sergei
Krishna, Gopal
Kriss, Gerard
Kristiansen, Jostein
Kriwattanawong, Wichean
Kron, Richard
Kroupa, Pavel
Krumholz, Mark
Kudritzki, Rolf-Peter
Kudrya, Yury
Kudryavtseva, Nadezhda (Nadia)
Kuijken, Koen
Kuin, Paul
Kulkarni, Prabhakar
KUMAI, Yasuki
Kunchev, Peter
Kunert-Bajraszewska, Magdalena
KUNO, Nario
Kunth, Daniel
Kuntschner, Harald
Kunz, Martin
Kurtanidze, Omar
Kurtz, Michael
KUSAKABE, Motohiko
KUSUNOSE, Masaaki
Kutner, Marc
Kuzio de Naray, Rachel
La Barbera, Francesco
Labbe, Ivo
Labeyrie, Antoine
Lacey, Cedric
Lachieze-Rey, Marc
La Franca, Fabio
Lagache, Guilaine
Lahav, Ofer
Lähteenmäki, Anne
Lake, George
Lake, Kayll
Lal, Dharam
La Mura, Giovanni
Lamy, Philippe
Lançon, Ariane
Landolt, Arlo
Lanfranchi, Gustavo
Lang, Kenneth
Laor, Ari
Lapi, Andrea
Larionov, Mikhail
Larsen, Søren
Larson, Richard
Lasota-Hirszowicz, Jean-Pierre
Laurikainen, Eija
Layzer, David

Leacock, Robert
Lêo, Joao Rodrigo
Le Borgne, Jean-Francois
Lebreton, Yveline
Lebrón, Mayra
Lee, Hyun-chul
Lee, Joon Hyeop
LEE, Myung Gyoon
LEE, Sang-Sung
LEE, Wo-Lung
Leeuw, Lerothodi
Le Fèvre, Olivier
Leger, Alain
Lehnert, Matthew
Lehto, Harry
Leibundgut, Bruno
Leinert, Christoph
Leitherer, Claus
Leloudas, Georgios
Lemke, Dietrich
Lenc, Emil
Lequeux, James
Leubner, Manfred
Leung, Kam
Levasseur-Regourd, Anny-Chantal
Levenson, Nancy
Levin, Yuri
Levine, Robyn
Lewis, Geraint
LI, Ji
LI, Li-Xin
LI, Qibin
LIANG, Yanchun
Liddle, Andrew
Liebscher, Dierck-E
Lilje, Per
Lillie, Charles
Lilly, Simon
LIM, Jeremy
Lima Neto, Gastao
LIN, Lihwai
LIN, Weipeng
LIN, Yen-Ting
Lindblad, Per
Linden-Vørnle, Michael
Lingenfelter, Richard
Lintott, Chris
LIOU, Guo Chin
LipunOv, Vladimir
LIU, Fukun
LIU, Xiang
LIU, Yongzhen
LIU, Yu
Lo, Fred K. Y.
Lobanov, Andrei
Lobo, Catarina
Lockman, Felix
Loiseau, Nora
łokas, Ewa
Lombardi, Marco
Londrillo, Pasquale
Longair, Malcolm
Longo, Giuseppe
Lonsdale, Carol
Lopes, Paulo
Lopez, Ericson
Lopez, Sebastian
Lopez Aguerri, Jose Alfonso
Lopez-Corredoira, Martin
López Cruz, Omar
Lopez Gonzalez, Maria
Lopez Hermoso, Maria
Lopez Moreno, Jose
Lopez Puertas, Manuel
López-Sánchez, Angel
Lord, Steven
LOU, Yu-Qing
Loubser, Ilani
Loup, Cecile
Louys, Mireille
Loveday, Jon
Lowenthal, James
Lu, Limin
LU, Youjun
Lubin, Lori
Lucero, Danielle
Lugger, Phyllis
Lukash, Vladimir
Luminet, Jean-Pierre
Luna, Homero
LUO, Ali
Lutz, Dieter
Lynden-Bell, Donald
Lynds, Beverly
Lynds, Roger
Lytvynenko, Leonid
MA, Jun
Maartens, Roy
Macalpine, Gordon
Maccacaro, Tommaso
Maccagni, Dario
MacCallum, Malcolm
Maccarone, Thomas
Macchetto, Ferdinando
MACHIDA, Mami
Maciejewski, Witold
Mackey, Alasdair
Mackie, Glen
Macquart, Jean-Pierre
Macri, Lucas
Madau, Piero
Madden, Suzanne
Maddox, Stephen
Madore, Barry
Madsen, Jes
MAEDA, Kei-ichi
Magain, Pierre
Magorrian, Stephen
Magrini, Laura
Magris, Gladis
Maguire, Kate
Maharaj, Sunil
Mahtessian, Abraham
Maia, Marcio
Maier, Christian
MAIHARA, Toshinori
Mainieri, Vincenzo
Maiolino, Roberto
Majumdar, Subhabrata
Makarov, Dmitry
Makarova, Lidia
MAKINO, Fumiyoshi
Malagnini, Maria
Malesani, Daniele
Malhotra, Sageeta
Malkan, Matthew
Mallik, D.
Mamon, Gary
Manchanda, R.
Mandolesi, Nazzareno
Mangalam, Arun
Mangum, Jeffrey
Mann, Ingrid
Mann, Robert
Mannucci, Filippo
Manrique, Alberto
Mansouri, Reza
Mao, Shude

MAO, Rui-Qing
Maoz, Dan
Marano, Bruno
Marcelin, Michel
Marco, Olivier
Marconi, Alessandro
Mardirossian, Fabio
Marecki, Andrzej
Marek, John
Marín-Franch, Antonio
Marino, Antonietta
Maris, Michele
Markoff, Sera
Marleau, Francine
Marquez, Isabel
Marr, Jonathon
Marranghello, Guilherme
Marston, Anthony
Martin, Christopher
Martin, Crystal
Martin, Donn
Martin, Maria
Martin, Rene
Martín, Sergio
Martinet, Louis
Martínez, Vicent
Martínez-Delgado, David
Martinez-Gonzalez, Enrique
Martini, Paul
Martinis, Mladen
Martins, Carlos
Martos, Marco
Marulli, Federico
Marziani, Paola
Masegosa, Josefa
Massardi, Marcella
Massaro, Francesco
Masters, Karen
Mather, John
Mathews, William
Mathewson, Donald
Mathez, Guy
MATSUMOTO, Toshio
MATSUMURA, Tomotake
MATSUSHITA, Kyoko
Matsuura, Mikako
MATSUURA, SHUJI
Matthews, Lynn
Matzner, Richard
Maucherat, Jean
Mauersberger, Rainer
Maurice, Eric
Maurogordato, Sophie
Mavrides, Stamatia
Max, Claire
Mayya, Divakara
Mazzarella, Joseph
Mazzei, Paola
McBreen, Brian
McCall, Marshall
McCracken, Henry
McGaugh, Stacy
McGimsey Jr, Ben
McKean, John
McMahon, Richard
McMillan, Paul
McNeil, Stephen
Mediavilla, Evencio
Mehlert, Dörte
Meier, David
Meikle, William
Meiksin, A.
Meisenheimer, Klaus
Melbourne, Jason
Melia, Fulvio
Mellema, Garrelt
Mellier, Yannick
Melnick, Jorge
Mel'nik, Anna
Melnyk, Olga
Melott, Adrian
Mendes de Oliveira, Cláudia
Meneghetti, Massimo
Menéndez-Delmestre, Karin
Menon, T.
Mercurio, Amata
Merighi, Roberto
Merluzzi, Paola
Merrifield, Michael
Meszaros, Attila
Meszaros, Peter
Metcalfe, Leo
Metevier, Anne
Meusinger, Helmut
Meyer, Angela
Meyer, David
Meyer, Martin
Meylan, Georges
Meza, Andres
Mezzetti, Marino
Middelberg, Enno
Mihov, Boyko
Mikhail, Joseph
Miley, George
Miller, Eric
Miller, Hugh
Miller, Joseph
Miller, Neal
Miller, Richard
Milvang-Jensen, Bo
Minchin, Robert
Mingaliev, Marat
Minikulov, Nasridin
MINOWA, Yosuke
Mirabel, Igor
Miralda-Escudé, Jordi
Miralles, Joan-Marc
Miranda, Oswaldo
Miroshnichenko, Alla
Misawa, Toru
Misconi, Nebil
Misner, Charles
Mitra, Abhas
Mitton, Simon
Miyaji, Takamitsu
Miyazaki, Satoshi
Miyoshi, Shigeru
MIZUNO, Takao
Mo, Houjun
Möllenhoff, Claus
Møller, Palle
Mohr, Joseph
Moiseev, Alexei
Molaro, Paolo
Moles Villamate, Mariano
Molinari, Emilio
Molla, Mercedes
Momjian, Emmanuel
Monaco, Pierluigi
Monard, Libert
Moodley, Kavilan
Moody, Joseph
Moore, Ben
Moos, Henry
Moreau, Olivier
Morelli, Lorenzo
Moreno, Edmundo

MORI, Masao
MOROKUMA, Tomoki
Morris, Simon
Mörtsell, Edvard
Moscardini, Lauro
Mota, David
MOTOHARA, Kentaro
Motta, Veronica
Mould, Jeremy
Mourão, Ana Maria
Muanwong, Orrarujee
Mücket, Jan
Mueller, Volker
Müller, Andreas
Muinonen, Karri
Mújica, Raul
MUKAI, Tadashi
Mukherjee, Krishna
Mukhopadhyay, Banibrata
Mulchaey, John
Muller, Erik
Muller, Richard
Muller, Sebastien
Muñoz Tuñón, Casiana
MURAKAMI, Izumi
MURAOKA, Kazuyuki
MURAYAMA, Hitoshi
MURAYAMA, Takashi
Murphy, John
Murphy, Michael
Murray, Stephen
Murray, Stephen
Murthy, Jayant
Mushotzky, Richard
Musielak, Zdzislaw
Muzzio, Juan
Nagao, Tohru
NAGASHIMA, Masahiro
Nahar, Sultana
Nair, Sunita
NAKAI, Naomasa
NAKAMICHI, Akika
NAKAMURA, Akiko
NAKANISHI, Hiroyuki
NAKANISHI, Kouichiro
NAKATA, Fumiaki
Nakos, Theodoros
Namboodiri, P.
NAMBU, Yasusada
Namouni, Fathi
Naoz, Smadar
Napolitano, Nicola
Narasimha, Delampady
Narbutis, Donatas
NARDETTO, Nicolas
Narlikar, Jayant
Naselsky, Pavel
Nasr-Esfahani, Bahram
Navarro, Julio
Nawar, Samir
Nedialkov, Petko
Nemiroff, Robert
Nesvadba, Nicole
Netzer, Hagai
Neves de Araujo, Jose
Nguyễn, Lan
Nicoll, Jeffrey
Nikołajuk, Marek
Ninković, Slobodan
Nipoti, Carlo
NISHIDA, Minoru
Nishikawa, Ken-Ichi
Nishimura, Tetsuo
NISHIMURA, Jun
Nityananda, Rajaram
Noerdlinger, Peter
NOGUCHI, Masafumi
NOH, Hyerim
Noll, Stefan
Noonan, Thomas
Norman, Colin
Norman, Dara
Norris, Raymond
Noterdaeme, Pasquier
Nottale, Laurent
Novello, Mario
Novikov, Igor
Novosyadlyj, Bohdan
Novotný, Jan
Nozari, Kourosh
Nucita, Achille
Nulsen, Paul
Nuza, Sebastian
Obregón Díaz, Octavio
O'Connell, Robert
Ocvirk, Pierre
O'Dea, Christopher
O'Dell, Stephen
O'Donoghue, Aileen
Oemler Jr, Augustus
Oey, Sally
Ogando, Ricardo
OHTA, Kouji
OKA, Tomoharu
OKAMOTO, Takashi
OKAMURA, Sadanori
OKUDA, Haruyuki
Oliver, Sebastian
Olling, Robert
Olofsson, Kjell
Olowin, Ronald
Omizzolo, Alessandro
Omnes, Roland
Onken, Christopher
Onuora, Lesley
Oosterloo, Thomas
Oozeer, Nadeem
Orienti, Monica
Origlia, Livia
Orlov, Viktor
Oscoz, Alejandro
Osman, Anas
Osmer, Patrick
Östlin, Göran
Ostorero, Luisa
Ostriker, Eve
Ostriker, Jeremiah
Ostrowski, Michał
O'Sullivan, Créidhe
OTA, Naomi
OTSUKI, Kaori
Ott, Heinz-Albert
Ott, Juergen
OUCHI, Masami
Ovcharov, Evgeni
Owen, Frazer
Owers, Matthew
OYA, Shin
OYABU, Shinki
Ozsváth, Istvan
Pacholczyk, Andrzej
Padmanabhan, Thanu
Padovani, Paolo
Page, Don
Page, Mathew
PAK, Soojong
Palmer, Philip
Palouš, Jan
Paltani, Stéphane
Palumbo, Giorgio

Panagia, Nino
Panessa, Francesca
Pannuti, Thomas
Panov, Kiril
Pantoja, Carmen
Papaelias, Philip
Papayannopoulos, Theodoros
Paragi, Zsolt
Paresce, Francesco
PARK, Jang Hyun
Parker, Quentin
Parnovsky, Sergei
Partridge, Robert
Pastoriza, Miriani
Pati, Ashok
Patsis, Panos
Patton, David
Paturel, Georges
Paul, Jacques
Pavlidou, Vasiliki
Pazhouhesh, Reza
Pearce, Frazer
Pearson, Timothy
Peck, Alison
Pecker, Jean-Claude
Pedersen, Kristian
Pedlar, Alan
Pedrosa, Susana
Peebles, P.
Peel, Michael
Peimbert, Manuel
Peletier, Reynier
Pellegrini, Silvia
Pellizza, Leonardo
Pello, Roser
Pen, Ue-Li
Peng, Eric
PENG, Qingyu
Pentericci, Laura
Penzias, Arno
Perea-Duarte, Jaime
Perets, Hagai
Pérez, Enrique
Perez, Fournon
Pérez-Garrido, Antonio
Pérez-González, Pablo
Pérez Torres, Miguel
Perley, Richard
Peroux, Céline
Perrin, Jean-Marie
Perry, Peter
Perryman, Michael
Persides, Sotirios
Persson, Carina
Peters, William
Peterson, Bradley
Peterson, Bruce
Peterson, Charles
Petit, Jean-Marc
Petitjean, Patrick
Petrosian, Artaches
Petrosian, Vahe
Petrov, Georgi
Petuchowski, Samuel
Pfenniger, Daniel
Pfleiderer, Jorg
PHAM, Diep
Philipp-May, Sabine
Phillipps, Steven
Phillips, Mark
Pierre, Marguerite
Pihlström, Ylva
Pikichian, Hovhannes
Pilbratt, Göran
Pimbblet, Kevin
Piotrovich, Mikhail
Pipino, Antonio
Piran, Tsvi
Pires Martins, Lucimara
Pirzkal, Norbert
Pisano, Daniel
Pitjev, Nikolaj
Pitkin, Matthew
Pizzella, Alessandro
Plana, Henri
Planesas, Pere
Plionis, Manolis
Podolský, Jiri
Pogge, Richard
Poggianti, Bianca
Polletta, Maria del Carmen
Polyachenko, Evgeny
Pompei, Emanuela
Ponman, Trevor
Pooley, David
Pooley, Guy
Popescu, Cristina
Popescu, Nedelia
Popović, Luka
Porquet, Delphine
Portinari, Laura
Poveda, Arcadio
Pović, Mirjana
Power, Chris
Poznanski, Dovi
Prabhu, Tushar
Pracy, Michael
Prandoni, Isabella
Premadi, Premana
Press, William
Prevot-Burnichon, Marie-Louise
Prieto, Almudena
Pritchet, Christopher
Proctor, Robert
Pronik, Vladimir
Proust, Dominique
Prugniel, Philippe
Puech, Mathieu
Puerari, Ivânio
Puetter, Richard
Puetzfeld, Dirk
Puget, Jean-Loup
Pulatova, Nadiia
Puschell, Jeffery
Pushkarev, Alexander
Pustilnik, Simon
Puxley, Phil
Puy, Denis
Puzia, Thomas
Pye, John
QIN, Bo
QIN, Yi-Ping
QU, Qinyue
Quinn, Peter
Quintana, Hernan
Quirk, William
Quirós, Israel
Radford, Simon
Rafanelli, Piero
Rahvar, Sohrab
Raiteri, Claudia
Ramella, Massimo
Rampazzo, Roberto
Ranalli, Piero
Rand, Richard
Rasmussen, Jesper
Rassat, Anais
Rastegaev, Denis

Ratnatunga, Kavan
Rauch, Michael
Ravindranath, Swara
Raychaudhury, Somak
Reach, William
Read, Andrew
Read, Justin
Rebolo, Rafael
Reboul, Henri
Recchi, Simone
Rector, Travis
Reddy, Bacham
Rees, Martin
Reeves, Hubert
Reichert, Gail
Reid, Mark
Reiprich, Thomas
Reisenegger, Andreas
Reitze, David
Rejkuba, Marina
Rekola, Rami
Renard, Jean-Baptiste
Rengarajan, Thinniam
Rephaeli, Yoel
Reshetnikov, Vladimir
Reunanen, Juha
Revaz, Yves
Revnivtsev, Mikhail
Rey, Soo-Chang
Reyes, Reinabelle
Reynolds, Cormac
RHEE, Myung Hyun
Rhoads, James
Rhodes, Jason
Riazi, Nematollah
Riazuelo, Alain
Ribeiro, André Luis
Ribeiro, Marcelo
Rich, Robert
Richard, Johan
Richer, Harvey
Richer, Michael
Richstone, Douglas
Richter, Gotthard
Richter, Philipp
Ricker, Paul
Ricotti, Massimo
Ridgway, Susan
Rindler, Wolfgang
Risaliti, Guido
Rivolo, Arthur
Rix, Hans-Walter
Rizzi, Luca
Robe, Henri
Robert, Carmelle
Roberts, David
Roberts, Morton
Roberts, Timothy
Roberts Jr, William
Robin, Annie
Robinson, Andrew
Robinson, I.
Robley, Robert
Robson, Ian
Rocca-Volmerange, Brigitte
Rodrigo, Rafael
Rodrigues de Oliveira Filho, Irapuan
Rodriguez Espinosa, Jose
Roeder, Robert
Rödiger, Elke
Roeser, Hermann-Josef
Romano, Patrizia
Romano-Diaz, Emilio
Romeo, Alessandro
Romer, A.
Romero-Colmenero, Encarnacion
Roming, Peter
Roos, Matts
Roos, Nicolaas
Rosa, Michael
Rosado, Margarita
Rosa González, Daniel
Rose, James
Ros Ibarra, Eduardo
Rosquist, Kjell
Rothberg, Barry
Rots, Arnold
Röttgering, Huub
Rouan, Daniel
Rowan-Robinson, Michael
Roxburgh, Ian
Rozas, Maite
Rubin, Vera
Rubino-Martin, Jose Alberto
Rubio, Monica
Rudnick, Lawrence
Ruffini, Remo
Ruszkowski, Mateusz
Rôžička, Adam
Rybicki, George
Ryder, Stuart
RYU, Dongsu
Saar, Enn
Sabra, Bassem
Sackett, Penny
Sadat, Rachida
Sadler, Elaine
Sadun, Alberto
Saffari, Reza
Sage, Leslie
Saglia, Roberto
Saha, Abhijit
Saha, Prasenjit
Sahai, Raghvendra
Sahibov, Firuz
Sahlén, Martin
Sahni, Varun
SAITOH, Takayuki
Saiz, Alejandro
Sakai, Shoko
SAKAMOTO, Tsuyoshi
SAKON, Itsuki
Sala, Ferran
Salaris, Maurizio
Salo, Heikki
Salomé, Philippe
Salvador-Sole, Eduardo
Salzer, John
Samurovi'c, Srdjan
Sanahuja Parera, Blai
Sánchez, Francisco
Sanchez, Norma
Sánchez Almeida, Jorge
Sánchez-Blázquez, Patricia
Sanchez-Saavedra, M.
Sancisi, Renzo
Sanders, David
Sanders, Robert
Sanroma, Manuel
Sansom, Anne
Santiago, Basilio
Santos-Lleó, Maria
Sanz, Jose
Sapar, Arved
Sapre, Ashok
Saracco, Paolo
Sarazin, Craig

Saripalli, Lakshmi
Sasaki, Toshiyuki
SASAKI, Minoru
SASAKI, Misao
SASAKI, Shin
Saslaw, William
Sastry, Shankara
SATO, Humitaka
SATO, Katsuhiko
SATO, Shinji
SATO, Jun'ichi
Saucedo Morales, Julio
Savage, Ann
Saviane, Ivo
Savin, Daniel
SAWA, Takeyasu
SAWADA-SATOH, Satoko
Sawicki, Marcin
Saxena, P.
Sazhin, Mikhail
Scaramella, Roberto
Schade, David
Schaerer, Daniel
Schaffner-Bielich, Jurgen
Schartel, Norbert
Schaye, Joop
Schechter, Paul
Schild, Rudolph
Schilizzi, Richard
Schindler, Sabine
Schinnerer, Eva
Schlosser, Wolfhard
Schmidt, Brian
Schmidt, Maarten
Schmidt, Robert
Schmitt, Henrique
Schmitz, Marion
Schneider, Donald
Schneider, Jean
Schneider, Peter
Schneider, Raffaella
Schramm, Thomas
Schröder, Anja
Schuch, Nelson
Schucking, Engelbert
Schulte-Ladbeck, Regina
Schumacher, Gerard
Schwarz, Ulrich
Schwehm, Gerhard
Schweizer, François
Schwope, Axel
Scodeggio, Marco
Scorza, Cecilia
Scott, Douglas
Scoville, Nicholas
Segaluvitz, Alexander
Seielstad, George
Seigar, Marc
Seimenis, John
Sellwood, Jerry
Semelin, Benoit
Semerák, Oldrich
Sempere, Maria
SEON, Kwang il
Sergeev, Sergey
Sergijenko, Olga
Serjeant, Stephen
Serote Roos, Margarida
SETA, Masumichi
SETO, Naoki
Setti, Giancarlo
Severgnini, Paola
Seymour, Nicholas
Shaffer, David
SHAN, Hongguang
Shandarin, Sergei
Shane, William
Shanks, Thomas
SHAO, Zhengyi
Shapero, Donald
Shapovalova, Alla
Sharina, Margarita
Sharma, Prateek
Sharp, Nigel
Sharples, Ray
Shastri, Prajval
Shaver, Peter
Shaviv, Giora
Shaya, Edward
Shefov, Nikolaj
SHEN, Juntai
SHEN, Zhiqiang
Sher, David
Sherwood, William
SHI, Huoming
SHIBATA, Masaru
Shields, Gregory
Shields, Joseph
SHIMASAKU, Kazuhiro
Shimon, Meir
SHINN, Jong-Ho
SHIOYA, Yasuhiro
SHIRAHATA, Mai
SHIRASAKI, Yuji
Shivanandan, Kandiah
Shostak, G.
Shukurov, Anvar
Siebenmorgen, Ralf
Siebert, Arnaud
Signore, Monique
Sigurdsson, Steinn
Sil'chenko, Olga
Silk, Joseph
Sillanpaa, Aimo
Silva, David
Silva, Laura
Simic, Sasa
Simkin, Susan
Sinachopoulos, Dimitris
Singh, Kulinder Pal
Siopis, Christos
Siringo, Giorgio
Sironi, Giorgio
Sistero, Roberto
Sivakoff, Gregory
Skillen, Ian
Skillman, Evan
Slavcheva-Mihova, Lyuba
Sleath, John
Slezak, Eric
Slosar, Anze
Smail, Ian
Smecker-Hane, Tammy
Smette, Alain
Smirnova, Aleksandrina
Smith, Eric
Smith, Haywood
Smith, Howard
Smith, Linda
Smith, Malcolm
Smith, Niall
Smith, Nigel
Smith, Robert
Smith, Rodney
Smith Castelli, Analia
Smoot III, George
Soares, Domingos Savio
Soberman, Robert
Sobouti, Yousef
Sodré, Laerte

SOFUE, Yoshiaki
SOHN, Young Jong
Soker, Noam
Sokołowski, Lech
Solanes, Josep
Sollerman, Jesper
Sołtan, Andrzej
Soltynski, Maciej
SONG, Doo-Jong
SONG, Liming
SONG, Yong-Seon
SORAI, Kazuo
Soria, Roberto
Sotnikova, Natalia
Soucail, Geneviève
Souradeep, Tarun
Souriau, Jean-Marie
Spaans, Marco
Spallicci di Filottrano, Alessandro
Sparke, Linda
Sparks, William
Sparrow, James
Spergel, David
Spiegel, Edward
Spinoglio, Luigi
Spinrad, Hyron
Springel, Volker
Spurzem, Rainer
Spyrou, Nicolaos
Squires, Gordon
Srianand, Raghunathan
Sridhar, Seshadri
Srinivasan, Ganesan
Stacy, Athena
Stadel, Joachim
Stanga, Ruggero
Statler, Thomas
Staude, Hans
Staveley-Smith, Lister
Stecker, Floyd
Steenbrugge, Katrien
Steigman, Gary
Steiman-Cameron, Thomas
Steinbring, Eric
Steinmetz, Matthias
Stiavelli, Massimo
Stirpe, Giovanna
Stocke, John
Stockton, Alan
Stoehr, Felix
Stolyarov, Vladislav
Stonkutė, Rima
Storchi-Bergmann, Thaisa
Storey, John
Storm, Jesper
Storrie-Lombardi, Lisa
Stott, John
Straumann, Norbert
Strauss, Michael
Strigachev, Anton
Stritzinger, Maximilian
Strom, Richard
Strom, Robert
Strubbe, Linda
Struble, Mitchell
Struck, Curtis
Strukov, Igor
Stuchlík, Zdenek
Stuik, Remko
SU, Cheng-yue
Subrahmanya, C.
Subrahmanyan, Ravi
Subramaniam, Annapurni
Subramanian, Kandaswamy
SUGAI, Hajime
SUGINOHARA, Tatsushi
SUGIYAMA, Naoshi
Suhhonenko, Ivan
Sulentic, Jack
Sullivan, Mark
Sullivan, III, Woodruff
Sundin, Maria
Sunyaev, Rashid
Suran, Marian
Surdej, Jean
SUSA, Hajime
Sutherland, Ralph
Sutherland, William
SUTO, Yasushi
Swings, Jean-Pierre
Sygnet, Jean-Francois
Sykes, Mark
Szalay, Alex
Szydłowski, Marek
Tacconi, Linda
Tacconi-Garman, Lowell
Tagger, Michel
Tagliaferri, Gianpiero
TAGOSHI, Hideyuki
TAKADA, Masahiro
TAKAGI, Toshinobu
TAKAHARA, Fumio
TAKAHASHI, Rohta
Takalo, Leo
TAKASHI, Hasegawa
TAKATA, Tadafumi
TAKATO, Naruhisa
TAKEUCHI, Tsutomu
TAKIZAWA, Motokazu
Tamm, Antti
Tammann, Gustav
TANABE, Hiroyoshi
TANABE, Kenji
TANAKA, Ichi Makoto
TANAKA, Masayuki
TANAKA, Yutaka
Tandon, S.
TANIGUCHI, Yoshiaki
Tantalo, Rosaria
Tanzella-Nitti, Giuseppe
Tarter, Jill
TARUYA, Atsushi
TASHIRO, Makoto
TATEKAWA, Takayuki
Tateyama, Claudio
Tauber, Jan
Taylor, A.
Taylor, Angela
Taylor, James
Teerikorpi, Veli
Tektunali, H.
Telles, Eduardo
Tempel, Elmo
Temporin, Sonia
Teng, Stacy
Tenjes, Peeter
Tepper Garcia, Thorsten
Terlevich, Elena
Terlevich, Roberto
Terzian, Yervant
Teuben, Peter
Teyssier, Romain
Thanjavur, Karunananth
Theis, Christian
Theureau, Gilles
Thöne, Christina
Thomas, Peter
Thomasson, Magnus

Thomsen, Bjarne
Thonnard, Norbert
Thornley, Michele
Thuan, Trinh
Thurston, Mark
Tifft, William
Tikhonov, Nikolai
Tilanus, Remo
Tipler, Frank
Tissera, Patricia
Tisserand, Patrick
Titov, Oleg
Toffolatti, Luigi
Toft, Sune
Toller, Gary
Tolstoy, Eline
Tomasella, Lina
TOMIMATSU, Akira
Tomita, Akihiko
TOMITA, Kenji
Tonry, John
Toomre, Alar
Tormen, Giuseppe
Torres-Papaqui, Juan
TOSAKI, Tomoka
Tosi, Monica
TOTANI, Tomonori
Tovmassian, Hrant
TOYAMA, Kiyotaka
Tozzi, Paolo
Traat, Peeter
Trager, Scott
Trakhtenbrot, Benny
Tran, Kim-Vy
Tremaine, Scott
Tremonti, Christy
Trenti, Michele
Tresse, Laurence
Treu, Tommaso
Trevese, Dario
Trimble, Virginia
Trinchieri, Ginevra
TRIPPE, Sascha
Tristram, Konrad
Trotta, Roberto
Trujillo Cabrera, Ignacio
TSAI, An-Li
Tsamparlis, Michael
TSUCHIYA, Toshio
TSUNEMI, Hiroshi
Tsutsumi, Takahiro
Tsvetkov, Dmitry
Tsygankov, Sergey
Tuffs, Richard
Tugay, Anatoliy
Tully, Richard
Turner, Edwin
Turner, Michael
Turnshek, David
Tyson, John
Tytler, David
Tyul'bashev, Sergei
Tzanavaris, Panayiotis
Tzioumis, Anastasios
Udaya, Shankar
UEDA, Haruhiko
UENO, Munetaka
Ugolnikov, Oleg
Ulmer, Melville
Ulrich, Marie-Helene
UMEMURA, Masayuki
Urbanik, Marek
Urry, Claudia
Uslenghi, Michela
Uson, Juan
Usowicz, Jerzy
Utrobin, Victor
Vaccari, Mattia
Vagnetti, Fausto
Väisänen, Petri
Valcheva, Antoniya
Valdes Parra, Jose
Valdivielso, Luisa
Valentijn, Edwin
Väliviita, Jussi-Pekka
Vallée, Jacques
Vallenari, Antonella
Valls-Gabaud, David
Valluri, Monica
Valotto, Carlos
Valtchanov, Ivan
Valtonen, Mauri
Valyavin, Gennady
van Albada, Tjeerd
van den Bergh, Sidney
van der Hulst, Jan
van der Kruit, Pieter
van der Laan, Harry
van der Marel, Roeland
Vandervoort, Peter
van der Werf, Paul
van de Weygaert, Rien
van Driel, Wim
Van Dyk, Schuyler
van Eymeren, Janine
van Gorkom, Jacqueline
van Haarlem, Michiel
van Kampen, Eelco
van Moorsel, Gustaaf
Vansevicius, Vladas
van Woerden, Hugo
van Zee, Liese
Varela López, Jesús
Varma, Ram
Vasta, Magda
Vaughan, Simon
Vauglin, Isabelle
Vavilova, Iryna
Vazdekis, Alexandre
Vedel, Henrik
Vega, Olga
Veilleux, Sylvain
Vennik, Jaan
Venters, Tonia
Venturi, Tiziana
Vercellone, Stefano
Verdes-Montenegro, Lourdes
Verdoes Kleijn, Gijsbert
Vergani, Daniela
Verheijen, Marc
Verkhodanov, Oleg
Verma, Aprajita
Vermeulen, Rene
Verschuur, Gerrit
Vesperini, Enrico
Vestergaard, Marianne
Vettolani, Giampaolo
Viana, Pedro
Vidal-Madjar, Alfred
Viel, Matteo
Vignali, Cristian
Vigroux, Laurent
Viironen, Kerttu
Vilchez, José
Vilkoviskij, Emmanuil
Villata, Massimo
Vishniac, Ethan
Vishveshwara, C.
Vittorio, Nicola

Vivas, Anna
Vlahakis, Catherine
Vlasyuk, Valerij
Vogel, Stuart
Voges, Wolfgang
Voit, Gerard
Vollmer, Bernd
Volonteri, Marta
von Borzeszkowski, H.
Vrtilek, Jan
Vucetich, Héctor
WADA, Keiichi
WADA, Takehiko
Wadadekar, Yogesh
Waddington, Ian
Wagner, Alexander
Wagner, Robert
Wagner, Stefan
Wagoner, Robert
Wainwright, John
WAKAMATSU, Ken-Ichi
Wakker, Bastiaan
Walker, Constance
Walker, Mark
Wall, Jasper
Wallis, Max
Walter, Fabian
Walterbos, Rene
Wambsganß, Joachim
WANAJO, Shinya
Wanas, Mamdouh
WANG, Ding-Xiong
WANG, Hong-Guang
WANG, Huiyuan
WANG, Jingyu
WANG, Junzhi
WANG, Tinggui
WANG, Yiping
WANG, Yu
Ward, Martin
Wardle, John
Warmels, Rein
WATANABE, Junichi
Watson, Darach
Waxman, Eli
Wayth, Randall
Weaver, Kimberly
Webb, John
Webb, Tracy
Webster, Adrian
Webster, Rachel
Weedman, Daniel
Wegner, Gary
WEI, Erhu
WEI, Jianyan
Weigelt, Gerd
Weilbacher, Peter
Weiler, Kurt
Weinberg, Jerry
Weinberg, Steven
Weistrop, Donna
Welch, Douglas
Welch, Gary
Wendt, Martin
Wesson, Paul
West, Michael
Westera, Pieter
Westmeier, Tobias
Wheatley, Peter
Whitcomb, Stanley
White, Raymond
White, Richard
White, Simon
Whiting, Alan
Whiting, Matthew
Whitmore, Bradley
Widrow, Lawrence
Wielebinski, Richard
Wielen, Roland
Wiik, Kaj
Wiita, Paul
Wilcots, Eric
Wild, Wolfgang
Wilkes, Belinda
Will, Clifford
Williams, Barbara
Williams, Robert
Williams, Theodore
Willis, Anthony
Willis, Jon
Willner, Steven
Wills, Beverley
Wills, Derek
Wilson, Gillian
Wilson, P.
Wilson, Robert
Windhorst, Rogier
Winkler, Hartmut
Winter, Lisa
Wise, Michael
Wisotzki, Lutz
Witt, Adolf
Witten, Louis
Wofford, Aida
Wold, Margrethe
Wolleben, Maik
Wolstencroft, Ramon
Wolter, Anna
Woltjer, Lodewijk
Wong, Tony
Woo, Joanna
WOO, Jong-Hak
Woolfson, Michael
Woosley, Stanford
Worrall, Diana
Woszczyna, Andrzej
Woudt, Patrick
Wozniak, Hervé
Wright, Edward
Wrobel, Joan
Wu, Yanling
WU, Hong
WU, Jianghua
WU, Jiun-Huei
WU, Wentao
WU, Xiangping
WU, Xue-bing
WU, Xue-Feng
Wucknitz, Olaf
Wulandari, Hesti
Wünsch, Richard
Wyithe, Stuart
Wynn-Williams, Gareth
Xanthopoulos, Emily
XIA, Xiao-Yang
XIANG, Shouping
Xilouris, Emmanouel
XINLIAN, Luo
XU, Chongming
XU, Dawei
XUE, Suijian
YAGI, Masafumi
YAHAGI, Hideki
Yahil, Amos
Yakovleva, Valerija
YAMADA, Shimako
YAMADA, Toru
YAMADA, Yoshiyuki
YAMAGATA, Tomohiko
YAMAMOTO, Hiroaki

YAMAMOTO, Tetsuo
YAMASAKI, Noriko
YAMASHITA, Kojun
YAMAUCHI, Aya
YANG, Xiaohu
YASUDA, Naoki
Yee, Howard
Yegorova, Irina
Yeşilyurt, Ibrahim
YI, Sukyoung
YOICHIRO, Suzuki
YOKOYAMA, Jun'ichi
YONEHARA, Atsunori
YONETOKU, Daisuke
Yoon, Suk-Jin
YOSHIDA, Hiroshi
YOSHIDA, Michitoshi
YOSHII, Yuzuru
YOSHIKAWA, Kohji
YOSHIOKA, Satoshi
Young, Lisa
YU, Qingjuan
YUAN, Feng
Yun, Min
Yushchenko, Alexander
Zacchei, Andrea
Zaggia, Simone
Zainal Abidin, Zamri
Zamkotsian, Frederic
Zamorani, Giovanni
Zamorano, Jaime
Zanichelli, Alessandra
Zannoni, Mario
Zaritsky, Dennis
Zaroubi, Saleem
Zasov, Anatoly
Zavala, Robert
Zeilinger, Werner
Zensus, J-Anton
Zepf, Stephen
Zezas, Andreas
Zhang, Yuying
ZHANG, Bo
ZHANG, Fenghui
ZHANG, Jialu
ZHANG, JiangShui
ZHANG, Jingyi
ZHANG, Shuang Nan
ZHANG, Tong-Jie
ZHANG, Xiaolei
ZHANG, Yang
ZHANG, Yanxia
ZHAO, Donghai
ZHENG, XianZhong
ZHOU, Hongyan
ZHOU, Jianfeng
ZHOU, Jianjun
ZHOU, Xu
ZHOU, Youyuan
ZHU, Ming
ZHU, Xingfeng
Zhuk, Alexander
Zibetti, Stefano
Zięba, Stanisław
Ziegler, Bodo
Ziegler, Harald
Zinn, Robert
Zirm, Andrew
Zitelli, Valentina
Zitrin, Adi
ZOU, Zhenlong
Zucca, Elena
Zwaan, Martin

Transactions IAU, Volume XXIXB
Proceedings IAU Symposium No. Volume XXIXB, 2018
P. Benvenuti, ed.
© International Astronomical Union 2019
doi:10.1017/S1743921318004064

CHAPTER XI

COMMISSIONS MEMBERSHIP
(until 31 August 2015)

NOTE. This chapter gives the membership of the "old" Commissions (listed by Commission number), as they were re-affiliated to the new Divisions that were established at the XXVIII GA in 2012. These Commissions ceased to exist at the XXIX GA and replaced by the new Commissions (se Chapter IX of these *Transactions*). The membership of the new Commissions was highly provisional and incomplete at the time of the XXIX GA (August 2015), therefore it is not reported here and it will be published in the next issue of the *Transactions B*.

Division A - Commission 4 Ephemerides

President: Catherine Y. Hohenkerk
Vice-President: Jean-Eudes Arlot

Organizing Committee Members:

William Thuillot
Mitsuru Sôma
Agnès Fienga
Marina V. Lukashova
Jose Manuel Ferrándiz
John A. Bangert
William M. Folkner
Elena V. Pitjeva
Steven A. Bell
Sean E. Urban
Jean-Eudes Arlot
George H. Kaplan

Members:

Abalakin, Viktor
Acton, Charles
AHN, Youngsook
ARAKIDA, Hideyoshi
Bartlett, Jennifer
Brozovic, Marina
Brumberg, Victor
Bueno de Camargo, Julio
Capitaine, Nicole
Chapront, Jean
Chapront-Touze, Michelle
Cooper, Nicholas
Dickey, Jean
Dunham, David
Eroshkin, Georgij
Espenak, Fred
Fominov, Aleksandr
FU, Yanning
Giorgini, Jon
Glebova, Nina
Harper, David
Hilton, James
HOU, Xiyun
Howard, Sethanne
Husarik, Marek
Iorio, Lorenzo
Ivantsov, Anatoliy
Johnston, Kenneth
Jubier, Xavier
Kaplan, George
KINOSHITA, Hiroshi
Klepczynski, William
Kolaczek, Barbara
Lara, Martin
Laskar, Jacques
Lehmann, Marek
Lenhardt, Helmut
Lieske, Jay
Lopez Moratalla, Teodoro

Madsen, Claus
Majid, Abdul
Mallamaci, Claudio
MASAKI, Yoshimitsu
Morrison, Leslie
Mueller, Ivan
Newhall, X.
Noyelles, Benoît
O'Handley, Douglas
Ofek, Eran
Olivier, Enrico
Page, Gary
Pavlyuchenkov, Yaroslav
Reasenberg, Robert
Rodin, Alexander
Romero Perez, Maria
Rossello, Gaspar
Seidelmann, P.
Shapiro, Irwin
Shiryaev, Alexander
SHU, Fengchun
Shuygina, Nadia
Simon, Jean-Louis
Skripnichenko, Vladimir
Standish, E.
Stewart, Susan
Suli, Áron
Vilinga, Jaime
Vondrak, Jan
Wallace, Patrick
WANG, Xiao-bin
Weratschnig, Julia
Wielen, Roland
Wilkins, George
Williams, Carol
Williams, James
Winkler, Gernot
Wytrzyszczak, Iwona
XIE, Yi
Yallop, Bernard

Division B - Commission 5 Documentation & Astronomical Data

President: Robert J. Hanisch
Vice-President: Michael W. Wise

Organizing Committee Members:

Daniel Egret
Anja C. Schröder
Douglas Tody
R. Elizabeth M. Griffin
Ajit K. Kembhavi
Fabio Pasian
Tara Murphy
Marsha Bishop
Heinz J Andernach

Members:

Abalakin, Viktor
Abt, Helmut
Accomazzi, Alberto
Acharya, Bannanje
Adelman, Saul
Aerts, Conny
Agüeros, Marcel
A'Hearn, Michael
Aizenman, Morris
AK, Serap
Alexander, Paul
Allan, Alasdair
Allen, Lori
Alvarez, Pedro
Andernach, Heinz
Andreon, Stefano
Antonelli, Lucio Angelo
Anupama, G.
Arenou, Frédéric
Argyle, Robert
Arlot, Jean-Eudes
Armstrong, John
Aspin, Colin
Baffa, Carlo
Bagla, Jasjeet
Banhatti, Dilip
Barbieri, Cesare
Barbuy, Beatriz
Bartczak, Przemysław
Beasley, Anthony
Beckmann, Volker
Bell Burnell, Jocelyn
Benacchio, Leopoldo
Benetti, Stefano
Benn, Chris
Bentley, Robert
Bersier, David
Berthier, Jerôme
Bertout, Claude
Bessell, Michael
Bhat, Ramesh
Bignall, Hayley
Bishop, Marsha
Bolatto, Alberto
Bond, Howard
Bond, Ian
Borde, Suzanne
Borisova, Ana
Borne, Kirk
Bosken, Sarah
Bourgeois, P.
Bouton, Ellen
Boyce, Peter
Brammer, Gabriel

Brandt, William
Brescia, Massimo
Brinchmann, Jarle
Brosch, Noah
Brouw, Willem
Brown, Michael
Brunner, Robert
Bucciarelli, Beatrice
Calabretta, Mark
Campana, Riccardo
Cappellaro, Enrico
Caretta, Cesar
Catelan, Márcio
Cenko, Stephen
CHANG, Hong
CHANG, Hsiang-Kuang
Chapman, Jacqueline
CHEN, Xuelei
Cheung, Cynthia
Chiappetti, Lucio
Christlieb, Norbert
CHU, Yaoquan
Ciardi, David
Clayton, Geoffrey
Cogan, Bruce
Coletti, Donna
Coluzzi, Regina
Conti, Alberto
Corbally, Christopher
Corbin, Brenda
Cordes, James
Crézé, Michel
Cristiani, Stefano
Cristóbal, David
Csabai, Istvan
CUI, Chenzhou
Cunniffe, John
Dalla, Silvia
Davis, Morris
de Carvalho, Reinaldo
De Cuyper, Jean-Pierre
Depagne, Éric
De Rossi, María
Derriere, Sebastien
DeVorkin, David
Dickel, Helene
Dimitrijevic, Milan
Dixon, Robert
Djorgovski, Stanislav
Dluzhnevskaya, Olga
Dobrzycki, Adam
Downes Wallace, Juan
Drimmel, Ronald
Dubois, Pascal
Ducati, Jorge
Ducourant, Christine
Durand, Daniel
Ederoclite, Alessandro
Egret, Daniel
Eichhorn, Guenther
Ekström García Nombela, Sylvia
Elia, Davide
Ellingsen, Simon
Elyiv, Andrii
FAN, Yufeng
Feigelson, Eric
Folgueira, Marta
Forveille, Thierry
FOUCAUD, Sebastien
Fox-Machado, Lester
Fraix-Burnet, Didier
FUKUSHIMA, Toshio
Fyfe, Duncan
Gabriel, Carlos
Gallagher, Sarah
Gallagher III, John
Garaimov, Vladimir
Gastaldello, Fabio
Gehrels, Neil
Genova, Françoise
Gezari, Suvi
Golden, Aaron
Golev, Valeri
Gomez, Monique
González, J.
Goodman, Alyssa
Graham, Eric
Green, Daniel
Green, David
Gregory, Philip
Greisen, Eric
Griffin, R. Elizabeth
Griffin, Roger
Grindlay, Jonathan
Groot, Paul
Grosbøl, Preben
Guibert, Jean
Guinan, Edward
GUO, Hongfang
GUO, Jianheng
Hamuy, Mario
Hanisch, Robert
Harmer, Dianne
Hauck, Bernard
Hearn, Anthony
Heinrich, Inge
Heiser, Arnold
Helou, George
Henden, Arne
Hessman, Frederic
Hestroffer, Daniel
Hledík, Stanislav
Hodge, Paul
Hopkins, Andrew
Horne, Keith
Howell, Steve
HUANG, Bi-kun
Hudec, Rene
Hudkova, Ludmila
Hunstead, Richard
Ioannou, Zacharias
Ivezic, Zeljko
Jacoby, George
Jaschek, Carlos
Jauncey, David
Jenkner, Helmut
JIN, WenJing
Jones, Derek
Jordán, Andrés
Kalberla, Peter
Kazantseva, Liliya
Kedziora-Chudczer, Lucyna
Kelly, Brandon
Kembhavi, Ajit
Kemp, D.
Kent, Brian
Khrutskaya, Evgenia
Kimball, Amy
Kitaeff, Vyacheslav
Koen, Marthinus
Kolenberg, Katrien
Kolobov, Dmitri
KONG, Xu
Kopatskaya, Evgenia
Koribalski, Bärbel
Kovalev, Yuri
Kovaleva, Dana
Krishna, Gopal
Kubát, Jiri

Kudryavtseva, Nadezhda (Nadia)
Kuin, Paul
Kulkarni, Shrinivas
Kunz, Martin
Labbe, Ivo
Laloe, Suzanne
Larson, Stephen
Laurent, Michel
Lazio, Joseph
Lee, William
Leibundgut, Bruno
Lequeux, James
Lesteven, Soizick
LI, De-He
Lintott, Chris
Lister, Matthew
LIU, Jinming
LIU, Siming
LIU, Xiaoqun
Long, Knox
Longo, Giuseppe
Lonsdale, Carol
Lopes, Paulo
Lortet, Marie-Claire
Loup, Cecile
Louys, Mireille
LU, Fangjun
Lubowich, Donald
Macquart, Jean-Pierre
Madore, Barry
MAEDA, Keiichi
Magnier, Eugene
Malkov, Oleg
Mann, Robert
Mannucci, Filippo
Martínez, Vicent
Martynov, D.
Mason, Brian
MATSUNAGA, Noriyuki
Matz, Steven
McAteer, R. T. James
McLean, Brian
McMahon, Richard
McNally, Derek
Meadows, A.
Mein, Pierre
Merín Martín, Bruno
Mermilliod, Jean-Claude
Mickaelian, Areg
Mighell, Kenneth
Mink, Jessica
Minniti, Dante
Mitton, Simon
Moitinho, André
Monet, David
Montes, David
Mookerjea, Bhaswati
Morbidelli, Roberto
Morris, Rhys
Morrison, Nancy
Muiños Haro, José
Murphy, Tara
Murtagh, Fionn
Murtagh, Fionn
NAGATA, Tetsuya
NAKAJIMA, Koichi
Nefedyev, Yury
Nesci, Roberto
NISHIMURA, Shiro
Norris, Raymond
Ochsenbein, François
Oertel, Goetz
Ogando, Ricardo
Ogorodnikov, K.
OHISHI, Masatoshi
Ojha, Roopesh
Oluseyi, Hakeem
Orellana, Rosa
Osborn, Wayne
Ott, Juergen
Pakhomov, Yury
Pamyatnykh, Alexey
Panessa, Francesca
Pasian, Fabio
Pasinetti, Laura
Paturel, Georges
Pauwels, Thierry
Peck, Alison
Pecker, Jean-Claude
Pence, William
Pérez-González, Pablo
PHAM, Diep
Philip, A.G.
Piskunov, Anatolij
Pitkin, Matthew
Pizzichini, Graziella
Plavchan, Peter
Polechova, Pavla
Potter, Stephen
Protsyuk, Yuri
Prša, Andrej
Pucillo, Mauro
Pushkarev, Alexander
Puxley, Phil
QU, Jinlu
Quinn, Peter
Ratnatunga, Kavan
Ray, Alak
Reardon, Kevin
Remy Battiau, Liliane
Renson, P.
Reyes-Ruiz, Mauricio
Richards, Mercedes
Rickard, Lee
Ridgway, Stephen
Riegler, Guenter
Rizzi, Luca
Robinson, Edward
Rocha-Pinto, Hélio
Rodrigues, Claudia
Roman, Nancy
Romaniello, Martino
Rossi, Corinne
Rots, Arnold
Rudnick, Lawrence
Russo, Guido
Saar, Enn
Saha, Abhijit
Sahal-Bréchot, Sylvie
Samodurov, Vladimir
Sarasso, Maria
Savanevich, Vadim
Schade, David
Schilbach, Elena
Schmadel, Lutz
Schmitz, Marion
Schneider, Jean
Schreiber, Roman
Schröder, Anja
Seaman, Rob
SEKIGUCHI, Kazuhiro
Seymour, Nicholas
Shafter, Allen
Shakeshaft, John
Sharp, Nigel
Shastri, Prajval
Shaw, Richard
Shelton, Ian
SHIRASAKI, Yuji

Silva, David
Skoda, Petr
Smart, Richard
Smith, Randall
Smith, Robert
SONG, Liming
Soszyński, Igor
Spite, François
Srianand, Raghunathan
Sterken, Christiaan
Stickland, David
Stil, Jeroen
Stoehr, Felix
Strelnitski, Vladimir
Strom, Stephen
Stroobant, P.
SU, Cheng-yue
Sullivan, Mark
SUNG, Hyun-Il
Surkis, Igor
Sutton, Edmund
Sykes, J.
Szalay, Alex
Szkody, Paula
TANAKA, Ichi Makoto
Taylor, A.
Tedds, Jonathan
Templeton, Matthew
TERASHITA, Yoichi
Tessema, Solomon
Teuben, Peter
Titov, Vladimir
Tody, Douglas
Torres-Papaqui, Juan
Trimble, Virginia
Tritton, Susan
Tsvetkov, Milcho
Turatto, Massimo
Turner, Kenneth
Tyson, John
UESUGI, Akira
Ulrich, Roger
Urban, Sean
Valls-Gabaud, David
Vandenbussche, Bart
Varela López, Jesús
Vavrek, Roland
Veillet, Christian
Velikodsky, Yuri
Verkhodanov, Oleg
Viotti, Roberto
Vishniac, Ethan
Vollmer, Bernd
Wallace, Patrick
WANG, Jian-Min
WANG, Shiang-Yu
Warren Jr, Wayne
Weilbacher, Peter
Weiss, Werner
Weller, Charles
Wells, Donald
Whitelock, Patricia
Whiting, Matthew
Wicenec, Andreas
Wielen, Roland
Wilkins, George
Williams, Robert
Willis, Anthony
Wise, Michael
Worley, Charles
Woudt, Patrick
YAMADA, Shimako
YANG, Hong-Jin
YANG, Xiaohu
Zacchei, Andrea
Zacharias, Norbert
Zender, Joe
ZHANG, Shu
ZHANG, Yanxia
ZHANG, Zhibin
ZHAO, Jun Liang
ZHAO, Yongheng
ZHOU, Jianfeng

Division B - Commission 6 Astronomical Telegrams

President: Hitoshi YAMAOKA
Vice-President: Daniel W. E. Green

Organizing Committee Members:

Alan C. Gilmore
Nikolay N. Samus
Gareth V. Williams
Timothy B. Spahr
Kaare Aksnes
Jana Tichá
Syuichi NAKANO

Members:

Aksnes, Kaare
Allan, Alasdair
Apostolovska, Gordana
Baransky, Olexander
Bazzano, Angela
Bouchard, Antoine
Buchar, E.
Candy, Michael
CHEN, Xinyang
Coletti, Donna
Corbin, Brenda
Cracco, Valentina
D Ammando, Filippo
Esenoğlu, Hasan
Filippenko, Alexei

Gal-Yam, Avishay
Gilmore, Alan
Green, Daniel
Grindlay, Jonathan
Hogg, A.
Kaminker, Alexander
Kastel, Galina
Kouveliotou, Chryssa
LIU, Guoqing
LIU, Jinming
Mattila, Seppo
MOROKUMA, Tomoki
Mrkos, Antonin
NAKAMURA, Tsuko
NAKANO, Syuichi
Ofek, Eran
Paragi, Zsolt
Phillips, Mark
Poznanski, Dovi
Roemer, Elizabeth
Rushton, Anthony
Samus, Nikolay
Seaman, Rob
Simon, Paul
Sivakoff, Gregory
Spahr, Timothy
Sullivan, Mark
Thernöe, K.
Tholen, David
Tichá, Jana
Tsvetkov, Milcho
URATA, Yuji
Valeev, Azamat
West, Richard
Williams, Gareth
YAMAOKA, Hitoshi

Division A - Commission 7 Celestial Mechanics & Dynamical Astronomy

President: Alessandro Morbidelli
Vice-President: Cristian Beaugé

Organizing Committee Members:

Piet Hut
Alessandra Celletti
Nader Haghighipour
Fernando Virgilio Roig
Jacques Laskar
Seppo Mikkola

Members:

Abad Medina, Alberto
Abalakin, Viktor
Ahmed, Mostafa
Aksenov, E.
Aksnes, Kaare
Alexander, Murray
Andoyer, H.
Andrade, Manuel
Anosova, Joanna
Antonacopoulos, Gregory
ARAKIDA, Hideyoshi
Archinal, Brent
Athanassoula, Evangelie (Lia)
Augereau, Jean-Charles
Balmino, Georges
Barabanov, Sergey
Barberis, Bruno
Barbosu, Mihail
Barkin, Yuri
Bartczak, Przemysław
Beaugé, Cristian
Bec-Borsenberger, Annick
Benest, Daniel
Bettis, Dale
Beutler, Gerhard
Bhatnagar, K.
Boccaletti, Dino
Bois, Eric
Borczyk, Wojciech
Borderies, Nicole
Borisov, Borislav
Boss, Alan
Bouchard, Antoine
Bozis, George
Branham, Richard
Breiter, Sławomir
Brieva, Eduardo
Brookes, Clive
Brouwer, D.
Brož, Miroslav
Brumberg, Victor
Brunini, Adrian
CAI, Michael
Candy, Michael
Caranicolas, Nicholas
Carlin, Jeffrey
Carpino, Mario
Carruba, Valerio
Cefola, Paul
Celletti, Alessandra
Chakrabarty, Dalia
Chambers, John
Chapanov, Yavor
Chapront, Jean
Chapront-Touze, Michelle
CHEN, Zhen
CHOI, Kyu Hong
Christou, Apostolos
Cionco, Rodolfo
Clemence, G.
Colin, Jacques
Conrad, Albert

Contopoulos, George
Cooper, Nicholas
CUI, Douxing
Davis, Morris
Deleflie, Florent
Deprit, Andre
Descamps, Pascal
de Sitter, W.
Dikova, Smilyana
Di Sisto, Romina
DONG, Xiaojun
Dourneau, Gerard
Drożyner, Andrzej
DU, Lan
Duboshin, G.
Duriez, Luc
Dvorak, Rudolf
Eckert, W.
Edelman, Colette
Efroimsky, Michael
El, Bakkali
Elipe, Antonio
Emelianov, Nikolaj
Emel'yanenko, Vacheslav
Esguerra, Jose Perico
FENG, Chugang
Fernández, Silvia
Ferrari, Fabricio
Ferraz-Mello, Sylvio
Ferrer, Martinez
Finch, Charlie
Floria Peralta, Luis
Fouchard, Marc
Froeschle, Claude
FUKUSHIMA, Toshio
Galibina, Irini
Gaposchkin, Edward
GENG, Lihong
Giacaglia, Giorgio
Giordano, Claudia
Giuliatti Winter, Silvia
Goldreich, Peter
Gomes, Rodney
González Camacho, Antonio
Goodwin, Simon
Goudas, Constantine
GoÅºdziewski, Krzysztof
Granvik, Mikael
Greenberg, Richard
Gronchi, Giovanni
Gusev, Alexander
Haghighipour, Nader
Hallan, Prem
Hamid, S.
Hamilton, Douglas
Hanslmeier, Arnold
Hau, George
HE, Miao-fu
Heggie, Douglas
Hellali, Yhya
HERSANT, Franck
Heyl, Jeremy
HORI, Genichiro
Horner, Jonathan
HOU, Xiyun
Hsieh, Henry
HU, Xiaogong
HUANG, Cheng
HUANG, Tianyi
Hurley, Jarrod
Hut, Piet
Iorio, Lorenzo
Ipatov, Sergei
Ismail, Mohamed
ITO, Takashi
Ivanov, Pavel
Ivanova, Violeta
Ivantsov, Anatoliy
Izmailov, Igor
Jäggi, Adrian
Jakubík, Marian
Janiczek, Paul
Jefferys, William
JI, Jianghui
JIANG, Ing-Guey
Journet, Alain
Jupp, Alan
Kalvouridis, Tilemachos
Kammeyer, Peter
Kent, Brian
Kenworthy, Matthew
Kholshevnikov, Konstantin
Kim, Yoo Jea
KIM, Sungsoo
KING, Sun-Kun
King-Hele, Desmond
KINOSHITA, Hiroshi
Kitiashvili, Irina
Klioner, Sergei
Klocok, Lubomir
Klokocnik, Jaroslav
Knežević, Zoran
KOKUBO, Eiichiro
Korchagin, Vladimir
Kornoš, Leonard
Koshkin, Nikolay
Kouwenhoven, M.B.N. (Thijs)
Kovačević, Andjelka
Kovalevsky, Jean
KOZAI, Yoshihide
Krivov, Alexander
Kustaanheimo, Paul
Kuznetsov, Eduard
Lala, Petr
Lammers, Uwe
Laskar, Jacques
La Spina, Alessandra
Lazovic, Jovan
Lecavelier des Etangs, Alain
Lega, Elena
Lemaître, Gérard
Le Poncin-Lafitte, Christophe
Levine, Stephen
LI, Yuqiang
LIAO, Xinhao
Libert, Anne-Sophie
Lieske, Jay
Lin, Douglas
Lissauer, Jack
LIU, Chengzhi
LIU, Wenzhong
Livadiotis, George
LU, BenKui
Lucchesi, David
Lundquist, Charles
MA, Jingyuan
MA, Lihua
Makhlouf, Amar
Malhotra, Renu
Mapelli, Michela
Marchal, Christian
Markellos, Vassilis
Martinet, Louis
Martins, Roberto
MASAKI, Yoshimitsu
Matas, Vladimir

Mavraganis, Anastasios
Meire, Raphael
Melbourne, William
Merman, Hirsh G.
Métris, Gilles
Michel, Patrick
Migaszewski, Cezary
Mignard, François
Mikkola, Seppo
Milani Comparetti, Andrea
Minazzoli, Olivier
Montgomery, Michele
Moore-Weiss, John
Morando, Bruno
Morbidelli, Alessandro
Moreira Morais, Maria
Muzzio, Juan
Myachin, Vladimir
Mysen, Eirik
Nacozy, Paul
Namouni, Fathi
Naroenkov, Sergey
Navone, Hugo
Nefedyev, Yury
Ninković, Slobodan
Nobili, Anna
Noskov, Boris
Novaković, Bojan
Novoselov, Viktor
Noyelles, Benoît
O'Handley, Douglas
Ollongren, A.
Orellana, Rosa
Orlov, Viktor
Osório, Isabel
Osório, José
Page, Gary
Pál, András
Parv, Bazil
Pauwels, Thierry
Peale, Stanton
Perets, Hagai
Perozzi, Ettore
Petit, Jean-Marc
Petrov, Sergey
Petrovskaya, Margarita
Pilat-Lohinger, Elke
Polyakhova, Elena
Popovic, Bozidar
Puetzfeld, Dirk
Reyes-Ruiz, Mauricio
Richardson, Derek
Robinson, William
Rodin, Alexander
Rodriguez-Eillamil, R.
Rossi, Alessandro
Ryabov, Yurij
Saad, Abdel-naby
Sansaturio, Maria
Sari, Re'em
Scheeres, Daniel
Scholl, Hans
Schubart, Joachim
Sconzo, Pasquale
Šegan, Stevo
Seidelmann, P.
Sein-Echaluce, M.
Shakht, Natalia
Shankland, Paul
Shapiro, Irwin
Sharaf, Sh
Shevchenko, Ivan
Sidlichovsky, Milos
Sima, Zdislav
Simó, Carles
Simon, Jean-Louis
Sinclair, Andrew
Skripnichenko, Vladimir
Soffel, Michael
SOFIA LYKAWKA, Patryk
Sokolov, Leonid
Sorokin, Nikolaj
Souchay, Jean
Stadel, Joachim
Standish, E.
Stellmacher, Irène
Steves, Bonnie
Stewart, Susan
Süli, Áron
Sultanov, G.
SUN, Fuping
SUN, Yisui
Sweatman, Winston
Szebehely, Victor
Szenkovits, Ferenc
Taborda, Jose
TAO, Jin-he
Tatevyan, Suriya
Taylor, Donald
Thiry, Yves
Thuillot, William
Tiscareno, Matthew
Titov, Vladimir
Tommei, Giacomo
TONG, Fu
Tremaine, Scott
Trenti, Michele
Tsiganis, Kleomenis
Tsuchida, Masayoshi
Valsecchi, Giovanni
Valtonen, Mauri
Varvoglis, Harry
Vashkovyak, Sofja
Vassiliev, Nikolaj
Veillet, Christian
Vieira Neto, Ernesto
Vienne, Alain
Vilhena de Moraes, Rodolpho
Vinet, Jean-Yves
Virtanen, Jenni
Vokrouhlicky, David
Vondrák, Jan
Walch, Jean-Jacques
Walker, Ian
WANG, Guangli
WANG, Xiaoya
WATANABE, Noriaki
WEI, Erhu
Whipple, Arthur
Wiegert, Paul
Williams, Carol
Winter, Othon
Wnuk, Edwin
WU, Lianda
Wytrzyszczak, Iwona
XIA, Yi
XIAOSHENG, Wan
XIONG, Jianning
XU, Jihong
XU, Pinxin
YANG, Xuhai
YANO, Taihei
Yarov-Yarovoj, Mikhail
YI, Zhaohua
Yokoyama, Tadashi
YOSHIDA, Haruo
YOSHIDA, Junzo
Yseboodt, Marie

YUASA, Manabu
Zafiropoulos, Basil
Zagar, F.
Zare, Khalil
ZHANG, Sheng-Pan
ZHANG, Wei
ZHANG, Wei
ZHANG, Xiaoxiang
ZHANG, Yang
ZHAO, Changyin
ZHAO, Haibin
ZHAO, You
Zhdanov, Valery
ZHENG, Jia-Qing
ZHENG, Xuetang
ZHENG, Yong
ZHOU, Hongnan
ZHOU, Ji-Lin
ZHOU, Li-Yong
ZHU, Wenyao

Division A - Commission 8 Astrometry

President: Norbert Zacharias
Vice-President: Anthony G.A. Brown

Organizing Committee Members:

Ramachrisna Teixeira
Oleksandr V. Shulga
Li CHEN
Valeri Makarov
Jean Souchay
Stephen C. Unwin
Naoteru GOUDA

Members:

Abad Hiraldo, Carlos
Abbas, Ummi
Adams, A.
Ahmed, Abdel-aziz
Ammons, Stephen
Andrei, Alexandre
Andronova, Anna
Arenou, Frédéric
Argyle, Robert
Arias, Elisa
Arlot, Jean-Eudes
Assafin, Marcelo
BABA, Junichi
Babusiaux, Carine
Bacchus, Pierre
Badescu, Octavian
Bakhtigaraev, Nail
Ballabh, Goswami
Bangert, John
Barkin, Yuri
Bartczak, Przemysław
Bartlett, Jennifer
Bastian, Ulrich
Belizon, Fernando
Benedict, George
Benevides Soares, Paulo
Bien, Reinhold
Billaud, G.
Boboltz, David
Bouchard, Antoine
Bougeard, Mireille
Bradley, Arthur
Branham, Richard
Brosche, Peter
Brouw, Willem
Brown, Anthony
Bucciarelli, Beatrice
Bueno de Camargo, Julio
Capitaine, Nicole
Carlin, Jeffrey
Casetti, Dana
Chakrabarty, Dalia
Chapanov, Yavor
Chemin, Laurent
CHEN, Alfred
CHEN, Li
CHEN, Li
CHEN, Linfei
Chiumiento, Giuseppe
Cioni, Maria-Rosa
Cooper, Nicholas
Corbin, Thomas
Costa, Edgardo
Crézé, Michel
Crifo, Francoise
Crosta, Mariateresa
Cudworth, Kyle
Dahn, Conard
Damljanovic, Goran
Danylevsky, Vassyl
da Rocha-Poppe, Paulo
Day-Jones, Avril
d'E Atkinson, R.
de Bruijne, Jos
Dejaiffe, Rene
Deller, Adam
Delmas, Christian
Del Santo, Melania
Devyatkin, Aleksandr
Dick, Steven
Dick, Wolfgang
DU, Lan
Ducourant, Christine
Duma, Dmitrij
Einicke, Ole
Emilio, Marcelo
Evans, Dafydd
Fabricius, Claus

FAN, Yu
Feissel, Martine
Fernandes-Martin, Vera
Fey, Alan
Finch, Charlie
Firneis, Maria
Fomin, Valerij
Fomin, Valery
Fors, Octavi
Franz, Otto
Fredrick, Laurence
Fresneau, Alain
Froeschle, Michel
Frutos-Alfaro, Francisco
FUJISHITA, Mitsumi
FUKUSHIMA, Toshio
Gatewood, George
Gaume, Ralph
Gauss, Stephen
Gavras, Panagiotis
Geffert, Michael
Germain, Marvin
GILLES, Dominique
Goddi, Ciriaco
Gontcharov, George
GOUDA, Naoteru
Goyal, A.
Guibert, Jean
Guseva, Irina
Hajian, Arsen
HAN, Inwoo
Hanson, Robert
Hartkopf, William
Helmer, Leif
Hemenway, Paul
Hering, Roland
Heudier, Jean-Louis
Hill, Graham
Hobbs, David
Høg, Erik
Høg, E.
HONG, Zhang
Hugues, James
Ianna, Philip
Ilin, Alexey
Iorio, Lorenzo
Irwin, Michael
Ivantsov, Anatoliy
Izmailov, Igor
Jackson, J.
Jackson, Paul
Jacobs, Christopher
Jahreiss, Hartmut
Jefferys, William
JIA, Lei
JIN, WenJing
Johnston, Kenneth
Jones, Burton
Jones, Derek
Jordi, Carme
Journet, Alain
Kalomeni, Belinda
Kanayev, Ivan
Kaplan, George
Kazantseva, Liliya
Kharchenko, Nina
Khrutskaya, Evgenia
King, Ivan
Klemola, Arnold
Klioner, Sergei
Klock, Benny
Kolesnik, Yuri
Kovalevsky, Jean
Kuimov, Konstantin
Kumkova, Irina
KURAYAMA, Tomoharu
Kurzyńska, Krystyna
Lacroute, Pierre
Lammers, Uwe
Lattanzi, Mario
Latypov, A.
Lazorenko, Peter
Lenhardt, Helmut
Lepine, Sebastien
Le Poncin-Lafitte, Christophe
Le Poole, Rudolf
LI, Qi
LI, Zhigang
Lindegren, Lennart
LIU, Chengzhi
Lopez, Carlos
Lopez, Jose
Lu, Phillip
LU, Chunlin
MA, Wenzhang
MacConnell, Darrell
Magnier, Eugene
Maigurova, Nadiia
Makarov, Valeri
Malkin, Zinovy
Mallamaci, Claudio
Marschall, Laurence
Marshalov, Dmitriy
McAlister, Harold
McLean, Brian
Mignard, François
Mink, Jessica
Miyamoto, Masanori
Monet, David
Morbidelli, Roberto
Morgan, H.
Morrison, Leslie
Muiños Haro, José
Murray, Andrew
NAKAGAWA, Akiharu
NAKAJIMA, Koichi
Naroenkov, Sergey
Nefedyev, Yury
Nemiro, André
NIINUMA, Kotaro
Nikoloff, Ivan
NIWA, Yoshito
Noel, Fernando
Noyelles, Benoît
Nunez, Jorge
Ofek, Eran
OHNISHI, Kouji
Oja, Tarmo
Olive, Don
Olsen, Hans
Osborn, Wayne
Osório, José
Page, Gary
Pakvor, Ivan
Pannunzio, Renato
Pascu, Dan
Pauwels, Thierry
Penna, Jucira
Perryman, Michael
Petrov, Sergey
PING, Jinsong
Pinigin, Gennadiy
Platais, Imants
Polozhentsev, Dmitrij
Poma, Angelo
Popescu, Petre
Pourbaix, Dimitri
Protsyuk, Yuri
Proverbio, Edoardo

Prusti, Timo
Pugliano, Antonio
Quijano, Luis
Rafferty, Theodore
Reddy, Bacham
Reffert, Sabine
Requieme, Yves
Reynolds, John
Rizvanov, Naufal
Rodin, Alexander
Roemer, Elizabeth
Roeser, Siegfried
Russell, Jane
SAKAMOTO, Tsuyoshi
Sanchez, Manuel
Sanders, Walter
Santamaria, Raffaele
Sarasso, Maria
SATO, Koichi
Schilbach, Elena
Schildknecht, Thomas
Schmeidler, F.
Scholz, Ralf-Dieter
Schombert, J.
Schreiber, Karl
Schwan, Heiner
Schwekendiek, Peter
Scott, F.
Ségransan, Damien
Seidelmann, P.
Shelus, Peter
SHEN, Kaixian
SHEN, Zhiqiang
Shokin, Yurij
SHU, Fengchun
Shulga, Oleksandr
Sivakoff, Gregory
Slaucitajs, Serge
Smart, Richard
Smith Jr, Clayton
Soderhjelm, Staffan
Solarić, Nikola
SOMA, Mitsuru
Souchay, Jean
Sovers, Ojars
Sozzetti, Alessandro
Spoljaric, Drago
Standish, E.
Stavinschi, Magda
Stein, John
Steinmetz, Matthias
Stewart, Susan
SUGANUMA, Masahiro
SUN, Fuping
SUNG, Hyun-Il
Surkis, Igor
TANG, Zheng-Hong
TARIS, François
Tavastsherna, K.
Tedds, Jonathan
Teixeira, Paula
Teixeira, Ramachrisna
Teleki, George
ten Brummelaar, Theo
Thuillot, William
TSUJIMOTO, Takuji
Tucker, Roy
Turon, Catherine
UEDA, Haruhiko
Unwin, Stephen
Upgren, Arthur
Urban, Sean
Valbousquet, Armand
Vallejo, Miguel
van Altena, William
van Leeuwen, Floor
Vass, Gheorghe
Vertypolokh, Olexander
Vilkki, Erkki
Volyanska, Margaryta
Wallace, Patrick
WANG, Guangli
WANG, Jia-Ji
WANG, Xiaoya
WANG, Zhengming
Wasserman, Lawrence
Watts, C.
WEI, Erhu
White, Graeme
Wicenec, Andreas
Wielen, Roland
Wood, H.
WU, Zhen-Yu
XIA, Yifei
XIE, Yi
XU, Jiayan
XU, Tong-Qi
YAMADA, Yoshiyuki
YAN, Haojian
YANG, Tinggao
YANO, Taihei
YASUDA, Haruo
Yatsenko, Anatolij
Yatskiv, Yaroslav
YE, Shuhua
YOSHIZAWA, Masanori
Zacchei, Andrea
Zacharias, Norbert
ZHANG, Wei
ZHANG, Yong
ZHENG, Yong
ZHU, Zi
Zverev, M.

Division E - Commission 10 Solar Activity

President: Karel Schrijver
Vice-President: Lyndsay Fletcher

Organizing Committee Members:

Daniel O. Gómez
Paul S. Cally
Paul Charbonneau
S. Sirajul Hasan
Yihua YAN
Astrid Veronig
Ayumi ASAI
Sarah Gibson

Members:

Abbett, William
Abdelatif, Toufik
Aboudarham, Jean
Ábrahám, Péter
Abramenko, Valentina
Afram, Nadine
Ahluwalia, Harjit
AI, Guoxiang
Airapetian, Vladimir
AKITA, Kyo
Alissandrakis, Costas
Almleaky, Yasseen
Altrock, Richard
Altschuler, Martin
Altyntsev, Alexandre
Aly, Jean-Jacques
Ambastha, Ashok
Ambroz, Pavel
Anastasiadis, Anastasios
Andersen, Bo Nyborg
Andretta, Vincenzo
Andries, Jesse
Antiochos, Spiro
Antonucci, Ester
Anzer, Ulrich
Arregui, Inigo
ASAI, Ayumi
Aschwanden, Markus
Atac, Tamer
Aurass, Henry
Avignon, Yvette
Babin, Arthur
Bagala, Liria
Bagare, S.
Balasubramaniam, K.
Balikhin, Michael
Ballester, Jose
Banerjee, Dipankar
BAO, Shudong
Baranovsky, Edward
Barrantes, Marco
Barrow, Colin
Barta, Miroslav
Basu, Sarbani
Batchelor, David
Beckers, Jacques
Bedding, Timothy
Beebe, Herbert
Bell, Barbara
Bellot Rubio, Luis
Belvedere, Gaetano
Bemporad, Alessandro
Benevolenskaya, Elena
Benz, Arnold
Berger, Mitchell
Berghmans, David
Berrilli, Francesco
Bertello, Luca
Bewsher, Danielle
Bianda, Michele
Bingham, Robert
Bobylev, Vadim
Bocchia, Romeo
Bogdan, Thomas
Bommier, Veronique
Bondal, Krishna
Bornmann, Patricia
Botha, Gert
Bothmer, Volker
Bouchard, Antoine
Bougeret, Jean-Louis
Boyer, René
Brajša, Roman
Brandenburg, Axel
Brandt, Peter
Braun, Douglas
Bray, Robert
Brekke, Pål
Bromage, Barbara
Brooke, John
Brosius, Jeffrey
Brown, John
Browning, Philippa
Brun, Allan
Bruner, Marilyn
Bruno, Roberto
Bruns, Andrey
Brynildsen, Nils
Buccino, Andrea
Buchlin, Eric
Buecher, Alain
Buechner, Joerg
Bumba, Vaclav
Busa', Innocenza
Bushby, Paul
Čadež, Vladimir
CAI, Mingsheng
Cally, Paul
Cane, Hilary
Carbonell, Marc
Cargill, Peter
Cauzzi, Gianna
CHAE, Jongchul
Chambe, Gilbert
Chandra, Suresh

CHANG, Heon-Young
Channok, Chanruangrit
Chaplin, William
Chapman, Gary
Charbonneau, Paul
CHEN, Peng-Fei
CHEN, Zhencheng
CHEN, Zhiyuan
Chernov, Gennadij
Chertok, Ilya
Chertoprud, Vadim
Chiuderi-Drago, Franca
CHIUEH, Tzihong
CHO, Kyung Suk
CHOE, Gwangson
Choudhary, Debi Prasad
Choudhuri, Arnab
Chupp, Edward
Cliver, Edward
Coffey, Helen
Collados, Manuel
Conway, Andrew
Cora, Alberto
Correia, Emilia
Costa, Joaquim
Craig, Ian
Cramer, Neil
Crannell, Carol
Culhane, John
Curdt, Werner
Dalla, Silvia
Damé, Luc
Dasso, Sergio
Datlowe, Dayton
Davila, Joseph
d'Azambuja, L.
Dechev, Momchil
de Feiter, L.
De Groof, Anik
de Jager, Cornelis
Del Toro Iniesta, Jose
Démoulin, Pascal
DENG, YuanYong
Dennis, Brian
Dere, Kenneth
DeRosa, Marc
de Toma, Giuliana
Deubner, Franz-Ludwig
Dialetis, Dimitris
DING, Mingde
DING, Youji
Dinulescu, Simona
Dobler, Wolfgang
Dobrzycka, Danuta
Dorch, Søren Bertil
Dorotovic, Ivan
Doyle MRIA, John
Druckmüller, Miloslav
Dryer, Murray
Dubau, Jacques
Dubois, Marc
Duchlev, Peter
Duldig, Marcus
Dumitrache, Cristiana
Dwivedi, Bhola
Efimenko, Volodymyr
Ellison, M.
Elste, Günther
Emslie, Gordon
Engvold, Oddbjørn
Erdelyi, Robertus
Ermolli, Ilaria
Esenoğlu, Hasan
Falchi, Ambretta
Falciani, Roberto
Falewicz, Robert
FAN, Yuhong
FANG, Cheng
Fárník, Frantisek
Fernandes, Francisco
Ferreira, João
Ferriz Mas, Antonio
Feulner, Georg
Filippov, Boris
Fisher, George
Fludra, Andrzej
Fluri, Dominique
Foing, Bernard
Fokker, Aad
Fontenla, Juan
Forbes, Terry
Forgács-Dajka, Emese
Fossat, Eric
Foullon, Claire
FU, Hsieh-Hai
Gabriel, Alan
Gaizauskas, Victor
Galal, A.
Galloway, David
Galsgaard, Klaus
GAN, Weiqun
Garaimov, Vladimir
García de la Rosa, Ignacio
Gary, Gilmer
Georgoulis, Manolis
Gergely, Tomas
Getling, Alexander
Ghizaru, Mihai
Gibson, David
Gibson, Sarah
Gill, Peter
Gilliland, Ronald
Gilman, Peter
Giménez de Castro, Carlos
Glatzmaier, Gary
Gleisner, Hans
Goedbloed, Johan
Gokhale, Moreshwar
Gómez, Daniel
Gömöry, Peter
Gontikakis, Constantin
Goossens, Marcel
Gopasyuk, Olga
Graffagnino, Vito
Grandpierre, Attila
Grant, Athay
Grechnev, Victor
Gregorio, Anna
Grib, Sergei
GU, Xiaoma
Gudiksen, Boris
Guglielmino, Salvatore
Guhathakurta, Madhulika
Gunár, Stanislav
Gupta, Surendra
Gurman, Joseph
Gurtovenko, E.
Györi, Lajos
Haberreiter, Margit
Hagyard, Mona
Hammer, Reiner
Hanaoka, Yoichiro
Hanasz, Jan
Hannah, Iain
Hansen, Richard
Hanslmeier, Arnold
HARA, Hirohisa
Harra, Louise
Harvey, John
Hasan, S. Sirajul

Hathaway, David
Haugan, Stein Vidar
HAYASHI, Keiji
Hayward, John
He, Han
Heinzel, Petr
Henoux, Jean-Claude
Herdiwijaya, Dhani
Hermans, Dirk
HIEI, Eijiro
Hildebrandt, Joachim
Hildner, Ernest
Hochedez, Jean-François
Hoeksema, Jon
Hohenkerk, Catherine
Hollweg, Joseph
Holman, Gordon
Holzer, Thomas
HONG, Hyon
Hood, Alan
Houdebine, Eric
Howard, Robert
Hoyng, Peter
Hudson, Hugh
Hurford, Gordon
HWANG, Junga
Ioshpa, Boris
Ireland, Jack
ISHII, Takako
Ishitsuka, Mutsumi
Isliker, Heinz
Ivanchuk, Victor
Ivanov, Evgenij
Ivchenko, Vasily
Jackson, Bernard
Jacobs, Carla
Jain, Rajmal
Jakimiec, Jerzy
Janssen, Katja
Jardine, Moira
Jefferies, John
JI, Haisheng
JIANG, Aimin
JIANG, Yunchun
Jimenez, Mancebo
JING, Hairong
Jones, Harrison
Jordan, Stuart
Joselyn, Jo
Jubier, Xavier
Jurčák, Jan
KABURAKI, Osamu
Kahler, Stephen
Kallenbach, Reinald
Kálmán, Bela
Kaltman, Tatyana
Kane, Sharad
KANG, Jin Sok
Käpylä, Petri
Karami, Kayoomars
Karlick, Márian
Karoff, Christoffer
Karpen, Judith
Kasiviswanathan, Sankarasubramanian
Kašparová, Jana
Katsova, Maria
Kaufmann, Pierre
Keppens, Rony
Khan, J
Khodachenko, Maxim
Khomenko, Elena
Khumlemlert, Thiranee
KILCIK, Ali
Kim, Iraida
KIM, Kap sung
Kiplinger, Alan
KITAI, Reizaburo
Kitchatinov, Leonid
Kitiashvili, Irina
Kjeldseth-Moe, Olav
Klein, Karl
Kliem, Bernhard
Klimchuk, James
Klvana, Miroslav
Knoska, Stefan
Kolobov, Dmitri
Kołomański, Sylwester
Kondrashova, Nina
Kontar, Eduard
Kopylova, Yulia
Kostik, Roman
Kotrc, Pavel
Koutchmy, Serge
Kovacs, Agnes
Koza, Julius
Kozlovsky, Ben
Kramer, William
Krat, Vladimir
Kretzschmar, Matthieu
Krimigis, Stamatios
Krittinatham, Watcharawuth
Krucker, Sam
Kryshtal, Alexander
Kryvodubskyj, Valery
KUBOTA, Jun
Kučera, Aleš
Kulhánek, Petr
Kuperus, Max
Kurochka, Evgenia
KUROKAWA, Hiroki
KUSANO, Kanya
Kuznetsov, Vladimir
Labrosse, Nicolas
Landi, Simone
Landman, Donald
Lang, Kenneth
Lawrence, John
Lazrek, Mohamed
Leibacher, John
Leiko, Uliana
Leka, K.D.
Leroy, Bernard
Leroy, Jean-Louis
LI, Chun-Sheng
LI, Hui
LI, Kejun
LI, Wei
LI, Zhi
Lie-Svendsen, Oystein
Lima, Joao
Lin, Yong
LIN, Jun
LIN, Yuanzhang
Liritzis, Ioannis
LIU, Siming
LIU, Xinping
LIU, Yang
LIU, Yu
Livadiotis, George
Livshits, Moisey
Longbottom, Aaron
Lopez Fuentes, Marcelo
Loukitcheva, Maria
Low, Boon
Lozitskij, Vsevolod
Lundstedt, Henrik
LUO, Xianhan
Lustig, Guenter

MA, Guanyi
Machado, E.
Machado, Marcos
Mackay, Duncan
MacKinnon, Alexander
MacQueen, Robert
Madjarska, Maria
Makarov, Valentin
MAKITA, Mitsugu
Malara, Francesco
Malherbe, Jean-Marie
Malitson, Harriet
Malville, J.
MANABE, Seiji
Mandrini, Cristina
Mann, Gottfried
Marcu, Alexandru
Maričić, Darije
Marilena, Mierla
Maris, Georgeta
Mariska, John
Marková, Eva
Martens, Petrus
Mason, Glenn
MASUDA, Satoshi
Matsuura, Oscar
Matthews, Sarah
Mattig, W.
Maxwell, Alan
McAteer, R. T. James
McCabe, Marie
McIntosh, Patrick
McKenna-Lawlor, Susan
McLean, Donald
Mein, Pierre
Mel'nik, Valentin
Mendes, Da
Mendoza-Briceño, César
Messerotti, Mauro
Messmer, Peter
Mészárosová, Hana
Michałek, Grzegorz
Miesch, Mark
Miletsky, Eugeny
Miralles, Mari Paz
MITRA, Dhrubaditya
Mohan, Anita
Moreno-Insertis, Fernando
Morita, Satoshi
MORIYAMA, Fumio
Motta, Santo
Muller, Richard
Musielak, Zdzislaw
Nagasawa, Shingo
NAKAJIMA, Hiroshi
Nakariakov, Valery
Namba, Osamu
Nandi, Dibyendu
Narain, Udit
Neidig, Donald
Neukirch, Thomas
Neupert, Werner
Newkirk Jr., Gordon
Nickeler, Dieter
NING, Zongjun
Nocera, Luigi
Noens, Jacques-Clair
Noyes, Robert
Nozawa, Satoshi
Nussbaumer, Harry
Obridko, Vladimir
Ofman, Leon
OH, Suyeon
OHKI, Kenichiro
OKAMOTO, Takenori Joten
Okten, Adnan
Oliver, Ramón
Oluseyi, Hakeem
Önel, Hakan
Opara, Fidelix
Oraevsky, Victor
Orlando, Salvatore
Ortiz Carbonell, Ada
O'Shea, Eoghan
Ozguc, Atila
Oezisik, Tuncay
Padmanabhan, Janardhan
Paletou, Frédéric
Pallavicini, Roberto
Pallé, Pere
Pallé Bagó, Enric
Pal'uš, Pavel
PAN, Liande
Pap, Judit
Parenti, Susanna
Parfinenko, Leonid
Pariat, Etienne
PARK, Young Deuk
Parkinson, William
Parnell, Clare
Pasachoff, Jay
Paternò, Lucio
Peres, Giovanni
Peter, Hardi
Petrie, Gordon
Petrosian, Vahe
Petrov, Nikola
Petrovay, Kristof
Pevtsov, Alexei
Pflug, Klaus
PHAN, Dong
Phillips, Kenneth
Pick, Monique
Pipin, Valery
Plainaki, Christina
Podesta, John
Poedts, Stefaan
Pohjolainen, Silja
Poland, Arthur
Poquerusse, Michel
Preka-Papadema, Panagiota
Preś, Paweł
Priest, Eric
Proctor, Michael
Prokakis, Theodore
Pustil'nik, Lev
Raadu, Michael
Ramelli, Renzo
Rao, A.
Raoult, Antoinette
Raulin, Jean-Pierre
Rayrole, Jean
Reale, Fabio
Rees, David
Reeves, Edmond
Reeves, Hubert
Regnier, Stephane
Régulo, Clara
Reinard, Alysha
Rendtel, Juergen
Rengel, Miriam
Reshetnyk, Volodymyr
RI, Son Jae
Rieger, Erich
Riehokainen, Aleksandr
Rijnbeek, Richard
Riley, Pete
Roberts, Walter

Robinson Jr, Richard
Roca Cortés, Teodoro
Roemer, Max
Romano, Paolo
Romoli, Marco
Rompolt, Bogdan
Roša, Dragan
Rösch, Jean
Roudier, Thierry
Rovira, Marta
Roxburgh, Ian
Rozelot, Jean-Pierre
Rudawy, Paweł
Ruderman, Michael
Rüedi, Isabelle
Ruediger, Guenther
Ruffolo, David
Ruiz Cobo, Basilio
Rusin, Vojtech
Russell, Alexander
Rust, David
Rust, David
Ruždjak, Domagoj
Ruzickova-Topolova, B.
Rybák, Jan
Rybansky, Milan
Safari, Hossein
Sahal-Bréchot, Sylvie
Saiz, Alejandro
SAKAO, Taro
SAKURAI, Takashi
SAKURAI, Kunitomo
Sánchez Almeida, Jorge
Saniga, Metod
Sasso, Clementina
Sattarov, Isroil
Sawyer, Constance
Schindler, Karl
Schlichenmaier, Rolf
Schmahl, Edward
Schmelz, Joan
Schmieder, Brigitte
Schober, Hans
Schrijver, Karel
Schuessler, Manfred
Schwartz, Pavol
Schwenn, Rainer
Seaton, Daniel
Shea, Margaret
Sheeley, Neil
SHIBASAKI, Kiyoto
Shimizu, Toshifumi
Shimojo, Masumi
Shine, Richard
SHIN'ICHI, Nagata
Sigalotti, Leonardo
Simnett, George
Simon, Guy
Simon, Paul
Simunac, Kristin
Sinha, Krishnanand
Smaldone, Luigi
Smerd, S.
Smith, Dean
Smith, Henry
Smol'kov, Gennadij
Snegirev, Sergey
Snik, Frans
Sobotka, Michal
Socas-Navarro, Hector
Solanki, Sami
Soliman, Mohamed
Soloviev, Alexandr
Somov, Boris
Spadaro, Daniele
Spicer, Daniel
Spruit, Hendrik
Srivastava, Nandita
Steiner, Oskar
Stellmacher, Götz
Stenflo, Jan
Stepanian, Natali
Stepanov, Alexander
Stepanov, V.
Steshenko, N.
Stewart, Ronald
Stix, Michael
Stoker, Pieter
Strong, Keith
Struminsky, Alexei
Sturrock, Peter
Subramanian, K.
Subramanian, Prasad
Sudar, Davor
Sukartadiredja, Darsa
SUN, Kai
SUZUKI, Takeru
Švanda, Michal
Sylwester, Barbara
Sylwester, Janusz
Szalay, Alex
TAKAHASHI, Kunio
TAKAKURA, Tatsuo
TAKANO, Toshiaki
Talon, Raoul
TAMENAGA, Tatsuo
TAN, Baolin
Tanaka, Katsuo
TANG, Yuhua
Tapping, Kenneth
Tarashchuk, Vera
Ternullo, Maurizio
Teske, Richard
Thomas, John
Thomas, Roger
Tikhomolov, Evgeniy
Tlatov, Andrej
Tobias, Steven
Tomczak, Michał
trân, Ha
Trellis, Michel
Treumann, Rudolf
Tripathi, Durgesh
Tripathy, Sushanta
Tritakis, Basil
Trottet, Gerard
Tsap, Yuriy
Tsinganos, Kanaris
TSUBOTA, Yukimasa
Uddin, Wahab
Underwood, James
Usoskin, Ilya
Válio, Adriana
Valnicek, Boris
van den Oord, Bert
van der Heyden, Kurt
van der Linden, Ronald
Van Doorsselaere, Tom
Van Hoven, Gerard
van't Veer, Frans
Vaughan, Arthur
Veck, Nicholas
Vekstein, Gregory
Velli, Marco
Venkatakrishnan, P.
Ventura, Rita
Vergez, Madeleine
Verheest, Frank
Verma, V.
Veronig, Astrid

Verwichte, Erwin
Vial, Jean-Claude
Viall, Nicholeen
Vieytes, Mariela
Vilinga, Jaime
Vilmer, Nicole
Vinod, S.
Voitenko, Yuriy
Vršnak, Bojan
Walker, Simon
Walsh, Robert
Wang, Yi-ming
WANG, Dongguang
WANG, Haimin
WANG, Huaning
WANG, Jia-Long
WANG, Jingyu
WANG, Min
WANG, Shujuan
Webb, David
White, Stephen
Wiehr, Eberhard
Wikstol, Oivind
Wilcox, John
Wild, John
Winebarger, Amy
Wittmann, Axel
Woehl, Hubertus
Wolfson, Richard
Woltjer, Lodewijk
Wood, Brian
Wu, Shi
WU, De Jin
WU, Linxiang
WU, Mingchan
XIE, Xianchun
XU, Aoao
XU, Jun
XU, Zhi
YAN, Yihua
YANG, Hong-Jin
YANG, Lei
YANG, Zhiliang
YE, Shi-hui
Yeh, Tyan
Yeşilyurt, Ibrahim
YI, Yu
YOKOYAMA, Takaaki
YOSHIMURA, Hirokazu
YOU, Jianqi
YU, Cong
YU, Dai
Yun, Hong-Sik
Zachariadis, Theodosios
Zappalà, Rosario
Zelenka, Antoine
Zender, Joe
Zhang, Bai-Rong
Zhang, Jie
Zhang, Tielong
ZHANG, Heqi
ZHANG, Jun
ZHANG, Mei
Zhitnik, Igor
ZHOU, Daoqi
ZHOU, Guiping
Zhugzhda, Yuzef
Zhukov, Vladimir
Zlobec, Paolo
ZOU, Yixin

Division E - Commission 12 Solar Radiation & Structure

President: Gianna Cauzzi
Vice-President: Nataliia Gennadievna Shchukina

Organizing Committee Members:

Mingde DING
Stuart M. Jefferies
Natalie Krivova
Michele Bianda
Sergio Dasso
Dean-Yi CHOU
Axel Brandenburg
Fernando Moreno-Insertis

Members:

Abbett, William
Aboudarham, Jean
Acton, Loren
AI, Guoxiang
Aime, Claude
Alissandrakis, Costas
Altrock, Richard
Altschuler, Martin
Andersen, Bo Nyborg
Anderson, Jay
ANDO, Hiroyasu
Andretta, Vincenzo
Ansari, S.M.
Antia, H.
Artzner, Guy
ASAI, Ayumi
Asplund, Martin
Avrett, Eugene
Ayres, Thomas
Babayev, Elchin
Baliunas, Sallie
Balthasar, Horst
Banerjee, Dipankar
Barta, Miroslav
Basu, Sarbani
Baturin, Vladimir
Beckers, Jacques

Beckman, John
Beebe, Herbert
Beiersdorfer, Peter
Bemporad, Alessandro
Benford, Gregory
Bertello, Luca
Bhardwaj, Anil
BI, Shaolan
Bianda, Michele
Bingham, Robert
Blamont, Jacques-Emile
Bocchia, Romeo
Boehm, Karl-Heinz
Bohm-Vitense, Erika
Bommier, Veronique
Bonnet, Roger
Book, David
Bornmann, Patricia
Borovik, Valerya
Bougeret, Jean-Louis
Brandenburg, Axel
Brandt, Peter
Brault, James
Bray, Robert
Breckinridge, James
Brosius, Jeffrey
Brueckner, Guenter
Bruls, Jo
Brun, Allan
Bruner, Marilyn
Bruning, David
Buchlin, Eric
Bumba, Vaclav
Čadež, Vladimir
Cauzzi, Gianna
Cavallini, Fabio
Ceppatelli, Guido
Chambe, Gilbert
CHAN, Kwing
Chapman, Gary
CHEN, Biao
Chertok, Ilya
CHOE, Gwangson
CHOU, Dean-Yi
Christensen-Dalsgaard, Jørgen
Clark, Thomas
Clette, Frederic
Collados, Manuel
Collet, Remo
Couvidat, Sebastien
Craig, Ian
Cramer, Neil
Damé, Luc
Dara, Helen
Dasso, Sergio
Dechev, Momchil
Degenhardt, Detlev
de Jager, Cornelis
Delache, Philippe
Deliyannis, John
Del Toro Iniesta, Jose
Demarque, Pierre
Deming, Leo
DeRosa, Marc
de Toma, Giuliana
Deubner, Franz-Ludwig
Di Mauro, Maria Pia
DING, Mingde
Diver, Declan
Dogan, Nadir
Donea, Alina
Doyle MRIA, John
Dravins, Dainis
Druckmüller, Miloslav
Dumont, Simone
Duvall Jr, Thomas
Ehgamberdiev, Shuhrat
Einaudi, Giorgio
Elliott, Ian
Elste, Günther
Epstein, Gabriel
Ermolli, Ilaria
Esser, Ruth
Fabiani, Sergio
Falciani, Roberto
Falewicz, Robert
FANG, Cheng
Feldman, Uri
Fernandes, Francisco
Feulner, Georg
Fisher, George
Fleck, Bernhard
Fluri, Dominique
Fofi, Massimo
Fomichev, Valerij
Fontenla, Juan
Forgács-Dajka, Emese
Fossat, Eric
Foukal, Peter
Foullon, Claire
Frazier, Edward
Froehlich, Claus
Gabriel, Alan
Gaizauskas, Victor
GAN, Weiqun
Garcia, Rafael
García-Berro, Enrique
Georgoulis, Manolis
Getling, Alexander
Giovanelli, Ronald
Glatzmaier, Gary
Gnevyshev, Mstislav
Goldman, Martin
Gomez, Maria
Gömöry, Peter
Gopalswamy, Natchimuthuk
Grant, Athay
Grevesse, Nicolas
GU, Xiaoma
Guhathakurta, Madhulika
Gunár, Stanislav
Gurtovenko, E.
Haberreiter, Margit
Hagyard, Mona
Hale, G.
Hamedivafa, Hashem
Hammer, Reiner
Hannah, Iain
Harvey, John
Hein, Righini
Hejna, Ladislav
HIEI, Eijiro
Hildner, Ernest
Hill, Frank
Hoang, Binh
Hoeksema, Jon
Hotinli, Metin
House, Lewis
Howard, Robert
Hoyng, Peter
HUANG, Guangli
Hudson, Hugh
HWANG, Junga
Illing, Rainer
Ivanov, Evgenij
Jabbar, Sabeh
Jackson, Bernard
Jacobs, Carla

Janssen, Katja
Jefferies, Stuart
Jones, Harrison
Jordan, Carole
Jordan, Stuart
Joshi, G.
Jurčák, Jan
Kalkofen, Wolfgang
Kálmán, Bela
Kaltman, Tatyana
Käpylä, Petri
Karlick, Márian
Karoff, Christoffer
Karpen, Judith
Kasiviswanathan, Sankarasubramanian
Kaufmann, Pierre
Keil, Stephen
Khan, J
Khomenko, Elena
Khumlemlert, Thiranee
KILCIK, Ali
Kim, Iraida
KIM, Yong Cheol
Kitiashvili, Irina
Klein, Karl
Kneer, Franz
Knoelker, Michael
Kolobov, Dmitri
Kołomański, Sylwester
Kononovich, Edvard
Kopylova, Yulia
Kosovichev, Alexander
Kostik, Roman
Kotov, Valery
Kotrc, Pavel
Koutchmy, Serge
Koza, Julius
Kramer, William
Krittinatham, Watcharawuth
Krivova, Natalie
Kryvodubskyj, Valery
Kubičela, Aleksandar
Kučera, Aleš
Kulčár, Ladislav
Kuperus, Max
Kuznetsov, Vladimir
Labrosse, Nicolas
Labs, Dietrich
Landi Degl'Innocenti, Egidio
Landman, Donald
Landolfi, Marco
Lanzafame, Alessandro
Leenaarts, Jorrit
Leibacher, John
Leka, K.D.
Leroy, Jean-Louis
Li, Linghuai
LIN, Jun
LIN, Yuanzhang
Linsky, Jeffrey
Livadiotis, George
Livingston, William
Locke, Jack
Lopez Arroyo, M.
Lopez Fuentes, Marcelo
Loukitcheva, Maria
Lüst, Reimar
Lustig, Guenter
Madjarska, Maria
Makarov, Valentin
MAKITA, Mitsugu
Mandrini, Cristina
Maričić, Darije
Marilli, Ettore
Marmolino, Ciro
Martínez Pillet, Valentin
Matthews, Sarah
Mattig, W.
McAteer, R. T. James
McKenna-Lawlor, Susan
Mein, Pierre
Melrose, Donald
Mendoza-Briceño, César
Mészárosová, Hana
Meyer, Friedrich
Michard, Raymond
Miesch, Mark
Milkey, Robert
Monteiro, Mario Joao
Moore, Ronald
Morabito, David
Moreno-Insertis, Fernando
Morita, Satoshi
MORIYAMA, Fumio
Mouradian, Zadig
Muller, Richard
Munro, Richard
Namba, Osamu
Nandi, Dibyendu
Neckel, Heinz
Nesis, Anastasios
New, Roger
Nicolas, Kenneth
Nikolsky, G.
Nordlund, Aake
Noyes, Robert
Nozawa, Satoshi
Obridko, Vladimir
OH, Suyeon
Önel, Hakan
Ortiz Carbonell, Ada
O'Shea, Eoghan
Ossendrijver, Mathieu
Owocki, Stanley
Padmanabhan, Janardhan
Paletou, Frédéric
Pallé, Pere
Pal'uš, Pavel
Papathanasoglou, Dimitrios
Parenti, Susanna
Parkinson, William
Pasachoff, Jay
Pauluhn, Anuschka
Pecker, Jean-Claude
Petrie, Gordon
Petrov, Nikola
Petrovay, Kristof
Pevtsov, Alexei
Pflug, Klaus
Phillips, Kenneth
Picazzio, Enos
Poquerusse, Michel
Povel, Hanspeter
Priest, Eric
Prokakis, Theodore
QU, Zhongquan
Radick, Richard
Ramelli, Renzo
Raoult, Antoinette
Reardon, Kevin
Rees, David
Reeves, Edmond
Régulo, Clara
Rengel, Miriam
Riehokainen, Aleksandr
Roca Cortés, Teodoro
Roddier, Francois

Roth, Markus
Roudier, Thierry
Rouppe van der Voort, Luc
Rovira, Marta
Rozelot, Jean-Pierre
Rusin, Vojtech
Russell, Alexander
Rutten, Robert
Ryabov, Boris
Rybák, Jan
Rybansky, Milan
Safari, Hossein
SAKAI, Junichi
Samain, Denys
Sánchez Almeida, Jorge
Saniga, Metod
Sasso, Clementina
Sauval, A.
Scherrer, Philip
Schleicher, Helmold
Schmahl, Edward
Schmidt, Wolfgang
Schmieder, Brigitte
Schmitt, Dieter
Schober, Hans
Schou, Jesper
Schuessler, Manfred
Schwartz, Pavol
Schwartz, Steven
Schwenn, Rainer
Seaton, Daniel
Severino, Giuseppe
Shchukina, Nataliia
Sheeley, Neil
Sheminova, Valentina
SHEN, Longxiang
SHIBASAKI, Kiyoto
Shine, Richard
Sigalotti, Leonardo
Sigismondi, Costantino
Sigwarth, Michael
Simon, George
Simon, Guy
Simunac, Kristin
Singh, Jagdev
Sinha, Krishnanand
Sivaraman, Koduvayur
Skumanich, Andrew
Smith, Peter
Sobolev, Vladislav
Solanki, Sami
Soloviev, Alexandr
Spicer, Daniel
Srivastava, Nandita
Stathopoulou, Maria
Staude, Juergen
Steffen, Matthias
Steiner, Oskar
Stenflo, Jan
Štěpán, Jiří
Stepanian, Natali
Stix, Michael
ST John, C.
Stodilka, Myroslav
Straus, Thomas
SUEMATSU, Yoshinori
Suemoto, Zenzaburo
Svalgaard, Leif
Švanda, Michal
TAN, Baolin
Tarashchuk, Vera
Teplitskaya, Raisa
Thomas, John
Tlatov, Andrej
Torelli, M.
trân, Ha
Tripathi, Durgesh
Trujillo Bueno, Javier
Tsiklauri, David
Tsiropoula, Georgia
Usoskin, Ilya
Uus, Undo
Van Doorsselaere, Tom
Van Hoven, Gerard
Vaughan, Arthur
Venkatakrishnan, P.
Ventura, Paolo
Vial, Jean-Claude
Vilmer, Nicole
Voitsekhovska, Anna
Volonte, Sergio
von der Lühe, Oskar
Vukicevic-Karabin, Mirjana
WANG, Haimin
WANG, Jingxiu
WANG, Shujuan
Warren Jr, Wayne
Weiss, Nigel
Wiik Toutain, Jun
Wittmann, Axel
Woehl, Hubertus
Worden, Simon
WU, Hsin-Heng
WU, Linxiang
XIE, Xianchun
XU, Zhi
YOICHIRO, Suzuki
YOSHIMURA, Hirokazu
YOU, Jianqi
Youssef, Nahed
Yun, Hong-Sik
Zampieri, Luca
Zarro, Dominic
Zelenka, Antoine
Zhang, Jie
ZHANG, Jun
ZHANG, Mei
ZHAO, Junwei
ZHOU, Daoqi
Zhugzhda, Yuzef
Zirker, Jack
Zuccarello, Francesca

Division B - Commission 14 Atomic & Molecular Data

President: Lyudmila I. Mashonkina
Vice-President: Farid Salama

Organizing Committee Members:

France Allard
Paul S. Barklem
Helen J. Fraser
Gillian Nave
Peter Beiersdorfer
Hampus Nilsson

Members:

Adelman, Saul
Afram, Nadine
Aggarwal, Kanti
Allard, France
Allard, Nicole
Allende Prieto, Carlos
Allen Jr, John
Arduini-Malinovsky, Monique
Arion, Douglas
Artru, Marie-Christine
Babcock, H.
Baird, K.
Balanca, Christian
Banerjee, Dipankar
Barber, Robert
Barklem, Paul
Barnbaum, Cecilia
Barrell, H.
Bartaya, R.
Bautista, Manuel
Bayet, Estelle
Behar, Ehud
Beiersdorfer, Peter
Bely-Dubau, Francoise
Berrington, Keith
Bhardwaj, Anil
Biémont, Emile
Black, John
Blackwell-Whitehead, Richard
Bodewits, Dennis
Boechat-Roberty, Heloisa
Bommier, Veronique
Branscomb, L.
Brault, James
Bromage, Gordon
Buchlin, Eric
Carroll, P.
Casasola, Viviana
cazaux, Stephanie
Chance, Kelly
CHEN, Guoming
CHEN, Huei-Ru
Christlieb, Norbert
Cichowolski, Silvina
Corliss, C.
Cornille, Marguerite
Czyzak, Stanley
Dalgarno, Alexander
Davis, Sumner
de Frees, Douglas
de Kertanguy, Amaury
Delsemme, Armand
Depagne, Éric
Désesquelles, Jean
d'Hendecourt, Louis
Diercksen, Geerd
Dimitrijevic, Milan
Dubau, Jacques
Dufay, Maurice
Dulieu, Francois
Eidelsberg, Michele
Epstein, Gabriel
Faucher, Paul
Feautrier, Nicole
Federici, Luciana
Federman, Steven
Filacchione, Gianrico
Fillion, Jean-Hugues
Fink, Uwe
Flower, David
Fluri, Dominique
Fraser, Helen
Fuhr, Jeffrey
Gabriel, Alan
Gallagher III, John
Gargaud, Muriel
Garrido, César
Glagolevskij, Yurij
Glinski, Robert
Glover, Simon
Goddi, Ciriaco
Goldbach, Claudine
Graae Jørgensen, Aleksandra
Grant, Ian
Grevesse, Nicolas
Harrison, Dean
Hartman, Henrik
Heiter, Ulrike
Henning, Thomas
Hesser, James
Hillier, John
Hoang, Binh
Homeier, Derek
Horácek, Jiri
Hörandel, Jörg
House, Lewis
Huber, Martin
Huebner, Walter
Ignjatovi'c, Ljubinko
Iliev, Ilian
Irwin, Alan
Irwin, Patrick
Jamar, Claude
Joblin, Christine
Johnson, Donald

Johnson, Fred
Joly, Francois
Jordan, Carole
Jørgensen, Uffe
Jorissen, Alain
Kamp, Inga
Kanekar, Nissim
Kaňuchová, Zuzana
KATO, Takako
Kennedy, Eugene
Kerber, Florian
Kielkopf, John
Kim, Joo Hyeon
KIM, Sang Joon
KIM, Zong
Kingston, Arthur
Kipper, Tonu
Kirby, Kate
Kohl, John
Kramida, Alexander
Kroto, Harold
KUAN, Yi-Jehng
Kupka, Friedrich
Kurucz, Robert
Lambert, David
Landman, Donald
Lang, James
Langhoff, Stephanie
Launay, Françoise
Launay, Jean-Michel
Lawrence, G.
Layzer, David
Leach, Sydney
Le Bourlot, Jacques
Le Floch, André
Leger, Alain
Lemaire, Jean-louis
LIANG, Guiyun
Linnartz, Harold
Littlefield, T.
LIU, Sheng-Yuan
Lo, Wing-Chi Nadia
Lobel, Alex
Loulergue, Michelle
Lovas, Francis
Lucero, Danielle
Lutz, Barry
Maillard, Jean-Pierre
Mandel'shtam, S.
Mardones, Diego
Martín, Sergio
Mashonkina, Lyudmila
Mason, Helen
McCall, Benjamin
McWhirter, R.
Meggers, W.
Mickelson, Michael
Mihajlov, Anatolij
Mookerjea, Bhaswati
Moore-Sitterly, Ch.
Morris, Patrick
Morton, Donald
Mumma, Michael
Nahar, Sultana
Nave, Gillian
Newsom, Gerald
Nguyễn, Phuong
Nicholls, Ralph
Nielsen, Krister
Nilsson, Hampus
Nollez, Gerard
Nussbaumer, Harry
O'Brian, Thomas
Oetken, L.
Oka, Takeshi
Omont, Alain
Orton, Glenn
OTSUKA, Masaaki
OZEKI, Hiroyuki
Palmeri, Patrick
PARK, Yong Sun
Parkinson, William
Paron, Sergio
Peach, Gillian
PEI, Chunchuan
Petrini, Daniel
Pettini, Marco
Piacentini, Ruben
Pilling, Sergio
Pintado, Olga
Pradhan, Anil
Querci, Francois
Quinet, Pascal
Ralchenko, Yuri
Ramírez, Jose
Rastogi, Shantanu
Redman, Matthew
Redman, Stephen
Rogers, Forrest
Roncin, Jean-Yves
Ross, John
Rostas, François
Roueff, Evelyne
Ruder, Hanns
Ryabchikova, Tatiana
Sahal-Bréchot, Sylvie
Salama, Farid
Sarre, Peter
Savanov, Igor
Savin, Daniel
Schrijver, Johannes
Schultz, David
Semenov, Dmitry
Sharp, Christopher
SHI, Jianrong
Shore, Bruce
Simić, Zoran
Sinha, Krishnanand
Smith, Peter
Smith, William
Somerville, William
SONG, In-Ok
Sorensen, Gunnar
Spielfiedel, Annie
Stancil, Philip
Stark, Glenn
Stehle, Chantal
ST John, C.
Strachan, Leonard
Strelnitski, Vladimir
Summers, Hugh
Sutherland, Ralph
Swings, Jean-Pierre
TAKAYANAGI, Kazuo
Tatum, Jeremy
Tayal, Swaraj
Tchang-Brillet, Lydia
Tennyson, Jonathan
Thorne, Anne
Tousey, Richard
Tozzi, Gian
TranMinh, Nguyet
Trefftz, Eleonore
Tripathi, Durgesh
Ulyanov, Oleg
van Dishoeck, Ewine
van Rensbergen, Walter
Varshalovich, Dmitrij
Vasta, Magda
Vavrek, Roland

Vidali, Gianfranco
Völk, Heinrich
Volonte, Sergio
Vujnovic, Vladis
Wahlgren, Glenn
Wakelam, Valentine
WANG, Feilu
WANG, Junxian
WANG, Junzhi
Wiese, Wolfgang
Wunner, Guenter
XIAO, Dong
YANG, Changgen
Yoshino, Kouichi
Young, Louise
Yu, Yan
Zeippen, Claude
ZENG, Qin
ZHAO, Gang

Division F - Commission 15 Physical Study of Comets & Minor Planets

President: Dominique Bockelée-Morvan
Vice-President: Ricardo Alfredo Gil-Hutton

Organizing Committee Members:

Irina N. Belskaya
Hideyo KAWAKITA
Javier Licandro
Hajime YANO
Björn J.R. Davidsson
Elisabetta Dotto
Thais Mothé-Diniz
Diane H. Wooden
Alan Fitzsimmons

Members:

ABE, Shinsuke
AGATA, Hidehiko
A'Hearn, Michael
ALIBERT, Yann
Allegre, Claude
Altwegg, Kathrin
Anders, Edward
Angeli, Claudia
Archinal, Brent
Arpigny, Claude
Babadjanov, Pulat
Bailey, Mark
Baldet, F.
Barabanov, Sergey
Baransky, Olexander
Barber, Robert
Barker, Edwin
Bar-Nun, Akiva
Barriot, Jean-Pierre
Barucci, Maria
Bear, Ealeal
Bell, Jeffrey
Belskaya, Irina
Belton, Michael
Bemporad, Alessandro
Bendjoya, Philippe
Benkhoff, Johannes
Bhardwaj, Anil
Bingham, Robert
Binzel, Richard
Birch, Peter
Birlan, Mirel
Biver, Nicolas
Blamont, Jacques-Emile
Blanco, Armando
Bobrovnikoff, Nicholas
Bockelee-Morvan, Dominique
Bodewits, Dennis
Boehnhardt, Hermann
Boice, Daniel
Bonev, Tanyu
Borisov, Borislav
Borisov, Galin
Borysenko, Serhii
Bowell, Edward
Brandt, John
Brecher, Aviva
Britt, Daniel
Brown, Robert
Brownlee, Donald
Brozovic, Marina
Brunk, William
Bueno de Camargo, Julio
Buie, Marc
Buratti, Bonnie
Burlaga, Leonard
Burns, Joseph
Busarev, Vladimir
Butler, Bryan
Campins, Humberto
Campo Bagatin, Adriano
Capaccioni, Fabrizio
Capria, Maria
Carruba, Valerio
Carruthers, George
Carsenty, Uri
Carusi, Andrea
Carvano, Jorge
Cellino, Alberto
Cerroni, Priscilla

Chandrasekhar, Thyagarajan
Chapman, Clark
Chapman, Robert
CHEN, Daohan
Chubko, Larysa
Clairemidi, Jacques
Clayton, Donald
Clayton, Geoffrey
Clube, S.
Cochran, Anita
Cochran, William
Colom, Pierre
Combi, Michael
Connolly, Harold
Connors, Martin
Conrad, Albert
Consolmagno, Guy
Cosmovici, Cristiano
Cottin, Hervé
Cremonese, Gabriele
Crovisier, Jacques
Cruikshank, Dale
Cuypers, Jan
Danks, Anthony
da Silveira, Enio
Davidsson, Björn
Davies, John
de Almeida, Amaury
Debehogne, Henri
Delbo, Marco
Dell' Oro, Aldo
Delsanti, Audrey
Delsemme, Armand
de Pater, Imke
Dermott, Stanley
de Sanctis, Giovanni
de Sanctis, Maria
Deutschman, William
de Val-Borro, Miguel
Di Martino, Mario
Dobrovolsky, Oleg
Doressoundiram, Alain
Dossin, F.
Dotto, Elisabetta
Dryer, Murray
Duffard, Rene
Duncan, Martin
Durech, Josef
Dzhapiashvili, Victor
Encrenaz, Therese
Erard, Stéphane
Ershkovich, Alexander
Eviatar, Aharon
Farnham, Tony
Fechtig, Hugo
Feldman, Paul
Fernández, Julio
Fernández, Yanga
Ferrin, Ignacio
Filacchione, Gianrico
Fitzsimmons, Alan
Fletcher, Leigh
Fornasier, Sonia
Forti, Giuseppe
Foryta, Dietmar
Fouchard, Marc
Fraser, Helen
Froeschle, Christiane
FUJIWARA, Akira
Fulchignoni, Marcello
FURUSHO, Reiko
Gajdoš, Stefan
Galád, Adrián
Gammelgaard, Peter
GAO, Jian
Geiss, Johannes
Gerakines, Perry
Gerard, Eric
Gibson, James
Gil-Hutton, Ricardo
Giovane, Frank
Gounelle, Matthieu
Gradie, Jonathan
Grady, Monica
Granvik, Mikael
Green, Daniel
Green, Simon
Greenberg, Richard
Gronkowski, Piotr
Grossman, Lawrence
Grün, Eberhard
Grundy, William
Gustafson, Bo
Hadamcik, Edith
Halliday, Ian
Hanner, Martha
Hapke, Bruce
Harris, Alan
Hartmann, William
Harwit, Martin
HASEGAWA, Sunao
Haser, Leo
Haupt, Hermann
Howell, Ellen
Hsieh, Henry
Huebner, Walter
Hughes, David
Huntress, Wesley
Husárik, Marek
Ibadinov, Khursand
Ibadov, Subhon
IP, Wing-Huen
Irvine, William
Irwin, Patrick
Israelevich, Peter
Ivanova, Oleksandra
Ivanova, Violeta
Ivezic, Zeljko
Jackson, William
Jakubík, Marian
Jedicke, Robert
Jehin, Emmanuel
Jockers, Klaus
Johnson, Torrence
Jorda, Laurent
Kaasalainen, Mikko
Käufl, Hans Ulrich
Kaňuchová, Zuzana
Karatekin, Özgür
Kavelaars, JJ.
KAWAKITA, Hideyo
Kaydash, Vadym
Keay, Colin
Keil, Klaus
Keller, Horst
Kidger, Mark
Kim, Joo Hyeon
KING, Sun-Kun
Kiselev, Nikolai
Kiss, Csaba
Klacka, Jozef
Knacke, Roger
Knežević, Zoran
Knight, Matthew
Koeberl, Christian
Kohoutek, Lubos
Kornoš, Leonard
Korokhin, Viktor
Korsun, Pavlo

Koschny, Detlef
Koshkin, Nikolay
KOZASA, Takashi
Krimigis, Stamatios
Krishna, Swamy
Kristensen, Leif
Krugly, Yurij
Kryszczynska, Agnieszka
KUAN, Yi-Jehng
Küppers, Michael
Lagerkvist, Claes-Ingvar
Lamy, Philippe
Lancaster, Brown
Lane, Arthur
Lara, Luisa
Larson, Harold
Larson, Stephen
La Spina, Alessandra
Laufer, Diana
Lazzarin, Monica
Lazzaro, Daniela
Lebofsky, Larry
LEE, Thyphoon
Levasseur-Regourd, Anny-Chantal
Levin, B.
Li, Jian-Yang
Licandro, Javier
Liller, William
Lillie, Charles
Lindsey, Charles
Lipschutz, Michael
Lissauer, Jack
Lisse, Carey
LIU, Xiaoqun
Lo Curto, Gaspare
Lodders, Katharina
Lopes, Rosaly
Lukyanyk, Igor
Lutz, Barry
Luu, Jane
Lyon, Ian
Magee-Sauer, Karen
Magnusson, Per
Mainzer, Amy
Makalkin, Andrei
Malaise, Daniel
Maran, Stephen
Marchi, Simone
Marcialis, Robert
Marciniak, Anna
Maris, Michele
Marzari, Francesco
Masiero, Joseph
Matson, Dennis
Matsuura, Oscar
Mazzotta Epifani, Elena
McCord, Thomas
McDonnell, J.
McFadden, Lucy
McKellar, A.
McKenna-Lawlor, Susan
Meech, Karen
Meisel, David
Mendillo, Michael
Mendis, Devamitta
Merline, William
Michałowski, Tadeusz
Michel, Patrick
Milani Comparetti, Andrea
Milet, Bernard
Millis, Robert
Moehlmann, Diedrich
MOON, Hong-Kyu
Moore, Elliott
Morrison, David
Mothé-Diniz, Thais
Mousis, Olivier
Mueller, Thomas
Muinonen, Karri
MUKAI, Tadashi
Mumma, Michael
NAKAMURA, Akiko
NAKAMURA, Tsuko
Napier, William
Nedelcu, Dan
Neff, John
Newburn Jr, Ray
Niedner, Malcolm
Ninkov, Zoran
Nolan, Michael
Noll, Keith
Novaković, Bojan
O'Dell, Charles
Ortiz, Jose
Pál, András
Paolicchi, Paolo
Parisot, Jean-Paul
Peixinho, Nuno
Pellas, Paul
Pendleton, Yvonne
Perez de Tejada, Hector
Picazzio, Enos
Piironen, Jukka
Pilcher, Carl
Pittich, Eduard
Pittichova, Jana
Prialnik, Dina
Proisy, Paul
Remy Battiau, Liliane
Reyes-Ruiz, Mauricio
Richardson, Derek
Rickman, Hans
Roemer, Elizabeth
Roig, Fernando
Rosenbush, Vera
Rossi, Alessandro
Rousselot, Philippe
Russel, Sara
Russell, Kenneth
Sagdeev, Roald
Salitis, Antonijs
Samarasinha, Nalin
Santos-Sanz, Pablo
Sasaki, Sho
SATO, Isao
Scaltriti, Franco
Schaller, Emily
Scheirich, Peter
Schleicher, David
Schloerb, F.
Schmidt, Maarten
Schober, Hans
Scholl, Hans
Sekanina, Zdenek
SEKIGUCHI, Tomohiko
Sen, Asoke
SEO, Haingja
Serra Ricart, Miquel
Shanklin, Jonathan
Sharma, A.
Sharp, Christopher
Shevchenko, Vasilij
SHIMIZU, Mikio
Shor, Viktor
Shoyoqubov, Shoayub
Shulman, Leonid
Sickafoose, Amanda
Simonia, Irakli
Sims, Mark

Sivaraman, Koduvayur
Sizonenko, Yuri
Smith, Bradford
Snyder, Lewis
SOFIA LYKAWKA, Patryk
Solc, Martin
Sosa, Andrea
Spinrad, Hyron
Steel, Duncan
Stern, S.
Stewart-Mukhopadhyay, Sarah
Surdej, Jean
Svoren, Jan
Swade, Daryl
Sykes, Mark
Szego, Karoly
Székely, Péter
Szutowicz, Sławomira
Tacconi-Garman, Lowell
TAKEDA, Hidenori
TANABE, Hiroyoshi
Tancredi, Gonzalo
Tanga, Paolo
TAO, Jun
Tarashchuk, Vera
Tatum, Jeremy
Tedesco, Edward
Terentjeva, Alexandra
Tholen, David
Thomas, Nicolas
TIAN, Feng
Tiscareno, Matthew
TOMITA, Koichiro
Tosi, Federico
Toth, Imre
Tozzi, Gian
Trujillo, Chadwick
Ugolnikov, Oleg
Valdés-Sada, Pedro
van Flandern, Tom
Vázquez, Roberto
Veeder, Glenn
Velikodsky, Yuri
Veverka, Joseph
Vilas, Faith
Voelzke, Marcos
Wallis, Max
WANG, Shiang-Yu
WANG, Xiao-bin
Wasson, John
WATANABE, Junichi
Wdowiak, Thomas
Weaver, Harold
Wehinger, Peter
Weidenschilling, S.
Weissman, Paul
Wells, Eddie
West, Richard
Wetherill, George
Wilkening, Laurel
Williams, Iwan
Wilson, Lionel
Womack, Maria
Wooden, Diane
Woolfson, Michael
Wurm, K.
Wyckoff, Susan
YABUSHITA, Shin
YANAGISAWA, Masahisa
YANG, Jongmann
YANG, Xiaohu
YANO, Hajime
Yavnel, Alexander
Yeomans, Donald
YI, Yu
YOSHIDA, Fumi
Zappalà, Vincenzo
Zarnecki, John
Zellner, Benjamin
ZHANG, Jun
ZHANG, Xiaoxiang
ZHAO, Haibin
ZHU, Jin

Division E - Commission 16 Physical Study of Planets & Satellites

President: Mark T. Lemmon

Organizing Committee Members:

Sang Joon KIM
Leonid V. Ksanfomality
Luisa M. Lara
David Morrison
Victor G. Tejfel
Padma A. Yanamandra-Fisher

Members:

AKABANE, Tokuhide
Akimov, Leonid
Alexandrov, Yuri
ALIBERT, Yann
Allison, Michael
Anders, Edward
Appleby, John
Archinal, Brent
Arthur, David
Atkinson, David
Atreya, Sushil
Balikhin, Michael

Barkin, Yuri
Barlow, Nadine
Barrow, Colin
Bartczak, Przemysław
Battaner, Eduardo
Bazilevsky, Alexandr
Beebe, Reta
Beer, Reinhard
Bell III, James
Belton, Michael
Bender, Peter
Ben-Jaffel, Lofti
Benkhoff, Johannes
Berge, Glenn
Bergstralh, Jay
Bertaux, Jean-Loup
Bezard, Bruno
Bhardwaj, Anil
Bhatia, R.
Billebaud, Francoise
Binzel, Richard
Blamont, Jacques-Emile
Blanco, Armando
Bobrov, M.
Bodewits, Dennis
Bondarenko, Lyudmila
Borisov, Borislav
Bosma, Pieter
Boss, Alan
Boyce, Peter
Brahic, André
Brecher, Aviva
Broadfoot, A.
Brown, Robert
Brunk, William
Buie, Marc
Buratti, Bonnie
Burba, George
Burns, Joseph
Busarev, Vladimir
Caldwell, John
Cameron, Winifred
Campbell, Donald
Campo Bagatin, Adriano
Capria, Maria
Carsmaru, Maria
Catalano, Santo
Cecconi, Baptiste
Chakrabarti, Supriya
Chapman, Clark
Chen, Dao-Han
CHEN, Daohan
Chevrel, Serge
Chung, Eduardo
Clairemidi, Jacques
Cochran, Anita
Combi, Michael
Connes, Janine
Cottin, Hervé
Cottini, Valeria
Coustenis, Athena
Cruikshank, Dale
Davies, Ashley
Davis, Gary
de Bergh, Catherine
DELEUIL, Magali
Demory, Brice-Olivier
DeNisco, Kenneth
de Pater, Imke
Dermott, Stanley
de Val-Borro, Miguel
Dickel, John
Dickey, Jean
Dlugach, Zhanna
Doressoundiram, Alain
Drake, Frank
Drossart, Pierre
Duffard, Rene
(Dunkin) Beardsley, Sarah
Durrance, Samuel
Dzhapiashvili, Victor
Ehrenreich, David
El- Baz, Farouk
Elston, Wolfgang
Encrenaz, Therese
Epishev, Vitali
Eshleman, Von
Esposito, Larry
Evans, Michael
Ferrari, Cécile
Ferrusca, Daniel
Feulner, Georg
Fielder, Gilbert
Filacchione, Gianrico
Fink, Uwe
Fletcher, Leigh
Fox, W.
FUJIWARA, Akira
Gautier, Daniel
Geiss, Johannes
Gerard, Jean-Claude
Giclas, Henry
Gierasch, Peter
Gillon, MichaÃ«l
Goldreich, Peter
Goldstein, Richard
Goody, Richard
Gorenstein, Paul
Gor'kavyi, Nikolai
Goudas, Constantine
Gounelle, Matthieu
Grav, Tommy
Green, Jack
Grieger, Bjoern
Grossman, Lawrence
Gulkis, Samuel
Gurshtein, Alexander
Hall, J.
Halliday, Ian
Hammel, Heidi
Hänninen, Jyrki
Harris, Alan
Harris, Alan
HASEGAWA, Ichiro
Helled, Ravit
Heng, Kevin
HERSANT, Franck
Hide, Raymond
Holberg, Jay
Horedt, Georg
Hovenier, J.
Hubbard, William
Hunt, Garry
HWANG, Junga
Irvine, William
Irwin, Patrick
IWASAKI, Kyosuke
Jin, Liping
Johnson, Torrence
Jordán, Andrés
Jurgens, Raymond
Käufl, Hans Ulrich
Kascheev, Rafael
Killen, Rosemary
Kim, Joo Hyeon
Kim, Yoo Jea
KIM, Sang Joon
KIM, Yongha
Kley, Wilhelm
Kostama, Veli-Petri

Kozak, Lyudmyla
Kraft, Ralph
Krimigis, Stamatios
Ksanfomality, Leonid
Ksanfomality, Leonid
Kuiper, G.
Kumar, Shiv
Kurt, Vladimir
Kuzmin, Arkadij
Lane, Arthur
Larson, Harold
Larson, Stephen
Laufer, Diana
LEE, Jun
Lellouch, Emmanuel
Lemmon, Mark
Levin, B.
Lewis, John
Li, Jian-Yang
Licandro, Javier
Lichtenegger, Herbert
Lineweaver, Charles
Lissauer, Jack
LIU, Chengzhi
LIU, Xiaoqun
Lockwood, G.
Lo Curto, Gaspare
Lodders, Katharina
Lopes, Rosaly
Lopez Moreno, Jose
Lopez Puertas, Manuel
Lopez Valverde, M.
Lutz, Barry
Luz, David
Lyon, Ian
Lyot, Bernard
Maggini, M.
Mainzer, Amy
Makalkin, Andrei
Marchi, Simone
Marcialis, Robert
Margot, Jean-Luc
Marov, Mikhail
Martinez-Frias, Jesus
Marzo, Giuseppe
Masursky, Harold
Matson, Dennis
MATSUI, Takafumi
Mazzotta Epifani, Elena
McCord, Thomas
McCullough, Peter
McElroy, Michael
McGrath, Melissa
McKinnon, William
Meadows, A.
Mendillo, Michael
Mickelson, Michael
Mikhail, Joseph
Millis, Robert
Mills, Franklin
Miyamoto, Syotaro
MIYAMOTO, Sigenori
Moehlmann, Diedrich
Molina, Antonio
Moncuquet, Michel
Montmessin, Franck
Moreno-Insertis, Fernando
Morozhenko, A.
Morrison, David
Mosser, Benoît
Mousis, Olivier
Mumma, Michael
Murphy, Robert
NAKAGAWA, Yoshitsugu
Nelson, Richard
Ness, Norman
Nixon, Conor
Noll, Keith
OHTSUKI, Keiji
Ortiz, Jose
Owen, Tobias
Pang, Kevin
Paolicchi, Paolo
PARK, Yong Sun
Pascu, Dan
Peixinho, Nuno
Perez, Mario
Petit, Jean-Marc
Pettengill, Gordon
PING, Jinsong
Pokorny, Zdenek
Politi, Romolo
Potter, Andrew
Psaryov, Volodymyr
Quanz, Sascha
Rao, M.
Rev. Phillips, T.
Rodionova, Zhanna
Rodrigo, Rafael
Roos-Serote, Maarten
Roques, Françoise
Rosenbush, Vera
Rossi, Alessandro
Runcorn, S.
Ruskol, Evgeniya
Russo, Pedro
Saissac, Joseph
Sampson, Russell
Sanchez-Lavega, Agustin
Sasaki, Sho
Schaller, Emily
Scheirich, Peter
Schleicher, David
Schloerb, F.
Schneider, Nicholas
SEO, Haingja
Sergis, Nick
Shapiro, Irwin
Sharonov, V.
Shevchenko, Vladislav
SHI, Huli
SHIMIZU, Mikio
SHIMIZU, Tsutomu
Shkuratov, Yurii
Sicardy, Bruno
Sims, Mark
Sjogren, William
Smith, Bradford
Snellen, Ignas
Soderblom, Larry
SOFIA LYKAWKA, Patryk
Sonett, Charles
Sprague, Ann
Stallard, Thomas
Stam, Daphne
Stern, S.
Sterzik, Michael
Stewart-Mukhopadhyay, Sarah
Stoev, Alexey
Stone, Edward
Strobel, Darrell
Strom, Robert
Synnott, Stephen
Tanga, Paolo
Taylor, Fredric
Tchouikova, Nadezhda
Tedds, Jonathan
Tejfel, Victor

Terrile, Richard
Tholen, David
Thomas, Nicolas
TIAN, Feng
Tiscareno, Matthew
Tosi, Federico
Trafton, Laurence
Trujillo, Chadwick
Tyler Jr, G.
van Flandern, Tom
Veiga, Carlos
Veverka, Joseph
Vidmachenko, Anatoliy
Walker, Alta
Walker, Simon
Wallace, Lloyd
Wannawichian, Suwicha
Wasserman, Lawrence
Wasson, John
Weidenschilling, S.
Wells, Eddie
Wetherill, George
Whitaker, Ewen
Wildt, R.
Williams, Iwan
Williams, James
Woolfson, Michael
Wu, Yanqin
Wurz, Peter
Yair, Yoav
Yanamandra-Fisher, Padma
YANG, Xiaohu
YI, Yu
Yoder, Charles
Young, Andrew
Young, Louise
Zarka, Philippe
Zhang, Tielong
ZHANG, Jun
ZHANG, Mian
Zharkov, Vladimir

Division A - Commission 19 Rotation of the Earth

President: Cheng-Li Huang
Vice-President: Richard Gross

Organizing Committee Members:

Vladimir E. Zharov
Daniela Thaller (Germany)
Harald Schuh
Oleg A. Titov
Zinovy M. Malkin
Bernd Richter
Benjamin F. Chao
David A. Salstein
Wiesław Kosek
Christian Bizouard

Members:

Abraham, Henry
Arabelos, Dimitrios
Archinal, Brent
Arias, Elisa
BANG, Yong
Banni, Aldo
Barkin, Yuri
Barlier, François
Bartlett, Jennifer
Beutler, Gerhard
Bizouard, Christian
Boehm, Johannes
Bolotin, Sergei
Bolotina, Olga
Boucher, Claude
Bougeard, Mireille
Bourda, Geraldine
Boytel, Jorge
Brentjens, Michiel
Brosche, Peter
Brzeziński, Aleksander
Capitaine, Nicole
Cazenave, Anny
Cecchini, G.
Chao, Benjamin
Chapanov, Yavor
CHEN, Wen Ping
Chiumiento, Giuseppe
Damljanovic, Goran
Débarbat, Suzanne
De Biasi, Maria
Defraigne, Pascale
Dehant, Véronique
Dejaiffe, Rene
Deleflie, Florent
de Viron, Olivier
Dick, Wolfgang
Dickey, Jean
Dickman, Steven
DU, Lan
El Shahawy, Mohamad
Enslin, Heinz
Eppelbaum, Lev
Escapa, Alberto
Feissel, Martine
FENG, Chugang
Fernández, Laura
Ferrándiz, Jose
Fliegel, Henry
Folgueira, Marta
FUJISHITA, Mitsumi
FUKUSHIMA, Toshio
Gambis, Daniel
GAO, Buxi

Gaposchkin, Edward
Gayazov, Iskander
Getino Fernández, Juan
Gozhy, Adam
Gross, Richard
Groten, Erwin
Guinot, Bernard
Gusev, Alexander
HAAS, RÜDIGER
HAN, Tianqi
HAN, Yanben
Hefty, Jan
Hobiger, Thomas
Huang, Cheng-Li
HUANG, Cheng
Hugentobler, Urs
IIJIMA, Shigetaka
Jäggi, Adrian
JIN, WenJing
Johnson, Thomas
Jubier, Xavier
KAKUTA, Chuichi
KAKUTA, Chuichi
KAMEYA, Osamu
Khoda, Oleg
Kimura, H.
Klepczynski, William
Knowles, Stephen
Kołaczek, Barbara
Korsun, Alla
Kosek, Wiesław
Kosteleck, Jan
Kouba, Jan
LEE, Jun
Lehmann, Marek
LI, Jinling
LI, Yong
LI, Zhengxin
LI, Zhian
LIAO, Dechun
Lieske, Jay
LIU, Chengzhi
LIU, Ciyuan
LUO, Dingjiang
LUO, Shi-Fang
Luzum, Brian
Ma, Chopo
MA, Lihua
Malkin, Zinovy
MANABE, Seiji
Markowitz, William
Marshalov, Dmitriy
MASAKI, Yoshimitsu
McCarthy, Dennis
Meinig, Manfred
Melbourne, William
Merriam, James
Milovanović, V.
Minazzoli, Olivier
Mironov, Nikolay
Monet, Alice
Morgan, Peter
Morrison, Leslie
Mueller, Ivan
Müller, Jürgen
Mysen, Eirik
Nastula, Jolanta
Naumov, Vitalij
Navarro, Julio
Newhall, X.
Niemi, Aimo
Nilsson, Tobias
Nothnagel, Axel
OOE, Masatsugu
Panafidina, Natalia
Paquet, Paul
PARK, Pilho
Pejović, Nadezda
Pešek, Ivan
Petit, Gérard
Petrov, Sergey
Picca, Domenico
Pilkington, John
Poma, Angelo
Popelar, Josef
Proverbio, Edoardo
Pugliano, Antonio
Ray, James
Richter, Bernd
Robertson, Douglas
Rochester, Michael
Rogister, Yves
Ron, Cyril
Roosbeek, Fabian
Rothacher, Markus
Ruder, Hanns
Rusu, I.
Rykhlova, Lidiya
Salstein, David
Sanchez, Manuel
Sansaturio, Maria
SASAO, Tetsuo
SATO, Koichi
Schillak, Stanisław
Schreiber, Karl
Schuh, Harald
Schutz, Bob
Seitz, Florian
SEKIGUCHI, Naosuke
Sevilla, Miguel
Seyed-Mahmoud, Behnam
Shapiro, Irwin
Shelus, Peter
SHI, Huli
SHU, Fengchun
Shuygina, Nadia
Sidore'nkov, Nikolay
Skurikhina, Elena
Soffel, Michael
Sollenberger, P.
Souchay, Jean
Steigenberger, Peter
Stephenson, F.
SUN, Fuping
Surkis, Igor
Tanner, Richard
Tapley, Byron
Tarady, Vladimir
Teleki, George
Thaller, Daniela
Thomas, Maik
Titov, Oleg
TSAO, Mo
Veillet, Christian
Vicente, Raimundo
Vondrák, Jan
Wallace, Patrick
WANG, Guangli
WANG, Kemin
WANG, Xiaoya
WANG, Zhengming
Weber, Robert
WEI, Erhu
Williams, James
Wilson, P.
Winkler, Gernot
Wooden, William
WU, Bin
WU, Shouxian
Wünsch, Johann

XIAO, Naiyuan
XU, Jiayan
XU, Tong-Qi
YANG, Fumin
Yatskiv, Yaroslav
YE, Shuhua
YOKOYAMA, Koichi
YU, Nanhua
ZHANG, Guo-Dong
ZHANG, Zhongping
ZHAO, Ming
Zharov, Vladimir
ZHENG, Yong
ZHONG, Min
ZHOU, Yonghong
ZHU, Yaozhong
ZHU, Yonghe

Division F - Commission 20 Positions & Motions of Minor Planets, Comets & Satellites

President: Steven R. Chesley
Vice-President: Daniela Lazzaro

Organizing Committee Members:

Robert Jedicke
Mikael Granvik
Shinsuke ABE
Jana Tichá
Timothy B. Spahr
Jin ZHU
Alan C. Gilmore
Petr Pravec

Members:

Abalakin, Viktor
ABE, Shinsuke
A'Hearn, Michael
Aikman, G.
Aksnes, Kaare
Arlot, Jean-Eudes
Babadjanov, Pulat
Baggaley, William
Bailey, Mark
Baransky, Olexander
Bec-Borsenberger, Annick
Behrend, Raoul
Benest, Daniel
Benkhoff, Johannes
Bernardi, Fabrizio
Berthier, JerÃ´me
Bien, Reinhold
Blanco, Carlo
Boerngen, Freimut
Borisov, Borislav
Borisov, Galin
Bowell, Edward
Branham, Richard
Brouwer, D.
Bueno de Camargo, Julio
Burns, Joseph
Candy, Michael
Carpino, Mario
Carusi, Andrea
Chang, Y.
Chapront-Touze, Michelle
Chesley, Steven
CHIO, Chol
Chodas, Paul
Churyumov, Klim
Cooper, Nicholas
DARHMAOUI, Hassane
Debehogne, Henri
Delbo, Marco
Delsemme, Armand
de Sanctis, Giovanni
Di Sisto, Romina
Donnison, John
Dourneau, Gerard
Doval, Jorge M.
Dunham, David
Dvorak, Rudolf
Dybczyński, Piotr
Edelman, Colette
Elst, Eric
Emelianov, Nikolaj
Emel'yanenko, Vacheslav
Epishev, Vitali
Evans, Michael
Fernández, Julio
Fernández, Yanga
Ferraz-Mello, Sylvio
Ferreri, Walter
Fors, Octavi
Forti, Giuseppe
Franklin, Fred
Fraser, Brian
Froeschle, Claude
FUSE, Tetsuharu
Gibson, James
Gilmore, Alan
Giorgini, Jon
Gomez, Edward
Gorshanov, Denis
Granvik, Mikael
Green, Daniel
Greenberg, Richard
Hahn, Gerhard
Hainaut, Olivier
Harper, David
Harris, Alan
HASEGAWA, Ichiro
Haupt, Hermann

HE, Miao-fu
Helin, Eleanor
Hemenway, Paul
Herget, Paul
Heudier, Jean-Louis
Hirose, Hideo
Hönig, Sebastian
Hol, Pedro
HOU, Xiyun
Hsieh, Henry
Hudkova, Ludmila
Hurnik, Hieronim
HURUKAWA, Kiichirō
Husárik, Marek
Ianna, Philip
Ivanova, Violeta
Ivantsov, Anatoliy
Izmailov, Igor
Jacobson, Robert
Jakubík, Marian
Jedicke, Robert
Kablak, Nataliya
Kazantsev, Anatolii
Kazimirchak-Polonskaya, E.
Khrutskaya, Evgenia
Kilmartin, Pamela
KIM, Sang Joon
KINOSHITA, Hiroshi
Kisseleva, Tamara
Klemola, Arnold
Knežević, Zoran
Knight, Matthew
Kohoutek, Lubos
KOSAI, Hiroki
Koschny, Detlef
Koshkin, Nikolay
KOZAI, Yoshihide
Kristensen, Leif
Królikowska-Sołtan, Małgorzata
Krugly, Yurij
Kulikova, Nelly
Lagerkvist, Claes-Ingvar
Larsen, Jeffrey
Laurin, Denis
Lazzaro, Daniela
Lemaître, Anne
Leuschner, A.
Li, Jian-Yang
LI, Guangyu
LI, Yong
Lieske, Jay
Lomb, Nicholas
Lovas, Miklos
MA, Guanyi
Mainzer, Amy
Makover, S.
Manara, Alessandro
Mancìni, Dario
Mardsen, Brian
Martins, Roberto
Matese, John
Maury, Alain
McMillan, Robert
McNaught, Robert
Medvedev, Yurij
Melita, Mario
Milet, Bernard
Millis, Robert
Mintz Blanco, Betty
Monet, Alice
Moore-Weiss, John
Morando, Bruno
Moravec, Zdeněk
Morris, Charles
Murray, Carl
Nacozy, Paul
NAKAMURA, Tsuko
NAKANO, Syuichi
Nedelcu, Dan
Neslusan, Lubos
Nguyễn, Phuong
Nobili, Anna
Novaković, Bojan
Noyelles, Benoît
Olive, Don
Owen Jr, William
Oezisik, Tuncay
Page, Gary
Pandey, A.
Pascu, Dan
Pat-El, Igal
Pauwels, Thierry
Perozzi, Ettore
Pierce, David
PING, Jinsong
Pittich, Eduard
Polyakhova, Elena
Porubcan, Vladimir
Pozhalova, Zhanna
Pravec, Petr
QIAO, Rongchuan
QINCHANG, Lin
Rajamohan, R.
Raju, Vasundhara
Rapaport, Michel
Reitsema, Harold
Rekola, Rami
Richardson, Derek
Rickman, Hans
Robertson, William
Roemer, Elizabeth
Roeser, Siegfried
Rossi, Alessandro
RUI, Qi
Russell, Kenneth
Santos-Sanz, Pablo
SATO, Isao
Savanevich, Vadim
Scheirich, Peter
Schmadel, Lutz
Schober, Hans
Scholl, Hans
Schubart, Joachim
Schuster, William
Seidelmann, P.
Sekanina, Zdenek
SEO, Haingja
Shanklin, Jonathan
Shelus, Peter
SHEN, Kaixian
SHI, Huli
Shor, Viktor
Sinclair, Andrew
Sitarski, Grzegorz
SOFIA LYKAWKA, Patryk
Solovaya, Nina
SOMA, Mitsuru
Sosa, Andrea
Spahr, Timothy
Standish, E.
Steel, Duncan
Stellmacher, Irène
Stokes, Grant
Süli, Áron
Sultanov, G.
Svoren, Jan
Sybiryakova, Yegeniya

Synnott, Stephen
Szabó, Gyula
Szutowicz, Sławomira
Tancredi, Gonzalo
Tatum, Jeremy
Taylor, Donald
Tholen, David
Thuillot, William
Tichá, Jana
Tiscareno, Matthew
Titov, Vladimir
TOMITA, Koichiro
Trujillo, Chadwick
Tsiganis, Kleomenis
Tsuchida, Masayoshi
Tuccari, Gino
van Flandern, Tom
van Houten-Groeneveld, Ingrid
Veillet, Christian
Vienne, Alain
Virtanen, Jenni
Voelzke, Marcos
Vu, Duong Tuyen
WANG, Xiaoya
Wasserman, Lawrence
Weissman, Paul
West, Richard
Whipple, Arthur
Wild, Paul
Williams, Gareth
Williams, Iwan
Williams, James
Wood, Harley
XIONG, Jianning
YABUSHITA, Shin
Yakhontova, N.
Yeomans, Donald
YIM, Hong Suh
YOSHIKAWA, Makoto
YUASA, Manabu
Zadnik, Marjan
Zagretdinov, Renat
Zappalà, Vincenzo
Zhang, Yu-Zhe
ZHANG, Jiaxiang
ZHANG, Qiang
ZHANG, Wei
ZHANG, Xiaoxiang
ZHANG, Yang
ZHAO, Haibin
ZHU, Jin
Ziołkowski, Krzysztof

Division J - Commission 21 Galactic & Extragalactic Background Radiation

President: Jayant Murthy
Vice-President: Jayant Murthy

Organizing Committee Members:

Ingrid Mann
William J. Baggaley
Kalevi Mattila
Anny-Chantal Levasseur-Regourd
Junichi WATANABE
Eli Dwek

Members:

Angione, Ronald
Asaad, A.
ASANO, Katsuaki
Baggaley, William
Banos, Cosmas
Barbier, D.
Behar, Ehud
Belkovich, Oleg
Blamont, Jacques-Emile
Boquien, Médéric
Bot, Caroline
Bouwens, Rychard
Bowyer, C.
Broadfoot, A.
Caputi, Karina
Chernyakova, Maria
Chiang, Hsin
Clairemidi, Jacques
de Petris, Marco
Dermott, Stanley
d'Hendecourt, Louis
Dodonov, Sergej
Dole, Herve
Dubin, Maurice
Dufay, J.
Dufay, Maurice
Dumont, Rene
Dwek, Eli
Eichhorn, Guenther
Feldman, Paul
FUJIWARA, Akira
Giovane, Frank
Gounelle, Matthieu
Grün, Eberhard
Gustafson, Bo
Hanner, Martha
HAO, Lei
Harwit, Martin
Hauser, Michael

Hecht, James
Henry, Richard
Hernández-Monteagudo, Carlos
Hofmann, Wilfried
HONG, Seung-Soo
Horns, Dieter
Huruhata, Masaaki
Hurwitz, Mark
Ingham, M.
Ivanov-Kholodny, Gor
Jackson, Bernard
James, John
Jelić, Vibor
Johns, Bethany
Joubert, Martine
Karygina, Zoya
Keshet, Uri
KOBAYASHI, Masakazu
Kopylov, Alexander
Korpela, Eric
Koutchmy, Serge
Kozak, Lyudmyla
Kramer, Busaba
Krassovsky, V.
Kulkarni, Prabhakar
Lamy, Philippe
Lapi, Andrea
Lebedinskij, A.
Leger, Alain
Leinert, Christoph
Lemke, Dietrich
Lenc, Emil
Levasseur-Regourd, Anny-Chantal
LI, Li-Xin
Lillie, Charles
Link, F.
Lopez Gonzalez, Maria
Lopez Moreno, Jose
Lopez Puertas, Manuel
MAIHARA, Toshinori
Mann, Ingrid
Martin, Donn
Massaro, Francesco
Mather, John
MATSUMOTO, Toshio
Mattila, Kalevi
Maucherat, Jean
McDonnell, J.
Mikhail, Joseph
Misconi, Nebil
Moodley, Kavilan
Morgan, David
Muinonen, Karri
MUKAI, Sonoyo
MUKAI, Tadashi
Murthy, Jayant
NAKAMURA, Akiko
Nawar, Samir
Neizvestny, Sergei
Nishimura, Tetsuo
Paresce, Francesco
Pérez-González, Pablo
Perrin, Jean-Marie
Pfleiderer, Jorg
Rauch, Michael
Reach, William
Renard, Jean-Baptiste
Risaliti, Guido
Roach, Franklin
Robley, Robert
Rodrigo, Rafael
Rozhkovskij, Dimitrij
Sánchez, Francisco
Sanchez-Saavedra, M.
Saxena, P.
Schlosser, Wolfhard
Schuh, Harald
Schwehm, Gerhard
Seymour, Nicholas
Shefov, Nikolaj
SHIRAHATA, Mai
Siringo, Giorgio
Smith, Robert
Soberman, Robert
Sparrow, James
Staude, Hans
Sykes, Mark
TANABE, Hiroyoshi
Tepper Garcia, Thorsten
Toller, Gary
Tsygankov, Sergey
Tyson, John
UENO, Munetaka
Ugolnikov, Oleg
Vaccari, Mattia
Venters, Tonia
Vrtilek, Jan
Wagner, Robert
Wallis, Max
WATANABE, Junichi
Weinberg, Jerry
Wendt, Martin
Wesson, Paul
Wheatley, Peter
Wilson, P.
Witt, Adolf
Wolleben, Maik
Wolstencroft, Ramon
Woolfson, Michael
YAMAMOTO, Tetsuo
YAMASHITA, Kojun
YANG, Xiaohu
Zerull, Reiner
Zhang, Yuying

Division F - Commission 22 Meteors, Meteorites & Interplanetary Dust

President: Petrus Matheus Marie Jenniskens
Vice-President: JiříBorovička

Organizing Committee Members:

Jin ZHU
Galina O. Ryabova
Shinsuke ABE
Diego Janches
Masateru ISHIGURO
Jérémie J. Vaubaillon
Guy Joseph Consolmagno

Members:

ABE, Shinsuke
Alexandrov, Alexander
Anders, Edward
Apai, Daniel
Asher, David
Babadjanov, Pulat
Baggaley, William
Belkovich, Oleg
Benkhoff, Johannes
Bhandari, N.
Borovička, Jiří
Brown, Peter
Brownlee, Donald
Busarev, Vladimir
Campbell-Brown, Margaret
Čapek, David
Carusi, Andrea
Cevolani, Giordano
Clifton, Kenneth
Clube, S.
Consolmagno, Guy
Cook, A.
Cooper, Timothy
Denning, W.
Dieleman, Pieter
Djorgovski, Stanislav
Dubin, Maurice
Duffard, Rene
Elford, William
Ellyett, Clifton
Fechtig, Hugo
Fedynsky, V.
Fireman, Edward
Forti, Giuseppe
Fromang, Sebastien
Gajdoš, Stefan
Glass, Billy
Gorbanev, Jury
Goswami, J.
Gounelle, Matthieu
Grady, Monica
Granvik, Mikael
Grün, Eberhard
Guigay, G.
Gustafson, Bo
Guth, Vladimir
Hajdukova, Maria
Hajduková, Jr., Maria
Halliday, Ian
Hanner, Martha
Harvey, Gale
HASEGAWA, Ichiro
HASEGAWA, Sunao
Hawkes, Robert
Helin, Eleanor
Hemenway, Curtis
Hey, James
Hirose, Hideo
Hodge, Paul
HONG, Seung-Soo
Husárik, Marek
ISHIGURO, Masateru
Jakubík, Marian
Janches, Diego
Javnel', A.
Jehin, Emmanuel
Jenniskens, Petrus
Jones, James
Jopek, Tadeusz
Kalenichenko, Valentin
Kaňuchová, Zuzana
Karakas, Amanda
Kasuga, Toshihiro
Keay, Colin
Khovritchev, Maxim
Kikwaya Eluo, Jean-Baptiste
KIMURA, Hiroshi
Koeberl, Christian
Kokhirova, Gulchehra
Kolomiyets, Svitlana
Kornoš, Leonard
Koschny, Detlef
Kostama, Veli-Petri
Koten, Pavel
Kozak, Pavlo
Kramer, Kh
Kruchinenko, Vitaliy
Kulikova, Nelly
Lamy, Philippe
Lemaire, Joseph
Levasseur-Regourd, Anny-Chantal
Levin, B.
Lodders, Katharina
Lugaro, Maria
Lyon, Ian
Makalkin, Andrei
Mann, Ingrid
Maris, Michele
Martinez-Frias, Jesus
Marvin, Ursula

Mason, John
Mawet, Dimitri
McDonnell, J.
McIntosh, Bruce
Meisel, David
Miles, Howard
Misconi, Nebil
Moór, Attila
Murray, Andrew
Murray, Carl
NAGAHARA, Hiroko
NAKAMURA, Takuji
NAKAZAWA, Kiyoshi
Napier, William
Newburn Jr, Ray
Nuth, Joseph
Olivier, C.
Pecina, Petr
Pellinen-Wannberg, Asta
Plavec, Zdenka
Politi, Romolo
Polnitzky, Gerhard
Poole, Graham
Quesada, Vinicio
Rajchl, Jaroslav
Rendtel, Juergen
Rickman, Hans
Ripken, Hartmut
Roggemans, Paul
Ryabova, Galina
Santos-Sanz, Pablo
Sasaki, Sho
Sekanina, Zdenek
SHANG, Hsien
Shrben, Lukáš
Simek, Milos
Simonenko, Alla
Soberman, Robert
Spurny, Pavel
Steel, Duncan
Stewart-Mukhopadhyay, Sarah
Štohl, Jan
Svestka, Jiri
Svoren, Jan
Tatum, Jeremy
Taylor, Andrew
Tedesco, Edward
Terentjeva, Alexandra
TOMITA, Koichiro
Tosi, Federico
Toth, Juraj
Trigo-Rodríguez, Josep
Valsecchi, Giovanni
Vaubaillon, Jérémie
Voloshchuk, Yuri
WANG, Dechang
WATANABE, Junichi
Webster, Alan
Weinberg, Jerry
Williams, Iwan
Woolfson, Michael
Wu, Zidian
XU, Pinxin
Yair, Yoav
YAMAMOTO, Masayuki
Yavnel, Alexander
Yeomans, Donald
Zadnik, Marjan
Zender, Joe
ZHANG, Xiaoxiang
ZHAO, Haibin
ZHU, Jin
Zvolankova, Judita

Division B - Commission 25 Astronomical Photometry & Polarimetry

President: Alistair Walker
Vice-President: Saul J. Adelman

Organizing Committee Members:

Wen Ping CHEN
J. Allyn Smith
Steve B. Howell
Kevin Volk
Antonio Mario Magalhaes
Barbara J. Anthony-Twarog
Jens Kirkeskov Knude
Donald W. Kurtz
John W. Menzies
W. Schoeneich

Members:

Ables, Harold
Adelman, Saul
Ahumada, Javier
Aigrain, Suzanne
AKITAYA, Hiroshi
Albrecht, Rudolf
Alecian, Evelyne
Anandaram, Mandayam
Andreuzzi, Gloria
Angel, J.
Angione, Ronald
Anthony-Twarog, Barbara
Arsenijevic, Jelisaveta
Ashok, N.
Aspin, Colin

Aungwerojwit, Amornrat
Babu, G.S.D.
Baldinelli, Luigi
Baliyan, Kiran
Balona, Luis
Baran, Andrzej
Barnes III, Thomas
Barrett, Paul
Barrientos, Luis
Baume, Gustavo
Behr, Alfred
Bellazzini, Michele
Berdyugin, Andrei
Bessell, Michael
Birkmann, Stephan
Bjorkman, Jon
Blecha, Andre
Borgman, Jan
Borisova, Ana
Borra, Ermanno
Boyle, Richard
Braithwaite, Jonathan
Breger, Michel
Brown, Douglas
Brown, Thomas
Buser, Roland
CAI, Mingsheng
Cantiello, Michele
Canto Martins, Bruno
Carciofi, Alex
Carney, Bruce
Carter, Brian
Castelaz, Micheal
Cesetti, Mary
Chadid, Merieme
CHEN, An-Le
CHEN, Wen Ping
Chugajnov, P.
Cioni, Maria-Rosa
Clark, David
Clem, James
Clocchiatti, Alejandro
Connolly, Leo
Copin, Yannick
Corradi, Wagner
Coyne, S.J, George
Cramer, Noel
Crawford, David
Cuypers, Jan
Dachs, Joachim
Dahn, Conard
DAI, Zhibin
DAISAKU, Nogami
Danford, Stephen
Davidge, Timothy
Demory, Brice-Olivier
Deshpande, M.
Dolan, Joseph
DOU, Jiangpei
Dubout, Renee
Ducati, Jorge
Ducourant, Christine
Edwards, Paul
Eggen, O.
Elkin, Vladimir
Elmhamdi, Abouazza
Evans, Dafydd
Fabiani, Sergio
Fabregat, Juan
Fabrika, Sergei
Feinstein, Alejandro
Fernández Lajús, Eduardo
Fernie, J.
Fluri, Dominique
Forte, Juan
Franco, Gabriel Armando
Freyhammer, Lars
Galadi-Enriquez, David
Gallouet, Louis
Garrison, Robert
Gehrz, Robert
Genet, Russell
Gerbaldi, Michele
Ghosh, Swarna
Gilliland, Ronald
Gillon, MichaÃ«l
Giorgi, Edgard
Glass, Ian
Golay, Marcel
Goy, Gerald
Graham, John
Grauer, Albert
Greaves, W.
Grenon, Michel
Grewing, Michael
Grundahl, Frank
Guetter, Harry
Guglielmino, Salvatore
Gutierrez-Moreno, A.
Hackman, Thomas
Hall, Douglas
Hardie, R.
Hauck, Bernard
Hayes, Donald
Heck, Andre
Heiter, Ulrike
Hensberge, Herman
Hertzsprung, Ejnar
Hilditch, Ronald
Hiltner, A.
Howell, Steve
HUANG, Lin
Hubrig, Swetlana
Huovelin, Juhani
Hyland, Harry
Ioannou, Zacharias
Irwin, Alan
Ivezic, Zeljko
Iyengar, K.
Jeffers, Sandra
Jerzykiewicz, Mikołaj
JIANG, Zhibo
Jordi, Carme
Joshi, Umesh
Karoff, Christoffer
Kasiviswanathan, Sankarasubramanian
KAWARA, Kimiaki
Kazlauskas, Algirdas
Kebede, Legesse
Keller, Stefan
KENTARO, Matsuda
Kepler, S.
Kilkenny, David
KIM, Seung-Lee
King, Ivan
Knude, Jens
Kornilov, Victor
Kóspál, Ágnes
Kulkarni, Prabhakar
Kunkel, William
Kurtz, Donald
Kurucz, Robert
Labhardt, Lukas
Landolt, Arlo
Landstreet, John
Laskarides, Paul
Laugalys, Vygandas
Lazauskaite, Romualda
Lemke, Michael

Lenzen, Rainer
Leroy, Jean-Louis
LI, Min
LI, Qingkang
LI, Sin
Linde, Peter
Lockwood, G.
Lub, Jan
Luna, Homero
Magalhaes, Antonio Mario
Magnier, Eugene
Maitzen, Hans
Manfroid, Jean
Manset, Nadine
Markkanen, Tapio
Marraco, Hugo
Marsden, Stephen
Martinez, Peter
Martinez Roger, Carlos
Maslennikov, Kirill
Mason, Paul
Mathys, Gautier
Mayer, Pavel
McDavid, David
McLean, Ian
Mendoza, V.
Menzies, John
Metcalfe, Travis
Mianes, Pierre
Miller, Joseph
Milone, Eugene
Mintz Blanco, Betty
Mironov, Aleksey
Moffett, Thomas
Moitinho, André
Monaco, Lorenzo
Mourard, Denis
Mumford, George
Munari, Ulisse
Narbutis, Donatas
Naylor, Tim
Neiner, Coralie
Nicolet, Bernard
Nikoghosyan, Elena
Nikonov, V.
NOGUCHI, Kunio
Notni, Peter
Oblak, Edouard
Oestreicher, Roland
Orsatti, Ana
Osawa, Kiyoteru
PAK, Soojong
Parimucha, Stefan
Pavani, Daniela
Pedreros, Mario
Pel, Jan
Penny, Alan
Perrin, Marshall
Petit, Pascal
Pfau, Werner
Pfeiffer, Raymond
Philip, A.G.
Piirola, Vilppu
Platais, Imants
Pokrzywka, Bartłomiej
Pulone, Luigi
QIAN, Shengbang
Rank-Lueftinger, Theresa
Rao, Pasagada
Raveendran, A.
Rawlings, Mark
Reglero Velasco, Victor
Reshetnyk, Volodymyr
Rhee, Jaehyon
Richardson, Lorna
Robb, Russell
Robinson, Edward
Rodrigues, Claudia
Romanyuk, Yaroslav
Rostopchina, Alla
Rufener, Fredy
Sabin, Laurence
Samec, Ronald
Santos Agostinho, Rui
Santos-Sanz, Pablo
Sarma, M.
Schiller, Stephen
Schmidt, Edward
Schoeneich, W.
Schuster, William
Seares, F.
SEKIGUCHI, Kazuhiro
Sen, Asoke
Shankland, Paul
Shawl, Stephen
Shoyoqubov, Shoayub
Sickafoose, Amanda
Simons, Douglas
Smith, J.
Smyth, Michael
Snik, Frans
Snowden, Michael
Stagg, Christopher
Steinlin, Uli
Sterken, Christiaan
Stetson, Peter
Stil, Jeroen
Stockman Jr, Hervey
Stone, Remington
Stonkutė, Rima
Straižys, Vytautas
Stritzinger, Maximilian
Subramaniam, Annapurni
Sudzius, Jokubas
Sullivan, Denis
Szkody, Paula
Szymański, Michał
Tandon, S.
Tapia-Perez, Santiago
Taranova, Olga
Taş, Günay
Tedds, Jonathan
Thompson, Rodger
Thurston, Mark
Tokunaga, Alan
Tolbert, Charles
Townsend, Richard
Turcu, Vlad
Ueta, Toshiya
Ulrich, Bruce
UMEDA, Hideyuki
Ureche, Vasile
Uslenghi, Michela
Vaughan, Arthur
Verma, R.
Vidotto, Aline
Volk, Kevin
Volo'shina, Irina
Vrba, Frederick
Walker, Alistair
Walker, William
WANG, Chuanjin
Warren Jr, Wayne
Weiss, Werner
Weistrop, Donna
Wesselius, Paul
Wheatley, Peter
White, Nathaniel
Wielebinski, Richard
Willstrop, Roderick

Winiarski, Maciej
WOO, Jong
Wramdemark, Stig
YAMASHITA, Yasumasa
YAO, Yongqiang
YIN, Jisheng
Young, Andrew
Yudin, Ruslan
YUJI, Ikeda
Zdanavičius, Justas
ZHU, Liying
Zoccali, Manuela
Zucker, Shay
Zwintz, Konstanze

Division G - Commission 26 Double & Multiple Stars

President: Brian D. Mason
Vice-President: Yurij Yu Balega

Organizing Committee Members:

Bo Reipurth
Andrei A. Tokovinin
Patricia J C Lampens
Jose-Angel Docobo
Vakhtang S. Tamazian
Marco Scardia
Frédéric Arenou
Edouard Oblak
Theo A. ten Brummelaar

Members:

Abt, Helmut
Ahumada, Javier
Aitken, R.
AK, Tansel
Allen, Christine
Andrade, Manuel
Anosova, Joanna
Antokhina, Eleonora
Arenou, Frédéric
Argyle, P.
Argyle, Robert
Armstrong, John
Aungwerojwit, Amornrat
Bacchus, Pierre
Bagnuolo Jr., William
Bailyn, Charles
Balega, Yurij
Batten, Alan
Beavers, Willet
Beklen, Elif
Boden, Andrew
Bonneau, Daniel
Boyajian, Tabetha
Brandner, Wolfgang
Brosche, Peter
Brown, David
Budaj, Jan
Carciofi, Alex
Cester, Bruno
CHEN, Wen Ping
CHEN, Zhen
Clarke, Catherine
Culver, Roger
Cvetkovi'c, Zorica
Dadaev, Aleksandr
DAISAKU, Nogami
De Becker, MichaÃ«l
De Cat, Peter
de Mink, Selma
de Val-Borro, Miguel
Docobo, Jose-Angel
Dominis Prester, Dijana
Dukes Jr., Robert
Dunham, David
Eggen, O.
Eldridge, John
Elkin, Vladimir
Falceta-Goncalves, Diego
Fekel, Francis
Fernandes, Joao
Ferrer, Osvaldo
Fletcher, J.
Fors, Octavi
Fox-Machado, Lester
Fracastoro, Mario
Franz, Otto
Fredrick, Laurence
Freyhammer, Lars
Gandolfi, Davide
Gatewood, George
Gaudenzi, Silvia
Gavras, Panagiotis
Geller, Aaron
Genet, Russell
Geyer, Edward
Ghez, Andrea
Gonçalves, Denise
Goodwin, Simon
Grundstrom, Erika
Gün, Gulnur
Hakkila, Jon
Halbwachs, Jean-Louis
Harrington, Robert
Hartigan, Patrick
Hartkopf, William
HE, Jinhua
Heacox, William
Hershey, John
Hertzsprung, Ejnar
Hidayat, Bambang
Hill, Graham
Hillwig, Todd
Hindsley, Robert
Horch, Elliott

Hummel, Christian
Hummel, Wolfgang
Ianna, Philip
Ireland, Michael
Izmailov, Igor
Izzard, Robert
Jahreiss, Hartmut
Jassur, Davoud
JEON, Young Beom
Johnston, Helen
Jurdana-Šepić, Rajka
Kafka, Styliani (Stella)
Kazantseva, Liliya
Kilpio, Elena
Kisseleva-Eggleton, Ludmila
Kitsionas, Spyridon
Kley, Wilhelm
Köhler, Rainer
Kouwenhoven, M.B.N. (Thijs)
Kroupa, Pavel
Kubát, Jiri
Lampens, Patricia
Latham, David
Lattanzi, Mario
Lee, William
LEE, Chung-Uk
LEE, Jae Woo
Leinert, Christoph
Lepine, Sebastien
Levato, Orlando
LIM, Jeremy
Ling, Josefina
Lippincott Zimmerman, Sarah
LIU, Michael
Lodén, Kerstin
Loden, Lars
Lyubchik, Yuri
Maddison, Sarah
Maíz Apellániz, Jesús
Malkov, Oleg
Malogolovets, Evgeny
Marsakova, Vladyslava
Martayan, Christophe
Martín, Eduardo
Mason, Brian
Mathieu, Robert
Mawet, Dimitri
McAlister, Harold
McBride, Vanessa
McDavid, David
Mennickent, Ronald
Meyer, Claude
Middleton, Christopher
Mikkola, Seppo
Mikołajewski, Maciej
Millour, Florentin
Milone, Eugene
Mohan, Chander
Morbey, Christopher
Morbidelli, Roberto
Morel, Pierre-Jacques
Morrell, Nidia
Negueruela, Ignacio
Neuhaeuser, Ralph
Nitschelm, Christian
Nürnberger, Dieter
Oblak, Edouard
Orlov, Viktor
Oswalt, Terry
Pannunzio, Renato
Parimucha, Stefan
Pauls, Thomas
Pereira, Claudio
Perets, Hagai
Peterson, Deane
Petr-Gotzens, Monika
Pietrukowicz, Paweł
Pluzhnik, Eugene
Pollacco, Don
Popovic, Georgije
Pourbaix, Dimitri
Poveda, Arcadio
Prieto, Cristina
Prieur, Jean-Louis
Prša, Andrej
Rastegaev, Denis
Reipurth, Bo
Roberts Jr, Lewis
Rodrigues de Oliveira Filho, Irapuan
Russell, Jane
Ruždjak, Domagoj
Sagar, Ram
Salukvadze, G.
Scardia, Marco
Scarfe, Colin
Schmidtke, Paul
Schöller, Markus
Shakht, Natalia
Shatsky, Nicolai
Simon, Michal
Sinachopoulos, Dimitris
Skokos, Charalampos
Smak, Józef
Smith, J.
Soderhjelm, Staffan
Sowell, James
Stein, John
Sterzik, Michael
Sudar, Davor
Szabados, Laszlo
Tamazian, Vakhtang
Tango, William
Tarasov, Anatolii
Teixeira, Paula
ten Brummelaar, Theo
Terquem, Caroline
Titov, Vladimir
Tokovinin, Andrei
Tomasella, Lina
Torres, Guillermo
Trimble, Virginia
TSAY, Wean-Shun
Tsygankov, Sergey
Turner, Nils
Udry, Stephane
Upgren, Arthur
Valbousquet, Armand
Valtonen, Mauri
van Altena, William
Van de Kamp, Peter
Van den Bos, H.
van der Bliek, Nicole
van der Hucht, Karel
van Dessel, Edwin
Vaňko, Martin
Vaz, Luiz Paulo
Vennes, Stephane
WANG, Jia-Ji
Weis, Edward
WEN, Linqing
Worley, Charles
YAN, Lin-shan
Zasche, Petr
Zavala, Robert
Zheleznyak, Alexander
ZHU, Liying
Zinnecker, Hans
Zucker, Shay

Division G - Commission 27 Variable Stars

President: Karen Pollard
Vice-President: Christopher Simon Jeffery

Organizing Committee Members:

Klaus G. Strassmeier
Dennis Stello
S. O. Kepler
Saskia Hekker
Katrien Uytterhoeven
Katrien Kolenberg
David E. Mkrtichian
Laurent Eyer
Márcio Catelan

Members:

Aerts, Conny
Aigrain, Suzanne
Airapetian, Vladimir
Aizenman, Morris
AK, Tansel
Albinson, James
Albrow, Michael
Alencar, Silvia
Alfaro, Emilio
Allan, David
Alpar, Mehmet
Amado, Pedro
ANDO, Hiroyasu
Andrievsky, Sergei
Andronov, Ivan
Antipin, Sergei
Antipova, Lyudmila
Antonello, Elio
Antonyuk, Kirill
Antov, Alexandar
Arellano Ferro, Armando
Arentoft, Torben
Arias, Maria
Arkhipova, Vera
Arsenijevic, Jelisaveta
Asteriadis, Georgios
Aungwerojwit, Amornrat
Avgoloupis, Stavros
Baade, Dietrich
Baglin, Annie
Balman, Solen
Balona, Luis
Baran, Andrzej
Baransky, Olexander
Barban, Caroline
Barnes III, Thomas
Bartolini, Corrado
Barway, Sudhashu
Barwig, Heinz
Baskill, Darren
Bastien, Pierre
Bath, Geoffrey
Bauer, Wendy
Bazot, Michael
Beaulieu, Jean-Philippe
Bedding, Timothy
Bedogni, Roberto
Belkacem, Kevin
Belmonte Aviles, Juan Antonio
Belserene, Emilia
Belvedere, Gaetano
Benkő, Jozsef
Benson, Priscilla
Berdnikov, Leonid
Berger, Mitchell
Bersier, David
Berthomieu, Gabrielle
Bessell, Michael
Bianchini, Antonio
Bjorkman, Karen
Bolton, Charles
Bond, Howard
Bopp, Bernard
Borczyk, Wojciech
Borisova, Ana
Bortoletto, Alexandre
Boyarchuk, Alexander
Boyarchuk, Margarita
Boyd, David
Bradley, Paul
Breger, Michel
Briquet, Maryline
Brown, Douglas
Bruntt, Hans
Buccino, Andrea
Burki, Gilbert
Burwitz, Vadim
Busa', Innocenza
Busko, Ivo
Butkovskaya, Varvara
Butler, Christopher
Butler, Dennis
Buzasi, Derek
Cacciari, Carla
Caldwell, John
Cameron, Andrew
Canavaggia, Renée
CAO, Huilai
Carciofi, Alex
Carrier, Fabien
Casares, Jorge
Catchpole, Robin
Catelan, Márcio
Cenko, Stephen
Chadid, Merieme
Chaplin, William
CHEN, Alfred
CHEN, An-Le
Cherchneff, Isabelle
Cherepashchuk, Anatolij
CHOU, Yi
Christensen-Dalsgaard, Jørgen
Christie, Grant

Cioni, Maria-Rosa
Clement, Christine
Clementini, Gisella
Clocchiatti, Alejandro
Cohen, Martin
Connolly, Leo
Contadakis, Michael
Cook, Kem
Corral, Luis
Córsico, Alejandro
Costa, Vitor
Cottrell, Peter
Coulson, Iain
Crause, Lisa
Creech-Eakman, Michelle
Cunha, Margarida
Cutispoto, Giuseppe
Cuypers, Jan
DAI, Zhibin
DAISAKU, Nogami
Dall'Ora, Massimo
D'Amico, Nicolo
Danford, Stephen
Daszynska-Daszkiewicz, Jadwiga
De Becker, MichaÃ«l
De Cat, Peter
de Groot, Mart
Delgado, Antonio
Demers, Serge
DENG, LiCai
Depagne, Éric
Derekas, Aliz
de Ridder, Joris
Deupree, Robert
Dickens, Robert
Diethelm, Roger
Di Mauro, Maria Pia
Donahue, Robert
Downes, Ronald
Dugan, R.
Dukes Jr., Robert
Dunlop, Storm
Dupuy, David
Dziembowski, Wojciech
Ederoclite, Alessandro
Edwards, Paul
Edwards, Suzan
Efremov, Yurij
Eggen, O.
Eggenberger, Patrick
El Basuny, Ahmed
Elkin, Vladimir
Engelbrecht, Chris
Esenoğlu, Hasan
Eskioğlu, A.
Evans, Aneurin
Evans, Nancy
Evren, Serdar
Eyer, Laurent
Fadeyev, Yurij
Feast, Michael
Ferland, Gary
Fernández Lajús, Eduardo
Fernie, J.
Figer, Donald
Fokin, Andrei
Formiggini, Lilliana
Fox-Machado, Lester
Frew, David
Freyhammer, Lars
Fromang, Sebastien
FU, Hsieh-Hai
FU, Jian-Ning
FUJIWARA, Tomoko
Gahm, Goesta
Gális, Rudolf
Gal-Yam, Avishay
Gameiro, Jorge
Gamen, Roberto
Garrido, Rafael
Gasparian, Lazar
Gavras, Panagiotis
Gay, Pamela
Genet, Russell
Gershberg, R.
Geyer, Edward
Gieren, Wolfgang
Gies, Douglas
Gillet, Denis
Glagolevskij, Yurij
Gondoin, Philippe
Gosset, Eric
Gough, Douglas
Goupil, Marie-Jose
Graham, John
Green, Daniel
Grinin, Vladimir
Groenewegen, Martin
Groh, Jose
Grouiller, H.
Grundstrom, Erika
Grygar, Jiri
Guerrero, Gianantonio
Guinan, Edward
Gün, Gulnur
Günthardt, Guillermo
GUO, Jianheng
Gurm, Hardev
Guzik, Joyce
Haas, Martin
Hackman, Thomas
Hackwell, John
Haefner, Reinhold
Haisch, Bernard
Halbwachs, Jean-Louis
Hall, Douglas
Hallinan, Gregg
Hamdy, M.
Handler, Gerald
HAO, Jinxin
Harmanec, Petr
Hawley, Suzanne
Heiser, Arnold
Hekker, Saskia
Hempelmann, Alexander
Henden, Arne
Hesser, James
Hill, Henry
Hintz, Eric
Hojaev, Alisher
Horner, Scott
Houdek, Gunter
Houk, Nancy
Howell, Steve
Huenemoerder, David
Humphreys, Elizabeth
Hutchings, John
Iben Jr, Icko
Iijima, Takashi
Ikonnikova, Natalia
Ireland, Michael
Ishida, Toshihito
Ismayilov, Nariman
ITA, Yoshifusa
Ivezic, Zeljko
Jablonski, Francisco
Janík, Jan
Jankov, Slobodan
Jeffers, Sandra

Jeffery, Simon
JEON, Young Beom
Jerzykiewicz, Mikołaj
Jetsu, Lauri
Jewell, Philip
Jha, Saurabh
JIANG, Biwei
JIANG, Shi-Yang
JIN, Zhenyu
Joner, Michael
Jones, Albert
Jurcsik, Johanna
Kadouri, Talib
Käufl, Hans Ulrich
Kafka, Styliani (Stella)
Kalomeni, Belinda
KAMBE, Eiji
KANAMITSU, Osamu
Kanbur, Shashi
Kanyo, Sandor
Karitskaya, Evgeniya
Karovska, Margarita
Karp, Alan
Katsova, Maria
Kaufer, Andreas
Kawaler, Steven
Kaye, Anthony
Kazarovets, Elena
Keller, Stefan
Kepler, S.
Kervella, Pierre
Kholopov, P.
Kilkenny, David
KIM, Chulhee
KIM, Seung-Lee
KIM, Tu Whan
KIM, Young-Soo
Kiplinger, Alan
Kippenhahn, Rudolf
Kiss, László
Kjeldsen, Hans
Kjurkchieva, Diana
Kochukhov, Oleg
Koen, Marthinus
Kolenberg, Katrien
Kolláth, Zoltan
Komžík, Richard
Konstantinova-Antova, Renada
Kopacki, Grzegorz
Korhonen, Heidi
Kóspál, Ágnes
Kővári, Zsolt
Kraft, Robert
Krautter, Joachim
Kreiner, Jerzy
Krisciunas, Kevin
Krzeminski, Wojciech
Krzesiński, Jerzy
Kubiak, Marcin
Kudryavtseva, Nadezhda (Nadia)
Kuhi, Leonard
Kunjaya, Chatief
Kunkel, William
KURAYAMA, Tomoharu
Kurtz, Donald
Kwee, K.
Lago, Maria
Lampens, Patricia
Landolt, Arlo
Laney, Clifton
Lanning, Howard
Lanza, Antonino
Larionov, Valeri
Laskarides, Paul
Lawlor, Timothy
Lawson, Warrick
Lázaro Hernando, Carlos
Le Bertre, Thibaut
Lebzelter, Thomas
Lee, Hyun-chul
LEE, Jae Woo
LEE, Jae Woo
LEE, Myung Gyoon
Leite, Scheid
Letarte, Bruno
Leung, Kam
LI, Yan
LI, Zhiping
Liermann, Adriane
Little-Marenin, Irene
LIU, Jifeng
Lloyd, Christopher
Lockwood, G.
Longmore, Andrew
Lopez, De
Lorenz-Martins, Silvia
Lub, Jan
Machado Folha, Daniel
Macri, Lucas
Madore, Barry
Maeder, Andre
Mahmoud, Farouk
Malov, Igor
Mannino, Giuseppe
Mantegazza, Luciano
Marchev, Dragomir
Marconi, Marcella
Margrave Jr, Thomas
Markoff, Sera
Marsakova, Vladyslava
Martayan, Christophe
Martic, Milena
Martinez, Peter
Mason, Paul
Mateu, Cecilia
Mathias, Philippe
MATSUMOTO, Katsura
MATSUNAGA, Noriyuki
Matthews, Jaymie
Mauche, Christopher
Mazumdar, Anwesh
McGraw, John
McSaveney, Jennifer
Melikian, Norair
Mennickent, Ronald
Messina, Sergio
Michel, Eric
Miglio, Andrea
Mikołajewski, Maciej
Millour, Florentin
Milone, Eugene
Milone, Luis
Minikulov, Nasridin
Mkrtichian, David
Moffett, Thomas
Mohan, Chander
Molenda-Å»akowicz, Joanna
Monard, Libert
Montalb'an, Josefina
Monteiro, Mario Joao
Montgomery, Michele
Morales Rueda, Luisa
Morel, Thierry
Morrison, Nancy
Moskalik, Paweł
Mosoni, Laszlo
Mowlavi, Nami

Mukai, Koji
Mumford, George
Murdin, Paul
NARDETTO, Nicolas
Naylor, Tim
Naze, Yael
Neff, John
Neiner, Coralie
Neustroev, Vitaly
NGEOW, Chow Choong
Niarchos, Panagiotis
Niemczura, Ewa
NIINUMA, Kotaro
Nijland, A.
Nikoghosyan, Elena
Nikolov, Andrej
Nitschelm, Christian
NITTA, Atsuko
Nota, Antonella
Nugis, Tiit
Odgers, Graham
O'Donoghue, Darragh
Ogłoza, Waldemar
Oláh, Katalin
Oliveira, Alexandre
Olivier, Enrico
Oluseyi, Hakeem
Østensen, Roy
Oswalt, Terry
O'Toole, Simon
OTSUKI, Kaori
Pál, András
Panei, Jorge
Papaloizou, John
Paparo, Margit
Papousek, Jiri
Parimucha, Stefan
PARK, Byeong-Gon
Parsamyan, Elma
Parthasarathy, Mudumba
Patat, Ferdinando
Pat-El, Igal
Paternò, Lucio
Pavlovski, Kresimir
Pazhouhesh, Reza
Pearson, Kevin
Penny, Matthew
Percy, John
Pérez Hernández, Fernando
Petersen, J.
Petit, Pascal
Petrov, Peter
Pettersen, Bjørn
Pietrukowicz, Paweł
Piirola, Vilppu
Pijpers, Frank
Plachinda, Sergei
Plavchan, Peter
Pollacco, Don
Pollard, Karen
Pont, Frédéric
Pop, Alexandru
Pop, Vasile
Price, Charles
Pricopi, Dumitru
Pringle, James
Pritzl, Barton
Provost, Janine
Pugach, Alexander
Pustõnski, Vladislav-Veniamin
QIAN, Shengbang
Rank-Lueftinger, Theresa
Ransom, Scott
Rao, N.
Ratcliff, Stephen
Reale, Fabio
Reiners, Ansgar
Reinsch, Klaus
Renson, P.
Rey, Soo-Chang
Rivinius, Thomas
Robinson, Edward
Rodriguez, Eloy
Romanov, Yuri
Rosenbush, Alexander
Rosino, Leonida
Rountree, Janet
Russev, Ruscho
Ruždjak, Domagoj
Sachkov, Mikhail
Sadik, Aziz
Safari, Hossein
Saha, Abhijit
SAKAMOTO, Tsuyoshi
SAKON, Itsuki
Samus, Nikolay
Sandmann, William
Sanyal, Ashit
Sareyan, Jean-Pierre
Sarma, M.
Sasselov, Dimitar
SATO, Naonobu
Schaefer, Bradley
Schlegel, Eric
Schmidt, Edward
Schmidtobreick, Linda
Schuh, Sonja
Schwartz, Philip
Schwarzenberg-Czerny, Alex
Schwope, Axel
Scuflaire, Richard
Seeds, Michael
Selam, Selim
Shahbaz, Tariq
Shahul Hameed, Mohin
Shakhovskaya, Nadejda
Shapley, H.
Shara, Michael
Sharma, Dharma
Shenavrin, Victor
Sherwood, William
Silvotti, Roberto
Singh, Harinder
Sinvhal, Shambhu
Sivakoff, Gregory
Skinner, Stephen
Smak, Józef
Smeyers, Paul
Smirnova, Olesja
Smit, Jan
Smith, Horace
Smith, Myron
Smolec, Radosław
Sódor, Ádám
Soliman, Mohamed
Somasundaram, Seetha
Soszyński, Igor
Southworth, John
Srivastava, Ram
Stachowski, Grzegorz
Starrfield, Sumner
Stellingwerf, Robert
Stello, Dennis
Stępień, Kazimierz
Sterken, Christiaan
Stoyanov, Kiril
Strassmeier, Klaus
Stringfellow, Guy

Strom, Karen
Strom, Stephen
Sudar, Davor
Szabados, Laszlo
Szabo, Robert
Szatmary, Karoly
Székely, Péter
Szkody, Paula
TAKATA, Masao
TAKEUTI, Mine
Tammann, Gustav
TAMURA, Shin'ichi
Tarasova, Taya
Taş, Günay
Teixeira, Paula
Teixeira, Teresa
Templeton, Matthew
Terzan, Agop
Tessema, Solomon
Thurston, Mark
Tjin-a-Djie, Herman
TOMINAGA, Nozomu
Tomov, Toma
Torres, Carlos Alberto
Townsend, Richard
Traulsen, Iris
Tremko, Jozef
Tsvetkov, Milcho
Tsvetkova, Katja
Tsygankov, Sergey
Turcu, Vlad
Turner, David
Tutukov, Aleksandr
Tylenda, Romuald
Udovichenko, Sergei
UEMURA, Makoto
UKITA, Nobuharu
Ulla Miguel, Ana
Usher, Peter
Uslenghi, Michela
Utrobin, Victor
Uytterhoeven, Katrien
Vaccaro, Todd
Valeev, Azamat
Valtier, Jean-Claude
Van Doorsselaere, Tom
van Genderen, Arnoud
Van Hoolst, Tim
Vaz, Luiz Paulo
Ventura, Rita
Viotti, Roberto
Vivas, Anna
Vogt, Nikolaus
Volo'shina, Irina
von Braun, Kaspar
Votruba, Viktor
Waelkens, Christoffel
Walker, Edward
Walker, Merle
Walker, William
Wallerstein, George
WANG, Xunhao
Warner, Brian
Watson, Robert
Webbink, Ronald
Wehlau, Amelia
Weis, Kerstin
Weiss, Werner
Welch, Douglas
Wenzel, Wolfgang
Wesson, Roger
Wheatley, Peter
Whitelock, Patricia
Williamon, Richard
Willson, Lee Anne
Wing, Robert
Wittkowski, Markus
Wood, Peter
Worters, Hannah
Woudt, Patrick
XIONG, Da Run
Yüce, Kutluay
YUJI, Ikeda
Zamanov, Radoslav
Zejda, Miloslav
ZHANG, Chengmin
ZHANG, Xiaobin
ZHU, Liying
Zijlstra, Albert
Zoła, Stanisław
Zsoldos, Endre
Zuckerman, Benjamin
Zwintz, Konstanze

Division J - Commission 28 Galaxies

President: John S. Gallagher III

Organizing Committee Members:

Stéphane J. Courteau
Monica Rubio
Marijn Franx
Avishai Dekel
Elena Terlevich
Chanda J. Jog
Naomasa NAKAI
Linda J. Tacconi
Shardha Jogee

Members:

Aalto, Susanne
Ables, Harold
Abrahamian, Hamlet
Adelman, Saul
Adler, David
Afanas'ev, Viktor
Aghaee, Alireza
Aguero, Estela
Aguilar, Luis
Aharonian, Felix
Ahmad, Farooq
Ajhar, Edward

Akashi, Muhammad
AKIYAMA, Masayuki
Alcaino, Gonzalo
Aldaya, Victor
Alexander, Tal
Alladin, Saleh
Allen, Ronald
Allington-Smith, Jeremy
Alloin, Danielle
Almaini, Omar
Aloisi, Alessandra
Alonso, Maria
Alonso, Maria
Alonso-Herrero, Almudena
Ammons, Stephen
Amram, Philippe
Andernach, Heinz
Anderson, Joseph
Andrillat, Yvette
ANN, Hong-Bae
Anosova, Joanna
Anton, Sonia
Antonelli, Lucio Angelo
Antoniou, Vallia
Aoki, Kentaro
Aparicio, Antonio
Aragón-Salamanca, Alfonso
Ardeberg, Arne
Aretxaga, Itziar
Argo, Megan
Arkhipova, Vera
Arnaboldi, Magda
Artamonov, Boris
Athanassoula, Evangelie (Lia)
Aussel, Herve
Avila-Reese, Vladimir
AYANI, Kazuya
Azzopardi, Marc
Baade, W.
BABA, Junichi
Bachev, Rumen
Baddiley, Christopher
Baes, Maarten
BAI, Jinming
Bailey, Mark
Bajaja, Esteban
Baker, Andrew
Baldwin, Jack
Balkowski-Mauger, Chantal
Ballabh, Goswami
Balogh, Michael
Bamford, Steven
Banhatti, Dilip
Bannikova, Elena
Barbon, Roberto
Barcons, Xavier
Barkhouse, Wayne
Barnes, David
Barr, Jordi
Barrientos, Luis
Barth, Aaron
Barthel, Peter
Barton, Elizabeth
Barway, Sudhashu
Bassino, Lilia
Basu, Baidyanath
Battaner, Eduardo
Battinelli, Paolo
Bauer, Amanda
Baum, Stefi
Bautista, Manuel
Bayet, Estelle
Beaulieu, Sylvie
Beck, Rainer
Beck, Sara
Beckmann, Volker
Begeman, Kor
Bender, Ralf
Benedict, George
Benetti, Stefano
Benítez, Erika
Bensby, Thomas
Bentz, Misty
Berczik, Peter
Bergeron, Jacqueline
Berkhuijsen, Elly
Berman, Vladimir
Berta, Stefano
Bertola, Francesco
Bettoni, Daniela
BIAN, Yulin
Bianchi, Simone
Biermann, Peter
Bignall, Hayley
Bijaoui, Albert
Binette, Luc
Binggeli, Bruno
Binney, James
Biretta, John
Birkinshaw, Mark
Björnsson, Claes-Ingvar
Blakeslee, John
Bland-Hawthorn, Jonathan
Blitz, Leo
Block, David
Blumenthal, George
Bodaghee, Arash
Böker, Torsten
Boissier, Samuel
Boisson, Catherine
Boksenberg, Alec
Boles, Thomas
Bolton, J.
Bolzonella, Micol
Bomans, Dominik
Bon, Natasa
Bongiovanni, Angel
Boquien, Médéric
Borchkhadze, Tengiz
Borne, Kirk
Bosma, Albert
Bot, Caroline
Bottinelli, Lucette
Bouwens, Rychard
Bowen, David
Bower, Gary
Braine, Jonathan
Braithwaite, Jonathan
Brammer, Gabriel
Braun, Robert
Bravo-Alfaro, Hector
Brecher, Kenneth
Bressan, Alessandro
Bridges, Terry
Briggs, Franklin
Brinchmann, Jarle
Brinkmann, Wolfgang
Brinks, Elias
Brodie, Jean
Bromberg, Omer
Brosch, Noah
Brosche, Peter
Brough, Sarah
Brouillet, Nathalie
Brown, Michael
Brown, Thomas
Brunner, Robert
Bruzual, Gustavo
Bryant, Julia

Buat, Véronique
Buote, David
Burbidge, Eleanor
Bureau, Martin
Burgarella, Denis
Burkert, Andreas
Burns, Jack
Busarello, Giovanni
Buta, Ronald
Butcher, Harvey
Byrd, Gene
BYUN, Yong Ik
Cabanac, Remi
CAI, Michael
Calderón, Jesús
Calura, Francesco
Calzetti, Daniela
Campusano, Luis
Cannon, John
Cannon, Russell
Cantiello, Michele
Canzian, Blaise
CAO, Li
CAO, Xinwu
Caon, Nicola
Capaccioli, Massimo
Cappellari, Michele
Caproni, Anderson
Caputi, Karina
Caretta, Cesar
Carigi, Leticia
Carollo, Marcella
Carrillo, Rene
Carswell, Robert
Carter, David
Casasola, Viviana
Casoli, Fabienne
Cattaneo, Andrea
Cayatte, Veronique
Cellone, Sergio
Cepa, Jordi
Cesetti, Mary
CHA, Seung-Hoon
CHAE, Kyu Hyun
Chakrabarti, Sandip
Chakrabarty, Dalia
Chamaraux, Pierre
CHANG, Ruixiag
Charmandaris, Vassilis
Chatterjee, Tapan
Chatzichristou, Eleni
Chavushyan, Vahram
Chelliah Subramonian, Stalin
Chelouche, Doron
Chemin, Laurent
Chen, Yanmei
CHEN, Dongni
CHEN, Jiansheng
CHEN, Lin-wen
CHEN, Yang
CHEN, Zhencheng
Chiappini, Cristina
CHIBA, Masashi
Chincarini, Guido
CHOU, Chih-Kang
CHOU, Mei-Yin
Choudhury, Tirthankar
CHU, Yaoquan
Chugai, Nikolaj
CHUN, Sun
Cid Fernandes, Roberto
Cinzano, Pierantonio
Cioni, Maria-Rosa
Ciotti, Luca
Ciroi, Stefano
Clark, David
Clavel, Jean
Clementini, Gisella
Cohen, Ross
Colbert, Edward
Colina, Luis
Combes, Françoise
Comerón, Sébastien
Comte, Georges
Conselice, Christopher
Conti, Alberto
Contopoulos, George
Cook, Kem
Corbett, Elizabeth
Corbin, Michael
Corsini, Enrico
Corwin Jr, Harold
CÃ´te, Patrick
CÃ´té, Stéphanie
Couch, Warrick
Courbin, Frederic
Courteau, Stéphane
Courtes, Georges
Courtois, Helene
Courvoisier, Thierry
Couto da Silva, Telma
Cowsik, Ramanath
Coziol, Roger
Cracco, Valentina
Crane, Philippe
Crawford, Carolin
Cress, Catherine
Cristóbal, David
Croston, Judith
Croton, Darren
Csabai, Istvan
CUI, Wenyuan
Cunniffe, John
Cunow, Barbara
Cypriano, Eduardo
da Costa, Luiz
DAISUKE, Iono
Dalla Bontà, Elena
Dallacasa, Daniele
Danks, Anthony
Dannerbauer, Helmut
Dantas, Christine
Da Rocha, Cristiano
Dasyra, Kalliopi
Davidge, Timothy
Davies, Benjamin
Davies, Jonathan
Davies, Rodney
Davies, Roger
Davis, Marc
Davis, Timothy
De Blok, Erwin
de Boer, Klaas
de Bruyn, A.
de Carvalho, Reinaldo
de Diego Onsurbe, Jose
de Grijs, Richard
de Jong, Roelof
Dejonghe, Herwig
Dekel, Avishai
De Lucia, Gabriella
Demarco, Ricardo
de Mello, Duilia
Demers, Serge
DENG, Zugan
Dennefeld, Michel
de Propris, Roberto
de Rijcke, Sven
de Silva, Gayandhi

Dessauges-Zavadsky, Miroslava
de Swardt, Bonita
Dettmar, Ralf-Juergen
Devost, Daniel
de Zeeuw, Pieter
Diaferio, Antonaldo
Diaz, Angeles
Diaz, Ruben
Díaz-Santos, Tanio
Dibay, E.
Dickey, John
Dietrich, Matthias
d'Odorico, Sandro
DOI, Mamoru
Dokuchaev, Vyacheslav
Dole, Herve
Domínguez, Mariano
Dominis Prester, Dijana
Donas, Jose
Donea, Alina
Dong, Xiao-Bo
Donner, Karl
D'Onofrio, Mauro
Donzelli, Carlos
Dopita, Michael
Doroshenko, Valentina
Dottori, Horacio
Dovciak, Michal
Doyon, Rene
Dressel, Linda
Dressler, Alan
Drinkwater, Michael
Driver, Simon
Duc, Pierre-Alain
Dufour, Reginald
Dultzin-Hacyan, Deborah
Dumont, Anne-Marie
Dunne, Loretta
Durret, Florence
Duval, Marie-France
Eales, Stephen
Edelson, Rick
Edmunds, Michael
Edwards, Louise
Efstathiou, George
Ehle, Matthias
Einasto, Jaan
Ekers, Ronald
Ellis, Richard
Ellis, Simon
Ellison, Sara
Elmegreen, Debra
Elvis, Martin
Elvius, Aina
Elyiv, Andrii
Emsellem, Eric
English, Jayanne
ENOKI, Motohiro
Espey, Brian
Evans, David
Evans, Robert
Fabbiano, Giuseppina
Faber, Sandra
Fabricant, Daniel
Fairall, Anthony
Falceta-Goncalves, Diego
Falco, Emilio
Falcón Barroso, Jesus
Fall, S.
Famaey, Benoit
FAN, Junhui
Farrell, Sean
Fasano, Giovanni
Fathi, Kambiz
Faure, Cécile
Feain, Ilana
Feast, Michael
Feinstein, Carlos
Feitzinger, Johannes
Ferguson, Annette
Ferland, Gary
Ferrarese, Laura
Ferrari, Fabricio
Ferreras, Ignacio
Ferrini, Federico
Field, George
Filippenko, Alexei
Firmani, Claudio
Fletcher, Andrew
Flin, Piotr
Florido, Estrella
Florsch, Alphonse
Floyd, David
Foltz, Craig
Forbes, Duncan
Ford, Holland
Ford Jr, W.
Foschini, Luigi
FOUCAUD, Sebastien
Fouqué, Pascal
Fraix-Burnet, Didier
Francis, Paul
Franx, Marijn
Freedman, Wendy
Freeman, Kenneth
Fricke, Klaus
Fried, Josef
Fritze, Klaus
Fritze-von Alvensleben, Uta
Frogel, Jay
Fuchs, Burkhard
FUJITA, Yutaka
FUKUGITA, Masataka
FUNATO, Yoko
Funes, José
Furlanetto, Steven
Gadotti, Dimitri
Gaensler, Bryan
Gallagher, Sarah
Gallagher III, John
Gallart, Carme
Gallazzi, Anna
Gallego, Jesús
Galletta, Giuseppe
Gallimore, Jack
Gamaleldin, Abdulla
Ganguly, Rajib
GAO, Jian
GAO, Yu
Garcia-Lorenzo, Maria
Gardner, Jonathan
Garilli, Bianca
Gavignaud, Isabelle
Gay, Pamela
Gelderman, Richard
Geller, Margaret
Gentile, Gianfranco
Georgiev, Tsvetan
Gerhard, Ortwin
Ghigo, Francis
Ghosh, P.
Giacani, Elsa
Giani, Elisabetta
Gibson, Brad
Gigoyan, Kamo
Gilbank, David
Giovanardi, Carlo
Giovanelli, Riccardo

Giroletti, Marcello
Gitti, Myriam
Glass, Ian
Glazebrook, Karl
Godłowski, Włodzimierz
Goicoechea, Luis
Gomez, Haley
Gonzalez Delgado, Rosa
Gonzalez-Serrano, Jose Ignacio
Gonzalez-Solares, Eduardo
Goodrich, Robert
Gordon, Karl
Gorgas, Garcia
Goss, W. Miller
GOTO, Tomotsugu
Gottesman, Stephen
Gouguenheim, Lucienne
Graham, Alister
Graham, John
Granato, Gian Luigi
Gray, Meghan
Grebel, Eva
Gregg, Michael
Greve, Thomas
Griffiths, Richard
Grillmair, Carl
Griv, Evgeny
Gronwall, Caryl
Grove, Lisbeth
Grupe, Dirk
GU, Minfeng
GU, Qiusheng
Gunn, James
Günthardt, Guillermo
Guseva, Natalia
Gutiérrez, Carlos
Gyulbudaghian, Armen
Haas, Martin
Hagen-Thorn, Vladimir
Haghi, Hosein
Hakopian, Susanna
HAMABE, Masaru
Hambaryan, Valeri
Hammer, François
HAN, Cheongho
HANAMI, Hitoshi
HAO, Lei
HARA, Tetsuya
Hardy, Eduardo
Harms, Richard
Harnett, Julienne
Hasan, Hashima
Hashimoto, Yasuhiro
HATTORI, Takashi
HATTORI, Makoto
Hatziminaoglou, Evanthia
Hau, George
Haugbølle, Troels
Hayes, Matthew
HE, XiangTao
Heald, George
Heckman, Timothy
Heidmann, Jean
Heidt, Jochen
Heinz, Sebastian
Held, Enrico
Helou, George
Henning, Patricia
Henry, Richard
Hensler, Gerhard
Héraudeau, Philippe
Hernández, Xavier
Hess, Kelley
Hewitt, Adelaide
Hewitt, Anthony
Hicks, Amalia
Hickson, Paul
Hintzen, Paul
HIRASHITA, Hiroyuki
Hjalmarson, Ake
Hjorth, Jens
Ho, Luis
Hodge, Paul
Hoekstra, Hendrik
Hönig, Sebastian
Holz, Daniel
Hopkins, Andrew
Hopp, Ulrich
Horellou, Cathy
Hornschemeier, Ann
Hornstrup, Allan
HOU, Jinliang
Houdashelt, Mark
Hough, James
HU, Fuxing
Hua, Chon Trung
Huang, Jiasheng
HUANG, Keliang
Huchtmeier, Walter
Huertas-Company, Marc
Huettemeister, Susanne
Hughes, David
Humphreys, Elizabeth
Humphreys, Roberta
Hunstead, Richard
Hunt, Leslie
Hunter, James
Huynh, Minh
HWANG, Chorng-Yuan
Ibata, Rodrigo
ICHIKAWA, Shin-ichi
ICHIKAWA, Takashi
Idiart, Thais
Ilić, Dragana
Iliev, Ilian
Illingworth, Garth
IM, Myungshin
IMANISHI, Masatoshi
Impey, Christopher
Infante, Leopoldo
INOUE, Akio
Irwin, Judith
Isaak, Kate
ISHIMARU, Yuhri
Israel, Frank
Issa, Issa
Ivezic, Zeljko
Ivison, Robert
IWAMURO, Fumihide
IWATA, Ikuru
IYE, Masanori
Izotov, Yuri
Izotova, Iryna
Jablonka, Pascale
Jáchym, Pavel
Jaffe, Walter
Jahnke, Knud
JANG, Minwhan
Jarrett, Thomas
Jerjen, Helmut
JIANG, Ing-Guey
Jiménez-Vicente, Jorge
Jog, Chanda
Johansson, Peter
Johnston, Helen
Johnston-Hollitt, Melanie
Joly, Monique
Jones, Christine
Jones, Paul

Jones, Thomas
Jordán, Andrés
Jorgensen, Inger
Joshi, Umesh
Jovanović, Predrag
Joy, Marshall
Jungwiert, Bruno
Junkes, Norbert
Junkkarinen, Vesa
Junor, William
Kaisin, Serafim
KAJISAWA, Masaru
Kalloglian, Arsen
KAMENO, Seiji
Kandalyan, Rafik
Kanekar, Nissim
KANEKO, Noboru
KANG, Xi
Karachentsev, Igor
Karachentseva, Valentina
KAROJI, Hiroshi
Karouzos, Marios
KASHIKAWA, Nobunari
Kaspi, Shai
Kassim, Namir
Katgert, Peter
Katsiyannis, Athanassios
Kauffmann, Guinevere
Kaufman, Michele
Kaviraj, Sugata
Kawada, Mitsunobu
KAWAKATU, Nozomu
Keel, William
Keenan, Ryan
Keeney, Brian
Kellermann, Kenneth
Kelly, Brandon
Kemp, Simon
Kennicutt, Robert
Kent, Brian
Keshet, Uri
Khachikian, Edward
Khanna, Ramon
Khare, Pushpa
Khosroshahi, Habib
Kilborn, Virginia
KIM, Dong Woo
KIM, Ji Hoon
KIM, Minsun
KIM, Sang Chul
KIM, Sungsoo
KIM, Woong-Tae
Kimball, Amy
King, Ivan
Kinman, Thomas
Kirshner, Robert
Kissler-Patig, Markus
Klein, Ulrich
Knapen, Johan
Knapp, Gillian
Knezek, Patricia
Kniazev, Alexei
Knudsen, Kirsten
KO, Chung-Ming
Kobayashi, Chiaki
KOBAYASHI, Masakazu
Koch, Andreas
Kochhar, Rajesh
KODAIRA, Keiichi
KODAMA, Tadayuki
Kogoshvili, Natela
Kollatschny, Wolfram
KOMIYAMA, Yutaka
KONG, Xu
Kontizas, Evangelos
Kontorovich, Victor
Koo, David
Koopmans, Leon
Koratkar, Anuradha
Koribalski, Bärbel
Kormendy, John
Kotilainen, Jari
Kotulla, Ralf
Kovačević, Jelena
KOZASA, Takashi
Kozłowski, Szymon
Kraan-Korteweg, Renée
Krabbe, Angela
Kraft, Ralph
Krajnović, Davor
Krause, Marita
Krishna, Gopal
Kriwattanawong, Wichean
Kron, Richard
Krumholz, Mark
Kudryavtseva, Nadezhda (Nadia)
KUMAI, Yasuki
Kunchev, Peter
Kunert-Bajraszewska, Magdalena
KUNO, Nario
Kunth, Daniel
Kuntschner, Harald
Kunz, Martin
Kuzio de Naray, Rachel
La Barbera, Francesco
Labbe, Ivo
La Franca, Fabio
Lagache, Guilaine
Lake, George
Lal, Dharam
La Mura, Giovanni
Lançon, Ariane
Lanfranchi, Gustavo
Lapi, Andrea
Larsen, Søren
Larson, Richard
Laurikainen, Eija
Layzer, David
Leacock, Robert
Lĕo, Joao Rodrigo
Lebrón, Mayra
Lee, Hyun-chul
Lee, Joon Hyeop
LEE, Myung Gyoon
LEE, Sang-Sung
Leeuw, Lerothodi
Le Fèvre, Olivier
Lehnert, Matthew
Lehto, Harry
Leibundgut, Bruno
Lenc, Emil
Lequeux, James
Levenson, Nancy
Levine, Robyn
LI, Ji
LI, Jing
LI, Qibin
LI, Xiaoqing
LIANG, Yanchun
Lilly, Simon
LIM, Jeremy
Lima Neto, Gastao
LIN, Lihwai
LIN, Weipeng
LIN, Yen-Ting
Lindblad, Per
Linden-Vørnle, Michael

Lintott, Chris
Lo, Fred K. Y.
Lobo, Catarina
Łokas, Ewa
Londrillo, Pasquale
Longo, Giuseppe
Lopes, Paulo
Lopez, Ericson
Lopez Aguerri, Jose Alfonso
López Cruz, Omar
Lopez Hermoso, Maria
López-Sánchez, Angel
Lord, Steven
Loubser, Ilani
Loup, Cecile
Low, Frank
Lowenthal, James
Lu, Limin
LU, Youjun
Lucero, Danielle
Lugger, Phyllis
Luminet, Jean-Pierre
LUO, Ali
Lutz, Dieter
Lynden-Bell, Donald
Lynds, Beverly
Lynds, Roger
MA, Jun
Macalpine, Gordon
Maccagni, Dario
Maccarone, Thomas
Macchetto, Ferdinando
Maciejewski, Witold
Mackey, Alasdair
Mackie, Glen
Macquart, Jean-Pierre
Madden, Suzanne
Madore, Barry
Magorrian, Stephen
Magrini, Laura
Magris, Gladis
Mahtessian, Abraham
Maier, Christian
Mainieri, Vincenzo
Maiolino, Roberto
Makarov, Dmitry
Makarova, Lidia
Malagnini, Maria
Malhotra, Sageeta
Mann, Robert
Mannucci, Filippo
Marcelin, Michel
Marco, Olivier
Marconi, Alessandro
Marín-Franch, Antonio
Marino, Antonietta
Markarjan, B.
Markoff, Sera
Marleau, Francine
Marquez, Isabel
Marr, Jonathon
Marston, Anthony
Martin, Crystal
Martin, Maria
Martin, Rene
Martín, Sergio
Martinet, Louis
Martínez, Vicent
Martini, Paul
Marziani, Paola
Masegosa, Josefa
Massardi, Marcella
Masters, Karen
Mathewson, Donald
MATSUMURA, Tomotake
MATSUSHITA, Kyoko
Matthews, Lynn
Mattila, Seppo
Mauersberger, Rainer
Maurice, Eric
Mayall, Nicholas
Mayya, Divakara
Mazzarella, Joseph
McBreen, Brian
McCracken, Henry
McGaugh, Stacy
McKean, John
McMillan, Paul
McNeil, Stephen
McVittie, George
Mediavilla, Evencio
Mehlert, Dörte
Meier, David
Meikle, William
Meisenheimer, Klaus
Melbourne, Jason
Mel'nik, Anna
Melnyk, Olga
Mendes de Oliveira, Cláudia
Menon, T.
Mercurio, Amata
Merluzzi, Paola
Merrifield, Michael
Metevier, Anne
Meusinger, Helmut
Meyer, Angela
Meyer, Martin
Meza, Andres
Mihov, Boyko
Miley, George
Miller, Eric
Miller, Hugh
Miller, Joseph
Miller, Neal
Miller, Richard
Mills, Bernard
Milvang-Jensen, Bo
Minkowski, R.
MINOWA, Yosuke
Mirabel, Igor
Miroshnichenko, Alla
Misawa, Toru
MIZUNO, Takao
Moiseev, Alexei
Moles Villamate, Mariano
Molinari, Emilio
Molla, Mercedes
Monaco, Pierluigi
Monard, Libert
Moodley, Kavilan
Moody, Joseph
Morelli, Lorenzo
MORI, Masao
MOROKUMA, Tomoki
MOTOHARA, Kentaro
Mould, Jeremy
Mourão, Ana Maria
Mueller, Volker
Müller, Andreas
Mújica, Raul
Mukhopadhyay, Banibrata
Mulchaey, John
Muller, Erik
Muller, Sebastien
Muñoz Tuñón, Casiana
MURAOKA, Kazuyuki
Muratorio, Gerard

MURAYAMA, Takashi
Murphy, Michael
Murray, Stephen
Mushotzky, Richard
Muzzio, Juan
Nagao, Tohru
NAGASHIMA, Masahiro
Nair, Sunita
NAKAI, Naomasa
NAKANISHI, Hiroyuki
NAKANISHI, Kouichiro
NAKATA, Fumiaki
Nakos, Theodoros
Namboodiri, P.
Napolitano, Nicola
Narbutis, Donatas
Narlikar, Jayant
Navarro, Julio
Nedialkov, Petko
Nesvadba, Nicole
Nguyễn, Lan
Nikołajuk, Marek
Ninković, Slobodan
Nipoti, Carlo
Nishikawa, Ken-Ichi
Nityananda, Rajaram
NOGUCHI, Masafumi
Noll, Stefan
Noonan, Thomas
Norman, Colin
Noterdaeme, Pasquier
Nucita, Achille
Nulsen, Paul
Nuza, Sebastian
O'Connell, Robert
Ocvirk, Pierre
O'Dea, Christopher
Oemler Jr, Augustus
Oey, Sally
Ogando, Ricardo
OHTA, Kouji
OKA, Tomoharu
OKAMOTO, Takashi
OKAMURA, Sadanori
Olling, Robert
Olofsson, Kjell
Omizzolo, Alessandro
Oosterloo, Thomas
Orienti, Monica
Origlia, Livia
Osman, Anas
Östlin, Göran
Ostorero, Luisa
Ostriker, Eve
OTA, Naomi
OTSUKI, Kaori
Ott, Juergen
OUCHI, Masami
Ovcharov, Evgeni
Owers, Matthew
OYA, Shin
OYABU, Shinki
Pacholczyk, Andrzej
Page, Mathew
PAK, Soojong
Palmer, Philip
Palumbo, Giorgio
Panessa, Francesca
Pannuti, Thomas
Pantoja, Carmen
Papayannopoulos, Theodoros
PARK, Jang Hyun
Parker, Quentin
Pastoriza, Miriani
Patton, David
Paturel, Georges
Pearce, Frazer
Pedrosa, Susana
Peel, Michael
Peimbert, Manuel
Peletier, Reynier
Pellegrini, Silvia
Pello, Roser
Peng, Eric
PENG, Qingyu
Pentericci, Laura
Perea-Duarte, Jaime
Perez, Fournon
Pérez-González, Pablo
Pérez Torres, Miguel
Peroux, Céline
Perry, Judith
Peters, William
Peterson, Bradley
Peterson, Charles
Petit, Jean-Marc
Petrosian, Artaches
Petrov, Georgi
Petuchowski, Samuel
Pfenniger, Daniel
Philipp-May, Sabine
Phillipps, Steven
Phillips, Mark
Pihlström, Ylva
Pikichian, Hovhannes
Piotrovich, Mikhail
Pipino, Antonio
Pires Martins, Lucimara
Pirzkal, Norbert
Pisano, Daniel
Pizzella, Alessandro
Plana, Henri
Pogge, Richard
Poggianti, Bianca
Polletta, Maria del Carmen
Polyachenko, Evgeny
Pompei, Emanuela
Pooley, David
Popescu, Cristina
Popović, Luka
Portinari, Laura
Poveda, Arcadio
Pović, Mirjana
Prabhu, Tushar
Pracy, Michael
Prandoni, Isabella
Press, William
Prevot-Burnichon, Marie-Louise
Prieto, Almudena
Pritchet, Christopher
Proctor, Robert
Pronik, Vladimir
Proust, Dominique
Prugniel, Philippe
Puech, Mathieu
Puerari, Ivânio
Pulatova, Nadiia
Pustilnik, Simon
Puxley, Phil
Puzia, Thomas
QIN, Yi-Ping
Quinn, Peter
Quintana, Hernan
Rafanelli, Piero
Raiteri, Claudia
Rampazzo, Roberto
Ranalli, Piero
Rand, Richard

Rasmussen, Jesper
Rauch, Michael
Ravindranath, Swara
Raychaudhury, Somak
Read, Andrew
Read, Justin
Reboul, Henri
Recchi, Simone
Rector, Travis
Reddy, Bacham
Reichert, Gail
Rejkuba, Marina
Rekola, Rami
Rephaeli, Yoel
Reshetnikov, Vladimir
Reunanen, Juha
Revaz, Yves
Revnivtsev, Mikhail
Rey, Soo-Chang
Reyes, Reinabelle
Reynolds, Cormac
Ribeiro, André Luis
Richard, Johan
Richer, Harvey
Richstone, Douglas
Richter, Gotthard
Richter, Philipp
Ridgway, Susan
Risaliti, Guido
Rix, Hans-Walter
Rizzi, Luca
Robert, Carmelle
Roberts, Morton
Roberts, Timothy
Roberts Jr, William
Rodrigues de Oliveira Filho, Irapuan
Rödiger, Elke
Roeser, Hermann-Josef
Romano, Patrizia
Romeo, Alessandro
Romero-Colmenero, Encarnacion
Roos, Nicolaas
Rosa, Michael
Rosado, Margarita
Rosa González, Daniel
Rose, James
Rothberg, Barry
Rots, Arnold
Rozas, Maite
Rubin, Vera
Rubio, Monica
RÅ¯žička, Adam
Ryder, Stuart
Sackett, Penny
Sadat, Rachida
Sadler, Elaine
Sadun, Alberto
Sahibov, Firuz
SAITOH, Takayuki
Saiz, Alejandro
Sakai, Shoko
SAKAMOTO, Kazushi
SAKON, Itsuki
Sala, Ferran
Salvador-Sole, Eduardo
Samland, Markus
Samurovi'c, Srdjan
Sanahuja Parera, Blai
Sánchez-Blázquez, Patricia
Sancisi, Renzo
Sandage, A.
Sanders, David
Sanders, Robert
Sanroma, Manuel
Sansom, Anne
Santiago, Basilio
Santos-Lleó, Maria
Sapre, Ashok
Saracco, Paolo
Sarazin, Craig
Sasaki, Toshiyuki
SASAKI, Minoru
Saslaw, William
Sastry, Shankara
Saucedo Morales, Julio
Savage, Ann
Saviane, Ivo
SAWA, Takeyasu
Sawicki, Marcin
Scaramella, Roberto
Schaerer, Daniel
Schaye, Joop
Schechter, Paul
Schmidt, Maarten
Schmitt, Henrique
Schmitz, Marion
Schröder, Anja
Schucking, Engelbert
Schwarz, Ulrich
Schweizer, François
Scodeggio, Marco
Scorza, Cecilia
Scoville, Nicholas
Seigar, Marc
Sellwood, Jerry
Semelin, Benoit
Sempere, Maria
SEON, Kwang il
Sergeev, Sergey
Serjeant, Stephen
Serote Roos, Margarida
Sĕrsic, J.
Setti, Giancarlo
Severgnini, Paola
SHAN, Hongguang
Shapley, H.
Shapovalova, Alla
Sharma, Prateek
Sharp, Nigel
Sharples, Ray
Shastri, Prajval
Shaver, Peter
Shaya, Edward
SHEN, Juntai
SHEN, Zhiqiang
Sherwood, William
Shields, Gregory
Shields, Joseph
SHIMASAKU, Kazuhiro
SHIRAHATA, Mai
SHIRASAKI, Yuji
Shostak, G.
Shukurov, Anvar
Siebenmorgen, Ralf
Siebert, Arnaud
Sigurdsson, Steinn
Sil'chenko, Olga
Sillanpaa, Aimo
Silva, David
Silva, Laura
Simic, Sasa
Simien, François
Simkin, Susan
Singh, Kulinder Pal
Siopis, Christos
Sivakoff, Gregory
Skillman, Evan
Slavcheva-Mihova, Lyuba

Slezak, Eric
Slipher, V.
Smail, Ian
Smecker-Hane, Tammy
Smirnova, Aleksandrina
Smith, Eric
Smith, Haywood
Smith, Malcolm
Soares, Domingos Savio
Sobouti, Yousef
SOHN, Young Jong
Sołtan, Andrzej
SONG, Guo Xuan
SONG, Liming
SORAI, Kazuo
Soria, Roberto
Sparks, William
Spinoglio, Luigi
Spinrad, Hyron
Sridhar, Seshadri
Srinivasan, Ganesan
Stacy, Athena
Stadel, Joachim
Statler, Thomas
Staveley-Smith, Lister
Steenbrugge, Katrien
Steiman-Cameron, Thomas
Steinbring, Eric
Stepanian, Jivan
Stiavelli, Massimo
Stirpe, Giovanna
Stoehr, Felix
Stone, Remington
Stonkutė, Rima
Storchi-Bergmann, Thaisa
Stott, John
Strauss, Michael
Strom, Richard
Strom, Robert
Strubbe, Linda
Stuik, Remko
SU, Cheng-yue
Subrahmanyam, P.
Subramaniam, Annapurni
SUGAI, Hajime
Sulentic, Jack
Sullivan, Mark
Sullivan, III, Woodruff
Sundin, Maria
SUSA, Hajime
Sutherland, Ralph
Tacconi, Linda
Tacconi-Garman, Lowell
Tagger, Michel
TAKADA, Masahiro
TAKAGI, Toshinobu
TAKAHASHI, Rohta
TAKASHI, Hasegawa
TAKATA, Tadafumi
TAKATO, Naruhisa
TAKEUCHI, Tsutomu
TAKIZAWA, Motokazu
Tamm, Antti
Tammann, Gustav
TANAKA, Ichi Makoto
TANAKA, Masayuki
TANAKA, Yutaka
TANIGUCHI, Yoshiaki
Tantalo, Rosaria
TASHIRO, Makoto
Taylor, James
Telles, Eduardo
Tempel, Elmo
Temporin, Sonia
Teng, Stacy
Tenjes, Peeter
Tepper Garcia, Thorsten
Terlevich, Elena
Terlevich, Roberto
Terzian, Yervant
Teuben, Peter
Teyssier, Romain
Thackeray, A.
Thakur, Ratna
Theis, Christian
Thöne, Christina
Thomasson, Magnus
Thonnard, Norbert
Thornley, Michele
Thuan, Trinh
Tiersch, Heinz
Tifft, William
Tikhonov, Nikolai
Tilanus, Remo
Tissera, Patricia
Tisserand, Patrick
Toft, Sune
Tolstoy, Eline
Tomita, Akihiko
TONG, Yi
Toomre, Alar
Torres-Papaqui, Juan
Tovmassian, Hrant
TOYAMA, Kiyotaka
Traat, Peeter
Trager, Scott
Tremaine, Scott
Tremonti, Christy
Trenti, Michele
Tresse, Laurence
Treu, Tommaso
Trimble, Virginia
Trinchieri, Ginevra
TRIPPE, Sascha
Trujillo Cabrera, Ignacio
TSAI, An-Li
TSUCHIYA, Toshio
Tsvetkov, Dmitry
Tuffs, Richard
Tugay, Anatoliy
Tully, Richard
Turner, Edwin
Tyson, John
Tyul'bashev, Sergei
Tzanavaris, Panayiotis
Ulrich, Marie-Helene
Urbanik, Marek
Uslenghi, Michela
Utrobin, Victor
Vaccari, Mattia
Valcheva, Antoniya
Valdes Parra, Jose
Valentijn, Edwin
Vallenari, Antonella
Valluri, Monica
Valotto, Carlos
Valtchanov, Ivan
Valtonen, Mauri
van Albada, Tjeerd
van den Bergh, Sidney
van der Hulst, Jan
van der Kruit, Pieter
van der Laan, Harry
van der Marel, Roeland
van Driel, Wim
van Eymeren, Janine
van Gorkom, Jacqueline
van Kampen, Eelco
van Moorsel, Gustaaf
Vansevicius, Vladas

van Woerden, Hugo
van Zee, Liese
Varela López, Jesús
Varma, Ram
Vasta, Magda
Vaughan, Simon
Vauglin, Isabelle
Vavilova, Iryna
Vazdekis, Alexandre
Vega, Olga
Veilleux, Sylvain
Venters, Tonia
Vercellone, Stefano
Verdes-Montenegro, Lourdes
Verdoes Kleijn, Gijsbert
Vergani, Daniela
Verma, Aprajita
Vermeulen, Rene
Véron, Marie-Paule
Viel, Matteo
Vigroux, Laurent
Viironen, Kerttu
Villata, Massimo
Vivas, Anna
Vlahakis, Catherine
Vlasyuk, Valerij
Voit, Gerard
Vollmer, Bernd
Volonteri, Marta
Vorontsov-Vel'yaminov, B.
Vrtilek, Jan
WADA, Keiichi
Wadadekar, Yogesh
Wagner, Alexander
Wagner, Stefan
WAKAMATSU, Ken-Ichi
Walker, Mark
Walter, Fabian
Walterbos, Rene
WANAJO, Shinya
WANG, Hong-Guang
WANG, Huiyuan
WANG, Junzhi
WANG, Tinggui
WANG, Yiping
WANG, Yu
Ward, Martin
Weedman, Daniel
WEI, Jianyan
Weilbacher, Peter
Weiler, Kurt
Welch, Gary
Westera, Pieter
Westerlund, Bengt
Westmeier, Tobias
White, Simon
Whiting, Matthew
Whitmore, Bradley
Wielebinski, Richard
Wielen, Roland
Wiita, Paul
Wilcots, Eric
Wild, Wolfgang
Williams, Barbara
Williams, Robert
Williams, Theodore
Willis, Jon
Wills, Beverley
Wills, Derek
Wilson, Andrew
Wilson, Gillian
Windhorst, Rogier
Winkler, Hartmut
Winter, Lisa
Wise, Michael
Wisotzki, Lutz
Wofford, Aida
Wold, Margrethe
Wong, Tony
WOO, Jong-Hak
Woosley, Stanford
Worrall, Diana
Woudt, Patrick
Wozniak, Hervé
Wrobel, Joan
Wu, Yanling
WU, Hong
WU, Jianghua
WU, Wentao
WU, Xue-bing
Wulandari, Hesti
Wünsch, Richard
Wynn-Williams, Gareth
Xanthopoulos, Emily
XIA, Xiao-Yang
Xilouris, Emmanouel
XU, Dawei
XUE, Suijian
YAGI, Masafumi
YAHAGI, Hideki
Yakovleva, Valerija
YAMADA, Shimako
YAMADA, Toru
YAMADA, Yoshiyuki
YAMAGATA, Tomohiko
YAMAUCHI, Aya
YI, Sukyoung
YONEHARA, Atsunori
YOSHIDA, Michitoshi
YOSHIKAWA, Kohji
Yun, Min
Zaggia, Simone
Zamorano, Jaime
Zaritsky, Dennis
Zaroubi, Saleem
Zasov, Anatoly
Zavatti, Franco
Zeilinger, Werner
Zepf, Stephen
Zezas, Andreas
Zhang, Yuying
ZHANG, Bo
ZHANG, Fenghui
ZHANG, JiangShui
ZHANG, Jingyi
ZHANG, Xiaolei
ZHANG, Yang
ZHENG, XianZhong
ZHOU, Hongyan
ZHOU, Jianjun
ZHOU, Xu
ZHOU, Youyuan
ZHU, Ming
Zibetti, Stefano
Ziegler, Bodo
Ziegler, Harald
Zinn, Robert
Zirm, Andrew
ZOU, Zhenlong
Zwaan, Martin

Division G - Commission 29 Stellar Spectra

President: Katia Cunha
Vice-President: David R. Soderblom

Organizing Committee Members:

Paul A Crowther
Vanessa M. Hill
Martin Asplund
Jorge Melendez
Nikolai E. Piskunov
Kim A. Venn
David Yong
Wako AOKI
Kenneth G. Carpenter

Members:

Abia, Carlos
Abt, Helmut
Adelman, Saul
Afram, Nadine
Aikman, G.
Airapetian, Vladimir
Ake III, Thomas
Alcalá, Juan Manuel
Alecian, Georges
Alencar, Silvia
Allende Prieto, Carlos
Andretta, Vincenzo
Andreuzzi, Gloria
Andrillat, Yvette
Annuk, Kalju
Antoniou, Vallia
AOKI, Wako
Appenzeller, Immo
Ardila, David
Aret, Anna
Arias, Maria
Arkharov, Arkadij
Artru, Marie-Christine
Asplund, Martin
Atac, Tamer
Audard, Marc
Aufdenberg, Jason
Baade, Dietrich
Bagnulo, Stefano
Bakker, Eric
Baliunas, Sallie
Ballereau, Dominique
Balman, Solen
Banerjee, Dipankar
Baratta, Giovanni
Barber, Robert
Barbuy, Beatriz
Barklem, Paul
Baron, Edward
Basri, Gibor
Batalha, Celso
Bauer, Wendy
Beckman, John
Beers, Timothy
Beiersdorfer, Peter
Bellas-Velidis, Ioannis
Bensby, Thomas
Berger, Jacques
Bertone, Emanuele
Bessell, Michael
Biazzo, Katia
Bikmaev, Ilfan
Boehm, Torsten
Boesgaard, Ann
Boggess, Albert
Bohlender, David
Bon, Natasa
Bond, Howard
Bonifacio, Piercarlo
Bonsack, Walter
Bopp, Bernard
Borczyk, Wojciech
Bouvier, JerÃ´me
Boyarchuk, Alexander
Bragaglia, Angela
Brandi, Elisande
Breysacher, Jacques
Brickhouse, Nancy
Briot, Danielle
Brown, Douglas
Brown, Paul
Bruhweiler, Frederick
Bruning, David
Bruntt, Hans
Bues, Irmela
Burkhart, Claude
Busa', Innocenza
Butkovskaya, Varvara
Butler, Keith
Canto Martins, Bruno
Carlin, Jeffrey
Carney, Bruce
Carpenter, Kenneth
Carretta, Eugenio
Carter, Bradley
casassus, simon
Cassatella, Angelo
Catala, Claude
Catalano, Santo
Catanzaro, Giovanni
Catchpole, Robin
Cayrel, Roger
Cesetti, Mary
Chadid, Merieme
Chavez-Dagostino, Miguel
CHEN, Alfred
CHEN, Yuqin
CHOU, Mei-Yin
Cidale, Lydia
Claudi, Riccardo
Climenhaga, John
Cohen, David
Collet, Remo
Coluzzi, Regina
Conti, Peter

Corbally, Christopher
Cornide, Manuel
Cottrell, Peter
Cowley, Anne
Cowley, Charles
Crowther, Paul
CUI, Wenyuan
Cunha, Katia
Curé, Michel
Dav ci'c, Miodrag
Daflon, Simone
DAISAKU, Nogami
Dall, Thomas
Damineli Neto, Augusto
da Silva, Licio
Davies, Benjamin
de Castro, Elisa
de Groot, Mart
de Laverny, Patrick
DELEUIL, Magali
del Peloso, Eduardo
Depagne, Éric
Derekas, Aliz
Divan, Lucienne
Doazan, Vera
Dolidze, Madona
Doppmann, Gregory
Dougados, Catherine
Dragunova, Alina
Drake, Natalia
Dufour, Patrick
Duncan, Douglas
Dworetsky, Michael
Edwards, Suzan
Elkin, Vladimir
Elmhamdi, Abouazza
Faraggiana, Rosanna
Feast, Michael
Felenbok, Paul
Fernandez-Figueroa, M.
Figer, Donald
Fitzpatrick, Edward
Floquet, Michele
Foing, Bernard
Foy, Renaud
Franchini, Mariagrazia
Francois, Patrick
Frandsen, Soeren
Freire Ferrero, Rubens
Freyhammer, Lars
Friel, Eileen
Fulbright, Jon
Fullerton, Alexander
Gamen, Roberto
García García, Miriam
Garcia-Hernandez, Domingo
García López, Ramón
Garmany, Katy
Garrison, Robert
Gautier, Daniel
Gehren, Thomas
Gerbaldi, Michele
Gershberg, R.
Gęsicki, Krzysztof
Ghosh, Kajal
Giampapa, Mark
Gilra, Daya
Giovannelli, Franco
Glagolevskij, Yurij
Glazunova, Ljudmila
Goebel, John
Gomboc, Andreja
Gonzalez, Guillermo
Gopka, Vera
Gorbaneva, Tatyana
Goswami, Aruna
Grady, Carol
Gratton, Raffaele
Gray, David
Griffin, R. Elizabeth
Griffin, Roger
Grundstrom, Erika
GU, Shenghong
Gustafsson, Bengt
Guthrie, Bruce
Hackman, Thomas
HAN, Inwoo
Hanson, Margaret
Hanuschik, Reinhard
Harmer, Charles
Harmer, Dianne
Hartman, Henrik
Hartmann, Lee
HASHIMOTO, Osamu
Hearnshaw, John
Heber, Ulrich
Heiter, Ulrike
Henrichs, Hubertus
Heske, Astrid
Hessman, Frederic
Hill, Grant
Hill, Vanessa
Hillier, John
Hinkle, Kenneth
HIRAI, Masanori
HIRATA, Ryuko
Hoeflich, Peter
Honda, Satoshi
HORAGUCHI, Toshihiro
Houk, Nancy
Houziaux, Leo
Hron, Josef
HU, Zhong wen
HUANG, Changchung
Hubert-Delplace, Anne-Marie
Hubrig, Swetlana
Huenemoerder, David
Hyland, Harry
Israelian, Garik
Ivans, Inese
IZUMIURA, Hideyuki
Jankov, Slobodan
Jaschek, Carlos
Jehin, Emmanuel
JIANG, Shi-Yang
Johnson, Christian
Johnson, Hollis
Johnson, Jennifer
Jordan, Carole
Josselin, Eric
Käufl, Hans Ulrich
Kawka, Adela
Kipper, Tonu
Kitchin, Christopher
Klochkova, Valentina
Koch, Andreas
Kochukhov, Oleg
KODAIRA, Keiichi
KOGURE, Tomokazu
Kolka, Indrek
Kordi, Ayman
Korn, Andreas
Korotin, Sergey
Kotnik-Karuza, Dubravka
Koubsky, Pavel
Kovachev, Bogomil
Kovács, József
Kovtyukh, Valery

Kraft, Robert
Krempeć-Krygier, Janina
Kučinskas, Arunas
Kwok, Sun
Lago, Maria
Lagrange, Anne-Marie
Laird, John
Lambert, David
Lamers, Henny
Lamontagne, Robert
Landstreet, John
Lanz, Thierry
Lĕo, Joao Rodrigo
Lèbre, Agnes
Leckrone, David
Le Contel, Jean-Michel
Lee, Hyun-chul
LEE, Chung-Uk
LEE, Jae Woo
Leedjarv, Laurits
Lester, John
Letarte, Bruno
Leushin, Valerij
Levato, Orlando
LI, Ji
LIANG, Guiyun
LIANG, Yanchun
Liebert, James
Little-Marenin, Irene
LIU, Guoqing
LIU, Jifeng
LIU, Michael
LIU, Yujuan
Lobel, Alex
Lodders, Katharina
Lopes, Dalton
Lubowich, Donald
Lucatello, Sara
Luck, R.
Lugaro, Maria
Lundstrom, Ingemar
LUO, Ali
Lyubimkov, Leonid
Magain, Pierre
Magazzu, Antonio
Magrini, Laura
Maillard, Jean-Pierre
Mainzer, Amy
Maitzen, Hans
Malaroda, Stella
Manteiga Outeiro, Minia
Marilli, Ettore
Marsden, Stephen
Martinez Fiorenzano, Aldo
Massey, Philip
Mathys, Gautier
Matsuura, Mikako
Mazzali, Paolo
McDavid, David
McGregor, Peter
McSaveney, Jennifer
McSwain, Mary
Mégessier, Claude
Melendez, Jorge
Melnikov, Oleg
Melo, Claudio
Merlo, David
Mickaelian, Areg
Mikulášek, Zdeněk
Moffat, Anthony
Molaro, Paolo
Monaco, Lorenzo
Monin, Dmitry
Montes, David
Moos, Henry
Morel, Thierry
Morossi, Carlo
Morrison, Nancy
Morton, Donald
Napiwotzki, Ralf
Nazarenko, Victor
Naze, Yael
Neckel, Heinz
Neiner, Coralie
Nicholls, Ralph
Niedzielski, Andrzej
Nielsen, Krister
Niemczura, Ewa
Nieva, Maria
Nilsson, Hampus
NISHIMURA, Shiro
Nissen, Poul
Norris, John
North, Pierre
Nugis, Tiit
OKAZAKI, Atsuo
O'Neal, Douglas
O'Toole, Simon
OTSUKA, Masaaki
OTSUKI, Kaori
Oudmaijer, Rene
Owocki, Stanley
Pakhomov, Yury
Pallavicini, Roberto
Parsons, Sidney
Parthasarathy, Mudumba
Pasinetti, Laura
Pavani, Daniela
Pavlenko, Yakov
Pedoussaut, André
Perrin, Marie-Noel
Peters, Geraldine
Peterson, Ruth
Petit, Pascal
Pilachowski, Catherine
Pintado, Olga
Pires Martins, Lucimara
Piskunov, Nikolai
Plavec, Mirek
Plez, Bertrand
Polcaro, V.
Polidan, Ronald
Polosukhina-Chuvaeva, Nina
Pompeia, Luciana
Porto de Mello, Gustavo
Primas, Francesca
Prinja, Raman
Przybylski, Antoni
Querci, Francois
Querci, Monique
Raassen, Ion
Rank-Lueftinger, Theresa
Rao, N.
Rastogi, Shantanu
Rautela, B.
Rauw, Gregor
Rawlings, Mark
Rebolo, Rafael
Reddy, Bacham
Rego, Fernandez
Reid, Warren
Reimers, Dieter
Reiners, Ansgar
Rettig, Terrence
Rhee, Jaehyon
Ringuelet, Adela
Rivinius, Thomas
Romanyuk, Iosif
Rose, James

Rossi, Corinne
Rossi, Lucio
Rossi, Silvia
Russell, H.
Rutten, Robert
Ryan, Sean
Ryde, Nils
Sachkov, Mikhail
SADAKANE, Kozo
Saffe, Carlos
SAKON, Itsuki
Sánchez Almeida, Jorge
Sanwal, Basant
Sareyan, Jean-Pierre
Sarre, Peter
Sasso, Clementina
Sbordone, Luca
Schild, Rudolph
Scholz, Gerhard
Schroeder, Klaus
Schuh, Sonja
Schuler, Simon
Seggewiss, Wilhelm
Selam, Selim
Shetrone, Matthew
SHI, Huoming
SHI, Jianrong
Shimansky, Vladislav
Sholukhova, Olga
Shore, Steven
Simić, Zoran
Simon, Theodore
Simón-Díaz, Sergio
Singh, Harinder
Singh, Mahendra
Sinnerstad, Ulf
Šlechta, Miroslav
Smalley, Barry
Smith, Graeme
Smith, Myron
Smith, Verne
Snow, Theodore
Soderblom, David
Sonneborn, George
Sonti, Sreedhar
Spite, François
Spite, Monique
Stalio, Roberto
Stateva, Ivanka
Stathakis, Raylee
Stawikowski, Antoni
Stecher, Theodore
Steffen, Matthias
Stencel, Robert
St-Louis, Nicole
Sundqvist, Jon
Suntzeff, Nicholas
Svolopoulos, Sotirios
Swings, Jean-Pierre
Szeifert, Thomas
TAKADA-HIDAI, Masahide
TAKAHASHI, Hidenori
TAKASHI, Hasegawa
Talavera, Antonio
Tantalo, Rosaria
Tautvaisiene, Gražina
Thevenin, Frederic
Todt, Helge
Tomasella, Lina
Tomov, Toma
José Miguel Torrejón, Jose Miguel
Torres-Papaqui, Juan
Tripathi, Durgesh
Ulla Miguel, Ana
Ulyanov, Oleg
Usenko, Igor
Utrobin, Victor
UTSUMI, Kazuhiko
Valdivielso, Luisa
Valeev, Azamat
Valenti, Jeff
Valtier, Jean-Claude
Valyavin, Gennady
van der Hucht, Karel
Van Doorsselaere, Tom
van Eck, Sophie
van't Veer-Menneret, Claude
Van Winckel, Hans
Vasta, Magda
Vasu-Mallik, Sushma
Venn, Kim
Vennes, Stephane
Verdugo, Eva
Verheijen, Marc
Vilhu, Osmi
Viotti, Roberto
Vladilo, Giovanni
Vogt, Nikolaus
Vogt, Steven
Vreux, Jean
Wade, Gregg
Wahlgren, Glenn
Wallerstein, George
WANG, Feilu
Waterworth, Michael
Wegner, Gary
Wehinger, Peter
Whelan, Emma
Williams, Peredur
Wilson, Olin
Wing, Robert
Wolff, Sidney
Wood, Brian
Wood, H.
Worters, Hannah
Wright, Nicholas
Wyckoff, Susan
YAMASHITA, Yasumasa
Yong, David
YOSHIOKA, Kazuo
Yüce, Kutluay
Yushkin, Maxim
Zaggia, Simone
Zapatero-Osorio, Maria Rosa
ZHANG, Bo
ZHANG, Haotong
ZHANG, Huawei
ZHANG, Yanxia
ZHAO, Gang
ZHU, Zhenxi
Zoccali, Manuela
Zorec, Juan
Zverko, Juraj

Division B - Commission 30 Radial Velocities

President: Dimitri Pourbaix
Vice-President: Tomaž Zwitter

Organizing Committee Members:

Alain Jorissen
David A. Katz
Matthias Steinmetz
Alceste Z. Bonanos
Tsevi Mazeh
Dante Minniti
Francesco A. Pepe

Members:

Abt, Helmut
Al-Malki, Mohammed
Andersen, Johannes
Arnold, Richard
Balona, Luis
Barbier-Brossat, Madeleine
Batten, Alan
Beavers, Willet
Beers, Timothy
Beuzit, Jean-Luc
Bonanos, Alceste
Borczyk, Wojciech
Bouigue, Roger
Boyajian, Tabetha
Breger, Michel
Buchhave, Lars
Burki, Gilbert
Burnage, Robert
Butkovskaya, Varvara
Butler, Paul
Campbell, Bruce
Cardoso Santos, Nuno
Carney, Bruce
Carquillat, Jean-Michel
Chadid, Merieme
Chemin, Laurent
CHEN, Yuqin
Cochran, William
Couto da Silva, Telma
Crampton, David
Crifo, Francoise
da Costa, Luiz
Davis, Marc
Davis, Robert
de Jonge, J.
de Medeiros, Jose
Derekas, Aliz
De Souza Pellegrini, Paulo
Douglas, Nigel
Dravins, Dainis
Dubath, Pierre
Duflot, Marcelle
Eaton, Joel
Edmondson, Frank
Elkin, Vladimir
Evans, David
Fairall, Anthony
FAN, Yufeng
Fekel, Francis
Fletcher, J.
Florsch, Alphonse
Foltz, Craig
Forveille, Thierry
Gamen, Roberto
Gandolfi, Davide
Garcia, Beatriz
Geller, Aaron
Georgelin, Yvon
Gilmore, Gerard
Giovanelli, Riccardo
Gnedin, Yurij
Gollnow, H.
Gonzalez, Jorge
Gouguenheim, Lucienne
Gray, David
Griffin, Roger
Hakopian, Susanna
Halbwachs, Jean-Louis
HAN, Inwoo
Hearnshaw, John
Hewett, Paul
Hilditch, Ronald
Hill, Graham
Holmberg, Johan
Howard, Andrew
Hrivnak, Bruce
HU, Zhong wen
HUANG, Changchung
Hube, Douglas
Hubrig, Swetlana
Ibata, Rodrigo
Imbert, Maurice
Irwin, Alan
Jasniewicz, Gerard
Jorissen, Alain
Kadouri, Talib
Karachentsev, Igor
Katz, David
Khalesseh, Bahram
Konacki, Maciej
Kovács, József
Krabbe, Angela
Kraft, Robert
Latham, David
Levato, Orlando
Lewis, Brian
Lindgren, Harri
LIU, Yujuan
Lo Curto, Gaspare
Łokas, Ewa
Lovis, Christophe
Marschall, Laurence
Martinez Fiorenzano, Aldo
Maurice, Eric
Mayor, Michel

Mazeh, Tsevi
McClure, Robert
McMillan, Robert
Meibom, Soren
Melnick, Gary
Mermilliod, Jean-Claude
Meylan, Georges
Mink, Jessica
Minniti, Dante
Missana, Marco
Mkrtichian, David
Monaco, Lorenzo
Morbey, Christopher
Morrell, Nidia
Naef, Dominique
Napolitano, Nicola
NARITA, Norio
Nordström, Birgitta
Oetken, L.
Ogando, Ricardo
Pedoussaut, André
Pepe, Francesco
Perrier-Bellet, Christian
Peterson, Ruth
Philip, A.G.
Plaskett, J.
Popov, Viktor
Pourbaix, Dimitri
Preston, George
Prevot, Louis
Pribulla, Theodor
Quintana, Hernan
Rastorguev, Alexey
Ratnatunga, Kavan
Reid, Warren
Romanov, Yuri
Royer, Frédéric
Rubenstein, Eric
Rubin, Vera
Sachkov, Mikhail
SAKAMOTO, Tsuyoshi
Samus, Nikolay
Sanwal, N.
Scarfe, Colin
Scholz, Gerhard
Schröder, Anja
Siebert, Arnaud
Sivan, Jean-Pierre
Skuljan, Jovan
Smith, Myron
Solivella, Gladys
Soubiran, Caroline
Stefanik, Robert
Steinmetz, Matthias
Stickland, David
Strauss, Michael
Suntzeff, Nicholas
Szabados, Laszlo
Thackeray, A.
Tokovinin, Andrei
Tomasella, Lina
Tonry, John
Udry, Stephane
van Dessel, Edwin
Verschueren, Werner
József Vinkó, Jozsef
Walker, Gordon
Wegner, Gary
Willstrop, Roderick
Wittenmyer, Robert
XIAO, Dong
YANG, Stephenson
Zaggia, Simone
ZHANG, Haotong
Zucker, Shay
Zwitter, Tomaž

Division A - Commission 31 Time

President: Mizuhiko HOSOKAWA
Vice-President: Elisa Felicitas Arias

Organizing Committee Members:

Demetrios N. Matsakis
Vladimir E. Zharov
Shougang ZHANG
Philip Tuckey
William Markowitz

Members:

Abele, Maris
AHN, Youngsook
Allan, David
Alley, Carrol
ARAKIDA, Hideyoshi
Archinal, Brent
Arias, Elisa
Becker, Werner
Belocerkovskij, D.
Benavente, Jose
Beutler, Gerhard
Boehm, Johannes
Bonanomi, J.
Breakiron, Lee
Brentjens, Michiel
Brumberg, Victor
Bruyninx, Carine
CAI, Yong
Carter, William
CHOU, Yi
Costain, C.
Damljanovic, Goran
Defraigne, Pascale
Dehant, Véronique

Dick, Wolfgang
Dickey, Jean
DONG, Shaowu
Douglas, R.
DU, Lan
Enslin, Heinz
Essen, L.
Fallon, Frederick
Feissel, Martine
Fliegel, Henry
Foschini, Luigi
FUJIMOTO, Masa-Katsu
Gaignebet, Jean
Gambis, Daniel
GAO, Yuping
Gökmen, Tarik
González, Gabriela
Granveaud, Michel
Guinot, Bernard
GUO, Ji
HAN, Tianqi
HANADO, Yuko
Hobbs, George
Hobiger, Thomas
HOSOKAWA, Mizuhiko
HU, Yonghui
HUA, Yu
IIJIMA, Shigetaka
Ivanov, Dmitrii
Jäggi, Adrian
JIN, WenJing
KAKUTA, Chuichi
Klepczynski, William
Kołaczek, Barbara
Koshelyaevsky, Nikolay
Kovalevsky, Jean
KOYAMA, Yasuhiro
Kwok, Sun
Lammers, Uwe
Le Poncin-Lafitte, Christophe
LI, xiaohui
LIANG, Zhonghuan
Lieske, Jay
LIU, Jinming
LIU, Tao
LU, BenKui
LU, Xiaochun
Luck, John
LUO, Dingchang
LUO, Shi-Fang
MA, Lihua
MA, Zhenguo
Maciesiak, Krzysztof
Malkin, Zinovy
Manchester, Richard
Markowitz, William
Mathur, B.
Matsakis, Demetrios
McCarthy, Dennis
Meinig, Manfred
Melbourne, William
Mendes, Virgilio
MIAO, Yongrui
Mikhailov, M.
Minazzoli, Olivier
Miyadi, Masasi
Morgan, Peter
Mueller, Ivan
Naumov, Vitalij
Newhall, X.
Nice, David
Noel, Fernando
Nörlund, N.
Paquet, Paul
Pavlov, Nikolaj
Penny, C.
Perrier, Général
Petit, Gérard
Petrov, Sergey
Pilkington, John
Pineau des Forets, Guillaume
Popelar, Josef
Potapov, Vladimir
Proverbio, Edoardo
Pugliano, Antonio
Pushkin, Sergej
QI, Guanrong
Ray, James
Ray, Paul
Robertson, Douglas
Rodin, Alexander
Rushton, Anthony
Sampson, R.
Schuler, Walter
Seaman, Rob
SEKIDO, Mamoru
Sheikh, Suneel
Smylie, Douglas
SONG, Jinan
Stairs, Ingrid
Stappers, Benjamin
Stoyko, N.
SUN, Fuping
Thomas, Claudine
Thomson, M.
TSUCHIYA, Atsushi
Tuckey, Philip
van Leeuwen, Joeri
Vannier, J.
Veillet, Christian
Vernotte, François
Vicente, Raimundo
Vilinga, Jaime
WANG, Yulin
Wilkins, George
Winkler, Gernot
Wooden, William
WU, Guichen
WU, Haitao
WU, Shouxian
XIE, Yi
XU, Bang-Xin
YANG, Ke-jun
YANG, Xuhai
Yatskiv, Yaroslav
YE, Shuhua
Zagar, F.
Zhai, Zaocheng
ZHANG, Haotong
ZHANG, Jintong
ZHANG, Shougang
ZHANG, Weiqun
Zharov, Vladimir
ZHENG, Ying
ZHENG, Yong
Zhuang, Qixiang

Division H - Commission 33 Structure & Dynamics of the Galactic System

President: Birgitta Nordström
Vice-President: Jonathan Bland-Hawthorn

Organizing Committee Members:

Evangelie (Lia) Athanassoula
Felix J. Lockman
Chanda J. Jog
Annie C. Robin
Dante Minniti
Sofia Feltzing
Jonathan Bland-Hawthorn

Members:

Aarseth, Sverre
Acosta Pulido, Jose
Adamson, Andrew
Afanas'ev, Viktor
Aguilar, Luis
AK, Serap
Alcobé, Santiago
Allende Prieto, Carlos
Altenhoff, Wilhelm
Ambastha, Ashok
Andersen, Johannes
Ardeberg, Arne
Ardi, Eliani
Arifyanto, Mochamad
Arnold, Richard
Asanok, Kitiyanee
Asplund, Martin
Asteriadis, Georgios
Athanassoula, Evangelie (Lia)
BABA, Junichi
Babusiaux, Carine
BAEK, Chang Hyun
Baier, Frank
Balazs, Lajos
Balbus, Steven
Balcells, Marc
Banhatti, Dilip
Barberis, Bruno
Bartašiute, Stanislava
Bash, Frank
Basu, Baidyanath
Baud, Boudewijn
Bauer, Amanda
Becker, Wilhelm
Bellazzini, Michele
Benjamin, Robert
Bensby, Thomas
Berkhuijsen, Elly
Bienaymé, Olivier
Binney, James
Bland-Hawthorn, Jonathan
Blitz, Leo
Bloemen, Hans
Blommaert, Joris
Bobylev, Vadim
Bodaghee, Arash
Bon, Edi
Borka Jovanović, Vesna
Brand, Jan
Bronfman, Leonardo
Brown, Warren
Burke, Bernard
Burton, W.
Butler, Ray
Caldwell, John
Cane, Hilary
CAO, Zhen
Caretta, Cesar
Carlin, Jeffrey
Carollo, Daniela
Carpintero, Daniel
Carrasco, Luis
Caswell, James
Cesarsky, Catherine
Cesarsky, Diego
Cesetti, Mary
CHA, Seung-Hoon
Chakrabarty, Dalia
Chapman, Jessica
Chemin, Laurent
CHEN, Li
CHEN, Li
CHEN, Yuqin
CHEN, Zhen
CHIBA, Masashi
CHOU, Mei-Yin
Christodoulou, Dimitris
Churchwell, Edward
Cincotta, Pablo
Cioni, Maria-Rosa
Clemens, Dan
Clube, S.
Cohen, Richard
Comerón, Sébastien
Comins, Neil
Contopoulos, George
Corradi, Romano
Costa, Edgardo
Courtes, Georges
Crampton, David
Crawford, David
Crézé, Michel
Cropper, Mark
Croton, Darren
Cubarsi, Rafael
Cudworth, Kyle
Cuperman, Sami
Dalla Bontà, Elena
Dambis, Andrei
Dauphole, Bertrand
Davies, Rodney

Dawson, Peter
de Jong, Teije
Dejonghe, Herwig
Dekel, Avishai
de Silva, Gayandhi
Diaferio, Antonaldo
Diaz, Ruben
Dickel, Helene
Dickel, John
Dickman, Robert
Dieter Conklin, Nannielou
Djorgovski, Stanislav
do Nascimento, José
Downes, Dennis
Drilling, John
Drimmel, Ronald
Ducati, Jorge
Ducourant, Christine
Dzigvashvili, R.
Edmondson, Frank
Efremov, Yurij
Egret, Daniel
Einasto, Jaan
Elmegreen, Debra
ESAMDIN, Ali
Esguerra, Jose Perico
ESIMBEK, Jarken
Evangelidis, E.
Evans, Wyn
Faber, Sandra
Fathi, Kambiz
Feast, Michael
Feitzinger, Johannes
Feltzing, Sofia
Fenkart, Rolf
Ferguson, Annette
Fernández, David
Ferrari, Fabricio
Figueras, Francesca
Flynn, Chris
Foster, Tyler
Freeman, Kenneth
Fuchs, Burkhard
FUJIMOTO, Masa-Katsu
FUJIMOTO, Mitsuaki
FUJIWARA, Takao
FUKUNAGA, Masataka
Fux, Roger
Ganguly, Rajib
Garzón, Francisco
Gemmo, Alessandra
Genkin, Igor
Genzel, Reinhard
Georgelin, Yvon
Georgelin, Yvonne
Gerhard, Ortwin
Gilmore, Gerard
Goldreich, Peter
Gomez, Ana
Gordon, Mark
Gottesman, Stephen
Grayzeck, Edwin
Green, Anne
Green, James
Grenon, Michel
Grillmair, Carl
Grindlay, Jonathan
GU, Minfeng
Gupta, Sunil
HABE, Asao
Habing, Harm
Haghi, Hosein
Hakkila, Jon
HAMAJIMA, Kiyotoshi
HANAMI, Hitoshi
Hanson, Margaret
Hartkopf, William
Hawkins, Michael
Hayli, Abraham
Haywood, Misha
Heiles, Carl
Helmi, Amina
Herbst, William
Herman, Jacobus
Hernández-Pajares, Manuel
Hetem Jr., Annibal
Heyl, Jeremy
Holmberg, Johan
HONMA, Mareki
HORI, Genichiro
HOZUMI, Shunsuke
Hron, Josef
Hulsbosch, A.
Humphreys, Elizabeth
Humphreys, Roberta
Ibata, Rodrigo
IGUCHI, Osamu
IKEDA, Norio
IKEUCHI, Satoru
ISHIHARA, Daisuke
Israel, Frank
Ivezic, Zeljko
Iwaniszewska, Cecylia
IYE, Masanori
Jablonka, Pascale
Jáchym, Pavel
Jackson, Peter
Jahreiss, Hartmut
Jalali, Mir Abbas
Jaschek, Carlos
Jasniewicz, Gerard
JIANG, Dongrong
JIANG, Ing-Guey
Jønch-Sørensen, Helge
Jog, Chanda
Johansson, Peter
Jones, Derek
Kaisin, Serafim
Kalnajs, Agris
KANG, Yong-Hee
Kasumov, Fikret
KATO, Shoji
Khovritchev, Maxim
Khrutskaya, Evgenia
KIM, Sang Chul
KIM, Sungsoo
KIM, Woong-Tae
King, Ivan
Kinman, Thomas
Klare, Gerhard
Knapp, Gillian
Korchagin, Vladimir
Kormendy, John
Krajnović, Davor
Kučinskas, Arunas
Kulsrud, Russell
Kutuzov, Sergei
Kuzmin, Grigori
Lafon, Jean-Pierre
Laloum, Maurice
Larson, Richard
Latham, David
LEE, Hyung-Mok
LEE, Kang Hwan
LEE, Myung Gyoon
LEE, Sang-Gak
Lepine, Sebastien
LI, Jing
LI, Jinzeng
Liebert, James

LIN, Qing
Lindblad, Per
Lockman, Felix
Lodén, Kerstin
Loden, Lars
Łokas, Ewa
LU, Youjun
LUO, Ali
Lynden-Bell, Donald
Lynga, Gosta
MacConnell, Darrell
Mackey, Alasdair
Maier, Christian
Majumdar, Subhabrata
Manchester, Richard
Mandel, Ilya
Marín-Franch, Antonio
Marochnik, Leonid
Martin, Christopher
Martínez-Delgado, David
Martos, Marco
Mateu, Cecilia
Mathewson, Donald
MATSUNAGA, Noriyuki
Matteucci, Francesca
Mayor, Michel
McBride, Vanessa
McClure-Griffiths, Naomi
McCuskey, S.
McGregor, Peter
McMillan, Paul
Meatheringham, Stephen
Mel'nik, Anna
Méndez Bussard, Rene
Merrifield, Michael
Migaszewski, Cezary
Mikkola, Seppo
Miller, Richard
Minniti, Dante
Mirabel, Igor
Mishurov, Yury
Miszalski, Brent
Miyamoto, Masanori
Moffat, Anthony
Mohammed, Ali
Moitinho, André
Monet, David
Monnet, Guy
Morales Rueda, Luisa
Moreno Lupiañez, Manuel
Morris, Mark
Morris, Rhys
Muench, Guido
NAKASATO, Naohito
Namboodiri, P.
Napolitano, Nicola
Narbutis, Donatas
Neckel, Th.
Nelemans, Gijs
Nelson, Alistair
Newberg, Heidi
Nikiforov, Igor
Ninković, Slobodan
NISHIDA, Minoru
NISHIDA, Mitsugu
Nordström, Birgitta
Norman, Colin
Nota, Antonella
Nuritdinov, Salakhutdin
Oblak, Edouard
Ocvirk, Pierre
Odenkirchen, Michael
Oey, Sally
Ogorodnikov, K.
OH, Kap Soo
Oja, Tarmo
Ojha, Devendra
OKA, Tomoharu
OKUDA, Haruyuki
Olano, Carlos
Ollongren, A.
Oluseyi, Hakeem
Orlov, Viktor
Ortiz, Roberto
Ostorero, Luisa
Ostriker, Eve
Ostriker, Jeremiah
Palmer, Patrick
Palouš, Jan
Pandey, A.
Pandey, Birendra
Papayannopoulos, Theodoros
PARK, Byeong-Gon
Parmentier, Geneviève
Patsis, Panos
Pauls, Thomas
Peimbert, Manuel
Peng, Eric
Perek, Luboš
Perets, Hagai
Perryman, Michael
Pesch, Peter
Philip, A.G.
Pier, Jeffrey
Pietrukowicz, Paweł
Pirzkal, Norbert
Polyachenko, Evgeny
Portinari, Laura
Price, R.
Priester, Wolfgang
Rabolli, Monica
Raharto, Moedji
Rastegaev, Denis
Ratnatunga, Kavan
Read, Justin
Recio-Blanco, Alejandra
Reid, Iain
Reif, Klaus
Reylé, Céline
Rhee, Jaehyon
Rho, Jeonghee
Rich, Robert
Richter, Philipp
Riegel, Kurt
Roberts, Morton
Roberts Jr, William
Robin, Annie
Rocha-Pinto, Hélio
Rodrigues de Oliveira Filho, Irapuan
Rohlfs, Kristen
RONG, Jianxiang
Rubin, Vera
Ruelas-Mayorga, R.
Ruiz, Maria Teresa
RÅ¯žička, Adam
Rybicki, George
Saar, Enn
SAKAMOTO, Tsuyoshi
Sakano, Masaaki
Sala, Ferran
Sánchez Doreste, Néstor
Sanchez-Saavedra, M.
Sandqvist, Aage
Santiago, Basilio
Santillán, Alfredo
Sanz, Jaume
Sargent, Annelia
Schechter, Paul

Schmidt, Maarten
Schmidt-Kaler, Theodor
Schödel, Rainer
Seggewiss, Wilhelm
Seimenis, John
Sellwood, Jerry
Serabyn, Eugene
SHAN, Hongguang
Shane, William
SHEN, Juntai
SHI, Huoming
SHIMIZU, Tsutomu
SHU, Frank
Siebert, Arnaud
Sigalotti, Leonardo
Simonson, S.
Sobouti, Yousef
Solomon, Philip
SONG, Guo Xuan
SONG, Liming
SONG, Qian
Sotnikova, Natalia
Soubiran, Caroline
Sparke, Linda
Spergel, David
Spiegel, Edward
Stecker, Floyd
Steiman-Cameron, Thomas
Steinlin, Uli
Steinmetz, Matthias
Stoehr, Felix
Stone, Jennifer
Strobel, Andrzej
Strubbe, Linda
Sturch, Conrad
SU, Cheng-yue
Subramaniam, Annapurni
SUMI, Takahiro
Surdin, Vladimir
Svolopoulos, Sotirios
Sygnet, Jean-Francois
TAKASHI, Hasegawa
Tammann, Gustav
TANAKA, Ichi Makoto
Tempel, Elmo
Terzides, Charalambos
Teyssier, Romain
Thé, Pik-Sin
Thielheim, Klaus
Thomas, Claudine
TIAN, Wenwu
Tinney, Christopher
Tobin, William
TOMISAKA, Kohji
TONG, Yi
Toomre, Alar
Toomre, Juri
Torra, Jordi
Torres-Papaqui, Juan
TOSA, Makoto
Trefzger, Charles
TSENG, Yao-Huan
TSUJIMOTO, Takuji
Turon, Catherine
Upgren, Arthur
Urquhart, James
Valluri, Monica
Valtonen, Mauri
van der Kruit, Pieter
Vandervoort, Peter
Van Rhijn, P.
van Woerden, Hugo
Varela Perez, Antonia
Vega, E.
Venugopal, V.
Vergne, María
Verschuur, Gerrit
Vetesnik, Miroslav
Villas da Rocha, Jaime
Vivas, Anna
Volkov, Evgeni
Volonteri, Marta
Wachlin, Felipe
Wagner, Alexander
Weaver, Harold
Weistrop, Donna
Westerlund, Bengt
Whiteoak, John
Whittet, Douglas
Wielebinski, Richard
Wielen, Roland
Woltjer, Lodewijk
Wong, Tony
Woodward, Paul
Wouterloot, Jan
Wramdemark, Stig
Wright, Nicholas
Wulandari, Hesti
Wünsch, Richard
Wyse, Rosemary
XU, Ye
YAMADA, Shimako
YAMAGATA, Tomohiko
YIM, Hong Suh
YOSHII, Yuzuru
Younis, Saad
YU, Qingjuan
Zachilas, Loukas
Zaggia, Simone
ZHANG, Bin
ZHANG, Haotong
ZHOU, Jianjun
ZHU, Qingfeng
Zoccali, Manuela

Division H - Commission 34 Interstellar Matter

President: Sun Kwok
Vice-President: Bon-Chul KOO

Organizing Committee Members:

Thomas Henning
Paola Caselli
Ji YANG
Mika J. Juvela
Laszlo Viktor Toth
Sylvie Cabrit
Elisabete M. de Gouveia
Dal Pino
Michal Różyczka
Dieter Breitschwerdt
Masato TSUBOI
Michael G. Burton
Susana Lizano

Members:

Aannestad, Per
Abgrall, Herve
Acker, Agnes
Adams, Fred
AIKAWA, Yuri
Aitken, David
AKABANE, Kenji
Akashi, Muhammad
Alcolea, Javier
Al-Mostafa, Zaki
Altenhoff, Wilhelm
Alves, João
Ambrocio-Cruz, Silvia
Andersen, Anja
Andersen, Morten
Andersson, B-G
Andrillat, Yvette
Andronov, Ivan
Anglada, Guillem
Arbutina, Bojan
Ardila, David
Arkhipova, Vera
Arny, Thomas
Arthur, Jane
ASANO, Katsuaki
Asanok, Kitiyanee
Audard, Marc
Azcarate, Diana
Baars, Jacob
Baart, Edward
BABA, Junichi
Babkovskaia, Natalia
Bachiller, Rafael
BAEK, Chang Hyun
Baker, Andrew
Ballesteros-Paredes, Javier
Balser, Dana
Baluteau, Jean-Paul
Bania, Thomas
Barlow, Michael
Barnes, Aaron
Baryshev, Andrey
Bash, Frank
Basu, Shantanu
Baudry, Alain
Bautista, Manuel
Bayet, Estelle
Becklin, Eric
Beckman, John
Beckwith, Steven
Bedogni, Roberto
Benaydoun, Jean-Jacques
Benisty, Myriam
Bergeron, Jacqueline
Bergin, Edwin
Bergman, Per
Bergström, Lars
Berkhuijsen, Elly
Bernat, Andrew
Bertout, Claude
Bhat, Ramesh
Bhatt, H.
Bianchi, Luciana
Bieging, John
Bignall, Hayley
Bignell, R.
Binette, Luc
Black, John
Blades, John
Blair, Guy
Blair, William
Bless, Robert
Blitz, Leo
Bloemen, Hans
Bobrowsky, Matthew
Bocchino, Fabrizio
Bochkarev, Nikolai
Bode, Michael
Bodenheimer, Peter
Boeshaar, Gregory
Boggess, Albert
Bohlin, Ralph
Boisse, Patrick
Boland, Wilfried
Bontemps, Sylvain
Boquien, Médéric
Bordbar, Gholam
Borgman, Jan
Borka Jovanović, Vesna
Borkowski, Kazimierz
Bot, Caroline
Boulanger, Francois
Boumis, Panayotis
Bourke, Tyler
Bouvier, JerÃ´me
Bowen, David
Brand, Jan
Brand, Peter
Breitschwerdt, Dieter

Briceño, Cesar
Brinkmann, Wolfgang
Bromage, Gordon
Brooks, Kate
Brouillet, Nathalie
Brown, Ronald
Bruhweiler, Frederick
Bujarrabal, Valentin
Burke, Bernard
Burton, Michael
Burton, W.
Bychkov, Konstantin
Bykov, Andrei
Bzowski, Maciej
Cabrit, Sylvie
CAI, Kai
Cambrésy, Laurent
Cami, Jan
Canto, Jorge
Caplan, James
Cappa de Nicolau, Cristina
Capriotti, Eugene
Capuzzo Dolcetta, Roberto
Carretti, Ettore
Carruthers, George
Casasola, Viviana
casassus, simon
Caselli, Paola
Castañeda, Héctor
Castelletti, Gabriela
Caswell, James
Cattaneo, Andrea
cazaux, Stephanie
Ceccarelli, Cecilia
Cecchi-Pestellini, Cesare
Centurión Martin, Miriam
Cernicharo, Jose
Cerruti Sola, Monica
Cersosimo, Juan
Cesarsky, Catherine
Cesarsky, Diego
CHA, Seung-Hoon
Chandra, Suresh
Chelouche, Doron
CHEN, Huei-Ru
CHEN, Xuefei
CHEN, Yafeng
CHEN, Yang
CHENG, Kwang
Cherchneff, Isabelle
Chevalier, Roger
CHIHARA, Hiroki
Chini, Rolf
Chopinet, Marguerite
Christopoulou, Panagiota-Eleftheria
Chu, You-Hua
Churchwell, Edward
Ciardullo, Robin
Cichowolski, Silvina
Ciroi, Stefano
Clark, Frank
Clarke, David
Clegg, Robin
Code, Arthur
Codella, Claudio
Coffey, Deirdre
Cohen, Marshall
Colangeli, Luigi
Collin, Suzy
Combes, Françoise
Corbelli, Edvige
Corradi, Romano
Corradi, Wagner
Costantini, Elisa
Costero, Rafael
Courtes, Georges
Cowie, Lennox
Cox, Donald
Cox, Pierre
Coyne, S.J, George
Cracco, Valentina
Crane, Philippe
Crawford, Ian
Crovisier, Jacques
Cruvellier, Paul
Cudaback, David
Cuesta Crespo, Luis
Cunningham, Maria
Czyzak, Stanley
Dahn, Conard
Dale, James
Dalgarno, Alexander
Danks, Anthony
Danly, Laura
Dannerbauer, Helmut
Davies, Jonathan
Davies, Rodney
Davis, Christopher
Davis, Timothy
de Almeida, Amaury
de Avillez, Miguel
De Bernardis, Paolo
de Boer, Klaas
De Buizer, James
Decourchelle, Anne
de Gouveia Dal Pino, Elisabete
de Gregorio-Monsalvo, Itziar
DEGUCHI, Shuji
Deharveng, Lise
Deiss, Bruno
de Jong, Teije
de La NoÃ«, Jerome
De Marco, Orsola
Dennefeld, Michel
Dent, William
Dewdney, Peter
d'Hendecourt, Louis
Dias da Costa, Roberto
Diaz, Ruben
Díaz-Santos, Tanio
Dib, Sami
Dickel, Helene
Dickel, John
Dickey, John
Dieleman, Pieter
Di Fazio, Alberto
Dinerstein, Harriet
Dinh, Trung
Dionatos, Odysseas
Disney, Michael
Djamaluddin, Thomas
Docenko, Dmitrijs
d'Odorico, Sandro
Dokuchaev, Vyacheslav
Dokuchaeva, Olga
Dominik, Carsten
Dopita, Michael
Dorschner, Johann
Dottori, Horacio
Dougados, Catherine
Downes, Dennis
Draine, Bruce
Dreher, John
Dubner, Gloria
Dubout, Renee
Dudorov, Aleksandr
Dufour, Reginald

Duley, Walter
Dunne, Loretta
Dupree, Andrea
Dutrey, Anne
Duvert, Gilles
Dwarkadas, Vikram
Dwek, Eli
Edwards, Suzan
Egan, Michael
Ehlerová, Soňa
Eisloeffel, Jochen
Elia, Davide
Elitzur, Moshe
Elliott, Kenneth
Ellison, Sara
Elmegreen, Bruce
Elmegreen, Debra
Elvius, Aina
Emerson, James
Encrenaz, Pierre
ESAMDIN, Ali
Escalante, Vladimir
ESIMBEK, Jarken
Esipov, Valentin
Esteban, César
Evans, Aneurin
Evans, Neal
Falceta-Goncalves, Diego
Falgarone, Edith
Falize, Emeric
Falk Jr, Sydney
Falle, Samuel
Federman, Steven
Feitzinger, Johannes
Felli, Marcello
Fendt, Christian
Ferland, Gary
Ferlet, Roger
Fernandes, Amadeu
Ferriere, Katia
Ferrini, Federico
Fesen, Robert
Fiebig, Dirk
Field, David
Field, George
Fierro, Julieta
Figer, Donald
Fischer, Jacqueline
Flannery, Brian
Fleck, Robert
Fletcher, Andrew
Florido, Estrella
Flower, David
Folini, Doris
Ford, Holland
Forster, James
Franco, Gabriel Armando
Franco, José
Fraschetti, Federico
Fraser, Helen
Freimanis, Juris
Frew, David
Fridlund, Malcolm
Frisch, Priscilla
Fromang, Sebastien
Fuente, Asuncion
Fukuda, Naoya
FUKUI, Yasuo
Fuller, Gary
Furniss, Ian
FURUYA, Ray
Gaensler, Bryan
Galli, Daniele
GAO, Jian
GAO, Yu
Garay, Guido
Garcia, Paulo
Garcia-Hernandez, Domingo
Garcia-Lario, Pedro
Garcia-Segura, Guillermo
Garnett, Donald
Gathier, Roel
Gaume, Ralph
Gaustad, John
Gay, Jean
Geballe, Thomas
Genzel, Reinhard
Georgelin, Yvon
Gerard, Eric
Gerin, Maryvonne
Gerola, Humberto
Gezari, Daniel
Ghanbari, Jamshid
Giacani, Elsa
Gibson, Steven
Gilra, Daya
Giovanelli, Riccardo
Glover, Simon
Goddi, Ciriaco
Goebel, John
Goldes, Guillermo
Goldreich, Peter
Goldsmith, Donald
Golovatyj, Volodymyr
Gomez, Gonzalez
Gonçalves, Denise
Gonzales-Alfonso, Eduardo
Goodman, Alyssa
Gordon, Courtney
Gordon, Karl
Gordon, Mark
Gosachinskij, Igor
Goss, W. Miller
Graham, David
Granato, Gian Luigi
Gredel, Roland
Green, Anne
Green, James
Gregorio-Hetem, Jane
Greisen, Eric
Grewing, Michael
Guelin, Michel
Guertler, Joachim
Guesten, Rolf
GUILLOTEAU, Stéphane
Gull, Theodore
Günthardt, Guillermo
GUO, Jianheng
Habing, Harm
Hackwell, John
Haisch Jr, Karl
HANAMI, Hitoshi
HAO, Lei
Hardebeck, Ellen
Harrington, J.
Harris, Alan
(Harris) Law, Stella
Harten, Ronald
Hartl, Herbert
Hartquist, Thomas
Harvey, Paul
Hatchell, Jennifer
Haverkorn, Marijke
Hayashi, Saeko
Haynes, Raymond
HE, Jinhua
Hébrard, Guillaume
Hecht, James
Heikkilä, Arto

Heiles, Carl
Hein, Righini
Heinz, Sebastian
Helfer, H.
Helmich, Frank
Helou, George
Heng, Kevin
Henkel, Christian
Henney, William
Henning, Thomas
Herbstmeier, Uwe
Hernández, Jesús
Herpin, Fabrice
HERSANT, Franck
Heydari-Malayeri, Mohammad
Heyer, Mark
Hidayat, Bambang
Higgs, Lloyd
Hildebrand, Roger
Hillenbrand, Lynne
Hippelein, Hans
HIRANO, Naomi
Hiriart, David
HIROMOTO, Norihisa
HIROSE, Shigenobu
Hjalmarson, Ake
Hobbs, Lewis
Höglund, Bertil
Hollenbach, David
Hollis, Jan
HONDA, Mitsuhiko
HONG, Seung-Soo
Hora, Joseph
Horácek, Jiri
Horns, Dieter
Houde, Martin
Houziaux, Leo
Hua, Chon Trung
Hudson, Reggie
Huggins, Patrick
Hulsbosch, A.
Hutchings, John
Hutsemekers, Damien
Hyung, Siek
IKEDA, Norio
Ikonnikova, Natalia
Il'in, Vladimir
INOUE, Akio
INUTSUKA, Shu-ichiro
Irvine, William
ISHIHARA, Daisuke
ISHII, Miki
Israel, Frank
Issa, Issa
ITOH, Hiroshi
ITOH, Yoichi
Iyengar, K.
Jabir, Niama
Jackson, James
Jacoby, George
Jacq, Thierry
Jaffe, Daniel
Jahnke, Knud
Jenkins, Edward
JIANG, Zhibo
Jiménez-Vicente, Jorge
JIN, Zhenyu
Joblin, Christine
Johnson, Fred
Johnston, Kenneth
Johnstone, Douglas
Jones, Christine
Jones, David
Jørgensen, Jes
Jørgensen, Uffe
Jourdain de Muizon, Marie
Jura, Michael
Just, Andreas
Justtanont, Kay
Juvela, Mika
Kafatos, Menas
Kaftan, May
KAIFU, Norio
Kalenskii, Sergei
Kaler, James
KAMAYA, Hideyuki
KAMAZAKI, Takeshi
KAMEGAI, Kazuhisa
KAMIJO, Fumio
Kamp, Inga
Kanekar, Nissim
Kantharia, Nimisha
Käpylä, Maarit
Karakas, Amanda
Kassim, Namir
Kaviraj, Sugata
Kawada, Mitsunobu
Keene, Jocelyn
Kegel, Wilhelm
Keheyan, Yeghis
Kemper, Francisca
Kennicutt, Robert
Khesali, Ali
KIM, Ji Hoon
KIM, Jongsoo
KIM, Woong-Tae
KIMURA, Hiroshi
KIMURA, Toshiya
King, David
Kirkpatrick, Ronald
Kirshner, Robert
Kiss, Csaba
Klessen, Ralf
Knacke, Roger
Knapp, Gillian
Knezek, Patricia
Knude, Jens
KO, Chung-Ming
KOBAYASHI, Naoto
Kohoutek, Lubos
KOIKE, Chiyoe
KONDO, Yoji
KONG, Xu
KOO, Bon-Chul
Koornneef, Jan
Korpela, Eric
Kóspál, Ágnes
KOZASA, Takashi
Krabbe, Angela
Krajnović, Davor
Kramer, Busaba
Krautter, Joachim
Kravchuk, Sergei
Kreysa, Ernst
Krishna, Swamy
Krumholz, Mark
KUAN, Yi-Jehng
KUDOH, Takahiro
Kuiper, Rolf
Kuiper, Thomas
Kulhánek, Petr
Kumar, C.
Kunth, Daniel
Kutner, Marc
Kwitter, Karen
Kwok, Sun
Kylafis, Nikolaos
Lada, Charles
Lafon, Jean-Pierre

LAI, Shih-Ping
Laloum, Maurice
Langer, William
Latter, William
Laureijs, Rene
Laurent, Claudine
Lauroesch, James
Lazarian, Alexandre
Lazio, Joseph
Lẽo, Joao Rodrigo
Lebrón, Mayra
Lee, Terence
LEE, Dae Hee
LEE, Hee Won
LEE, Jeong-Eun
LEE, Jung-Won
LEE, Myung Gyoon
Lefloch, Bertrand
Leger, Alain
Lehtinen, Kimmo
Leisawitz, David
Lépine, Jacques
Lequeux, James
Le Squeren, Anne-Marie
Leto, Giuseppe
Leung, Chun
LI, Jinzeng
LIANG, Yanchun
Ligori, Sebastiano
Likkel, Lauren
Liller, William
Limongi, Marco
LIN, Weipeng
Linke, Richard
Linnartz, Harold
Lis, Dariusz
Liseau, René
Liszt, Harvey
LIU, Sheng-Yuan
LIU, Xiaowei
Lizano, Susana
Lloyd, Myfanwy
Lo, Fred K. Y.
Lo, Wing-Chi Nadia
Lockman, Felix
Lodders, Katharina
Loinard, Laurent
López Garcia, Jose
Loren, Robert
Lortet, Marie-Claire
Louise, Raymond
Lovas, Francis
Low, Frank
Lozinskaya, Tatjana
Lucas, Robert
Lucero, Danielle
LUO, Shaoguang
Lynds, Beverly
Lyon, Ian
MA, Jun
Maciel, Walter
MacLeod, John
Mac Low, Mordecai-Mark
Madsen, Gregory
Magrini, Laura
Maier, Christian
MAIHARA, Toshinori
MAKIUTI, Sin'itirou
Malbet, Fabien
Mallik, D.
Mampaso, Antonio
Manchado, Arturo
Manchester, Richard
Manfroid, Jean
Mardones, Diego
Maret, Sébastien
Marleau, Francine
Marston, Anthony
Martin, Christopher
Martin, Peter
Martin, Robert
Martín, Sergio
Martin-Pintado, Jesus
Masson, Colin
Mather, John
Mathews, William
Mathewson, Donald
Mathis, John
MATSUHARA, Hideo
MATSUMOTO, Tomoaki
MATSUMURA, Tomotake
MATSUMURA, Masafumi
Matsuura, Mikako
Mattila, Kalevi
Mauersberger, Rainer
McCall, Benjamin
McCall, Marshall
McClure-Griffiths, Naomi
Mccombie, June
McCray, Richard
McGee, Richard
McGehee, Peregrine
McGregor, Peter
McKee, Christopher
McNally, Derek
Meaburn, John
Mebold, Ulrich
Mehringer, David
Meier, Robert
Meixner, Margaret
Mellema, Garrelt
Melnick, Gary
Mendez, Roberto
Mennella, Vito
Menon, T.
Menten, Karl
Menzies, John
Meszaros, Peter
Meyer, Martin
Millar, Thomas
Miller, Eric
Miller, Joseph
Milne, Douglas
Minier, Vincent
MINN, Young-Ki
Minter, Anthony
Miszalski, Brent
Mitchell, George
MIYAMA, Syoken
MIZUNO, Shun
Mo, Jinger
Monin, Jean-Louis
Montmerle, Thierry
Mookerjea, Bhaswati
Moór, Attila
Moore, Marla
Moreno-Corral, Marco
Morgan, David
Morris, Mark
Morris, Patrick
Morton, Donald
Mosoni, Laszlo
Mouschovias, Telemachos
Muench, Guido
Mufson, Stuart
Mulas, Giacomo
Muller, Erik
Muller, Sebastien
Murthy, Jayant
Myers, Philip

NAGAHARA, Hiroko
NAGATA, Tetsuya
NAKADA, Yoshikazu
NAKAGAWA, Takao
NAKAMOTO, Taishi
NAKAMURA, Fumitaka
NAKANO, Makoto
NAKANO, Takenori
Nammahachak, Suwit
Natta, Antonella
Naze, Yael
Nesvadba, Nicole
Nguyen-Quang, Rieu
Nikoghosyan, Elena
Nikoli'c, Silvana
NISHI, Ryoichi
NOMURA, Hideko
Nordh, Lennart
Norman, Colin
Noterdaeme, Pasquier
Nürnberger, Dieter
Nulsen, Paul
Nussbaumer, Harry
Nuth, Joseph
Oberst, Thomas
O'Dell, Charles
O'Dell, Stephen
Oey, Sally
OHISHI, Masatoshi
OHTANI, Hiroshi
OKA, Tomoharu
OKUDA, Haruyuki
OKUMURA, Shin-ichiro
Olofsson, Hans
Omont, Alain
OMUKAI, Kazuyuki
ONAKA, Takashi
Onello, Joseph
Opendak, Michael
Orlando, Salvatore
OSAKI, Toru
Osborne, John
Oskinova, Lidia
Ostriker, Eve
OTSUKA, Masaaki
Ott, Juergen
Oudmaijer, Rene
Pagani, Laurent
Pagano, Isabella
PAK, Soojong
Palla, Francesco
Palmer, Patrick
Palumbo, Maria Elisabetta
Panagia, Nino
Pandey, Birendra
Pankonin, Vernon
PARK, Yong Sun
Parker, Eugene
Parker, Quentin
Paron, Sergio
Parthasarathy, Mudumba
Pauls, Thomas
Pavlyuchenkov, Yaroslav
Pecker, Jean-Claude
Peeters, Els
Peimbert, Manuel
Pellegrini, Silvia
Pena, Miriam
Pendleton, Yvonne
PENG, Qingyu
Penzias, Arno
Pequignot, Daniel
Perault, Michel
Persi, Paolo
Persson, Carina
Peters, William
Petrosian, Vahe
Petuchowski, Samuel
Philipp-May, Sabine
Phillips, Thomas
Pihlström, Ylva
Pikel'ner, S.
Pilbratt, Göran
Pilling, Sergio
Pineau des Forets, Guillaume
Pinte, Christophe
Pittard, Julian
Plume, Rene
Podio, Linda
Pöppel, Wolfgang
Pongracic, Helen
Pontoppidan, Klaus
Porceddu, Ignazio
Pottasch, Stuart
Pound, Marc
Pouquet, Annick
Prasad, Sheo
Preite Martinez, Andrea
Price, Daniel
Price, R.
Prochaska, Jason
Prusti, Timo
Puget, Jean-Loup
QIN, Zhihai
Ramírez, Jose
Ramos-Larios, Gerardo
Ranalli, Piero
Rastogi, Shantanu
Ratag, Mezak
Rathborne, Jill
Rauch, Thomas
Rawlings, Jonathan
Rawlings, Mark
Raymond, John
Recchi, Simone
Redman, Matthew
Reid, Michael
Reipurth, Bo
Rengarajan, Thinniam
Rengel, Miriam
Reshetnyk, Volodymyr
Reyes, Rafael
Reynolds, Cormac
Reynolds, Ronald
Reynoso, Estela
Rho, Jeonghee
Richter, Philipp
Rickard, Lee
Roberge, Wayne
Roberts, Douglas
Roberts Jr, William
Robinson, Garry
Roche, Patrick
Rodrigues, Claudia
Rodriguez, Luis
Rodríguez, Monica
Rödiger, Elke
Roelfsema, Peter
Röser, Hans-peter
Roger, Robert
Rogers, Alan
Rohlfs, Kristen
Román-Zúñiga, Carlos
Rosa, Michael
Rosado, Margarita
Rosino, Leonida
Rosolowsky, Erik
Rouan, Daniel
Rowell, Gavin

Roxburgh, Ian
Rozhkovskij, Dimitrij
Różyczka, Michal
Ryabov, Michael
SABANO, Yutaka
Sabbadin, Franco
Sabin, Laurence
Sahu, Kailash
SAIGO, Kazuya
Sakano, Masaaki
SAKON, Itsuki
Salama, Farid
Salinari, Piero
Salomé, Philippe
Salpeter, Edwin
Salter, Christopher
Samodurov, Vladimir
Sánchez Doreste, Néstor
Sanchez-Saavedra, M.
Sancisi, Renzo
Sandell, Göran
Sandqvist, Aage
Sankrit, Ravi
Sarazin, Craig
Sargent, Annelia
Sarma, N.
Sarre, Peter
SATO, Fumio
SATO, Shuji
Savage, Blair
Savedoff, Malcolm
Scalo, John
Scappini, Flavio
Scherb, Frank
Schilke, Peter
Schlemmer, Stephan
Schmid-Burgk, J.
Schmidt-Kaler, Theodor
Schröder, Anja
Schulz, R.
Schure, Klara
Schwartz, Philip
Schwarz, Ulrich
Scott, Eugene
Scoville, Nicholas
SEKI, Munezo
Sellgren, Kristen
Sembach, Kenneth
Semenov, Dmitry
Sen, Asoke
SEON, Kwang il
Shadmehri, Mohsen
Shalchi, Andreas
Shane, William
Shapiro, Stuart
Sharma, Prateek
Sharpless, Stewart
Shaver, Peter
Shawl, Stephen
Shchekinov, Yuri
Shematovich, Valerij
Sherwood, William
Shields, Gregory
SHIMOIKURA, Tomomi
Shipman, Russell
Šmelds, Ivar
SHU, Frank
Shull, John
Shull, Peter
Shustov, Boris
Siebenmorgen, Ralf
Sigalotti, Leonardo
Silich, Sergey
Silk, Joseph
Silva, Laura
Silvestro, Giovanni
Simón-Díaz, Sergio
Simonia, Irakli
Simons, Stuart
Sitko, Michael
Sivan, Jean-Pierre
Skilling, John
Skulskyj, Mychajlo
Slane, Patrick
Sloan, Gregory
Smirnova, Tatiana
Smith, Craig
Smith, Michael
Smith, Peter
Smith, Randall
Smith, Robert
Smith, Robert
Smith, Tracy
Snell, Ronald
Snow, Theodore
Sobolev, Andrey
Sofia, Sabatino
Sofia, Ulysses
SOFUE, Yoshiaki
Solc, Martin
Solomon, Philip
Somerville, William
SONG, In-Ok
Spaans, Marco
Spitzer, L.
Stahler, Steven
Stanga, Ruggero
Stanghellini, Letizia
Stanimirovic, Snezana
Stapelfeldt, Karl
Stark, Ronald
Stasinska, Grazyna
Stebbins, Joel
Stecher, Theodore
Stecklum, Bringfried
Stenholm, Björn
Stenholm, Lars
Stone, James
Stone, Jennifer
Strom, Richard
Struve, M.
Suh, Kyung-Won
Suleymanova, Svetlana
SUN, Jin
SUNG, Hyun-Il
SUSA, Hajime
Sutherland, Ralph
SUZUKI, Tomoharu
Swade, Daryl
Sylvester, Roger
Szczerba, Ryszard
TACHIHARA, Kengo
Tafalla, Mario
TAKAHASHI, Hidenori
TAKAHASHI, Junko
TAKANO, Toshiaki
TAMURA, Motohide
TAMURA, Shin'ichi
TANAKA, Masuo
Tantalo, Rosaria
Tauber, Jan
Taylor, Kenneth
Teixeira, Paula
Tenorio-Tagle, Guillermo
Tepper Garcia, Thorsten
TERADA, Yukikatsu
Terzian, Yervant
Testi, Leonardo
Teyssier, Romain
Thaddeus, Patrick

Thé, Pik-Sin
Thöne, Christina
Thompson, A.
Thonnard, Norbert
Thronson Jr, Harley
Tilanus, Remo
Tokarev, Yurij
Torrelles, Jose Maria
Torres-Peimbert, Silvia
Tosi, Monica
Toth, Laszlo
Tothill, Nicholas
Townes, Charles
Trammell, Susan
Treffers, Richard
Tremonti, Christy
Trinidad, Miguel
TSAI, An-Li
TSUBOI, Masato
Turner, Kenneth
Tyul'bashev, Sergei
Ueta, Toshiya
Ulrich, Marie-Helene
Urov sevi'c, Dejan
Urquhart, James
USUDA, Tomonori
van den Ancker, Mario
VandenBout, Paul
van der Hulst, Jan
van der Laan, Harry
van der Tak, Floris
van de Steene, Griet
van Dishoeck, Ewine
van Gorkom, Jacqueline
van Loon, Jacco
van Woerden, Hugo
Varshalovich, Dmitrij
Vasta, Magda
Vavrek, Roland
Vázquez, Roberto
Vega, Olga
Velázquez, Pablo
Verdoes Kleijn, Gijsbert
Verheijen, Marc
Verma, Aprajita
Verner, Ekaterina
Verschuur, Gerrit
Viala, Yves
Viallefond, Francois
Vidal, Jean-Louis
Vidal-Madjar, Alfred
Viegas, Sueli
Viironen, Kerttu
Vijh, Uma
Vilchez, José
Villaver, Eva
Vink, Jacco
Viti, Serena
Vlahakis, Catherine
Vlemmings, Wouter
Vogel, Manfred
Voit, Gerard
Volk, Kevin
Vorobyov, Eduard
Voronkov, Maxim
Voshchinnikov, Nikolai
Vrba, Frederick
Wakelam, Valentine
Wakker, Bastiaan
Walker, Gordon
Walmsley, C.
Walsh, Andrew
Walsh, Wilfred
Walton, Nicholas
Wang, Q. Daniel
WANG, Chen
WANG, Hongchi
WANG, Hong-Guang
WANG, Jun-Jie
WANG, Junzhi
WANG, Min
Wannier, Peter
Wardle, Mark
Ward-Thompson, Derek
Watt, Graeme
Weaver, Harold
Weiler, Kurt
Weinberger, Ronald
Weisheit, Jon
Wendker, Heinrich
Wendt, Martin
Wesselius, Paul
Wesson, Roger
Weymann, Ray
Whelan, Emma
White, Glenn
White, Richard
Whitelock, Patricia
Whiteoak, John
Whittet, Douglas
Whitworth, Anthony
Wickramasinghe, N.
Wiebe, Dmitri
Wild, Wolfgang
Wilkin, Francis
Williams, David
Williams, Robert
Williams, Robin
Willis, Allan
Willner, Steven
Wilson, Christine
Wilson, Robert
Wilson, Thomas
Winnberg, Anders
Witt, Adolf
Wolff, Michael
Wolfire, Mark
Wolstencroft, Ramon
Wolszczan, Alexander
Woltjer, Lodewijk
Wong, Tony
Wood, Brian
Wood, Douglas
Woodward, Paul
Woolf, Neville
Wootten, Henry
Wouterloot, Jan
Wright, Edward
Wu, Chi
Wünsch, Richard
Wynn-Williams, Gareth
YABUSHITA, Shin
YAMADA, Masako
YAMAMOTO, Hiroaki
YAMAMOTO, Satoshi
YAMAMURA, Issei
YAMASHITA, Takuya
YAN, Jun
YANG, Ji
York, Donald
Yorke, Harold
YOSHIDA, Shigeomi
Younis, Saad
YUI, Yukari
Yun, Joao
Zavagno, Annie
Zealey, William
Zeilik, Michael
ZENG, Qin
ZHANG, Cheng-Yue

ZHANG, JiangShui
ZHANG, Jingyi
ZHOU, Jianjun
ZHOU, Zhen-pu
ZHU, Qingfeng
ZHU, Wenbai
Zibetti, Stefano
Zijlstra, Albert
Zinchenko, Igor
Zuckerman, Benjamin

Division G - Commission 35 Stellar Constitution

President: Marco Limongi
Vice-President: John C. Lattanzio

Organizing Committee Members:

Inma Dominguez
Giora Shaviv
Amanda I. Karakas
Marcella Marconi
Jacco Th. van Loon
Claus Leitherer
Jordi Isern

Members:

Adams, Mark
Aizenman, Morris
Angelov, Trajko
Antia, H.
Appenzeller, Immo
ARAI, Kenzo
Arentoft, Torben
Argast, Dominik
ARIMOTO, Nobuo
Arnett, W.
Arnould, Marcel
Audouze, Jean
Baglin, Annie
Barnes, Sydney
Basu, Sarbani
Baym, Gordon
Bazot, Michael
Beaudet, Gilles
Becker, Stephen
Belkacem, Kevin
Belmonte Aviles, Juan Antonio
Benz, Willy
Bergeron, Pierre
Bertelli, Gianpaolo
Berthomieu, Gabrielle
Bisnovatyi-Kogan, Gennadij
Blaga, Cristina
Bludman, Sidney
Bocchia, Romeo
Bodenheimer, Peter
Boehm, Karl-Heinz
Bombaci, Ignazio
Bono, Giuseppe
Boss, Alan
Braithwaite, Jonathan
Brassard, Pierre
Bravo, Eduardo
Bressan, Alessandro
Brown, David
Browning, Matthew
Brownlee, Robert
Bruenn, Stephen
Brun, Allan
Busso, Maurizio
Callebaut, Dirk
Caloi, Vittoria
Campbell, Simon
Canal, Ramon
Caputo, Filippina
Carson, T.
Castor, John
Chaboyer, Brian
Chabrier, Gilles
Chamel, Nicolas
Chan, Roberto
CHAN, Kwing
Chandrasekhar, S.
Charbonnel, Corinne
Charpinet, Stéphane
Chechetkin, Valerij
Chevalier, Claude
Chiosi, Cesare
Chitre, Shashikumar
Chkhikvadze, Iakob
Christensen-Dalsgaard, Jørgen
Cohen, Judith
Connolly, Leo
Córsico, Alejandro
Cowan, John
Cowling, Thomas
Cristallo, Sergio
Crowther, Paul
D'Antona, Francesca
Das, Mrinal
Daszynska-Daszkiewicz, Jadwiga
Davies, Benjamin
Davis Jr, Cecil
Dearborn, David
de Greve, Jean-Pierre
Deinzer, W.
de Jager, Cornelis
Deliyannis, Constantine
de Loore, Camiel

Demarque, Pierre
de Medeiros, Jose
de Mink, Selma
Denisenkov, Pavel
de Silva, Gayandhi
Deupree, Robert
Di Mauro, Maria Pia
Dluzhnevskaya, Olga
Domiciano de Souza, Armando
Dominguez, Inma
Dupuis, Jean
Durisen, Richard
Dziembowski, Wojciech
Edwards, Alan
Edwards, Suzan
Eggenberger, Patrick
Eggleton, Peter
Ekström García Nombela, Sylvia
Eldridge, John
Elmhamdi, Abouazza
Eminzade, T.
Endal, Andrew
Engelbrecht, Chris
Ergma, Ene
ERIGUCHI, Yoshiharu
Ezer-Eryurt, Dilhan
Fadeyev, Yurij
Faulkner, John
Flannery, Brian
Fontaine, Gilles
Forbes, J.
Forestini, Manuel
Fossat, Eric
Foukal, Peter
Fox-Machado, Lester
FUJIMOTO, Masayuki
Gabriel, Maurice
Gallino, Roberto
Garcia, Domingo
García García, Miriam
García-Berro, Enrique
Gautschy, Alfred
Georgy, Cyril
Geroyannis, Vassilis
Giannone, Pietro
Gimenez, Álvaro
Girardi, Leo
Giridhar, Sunetra
Glatzmaier, Gary
Goedhart, Sharmila
Gomez, Haley
Gonçalves, Denise
Goriely, Stephane
Goswami, Aruna
Gough, Douglas
Goupil, Marie-Jose
Graham, Eric
Greggio, Laura
Groh, Jose
Guenther, David
Gurm, Hardev
Guzik, Joyce
HACHISU, Izumi
Hammond, Gordon
HAN, Zhanwen
HASHIMOTO, Masa-aki
Heger, Alexander
Henry, Richard
Hernanz, Margarita
Hillier, John
Hirschi, Raphael
Hollowell, David
Homeier, Derek
Huggins, Patrick
Humphreys, Roberta
Iben Jr, Icko
Iliev, Ilian
Imbroane, Alexandru
Imshennik, Vladimir
Isern, Jordi
ISHIHARA, Daisuke
ISHIZUKA, Toshihisa
ITOH, Naoki
Ivanova, Natalia
IWAMOTO, Nobuyuki
Izzard, Robert
James, Richard
Jones, David
Jørgensen, Jes
José, Jordi
Kaehler, Helmuth
Kalirai, Jason
KAMINISHI, Keisuke
Kaminker, Alexander
Käpylä, Petri
Karakas, Amanda
KATO, Mariko
Kervella, Pierre
Khozov, Gennadij
KIGUCHI, Masayoshi
King, David
Kippenhahn, Rudolf
Kiziloglu, Nilgun
Knoelker, Michael
Kochhar, Rajesh
Koester, Detlev
Konar, Sushan
Kosovichev, Alexander
Kovetz, Attay
Kozłowski, Maciej
Kučinskas, Arunas
Kuiper, Rolf
Kumar, Shiv
Kupka, Friedrich
Kwok, Sun
Labay, Javier
Lamb, Susan
Lamb Jr, Donald
Lamzin, Sergei
Langer, Norbert
Larson, Richard
Laskarides, Paul
Lasota-Hirszowicz, Jean-Pierre
Lattanzio, John
Lebovitz, Norman
Lebreton, Yveline
Lee, William
LEE, Thyphoon
Leitherer, Claus
Lépine, Jacques
LI, Qingkang
LI, Zongwei
Liebendörfer, Matthias
Lignieres, François
Limongi, Marco
Linnell, Albert
Littleton, John
LIU, Guoqing
Livio, Mario
Lucatello, Sara
Lugaro, Maria
LUO, Guoquan
MAEDA, Keiichi
Maeder, Andre
Maheswaran, Murugesapillai
Mallik, D.

Marconi, Marcella
Marín-Franch, Antonio
Martayan, Christophe
Massevich, Alla
Mathis, Stephane
Matteucci, Francesca
Mazumdar, Anwesh
Mazurek, Thaddeus
Mazzitelli, Italo
McDavid, David
Melbourne, Jason
Melik-Alaverdian, Yu
Mendes, Luiz
Mestel, Leon
Meyer-Hofmeister, Emmi
Meynet, meynet
Michaud, Georges
Miszalski, Brent
Mitalas, Romas
MIYAJI, Shigeki
Möllenhoff, Claus
Mohan, Chander
Moiseenko, Sergey
Monaghan, Joseph
Monier, Richard
Montalb'an, Josefina
Monteiro, Mario Joao
Moore, Daniel
Morgan, John
Moskalik, Paweł
Moss, David
Mowlavi, Nami
Nadyozhin, Dmitrij
NAGATAKI, Shigehiro
NAKAMURA, Takashi
NAKANO, Takenori
NAKAZAWA, Kiyoshi
Narasimha, Delampady
NARITA, Shinji
Nelemans, Gijs
Newman, Michael
NISHIDA, Minoru
Noels, Arlette
NOMOTO, Ken'ichi
Odell, Andrew
OHYAMA, Noboru
OKAMOTO, Isao
Oliveira, Joana
OSAKI, Yoji
Ostriker, Jeremiah
Oswalt, Terry
O'Toole, Simon
Oudmaijer, Rene
Owocki, Stanley
PALACIOS, Ana
Pamyatnykh, Alexey
Pande, Girish
Panei, Jorge
Papaloizou, John
Pearce, Gillian
Phillips, Mark
Pines, David
Pinotsis, Antonis
Plavec, Mirek
Pongracic, Helen
Pontoppidan, Klaus
Porfir'ev, Vladimir
Poveda, Arcadio
Prentice, Andrew
Prialnik, Dina
Primas, Francesca
Proffitt, Charles
Provost, Janine
Pulone, Luigi
QU, Qinyue
Raedler, K.
Ramadurai, Souriraja
Rauscher, Thomas
Ray, Alak
Rayet, Marc
Reeves, Hubert
Reisenegger, Andreas
Renzini, Alvio
Reyniers, Maarten
Richard, Olivier
Ritter, Hans
Rizzi, Luca
Roca Cortés, Teodoro
Roxburgh, Ian
Ruiz-Lapuente, María
Russell, H.
Sackmann, Inge
SAIO, Hideyuki
SAKASHITA, Shiro
Salaris, Maurizio
Salpeter, Edwin
Santos, Filipe
Sarna, Marek
SATO, Katsuhiko
Savedoff, Malcolm
Savonije, Gerrit
Scalo, John
Schatten, Kenneth
Schild, Hansruedi
Schoenberner, Detlef
Schuler, Simon
Schutz, Bernard
Scuflaire, Richard
Sears, Richard
Seidov, Zakir
Sengbusch, Kurt
Shaviv, Giora
SHIBAHASHI, Hiromoto
SHIBATA, Yukio
Shustov, Boris
Siess, Lionel
Sigalotti, Leonardo
Signore, Monique
Sills, Alison
Silvestro, Giovanni
Simón-Díaz, Sergio
Sion, Edward
Smeyers, Paul
Smith, Robert
Smolec, Radosław
Sobouti, Yousef
Sofia, Sabatino
Sparks, Warren
Spiegel, Edward
Sreenivasan, S.
Stancliffe, Richard
Starrfield, Sumner
Stée, Philippe
Stellingwerf, Robert
Stergioulas, Nikolaos
Stringfellow, Guy
Strittmatter, Peter
Suda, Takuma
SUGIMOTO, Daiichiro
Sundqvist, Jon
Sweigart, Allen
Taam, Ronald
TAKAHARA, Mariko
Tassoul, Jean-Louis
Tassoul, Monique
Thielemann, Friedrich-Karl
Thomas, Hans
Tjin-a-Djie, Herman
Todt, Helge
Tohline, Joel

TOMINAGA, Nozomu
Toomre, Juri
Tornambe, Amedeo
Townsend, Richard
Trimble, Virginia
Truran, James
Tscharnuter, Werner
Turck-Chièze, Sylvaine
Tutukov, Aleksandr
Úbeda, Leonardo
UCHIDA, Juichi
ud-Doula, Asif
Ulrich, Roger
UNNO, Wasaburo
Utrobin, Victor
Uus, Undo
VandenBerg, Don
van den Heuvel, Edward
van der Borght, Rene
van der Raay, Herman
van Horn, Hugh
van Loon, Jacco
van Riper, Kenneth
Vauclair, Gérard
Ventura, Paolo
Vila, Samuel
Vilhu, Osmi
Vilkoviskij, Emmanuil
Vink, Jorick
Ward, Richard
Weaver, Thomas
Webbink, Ronald
Weiss, Achim
Weiss, Nigel
Wheeler, J.
Willson, Lee Anne
Wilson, Robert
Winkler, Karl-Heinz
Wittkowski, Markus
Wood, Matthew
Wood, Peter
Woosley, Stanford
XIONG, Da Run
YAMADA, Shimako
YAMAOKA, Hitoshi
YI, Sukyoung
Yorke, Harold
YOSHIDA, Shin'ichirou
YOSHIDA, Takashi
Yungelson, Lev
Yushkin, Maxim
Zahn, Jean-Paul
Zampieri, Luca
Ziółkowski, Janusz

Division G - Commission 36 Theory of Stellar Atmospheres

President: Joachim Puls
Vice-President: van Hubeny

Organizing Committee Members:

Rolf-Peter Kudritzki
Mats Carlsson
Tatiana A. Ryabchikova
France Allard
Martin Asplund
Carlos Allende Prieto
Thomas R. Ayres
Bengt Gustafsson

Members:

Abbott, David
Afram, Nadine
Allard, France
Altrock, Richard
Andretta, Vincenzo
Ardila, David
Aret, Anna
Arpigny, Claude
Asplund, Martin
Atanackovic, Olga
Auer, Lawrence
Aufdenberg, Jason
Auman, Jason
Avrett, Eugene
Ayres, Thomas
Baade, Dietrich
Baird, Scott
Baliunas, Sallie
Balona, Luis
Barber, Robert
Barbuy, Beatriz
Barklem, Paul
Baschek, Bodo
Basri, Gibor
Becker, Sylvia
Bennett, Philip
Berdyugina, Svetlana
Bernat, Andrew
Bertone, Emanuele
Bertout, Claude
Biazzo, Katia
Bingham, Robert
Blanco, Carlo
Bless, Robert
Blomme, Ronny
Bodo, Gianluigi
Boehm, Karl-Heinz
Boesgaard, Ann
Bohm-Vitense, Erika

Bonifacio, Piercarlo
Bopp, Bernard
Brown, Alexander
Brown, David
Brown, Douglas
Browning, Matthew
Buchlin, Eric
Bues, Irmela
Busa', Innocenza
Butkovskaya, Varvara
Cameron, Andrew
Carbon, Duane
Carlsson, Mats
Carson, T.
Cassinelli, Joseph
Castelli, Fiorella
Castor, John
Catala, Claude
Catalano, Franco
Catalano, Santo
Cayrel, Roger
Chalonge, Daniel
CHAN, Kwing
CHEN, Peisheng
Christlieb, Norbert
Chugai, Nikolaj
Cidale, Lydia
Cohen, David
Collet, Remo
Conti, Peter
Cowley, Charles
Cram, Lawrence
Crivellari, Lucio
Cruzado, Alicia
Cugier, Henryk
CUI, Wenyuan
Cuntz, Manfred
Cuny, Yvette
Curé, Michel
Dall, Thomas
Daszynska-Daszkiewicz, Jadwiga
Davis Jr, Cecil
Day-Jones, Avril
Decin, Leen
de Jager, Cornelis
de Koter, Alex
Deliyannis, Constantine
Depagne, Éric
Dimitrijevic, Milan
Doazan, Vera
Domke, Helmut
Donati, Jean-Francois
Doyle MRIA, John
Drake, Stephen
Dravins, Dainis
Dreizler, Stefan
Duari, Debiprosad
Dufton, Philip
Dupree, Andrea
Edvardsson, Bengt
Elmhamdi, Abouazza
Elste, Günther
Eriksson, Kjell
Evangelidis, E.
Faraggiana, Rosanna
Faurobert, Marianne
Feigelson, Eric
Ferreira, João
Firsch, H.
Fitzpatrick, Edward
Fluri, Dominique
Fontaine, Gilles
Fontenla, Juan
Forveille, Thierry
Foy, Renaud
Freire Ferrero, Rubens
Frémat, Yves
Freytag, Bernd
Friend, David
Frisch, Helene
Frisch, Uriel
Froeschle, Christiane
Gail, Hans-Peter
Gallino, Roberto
García García, Miriam
García López, Ramón
Gebbie, Katharine
Gęsicki, Krzysztof
Giampapa, Mark
Gigas, Detlef
Gonzalez, Jean-Francois
Gordon, Charlotte
Goswami, Aruna
Gough, Douglas
Grant, Athay
Grant, Ian
Gratton, Raffaele
Gray, David
Grevesse, Nicolas
Grinin, Vladimir
Groh, Jose
Güdel, Manuel
Gussmann, Ernst-August
Gustafsson, Bengt
Haberreiter, Margit
Haisch, Bernard
Hall, Douglas
Hamann, Wolf-Rainer
Harper, Graham
Hartmann, Lee
Harutyunian, Haik
Hearn, Anthony
Heasley, James
Heber, Ulrich
Heiter, Ulrike
Hempel, Marc
Heyl, Jeremy
Hill, Vanessa
Hillier, John
Ho, Wynn
Hoare, Melvin
Hoeflich, Peter
Höfner, Susanne
Holzer, Thomas
Homeier, Derek
Hotinli, Metin
House, Lewis
HUANG, He
Hubeny, Ivan
Hui bon Hoa, Alain
Hutchings, John
Ignace, Richard
Ignjatovi'c, Ljubinko
ITO, Yutaka
Ivanov, Vsevolod
Jahn, Krzysztof
Jankov, Slobodan
Jatenco-Pereira, Vera
Jefferies, John
Jevremovic, Darko
Johnson, Hollis
Jones, Carol
Jordan, Stefan
Judge, Philip
Kadouri, Talib
Kalkofen, Wolfgang
Kamp, Inga
Kamp, Lucas
Kandel, Robert

Käpylä, Petri
Karp, Alan
Kašparová, Jana
Katsova, Maria
Kervella, Pierre
Kiselman, Dan
Klein, Richard
KOBAYASHI, Masakazu
Kochukhov, Oleg
KODAIRA, Keiichi
Koester, Detlev
Kolesov, Aleksandr
KONDO, Yoji
Kontizas, Evangelos
Korcakova, Daniela
Korn, Andreas
KOZASA, Takashi
Kraus, Michaela
Krikorian, Ralph
Krishna, Swamy
Krticka, Jiri
Kubát, Jiri
Kučinskas, Arunas
Kudritzki, Rolf-Peter
Kuhi, Leonard
Kumar, Shiv
Kupka, Friedrich
Kurucz, Robert
Lambert, David
Lamers, Henny
Lanz, Thierry
LEE, Jae Woo
LEE, Jun
Leenaarts, Jorrit
Leibacher, John
Leitherer, Claus
LI, Ji
Liebert, James
Linnell, Albert
Linsky, Jeffrey
LIU, Caipin
LIU, Yujuan
Lobel, Alex
Lucatello, Sara
Luck, R.
Ludwig, Hans
Luo, Qinghuan
Luttermoser, Donald
Lyubimkov, Leonid
Machado, Maria
Madej, Jerzy
Magazzu, Antonio
Magnan, Christian
Maguire, Kate
Marlborough, J.
Marlborough, Michael
Marley, Mark
Martins, Fabrice
Mashonkina, Lyudmila
Massaglia, Silvano
Mathys, Gautier
MATSUMOTO, Masamichi
Mauas, Pablo
Medupe, Rodney
Merlo, David
Michaud, Georges
Mihajlov, Anatolij
MIYAMOTO, Sigenori
Mnatsakanian, Mamikon
Mohan Rao, Desaraju
Molaro, Paolo
Montalb'an, Josefina
Morel, Thierry
Morris, Patrick
Muench, Guido
MUKAI, Sonoyo
Musielak, Zdzislaw
Mutschlecner, Joseph
Nagirner, Dmitrij
Najarro de la Parra, Francisco
Narasimha, Delampady
NARDETTO, Nicolas
NARIAI, Kyoji
Neff, John
Niemczura, Ewa
Nieva, Maria
Nikoghossian, Arthur
NISHIMURA, Masayoshi
Nordlund, Aake
Nowotny-Schipper, Walter
Oskinova, Lidia
Owocki, Stanley
Oxenius, Joachim
Pacharin-Tanakun, P.
Pakhomov, Yury
Pallavicini, Roberto
Pandey, Birendra
Panek, Robert
Pannekoek, A.
Pasinetti, Laura
Pavlenko, Yakov
Pecker, Jean-Claude
Peraiah, Annamaneni
Peters, Geraldine
Pinsonneault, Marc
Pintado, Olga
Pinto, Philip
Piskunov, Nikolai
Plez, Bertrand
Pogodin, Mikhail
Pottasch, Stuart
Przybilla, Norbert
Puls, Joachim
Pustõnski, Vladislav-Veniamin
Puzia, Thomas
Querci, Francois
Querci, Monique
Ramsey, Lawrence
Randich, Sofia
Rangarajan, K.
Rauch, Thomas
Reale, Fabio
Reid, Warren
Reimers, Dieter
Reiners, Ansgar
Ross, John
Rostas, François
Rovira, Marta
Różańska, Agata
Rucinski, Slavek
Rutten, Robert
Ryabchikova, Tatiana
Rybicki, George
Sachkov, Mikhail
Saffe, Carlos
SAIO, Hideyuki
Sakhibullin, Nail
Sapar, Arved
Sapar, Lili
Sarre, Peter
Sasselov, Dimitar
Sasso, Clementina
Sauty, Christophe
Savanov, Igor
Sbordone, Luca
Schaerer, Daniel
Scharmer, Goeran
Schmalberger, Donald

Schmid-Burgk, J.
Schmutz, Werner
Schoenberner, Detlef
Scholz, M.
Schrijver, Karel
Sedlmayer, Erwin
Sengupta, Sujan
SHANG, Hsien
SHI, Jianrong
Shimansky, Vladislav
Shine, Richard
Shipman, Harry
Short, Christopher
Sigut, T. A. Aaron
Simon, Klaus
Simon, Theodore
Simón-Díaz, Sergio
Simonneau, Eduardo
Skumanich, Andrew
Sneden, Chris
Sobolev, Vladislav
Socas-Navarro, Hector
Soderblom, David
Spiegel, Edward
Spite, François
Spite, Monique
Spruit, Hendrik
Stalio, Roberto
Stauffer, John
Stée, Philippe
Steffen, Matthias
Stein, Robert
Stępień, Kazimierz
Stern, Robert
Strom, Stephen
Stuik, Remko
Sundqvist, Jon
Swann, Cornelis
TAKEDA, Yoichi
Tautvaisiene, Gražina
Thejll, Peter
Thomas, Richard
Thompson, Rodger
Todt, Helge
Toomre, Juri
Townsend, Richard
Traving, Gerhard
TSUJI, Takashi
UESUGI, Akira
Ulmschneider, Peter
UNNO, Wasaburo
Utrobin, Victor
Vakili, Farrokh
Valeev, Azamat
van't Veer, Frans
van't Veer-Menneret, Claude
Vardavas, Ilias
Vasu-Mallik, Sushma
Vaughan, Arthur
Velusamy, T.
Vennes, Stephane
Vidotto, Aline
Vieytes, Mariela
Viik, Tõnu
Vilhu, Osmi
Vink, Jorick
Walter, Frederick
WATANABE, Tetsuya
Waters, Laurens
Weber, Stephen
Werner, Klaus
White, Richard
Wickramasinghe, N.
Willson, Lee Anne
Wilson, S.
Woehl, Hubertus
Wolff, Sidney
Wrubel, M.
Yanovitskij, Edgard
Yengibarian, Norair
Yorke, Harold
Začs, Laimons
Zahn, Jean-Paul
ZHANG, Bo
ZHAO, Gang

Division H - Commission 37 Star Clusters & Associations

President: Giovanni Carraro
Vice-President: Richard de Grijs

Organizing Committee Members:

Peter B. Stetson
Douglas Paul Geisler
Simon P. Goodwin
Barbara J. Anthony-Twarog
Bruce G. Elmegreen
Dante Minniti

Members:

Aarseth, Sverre
Abou'el-ella, Mohamed
Ahumada, Andrea
Ahumada, Javier
Aigrain, Suzanne
Ajhar, Edward
AK, Serap
AK, Tansel
Akeson, Rachel
Alcaino, Gonzalo
Alfaro, Emilio
Alksnis, Andrejs

Allen, Christine
Allen, Lori
Alter, G.
Alves, Virgínia
Anderson, Joseph
Andreuzzi, Gloria
Anthony-Twarog, Barbara
Antoniou, Vallia
Aparicio, Antonio
Arifyanto, Mochamad
Armandroff, Taft
Auriere, Michel
Baade, W.
Bailyn, Charles
Balázs, Bela
Barkhatova, Klaudia
Barmby, Pauline
Barrado Navascués, David
Bartašiute, Stanislava
Bastian, Nathan
Baume, Gustavo
Baumgardt, Holger
Beck, Sara
Becker, Wilhelm
Bekki, Kenji
Bellazzini, Michele
Benacquista, Matthew
Biazzo, Katia
Bijaoui, Albert
Blum, Robert
Boily, Christian
Bonatto, Charles
Bosch, Guillermo
Bragaglia, Angela
Brandner, Wolfgang
Brown, Anthony
Buonanno, Roberto
Burderi, Luciano
Burkhead, Martin
Butler, Dennis
Butler, Ray
Buzzoni, Alberto
Byrd, Gene
Calamida, Annalisa
Callebaut, Dirk
Caloi, Vittoria
Campbell, Simon
Cannon, Russell
Cantiello, Michele
Canto Martins, Bruno
Caputo, Filippina
Capuzzo Dolcetta, Roberto
Carney, Bruce
Carraro, Giovanni
Carretta, Eugenio
Chaboyer, Brian
Charbonnel, Corinne
Chavarria-K, Carlos
CHEN, Huei-Ru
CHEN, Li
CHENG, Kwang
Chiosi, Cesare
Christian, Carol
Chryssovergis, Michael
CHUN, Mun-suk
Claria, Juan
Clark, David
Clementini, Gisella
Colin, Jacques
Corral, Luis
Covino, Elvira
Crause, Lisa
Cropper, Mark
Cudworth, Kyle
Da Costa, Gary
Dale, James
D'Amico, Nicolo
Danford, Stephen
D'Antona, Francesca
Dapergolas, Anastasios
Daube-Kurzemniece, Ilga
Davies, Benjamin
Davies, Melvyn
Dehghani, Mohammad
De Marchi, Guido
Demarque, Pierre
Demers, Serge
de Mink, Selma
DENG, LiCai
de Silva, Gayandhi
Díaz-Santos, Tanio
Dickens, Robert
Di Fazio, Alberto
Dionatos, Odysseas
Djupvik, Anlaug Amanda
Dluzhnevskaya, Olga
Dougados, Catherine
Downes Wallace, Juan
Drissen, Laurent
Durrell, Patrick
Eastwood, Kathleen
Eggen, O.
El Basuny, Ahmed
Eldridge, John
Elmegreen, Bruce
Fall, S.
Feinstein, Alejandro
Figer, Donald
Forbes, Douglas
Forte, Juan
Fox-Machado, Lester
Freeman, Kenneth
Freyhammer, Lars
Friel, Eileen
FUKUSHIGE, Toshiyuki
Fusi-Pecci, Flavio
Gandolfi, Davide
Garcia, Beatriz
Geffert, Michael
Geisler, Douglas
Geller, Aaron
Giersz, Mirosław
Giorgi, Edgard
Glushkova, Elena
Golay, Marcel
Goodwin, Simon
Gouliermis, Dimitrios
Gratton, Raffaele
Green, Elizabeth
Griffiths, William
Grillmair, Carl
Grundahl, Frank
Guetter, Harry
Haghi, Hosein
Haisch Jr, Karl
Hanes, David
Hanson, Margaret
Harris, Gretchen
Harris, Hugh
Harvel, Christopher
Hawarden, Timothy
Hayes, Matthew
Hazen, Martha
Heggie, Douglas
Herbst, William
Hesser, James
Heudier, Jean-Louis
Hilker, Michael
Hillenbrand, Lynne
Hills, Jack

Hodapp, Klaus
Hogg, A.
Hünsch, Matthias
Hut, Piet
Iben Jr, Icko
IKEDA, Norio
Illingworth, Garth
ITOH, Yoichi
Ivanova, Natalia
Janes, Kenneth
JEON, Young Beom
JIANG, Zhibo
Johnson, Christian
Johnson, H.
Jordi, Carme
Joshi, Umesh
Kaisin, Serafim
Kalirai, Jason
Kamp, Lucas
Karakas, Amanda
Kaviraj, Sugata
Kholopov, P.
Kilambi, G.
KIM, Sang Chul
KIM, Seung-Lee
KIM, Sungsoo
King, Ivan
Kitsionas, Spyridon
KO, Chung-Ming
Koch, Andreas
Kontizas, Evangelos
Kontizas, Mary
Kotulla, Ralf
Kouwenhoven, M.B.N. (Thijs)
Kraft, Robert
Kroupa, Pavel
Krumholz, Mark
Kun, Maria
Kundu, Arunav
Kurtev, Radostin
Lada, Charles
Landolt, Arlo
Lapasset, Emilio
Larsson-Leander, Gunnar
Laugalys, Vygandas
Laval, Annie
Lee, Hyun-chul
Lee, Young-Wook
LEE, Jae Woo
LEE, Kang Hwan
Leisawitz, David
Leonard, Peter
LI, Jinzeng
LIU, Michael
Lodieu, Nicolas
Loktin, Alexhander
Lu, Phillip
Lucatello, Sara
Lynden-Bell, Donald
Lynga, Gosta
Maccarone, Thomas
Mackey, Alasdair
Maeder, Andre
Magrini, Laura
Maíz Apellániz, Jesús
Makalkin, Andrei
MAKINO, Junichiro
Mamajek, Eric
Mandel, Ilya
Marco, Amparo
Mardling, Rosemary
Marín-Franch, Antonio
Marino, Antonietta
Markkanen, Tapio
Markov, Haralambi
Marraco, Hugo
Marsden, Stephen
Marshall, Kevin
Martinez Roger, Carlos
Martins, Donald
McGehee, Peregrine
Meibom, Soren
Mel'nik, Anna
Menon, T.
Menzies, John
Mermilliod, Jean-Claude
Meyer, Michael
Meylan, Georges
Milone, Eugene
Minniti, Dante
Moehler, Sabine
Mohan, Vijay
Moitinho, André
Monaco, Lorenzo
Montalb'an, Josefina
Moór, Attila
Mould, Jeremy
Muminov, Muydinjon
Murray, Andrew
Muzzio, Juan
Narbutis, Donatas
Navone, Hugo
Naylor, Tim
Nemec, James
Nesci, Roberto
Neuhaeuser, Ralph
Newell, Edward
Nikoghosyan, Elena
Ninkov, Zoran
Nissen, Poul
Nota, Antonella
Nürnberger, Dieter
Oey, Sally
OGURA, Katsuo
Oliveira, Joana
Origlia, Livia
Ortolani, Sergio
Oskinova, Lidia
Osman, Anas
OTSUKI, Kaori
Oudmaijer, Rene
PALACIOS, Ana
Pandey, A.
PARK, Byeong-Gon
Parmentier, Geneviève
Parsamyan, Elma
Patten, Brian
Paunzen, Ernst
Pavani, Daniela
Pedreros, Mario
Peng, Eric
Penny, Alan
Perez, Mario
Peterson, Charles
Petrovskaya, Margarita
Peykov, Zvezdelin
Phelps, Randy
Piatti, Andrés
Pietrukowicz, Paweł
Pilachowski, Catherine
Piskunov, Anatolij
Platais, Imants
Pooley, David
Porras Juárez, Bertha
Portegies Zwart, Simon
Poveda, Arcadio
Price, Daniel
Pritchet, Christopher
Pulone, Luigi

Puzia, Thomas
Quanz, Sascha
Raimondo, Gabriella
Rathborne, Jill
Ravindranath, Swara
Rebull, Luisa
Recillas Pishmish, Elsa
Renzini, Alvio
Rey, Soo-Chang
Rhee, Jaehyon
Richard, Olivier
Richer, Harvey
Richtler, Tom
Robberto, Massimo
Rodrigues de Oliveira Filho, Irapuan
Román-Zúñiga, Carlos
Rosino, Leonida
Rothberg, Barry
Rountree, Janet
Rowell, Gavin
Royer, Pierre
Ruprecht, Jaroslav
Russeva, Tatjana
Sagar, Ram
Salukvadze, G.
Samus, Nikolay
Sanchez Bejar, Victor
Sanders, Walter
Santiago, Basilio
Santos Jr., Joao
Sarajedini, Ata
Schild, Hansruedi
Schödel, Rainer
Scholz, Alexander
Schuler, Simon
Schweizer, François
Seitzer, Patrick
Semkov, Evgeni
Shawl, Stephen
Sher, David
SHI, Huoming
SHU, Chenggang
SIMODA, Mahiro
Skinner, Stephen
Smith, Graeme
Smith, J.
Song, Inseok
Southworth, John
Spurzem, Rainer
Stauffer, John
Stetson, Peter
Stringfellow, Guy
Stuik, Remko
Subramaniam, Annapurni
SUGIMOTO, Daiichiro
SUNG, Hwankyung
Suntzeff, Nicholas
Swope, Henrietta
Székely, Péter
Tadross, Ashraf
TAKAHASHI, Hidenori
TAKAHASHI, Koji
TAKASHI, Hasegawa
Taş, Günay
Tautvaisiene, Gražina
Terranegra, Luciano
Terzan, Agop
Thoul, Anne
Thurston, Mark
Tikhonov, Nikolai
Tornambe, Amedeo
Tosi, Monica
Tremonti, Christy
Trenti, Michele
Tripicco, Michael
Trullols, I.
Tsvetkov, Milcho
Tsvetkova, Katja
Turner, David
Twarog, Bruce
Úbeda, Leonardo
Upgren, Arthur
van Altena, William
van den Berg, Maureen
VandenBerg, Don
van den Bergh, Sidney
van der Bliek, Nicole
Vazquez, Ruben
Veltchev, Todor
Ventura, Paolo
Verschueren, Werner
Vesperini, Enrico
von Hippel, Theodore
Walker, Gordon
Walker, Merle
Warren Jr, Wayne
Weaver, Harold
Wehlau, Amelia
Wielen, Roland
Wofford, Aida
Worters, Hannah
Wramdemark, Stig
Wright, Nicholas
WU, Hsin-Heng
WU, Zhen-Yu
Wulandari, Hesti
Xiradaki, Evangelia
YI, Sukyoung
YIM, Hong Suh
Yong, David
YUMIKO, Oasa
Zaggia, Simone
Zakharova, Polina
Zapatero-Osorio, Maria Rosa
Zdanavičius, Justas
ZHANG, Fenghui
ZHAO, Jun Liang
ZHU, Qingfeng
Zinn, Robert
Zoccali, Manuela
Zwintz, Konstanze

Division B - Commission 40 Radio Astronomy

President: Jessica Mary Chapman
Vice-President: Gabriele Giovannini

Organizing Committee Members:

Monica Rubio
Raffaella Morganti
Justin L. Jonas
Ren-Dong NAN
Joseph Lazio
Prajval Shastri
Christopher L. Carilli
Hisashi HIRABAYASHI
Richard E. Hills

Members:

Abdulla, Shaker
Ábrahám, Péter
Ade, Peter
AKABANE, Kenji
Akujor, Chidi
Alberdi, Antonio
Alexander, Joseph
Alexander, Paul
Allen, Ronald
Aller, Hugh
Aller, Margo
Altenhoff, Wilhelm
Altunin, Valery
Ambrosini, Roberto
AN, TAO
Andernach, Heinz
Anglada, Guillem
Antonova, Antoaneta
Aparici, Juan
Argo, Megan
Arnal, Edmundo
Asanok, Kitiyanee
Asareh, Habibolah
Aschwanden, Markus
Assousa, George
Aubier, Monique
Augusto, Pedro
Aurass, Henry
Avignon, Yvette
Baan, Willem
Baars, Jacob
Baart, Edward
Bååth, Lars
Babkovskaia, Natalia
Bachiller, Rafael
Bagri, Durgadas
Bailes, Matthew
Bajaja, Esteban
Bajkova, Anisa
Baker, Andrew
Baker, Joanne
Balasubramanian, V.
Balasubramanyam, Ramesh
Ball, Lewis
Bally, John
Balonek, Thomas
Banhatti, Dilip
Barrett, Alan
Barrow, Colin
Bartel, Norbert
Barthel, Peter
Bartkiewicz, Anna
Barvainis, Richard
Baryshev, Andrey
Bash, Frank
Basu, Kaustuv
Batty, Michael
Baudry, Alain
Bauer, Amanda
Baum, Stefi
Bayet, Estelle
Beasley, Anthony
Beck, Rainer
Beck, Sara
Bekki, Kenji
Beklen, Elif
Benaglia, Paula
Benn, Chris
Bennett, Charles
Benson, Priscilla
Benz, Arnold
Berge, Glenn
Bergman, Per
Berkhuijsen, Elly
Bhandari, Rajendra
Bhat, Ramesh
Bhonsle, Rajaram
Bieging, John
Biermann, Peter
Bietenholz, Michael
Bigdeli, Mohsen
Biggs, James
Bignall, Hayley
Bignell, R.
Biraud, François
Biretta, John
Birkinshaw, Mark
Blair, David
Blandford, Roger
Bloemhof, Eric
Blum, Emile
Blundell, Katherine
Boboltz, David
Bock, Douglas
Bockelee-Morvan, Dominique
Boischot, André
Bolatto, Alberto
Bondi, Marco
Boonstra, Albert
Booth, Roy
Borisov, Galin
Borka Jovanović, Vesna
Bos, Albert
Bosch-Ramon, Valenti

Bot, Caroline
Bottinelli, Lucette
Bouton, Ellen
Bower, Geoffrey
Bowers, Phillip
Branchesi, Marica
Bregman, Jacob
Brentjens, Michiel
Breton, Rene
Bridle, Alan
Briggs, Franklin
Brinks, Elias
Britzen, Silke
Broderick, John
Bronfman, Leonardo
Brooks, Kate
Broten, Norman
Brouw, Willem
Brown, Joanna
Brown, Jo-Anne
Browne, Ian
Brunetti, Gianfranco
Brunthaler, Andreas
Bryant, Julia
Bujarrabal, Valentin
Burderi, Luciano
Burke, Bernard
Campbell, Robert
Campbell-Wilson, Duncan
Caproni, Anderson
Carilli, Christopher
Carlqvist, Per
Caroubalos, Constantinos
Carretti, Ettore
Carvalho, Joel
Casasola, Viviana
casassus, simon
Casoli, Fabienne
Cassano, Rossella
Castelletti, Gabriela
Castets, Alain
Caswell, James
Cawthorne, Timothy
Cecconi, Baptiste
Celotti, Anna Lisa
Cernicharo, Jose
CH, ISHWARA CHANDRA
CHAN, Kwing
Chandler, Claire
Chapman, Jessica
Charlot, Patrick
CHEN, Huei-Ru
CHEN, Xuefei
CHEN, Yongjun
CHEN, Zhijun
CHEN, Zhiyuan
Chengalur, Jayaram
CHIKADA, Yoshihiro
CHIN, Yi-nan
Chini, Rolf
CHO, Se Hyung
Choudhury, Tirthankar
Christiansen, Wayne
CHU, Hanshu
CHUNG, Hyun-Soo
Chyży, Krzysztof
Cichowolski, Silvina
Ciliegi, Paolo
Clark, Barry
Clark, David
Clark, Frank
Clegg, Andrew
Clemens, Dan
Cohen, Marshall
Cohen, Richard
Cole, Trevor
Coleman, Paul
Colomer, Francisco
Combes, Françoise
Combi, Jorge
Condon, James
Conklin, Edward
Contreras, Maria
Conway, John
Conway, Robin
Corbel, Stéphane
Cordes, James
Costa, Marco
Cotton Jr, William
Courtois, Helene
Crane, Patrick
Crawford, Fronefield
Croft, Steve
Croom, David
Croston, Judith
Crovisier, Jacques
Crutcher, Richard
Cudaback, David
Cunningham, Maria
Dagkesamanskii, Rustam
Daintree, Edward
DAISHIDO, Tsuneaki
DAISUKE, Iono
Dallacasa, Daniele
D'Amico, Nicolo
Dannerbauer, Helmut
Davies, Rodney
Davis, Michael
Davis, Richard
Davis, Robert
Davis, Timothy
de Bergh, Catherine
De Bernardis, Paolo
Degaonkar, S.
de Gregorio-Monsalvo, Itziar
de Jager, Cornelis
de Lange, Gert
Delannoy, Jean
de La NoÃ«, Jerome
Deller, Adam
Denisse, Jean-Francois
Dent, William
de Petris, Marco
de Ruiter, Hans
Deshpande, Avinash
Despois, Didier
de Vicente, Pablo
Dewdney, Peter
Dhawan, Vivek
Diamond, Philip
Dickel, Helene
Dickel, John
Dickey, John
Dickman, Robert
Dieter Conklin, Nannielou
Dionatos, Odysseas
Dixon, Robert
DOBASHI, Kazuhito
Dodson, Richard
Doubinskij, Boris
Dougherty, Sean
Douglas, James
Downes, Dennis
Downs, George
Drake, Frank
Drake, Stephen
Dravskikh, Aleksandr
Dreher, John

Dubner, Gloria
Duffett-Smith, Peter
Dulk, George
Dutrey, Anne
Dwarakanath, K.
Dyson, Freeman
Eales, Stephen
Edelson, Rick
Edwards, Philip
Ehle, Matthias
Ekers, Ronald
Elia, Davide
Ellingsen, Simon
Ellis, Graeme
Elsmore, Bruce
Emerson, Darrel
Emonts, Bjorn
Epstein, Eugene
Erickson, William
ESAMDIN, Ali
Eshleman, Von
ESIMBEK, Jarken
Evans, Kenton
Ewing, Martin
EZAWA, Hajime
Facondi, Silvia
Falcke, Heino
Fanaroff, Bernard
Fanti, Roberto
Farrell, Sean
Faulkner, Andrew
Feain, Ilana
Fedotov, Leonid
Feigelson, Eric
Feldman, Paul
Felli, Marcello
Feretti, Luigina
Fernandes, Francisco
Ferrari, Attilio
Ferrusca, Daniel
Fey, Alan
Field, George
Figueiredo, Newton
Filipovic, Miroslav
Fleischer, Robert
Fletcher, Andrew
Florkowski, David
Foley, Anthony
Fomalont, Edward
Fort, David
Forveille, Thierry
Fouqué, Pascal
Frail, Dale
Frater, Robert
Frey, Sandor
Friberg, Per
Fuerst, Ernst
FUKUI, Yasuo
Gabanyi, Krisztina
Gabuzda, Denise
Gaensler, Bryan
Gallego, Juan Daniel
Gallimore, Jack
Gangadhara, R.T.
GAO, Yu
Garaimov, Vladimir
Garay, Guido
Garrett, Michael
Garrington, Simon
Gasiprong, Nipon
Gaume, Ralph
Gawroński, Marcin
Gaylard, Michael
Geldzahler, Barry
GENG, Lihong
Gentile, Gianfranco
Genzel, Reinhard
Gerard, Eric
Gergely, Tomas
Gervasi, Massimo
Ghigo, Francis
Ghosh, Tapasi
Gimenez, Álvaro
Gioia, Isabella
Giovannini, Gabriele
Giroletti, Marcello
Gitti, Myriam
Goddi, Ciriaco
Goedhart, Sharmila
Goldwire, Jr., Henry
Gomez, Gonzalez
Gómez Fernández, Jose
GONG, Biping
Gonze, Roger
Gopalswamy, Natchimuthuk
Gordon, Chris
Gordon, Mark
Gorschkov, Aleksandr
Gosachinskij, Igor
Goss, W. Miller
Gottesman, Stephen
Gower, Ann
Gower, J.
Graham, David
Gratchev, Valerij
Green, Anne
Green, David
Green, James
Gregorini, Loretta
Gregorio-Hetem, Jane
Gregory, Philip
Grewing, Michael
GU, Minfeng
GU, Xuedong
Gubchenko, Vladimir
Guelin, Michel
Guesten, Rolf
Guidice, Donald
GUILLOTEAU, Stéphane
Gulkis, Samuel
Gull, Stephen
Gulyaev, Sergei
GUO, Jianheng
Gupta, Yashwant
Gurvits, Leonid
Gwinn, Carl
Haddock, Fred
Hall, Peter
Hallinan, Gregg
Hamilton, Phillip
HAN, JinLin
HAN, Wenjun
Hanasz, Jan
HANDA, Toshihiro
Hanisch, Robert
Hankins, Timothy
Hardee, Philip
Harnett, Julienne
Harris, Daniel
Harten, Ronald
Harvey-Smith, Lisa
Haschick, Aubrey
HASEGAWA, Tetsuo
Haslam, C.
Hatziminaoglou, Evanthia
Haverkorn, Marijke
Hayashi, Masahiko
Haynes, Martha
Haynes, Raymond

Hazard, Cyril
HAZUMI, Masashi
HE, Jinhua
Heald, George
Heeralall-Issur, Nalini
Heiles, Carl
Heinz, Sebastian
Helou, George
Henkel, Christian
Herpin, Fabrice
Heske, Astrid
Hess, Kelley
Hessels, Jason
Hewish, Antony
Hey, James
Hibbard, John
Higgs, Lloyd
Hills, Richard
HIRABAYASHI, Hisashi
HIRAMATSU, Masaaki
HIROTA, Tomoya
Hjalmarson, Ake
Ho, Paul
Ho, Wynn
Hoang, Binh
Hobbs, George
Höglund, Bertil
Hofner, Peter
Högbom, Jan
Hogg, David
Hojaev, Alisher
Hollis, Jan
HONG, Xiaoyu
Hopkins, Andrew
Horiuchi, Shinji
Hotan, Aidan
Howard III, William
HUANG, Fuquan
HUANG, Hui-Chun
Huchtmeier, Walter
Hughes, David
Hughes, Philip
Hulsbosch, A.
Humphreys, Elizabeth
Hunstead, Richard
Huynh, Minh
HWANG, Chorng-Yuan
Ibrahim, Zainol Abidin
IGUCHI, Satoru
IKEDA, Norio
Ikhsanov, Robert
Ikhsanova, Vera
Iliev, Ilian
Ilin, Gennadii
IMAI, Hiroshi
INATANI, Junji
INOUE, Makoto
Ipatov, Aleksandr
Irvine, William
ISHIGURO, Masato
Israel, Frank
Ivanov, Dmitrii
IWATA, Takahiro
Jáchym, Pavel
Jackson, Carole
Jackson, Neal
Jacq, Thierry
Jaffe, Walter
Jamrozy, Marek
Janssen, Michael
Jauncey, David
Jelić, Vibor
Jenkins, Charles
Jewell, Philip
JI, Shuchen
JIANG, Zhibo
JIN, Shengzeng
JIN, Zhenyu
Johnson, Donald
Johnston, Helen
Johnston, Kenneth
Johnston-Hollitt, Melanie
Joly, Francois
Jonas, Justin
Jones, Dayton
Jones, Paul
Joshi, Mohan
Josselin, Eric
JUNG, Jae-Hoon
Kaftan, May
Kaidanovski, Mikhail
KAIFU, Norio
KAKINUMA, Takakiyo
Kalberla, Peter
Kaltman, Tatyana
KAMAZAKI, Takeshi
KAMEGAI, Kazuhisa
KAMENO, Seiji
KAMEYA, Osamu
Kandalyan, Rafik
Kanekar, Nissim
KANG, Gon-Ik
Kapahi, V.
Kardashev, Nicolay
Karouzos, Marios
Kassim, Namir
KASUGA, Takashi
Kaufmann, Pierre
KAWABE, Ryohei
KAWAGUCHI, Kentarou
KAWAMURA, Akiko
Kedziora-Chudczer, Lucyna
Kellermann, Kenneth
Kent, Brian
Keshet, Uri
Kesteven, Michael
Khaikin, Vladimir
Khodachenko, Maxim
Kijak, Jarosław
Kilborn, Virginia
Killeen, Neil
KIM, Hyun-Goo
KIM, Ji Hoon
KIM, Kwang tae
KIM, Sang Joon
KIM, Tu Whan
Kimball, Amy
Kislyakov, Albert
Kitaeff, Vyacheslav
Klein, Karl
Klein, Ulrich
Knudsen, Kirsten
Ko, Hsien
KOBAYASHI, Hideyuki
Kocharovsky, Vitaly
KODA, Jin
KOHNO, Kotaro
KOJIMA, Masayoshi
Kolomiyets, Svitlana
Kondratiev, Vladislav
Konovalenko, Alexander
Kopylova, Yulia
Korpela, Eric
Korzhavin, Anatoly
Kovalev, Yuri
Kovalev, Yuri
KOYAMA, Yasuhiro
Kraft, Ralph
Kramer, Busaba
Kramer, Michael

Kreysa, Ernst
Krichbaum, Thomas
Krishna, Gopal
Krishnamohan, S.
Krishnan, Thiruvenkata
Kronberg, Philipp
Krügel, Endrik
KUAN, Yi-Jehng
Kudryavtseva, Nadezhda (Nadia)
Kuijpers, H.
Kuiper, Thomas
Kulkarni, Prabhakar
Kulkarni, Shrinivas
Kulkarni, Vasant
Kumkova, Irina
Kundt, Wolfgang
Kunert-Bajraszewska, Magdalena
KURAYAMA, Tomoharu
Kus, Andrzej
Kutner, Marc
Kuzmin, Arkadij
Kwok, Sun
Lada, Charles
La Franca, Fabio
LAI, Shih-Ping
Laing, Robert
Lal, Dharam
Landecker, Thomas
Landt, Hermine
Lang, Kenneth
Langer, William
Langston, Glen
LaRosa, Theodore
Lasenby, Anthony
Lawrence, Charles
Lazio, Joseph
Leahy, J.
Lebrón, Mayra
LEE, Chang Won
LEE, Jeong-Eun
LEE, Jung-Won
LEE, Sang-Sung
LEE, Yong Bok
LEE, Youngung
Lefloch, Bertrand
Lehnert, Matthew
Lenc, Emil
Lépine, Jacques
Lequeux, James
Lesch, Harald
Le Squeren, Anne-Marie
Lestrade, Jean-François
Leung, Chun
Levreault, Russell
Li, Hong-Wei
LI, Chun-Sheng
LI, Gyong
LI, Shiguang
LI, Zhi
Likkel, Lauren
Lilley, Edward
LIM, Jeremy
Lindqvist, Michael
Linke, Richard
Lis, Dariusz
Liseau, René
Lister, Matthew
LIU, Sheng-Yuan
LIU, Xiang
LIU, Yuying
Lo, Fred K. Y.
Lo, Wing-Chi Nadia
Locke, Jack
Lockman, Felix
Loiseau, Nora
Longair, Malcolm
Loren, Robert
Loubser, Ilani
Loukitcheva, Maria
Lovell, James
Lozinskaya, Tatjana
LU, Yang
Lubowich, Donald
Lucero, Danielle
Luks, Thomas
LUO, Xianhan
Luque-Escamilla, Pedro
Lyne, Andrew
Lytvynenko, Leonid
MA, Guanyi
Macchetto, Ferdinando
MacDonald, Geoffrey
MacDonald, James
Machalski, Jerzy
Maciesiak, Krzysztof
Mack, Karl-Heinz
MacLeod, John
MAEHARA, Hideo
Malofeev, Valery
Manchester, Richard
Mandolesi, Nazzareno
Mantovani, Franco
MAO, Rui-Qing
Maran, Stephen
Marcaide, Juan-Maria
Mardones, Diego
Mardyshkin, Vyacheslav
Marecki, Andrzej
Markoff, Sera
Marques, Dos
Marscher, Alan
Marshalov, Dmitriy
Martí, Josep
Martin, Christopher
Martin, Robert
Martín, Sergio
Martin-Pintado, Jesus
Marvel, Kevin
Masheder, Michael
Masłowski, Józef
Mason, Paul
Massardi, Marcella
Massaro, Francesco
Masson, Colin
Masters, Karen
Matheson, David
Matsakis, Demetrios
MATSUO, Hiroshi
MATSUSHITA, Satoki
Matthews, Brenda
Matthews, Henry
Mattila, Kalevi
Matveenko, Leonid
Mauersberger, Rainer
Maxwell, Alan
McAdam, Bruce
McConnell, David
McCulloch, Peter
McKean, John
McKenna-Lawlor, Susan
McLean, Donald
McMullin, Joseph
Mebold, Ulrich
Meeks, M.
Meier, David
Melikidze, Giorgi
Menon, T.
Menten, Karl

Meyer, Martin
Michalec, Adam
Migenes, Victor
Mikhailov, Andrey
Miley, George
Miller, Neal
Mills, Bernard
Milne, Douglas
Mirabel, Igor
Miroshnichenko, Alla
Mitchell, Kenneth
Miyawaki, Ryosuke
MIYAZAKI, Atsushi
MIYOSHI, Makoto
MIZUNO, Akira
MIZUNO, Norikazu
Moellenbrock III, George
Moffet, Alan
Moffett, David
Momjian, Emmanuel
MOMOSE, Munetake
Montmerle, Thierry
Morabito, David
Moran, James
Morgan, Lawrence
Morganti, Raffaella
Morison, Ian
MORITA, Kazuhiko
MORIYAMA, Fumio
Morras, Ricardo
Morris, David
Morris, Mark
Moscadelli, Luca
Mosoni, Laszlo
Muller, Erik
Muller, Sebastien
Mundy, Lee
MURAOKA, Kazuyuki
MURATA, Yasuhiro
Murdoch, Hugh
Murphy, Tara
Mutel, Robert
Muxlow, Thomas
Myers, Philip
Nadeau, Daniel
Nagnibeda, Valerij
NAKAJIMA, Junichi
NAKANO, Takenori
NAKASHIMA, Jun-ichi
Nammahachak, Suwit
NAN, Ren-Dong
Neeser, Mark
Neff, John
Nesvadba, Nicole
Nguyen-Quang, Rieu
Nicastro, Luciano
Nice, David
Nicholls, Jennifer
Nicolson, George
NIINUMA, Kotaro
Nikoli'c, Silvana
NISHIO, Masanori
Norris, Raymond
Nürnberger, Dieter
Nuza, Sebastian
O'Dea, Christopher
O'Donoghue, Aileen
OGAWA, Hideo
Ohashi, Nagayoshi
OHISHI, Masatoshi
Ojha, Roopesh
OKA, Tomoharu
OKUMURA, Sachiko
Olberg, Michael
Önel, Hakan
ONISHI, Toshikazu
Onuora, Lesley
Oozeer, Nadeem
Orchiston, Wayne
Orienti, Monica
O'Sullivan, John
Otmianowska-Mazur, Katarzyna
Ott, Juergen
Owen, Frazer
Özel, Mehmet
Özeren, Ferhat
Pacholczyk, Andrzej
Padman, Rachael
Palmer, Patrick
Panessa, Francesca
Pankonin, Vernon
Paragi, Zsolt
Paredes Poy, Josep
Parijskij, Yurij
PARK, Yong Sun
Parker, Edward
Parma, Paola
Paron, Sergio
Parrish, Allan
Pasachoff, Jay
Pashchenko, Mikhail
Patel, Nimesh
Pauliny, Toth
Pauls, Thomas
Pavlyuchenkov, Yaroslav
Pawsey, J.
Payne, David
Pearson, Timothy
Peck, Alison
Pedersen, Holger
Pedlar, Alan
Peel, Michael
PENG, Bo
PENG, Qingyu
Penzias, Arno
Perez, Fournon
Pérez Torres, Miguel
Perley, Richard
Perozzi, Ettore
Persson, Carina
Peters, William
Petrova, Svetlana
Pettengill, Gordon
PHAM, Diep
Philipp-May, Sabine
Phillips, Christopher
Phillips, Thomas
Pick, Monique
PING, Jinsong
Pisano, Daniel
Pitkin, Matthew
Planesas, Pere
Pogrebenko, Sergei
Polatidis, Antonios
Pompei, Emanuela
Ponsonby, John
Pooley, Guy
Porcas, Richard
Porras Juárez, Bertha
Potapov, Vladimir
Prandoni, Isabella
Preston, Robert
Preuss, Eugen
Price, R.
Priester, Wolfgang
Pshirkov, Maxim
Puschell, Jeffery
Pushkarev, Alexander
Puxley, Phil

QIU, Yuhai
Radford, Simon
Rahimov, Ismail
Ransom, Scott
Rao, A.
Raoult, Antoinette
Rathborne, Jill
Ray, Paul
Ray, Tom
Readhead, Anthony
Redman, Matthew
Reich, Wolfgang
Reid, Mark
Reif, Klaus
Reyes, Francisco
Reynolds, Cormac
Reynolds, John
RHEE, Myung Hyun
Ribes, Jean-Claude
Ribó, Marc
Richer, John
Rickard, Lee
Rickett, Barnaby
Ridgway, Susan
Riley, Julia
Rioja, Maria
Rizzo, Jose
Roberts, David
Roberts, Morton
Robertson, Douglas
Robertson, James
Robinson Jr, Richard
Rodriguez, Luis
Roeder, Robert
Roelfsema, Peter
Roennaeng, Bernt
Röser, Hans-peter
Roger, Robert
Rogers, Alan
Rogstad, David
Rohlfs, Kristen
Romanov, Andrey
Romero, Gustavo
Romney, Jonathan
Rosa González, Daniel
Rosolowsky, Erik
Rowson, Barrie
Rubio, Monica
Rubio-Herrera, Eduardo
Rudnick, Lawrence
Rudnitskij, Georgij
Rushton, Anthony
Russell, Jane
Rydbeck, Gustaf
Ryś, Stanisław
Sadler, Elaine
Saikia, Dhruba
SAKAMOTO, Seiichi
Salomé, Philippe
Salpeter, Edwin
Salter, Christopher
Samodurov, Vladimir
Sandell, Göran
Sanders, David
Sargent, Annelia
Saripalli, Lakshmi
Sarma, Anuj
Sarma, N.
Sastry, Ch.
SATO, Fumio
Saunders, Richard
Savage, Ann
Savolainen, Tuomas
Sawada, Tsuyoshi
SAWADA-SATOH, Satoko
Sawant, Hanumant
Scalise Jr, Eugenio
Schaal, Ricardo
Schilizzi, Richard
Schilke, Peter
Schlickeiser, Reinhard
Schmidt, Maarten
Schröder, Anja
Schuch, Nelson
Schulz, R.
Schure, Klara
Schwartz, Philip
Schwarz, Ulrich
Scott, John
Scott, Paul
Seaquist, Ernest
Seielstad, George
SEKIDO, Mamoru
SEKIMOTO, Yutaro
Semenov, Dmitry
Sese, Rogel Mari
SETA, Masumichi
Setti, Giancarlo
Seymour, Nicholas
Shaffer, David
Shakeshaft, John
SHANG, Hsien
Shaposhnikov, Vladimir
Shastri, Prajval
Shaver, Peter
SHEN, Zhiqiang
Shepherd, Debra
Shevgaonkar, R.
SHIBATA, Katsunori
Shimmins, Albert
SHIMOIKURA, Tomomi
Shinnaga, Hiroko
Šmelds, Ivar
Sholomitsky, Gennady
Shone, David
SHU, Fengchun
Shulga, Valerii
Sieber, Wolfgang
Singal, Ashok
Sinha, Rameshwar
Siringo, Giorgio
Skillman, Evan
Slade, Martin
Slee, O.
Slysh, Vyacheslav
Smerd, S.
Smirnova, Tatiana
Smith, Dean
Smith, Francis
Smith, Niall
Smolentsev, Sergej
Smol'kov, Gennadij
Snellen, Ignas
Sobolev, Yakov
Sodin, Leonid
SOFUE, Yoshiaki
Sokolov, Konstantin
SOMANAH, Radhakhrishna
SONG, Qian
Soria, Roberto
Sorochenko, Roman
Spencer, John
Spencer, Ralph
Richard, Richard
Sridharan, Tirupati
Stairs, Ingrid
Stanghellini, Carlo
Stanley, G.
Stannard, David

Stappers, Benjamin
Steffen, Matthias
Steinberg, Jean-Louis
Stewart, Paul
Stewart, Ronald
Stil, Jeroen
Stone, R.
Storey, Michelle
Strom, Richard
Strukov, Igor
Subrahmanya, C.
Subrahmanyan, Ravi
SUGITANI, Koji
Sukumar, Sundarajan
Suleymanova, Svetlana
Sullivan, III, Woodruff
SUNADA, Kazuyoshi
Surkis, Igor
Swarup, Govind
Swenson Jr., George
Szomoru, Arpad
Szymczak, Marian
TABARA, Hiroto
TAKABA, Hiroshi
TAKAGI, Kojiro
TAKAKURA, Tatsuo
TAKANO, Shuro
TAKANO, Toshiaki
Tammi, Joni
TAN, Baolin
Tanaka, Haruo
TANAKA, Riichiro
Tapping, Kenneth
Tarter, Jill
TATEMATSU, Ken'ichi
Tauber, Jan
Taylor, A.
te Lintel Hekkert, Peter
Tello Bohorquez, Camilo
Teng, Stacy
Terasranta, Harri
Terzian, Yervant
Theureau, Gilles
Thomasson, Peter
Thompson, A.
Thum, Clemens
TIAN, Wenwu
Tingay, Steven
Tiplady, Adrian
Tofani, Gianni
Tolbert, Charles
Tomasi, Paolo
Tornikoski, Merja
Torrelles, Jose Maria
TOSAKI, Tomoka
Tovmassian, Hrant
Townes, Charles
Trigilio, Corrado
Trinidad, Miguel
TRIPPE, Sascha
Tritton, Keith
Troland, Thomas
Truong, Bach
Trushkin, Sergey
TSAI, An-Li
TSUBOI, Masato
Tsutsumi, Takahiro
Tuccari, Gino
Turło, Zygmunt
Turner, Barry
Turner, Jean
Turner, Kenneth
Turtle, A.
Tyul'bashev, Sergei
Tzioumis, Anastasios
Udaya, Shankar
Ulrich, Bruce
Ulrich, Marie-Helene
Ulvestad, James
Ulyanov, Oleg
Umana, Grazia
UMEMOTO, Tomofumi
Unger, Stephen
Unwin, Stephen
Urama, Johnson
Urov sevi'c, Dejan
Urquhart, James
Uson, Juan
Vaccari, Mattia
Vakoch, Douglas
Vallée, Jacques
Valtaoja, Esko
Valtonen, Mauri
Val'tts, Irina
VandenBout, Paul
van der Hulst, Jan
van der Kruit, Pieter
van der Laan, Harry
van der Tak, Floris
van Driel, Wim
van Gorkom, Jacqueline
van Kampen, Eelco
van Langevelde, Huib
van Leeuwen, Joeri
van Woerden, Hugo
Vats, Hari
Vaughan, Alan
Velusamy, T.
Venturi, Tiziana
Venugopal, V.
Verheijen, Marc
Verkhodanov, Oleg
Vermeulen, Rene
Verschuur, Gerrit
Verter, Frances
Vestergaard, Marianne
Vilas, Faith
Vilas-Boas, José
Vitkevitch, V.
Vivekanand, M.
Vlahakis, Catherine
Vlemmings, Wouter
Vogel, Stuart
Volvach, Alexander
Voronkov, Maxim
Wadadekar, Yogesh
WAJIMA, Kiyoaki
Walker, Robert
Wall, Jasper
Wall, William
Walmsley, C.
Walsh, Andrew
Walsh, Wilfred
WAN, Tong-Shan
WANG, Chen
WANG, Guangli
WANG, Hong-Guang
WANG, Jingyu
WANG, Junzhi
WANG, Min
WANG, Na
WANG, Shouguan
WANG, Shujuan
Wannier, Peter
Wardle, John
Ward-Thompson, Derek
Warmels, Rein
Warner, Peter
Watson, Robert
Wayth, Randall

Wehrle, Ann
Wei, Mingzhi
WEI, Erhu
Weigelt, Gerd
Weiler, Edward
Weiler, Kurt
Welch, William
Weliachew, Leonid
Wellington, Kelvin
WEN, Linqing
Wendker, Heinrich
Wendt, Harry
WENLEI, Shan
Westmeier, Tobias
White, Glenn
Whiteoak, John
Whiting, Matthew
Wicenec, Andreas
Wickramasinghe, N.
Wielebinski, Richard
Wiik, Kaj
Wiklind, Tommy
Wild, John
Wild, Wolfgang
Wilkinson, Peter
Willis, Anthony
Wills, Beverley
Wills, Derek
Willson, Robert
Wilner, David
Wilson, Andrew
Wilson, Robert
Wilson, Thomas
Wilson, William
Windhorst, Rogier
Winnberg, Anders
Wise, Michael
Witzel, Arno
Woan, Graham
Wolleben, Maik
Wolszczan, Alexander
Woltjer, Lodewijk
Wong, Tony
Wood, Douglas
Woodsworth, Andrew
Wootten, Henry
Wright, Alan
Wrobel, Joan
Wu, Nailong
WU, Huai-Wei
WU, Xinji
WU, Yuefang
Wucknitz, Olaf
XIA, Zhiguo
XU, Pei-Yuan
XU, Zhi-Cai
YAMAMOTO, Hiroaki
YANG, Ji
YANG, Jian
YANG, Zhigen
YAO, Qijun
YE, Shuhua
Yin, Qi-Feng
YONEKURA, Yoshinori
Younis, Saad
YU, Zhiyao
Yusef-Zadeh, Farhad
Zabolotny, Vladimir
Zainal Abidin, Zamri
Zaitsev, Valerij
Zanichelli, Alessandra
Zannoni, Mario
Zarka, Philippe
Zavala, Robert
Zensus, J-Anton
ZHANG, Chengmin
ZHANG, Haiyan
ZHANG, Hongbo
ZHANG, Jian
ZHANG, JiangShui
ZHANG, Jin
ZHANG, Jingyi
ZHANG, Qizhou
ZHANG, Xizhen
ZHANG, Yong
ZHANG, Zhibin
ZHAO, Jun-Hui
Zheleznyak, Alexander
Zheleznyakov, Vladimir
ZHENG, Xiaonian
ZHENG, Xinwu
ZHOU, Jianfeng
ZHOU, Jianjun
ZHOU, Ti-jian
ZHOU, Xia
ZHU, LiChun
ZHU, Ming
ZHU, Qingfeng
ZHU, Wenbai
Zięba, Stanisław
Zinchenko, Igor
Zlobec, Paolo
Zlotnik, Elena
Zuckerman, Benjamin
Zwaan, Martin
Zylka, Robert

Division C - Commission 41 History of Astronomy

President: Brian Warner
Vice-President: Xiaochun Sun

Organizing Committee Members:

Eugene F. Milone
Juan Antonio Belmonte Aviles
Raymond P. Norris
Brenda G. Corbin
Mitsuru SOMA
Luisa Pigatto

Members:

Abalakin, Viktor
Abt, Helmut
Acharya, Bannanje
AHN, Youngsook
Alves, Virgínia
Andrews, David
Ansari, S.M.
Antonello, Elio
Arifyanto, Mochamad
Ashbrook, J.
Ashok, N.
Babu, G.S.D.
Babul, Arif
Badolati, Ennio
Bailey, Mark
Ball, Lewis
Ballabh, Goswami
Balyshev, Marat
Bandyopadhyay, A.
Baneke, David
Barlai, Katalin
Batten, Alan
Belmonte Aviles, Juan Antonio
Bennett, Jim
Benson, Priscilla
Berendzen, Richard
Bertola, Francesco
Bessell, Michael
Bhatia, Vishnu
Bhatt, H.
Bhattacharjee, Pijushpani
Bien, Reinhold
Bishop, Roy
BO, Shuren
Boccaletti, Dino
Boerngen, Freimut
Bon, Edi
Bònoli, Fabrizio
Botez, Elvira
Bougeret, Jean-Louis
Bouton, Ellen
Bowen, David
Brecher, Kenneth
Bretones, Paulo
Brooks, Randall
Brosche, Peter
Brouw, Willem
Brunet, Jean-Pierre
Burman, Ronald
CAI, Kai
Campana, Riccardo
Cannon, Russell
Caplan, James
Carlson, John
CHANG, Heon-Young
Chapman, Allan
Chapman, Jessica
CHEN, Meidong
CHIN, Yi-nan
Chinnici, Ileana
Choudhary, Debi Prasad
Chung, Eduardo
Clifton, Gloria
Corbin, Brenda
Cornejo, Alejandro
CUI, Shizhu
CUI, Zhenhua
Dadic, Zarko
Danezis, Emmanuel
Das, P.
Davies, Rodney
Davis, A. E. L.
Davoust, Emmanuel
Débarbat, Suzanne
de Jong, Teije
Dekker, E.
Denisse, Jean-Francois
DeVorkin, David
Dick, Steven
Dick, Wolfgang
Dingle, H.
Dluzhnevskaya, Olga
Dorschner, Johann
Duffard, Rene
Dumont, Simone
Dutil, Yvan
Dworetsky, Michael
Edmondson, Frank
Edwards, Philip
Ehgamberdiev, Shuhrat
Engels, Dieter
Espenak, Fred
Esteban, César
Evans, Robert
Ferlet, Roger
Fernie, J.
Feulner, Georg
Field, J. V.
Fierro, Julieta
Firneis, Maria
Flin, Piotr
Florides, Petros
Fluke, Christopher
Fodera, Serio
Forbes, Eric
Freeman, Kenneth
Frew, David
Funes, José
Gábor, Pavel
Gangui, Alejandro
Gavrilov, Mikhail
Geffert, Michael
Geyer, Edward
Gillingham, Peter
Gingerich, Owen
Glass, Ian
Goss, W. Miller
Graham, John
Green, Anne
Green, Daniel
Green, David
Griffin, R. Elizabeth
Gurshtein, Alexander
Gussmann, Ernst-August
Hadrava, Petr
HAN, Wonyong
Hasan, S. Sirajul
HASEGAWA, Ichiro
Haubold, Hans
Haupt, Hermann
Hayli, Abraham
Haynes, Raymond
Haynes, Roslynn
Hearnshaw, John
Heck, Andre
Heddle, Douglas
Hellman, C.
Helou, George
Hemenway, Mary
Herrmann, Dieter

Hidayat, Bambang
HIRAI, Masanori
Hockey, Thomas
Høg, Erik
Hollow, Robert
Holmberg, Gustav
Hopkins, Andrew
Horsky, J.
Hoskin, Michael
HU, Tiezhu
Huan, Nguyen
Hunstead, Richard
HURUKAWA, Kiichirō
HWANG, Chorng-Yuan
Hysom, Edmund
Hyung, Siek
Ibrahim, Alaa
Jafelice, Luiz
Jahreiss, Hartmut
Jauncey, David
JEONG, Jang-Hae
JIANG, Xiaoyuan
Jiménez-Vicente, Jorge
Jones, Paul
Kapoor, Ramesh
Keay, Colin
Keller, Hans-Ulrich
Kellermann, Kenneth
Kepler, S.
Kerschbaum, Franz
Kilambi, G.
KIM, Chun-Hwey
KIM, Sang Chul
KIM, SANG HYUK
KIM, Yong Cheol
KIM, Yonggi
KIM, Young-Soo
King, David
Kippenhahn, Rudolf
Knight, Matthew
Kochhar, Rajesh
Kollerstrom, Nicholas
Kolomiyets, Svitlana
Komonjinda, Siramas
Koribalski, Bärbel
Kosovichev, Alexander
Kovács, József
Krajnović, Davor
Kreiner, Jerzy
Krisciunas, Kevin
Krishnan, Thiruvenkata
Krupp, Edwin
Kunitzsch, Paul
Lanfranchi, Gustavo
Lang, Kenneth
LAS VERGNAS, Olivier
Launay, Françoise
LEE, Eun Hee
LEE, Ki-Won
LEE, Woo baik
LEE, Yong Bok
LEE, Yong Sam
Le Guet Tully, Françoise
Lerner, Michel-Pierre
Leung, Kam
Levy, Eugene
LI, Min
LI, Yong
LI, Zhisen
Liller, William
Liritzis, Ioannis
LIU, Ciyuan
Locher, Kurt
Lomb, Nicholas
Longo, Giuseppe
Lopes, Rosaly
López, Alejandro
Lopez, Carlos
Luminet, Jean-Pierre
Ma, Chunyu
Mahoney, Terence
Maíz Apellániz, Jesús
Malin, David
Mallamaci, Claudio
Malville, J.
Manchester, Richard
Marco, Olivier
Mason, Brian
Mathewson, Donald
McAdam, Bruce
McConnell, David
McKenna-Lawlor, Susan
McLean, Donald
Meadows, A.
Meech, Karen
Mendillo, Michael
Menon, T.
Mickaelian, Areg
Mickelson, Michael
Mikhail, Joseph
Milne, Douglas
Milone, Eugene
Moesgaard, Kristian
Molnar, Michael
Moore-Weiss, John
Mumford, George
Nadal, Robert
NAKAMURA, Tsuko
NAKAYAMA, Shigeru
Narlikar, Jayant
Naze, Yael
Nefedyev, Yury
Neugebauer, O.
NHA, Il Seong
Nicolaidis, Efthymios
NING, Xiaoyu
Nitschelm, Christian
Norris, Raymond
North, John
Nussbaumer, Harry
OH, Kyu-Dong
Ohashi, Nagayoshi
Ôhashi, Yukio
Olivier, Enrico
Olowin, Ronald
Oproiu, Tiberiu
Orchiston, Wayne
Osório, José
Oudet, J.
Pang, Kevin
Papathanasoglou, Dimitrios
Pasachoff, Jay
Pati, Ashok
Pecker, Jean-Claude
Pedersen, Olaf
Peterson, Charles
Pettersen, Bjørn
Pfleiderer, Jorg
PHAN, Dong
Pigatto, Luisa
Pilbratt, Göran
Pineda de Carias, Maria
Pinigin, Gennadiy
Polcaro, V.
Polozhentsev, Dmitrij
Polyakhova, Elena
Poulle, Emmanuel
Pozhalova, Zhanna
Preston, Robert
Prokakis, Theodore

Proverbio, Edoardo
Pustylnik, Izold
QUAN, Hejun
Rafferty, Theodore
Ray, Tom
Reboul, Henri
Robertson, James
Rubio-Herrera, Eduardo
Ruggles, Clive
Ryder, Stuart
Saucedo Morales, Julio
Sbirkova-Natcheva, Temenuzhka
Schaefer, Bradley
Schechner, Sara
Schmadel, Lutz
Schmeidler, F.
Schmidt, Maarten
Schnell, Anneliese
Schulz, R.
Seck, Friedrich
Seggewiss, Wilhelm
Serio, Salvatore
Shank, Michael
Shankland, Paul
Shaver, Peter
Shelton, Ian
Shingareva, Kira
Shore, Steven
Shukre, C.
Sibhardth, B.
Sigismondi, Costantino
Signore, Monique
Sima, Zdislav
Simonia, Irakli
Simpson, Allen
Sinachopoulos, Dimitris
Singh, Jagdev
Šlechta, Miroslav
Slee, O.
Slysh, Vyacheslav
Smith, Malcolm
Sobouti, Yousef
Solc, Martin
SOMA, Mitsuru
Soonthornthum, Boonrucksar
Souchay, Jean
Stathopoulou, Maria
Stavinschi, Magda
Stavinschi, Magda
Steel, Duncan
Steele, John
Steinle, Helmut
Stephenson, F.
Sterken, Christiaan
Stoev, Alexey
Storey, Michelle
Subramanian, K.
Sullivan, III, Woodruff
Sun, Xiaochun
Svolopoulos, Sotirios
Swarup, Govind
Swerdlow, Noel
Szabados, Laszlo
Szostak, Roland
Tammann, Gustav
TAMURA, Shin'ichi
TANIKAWA, Kiyotaka
Taub, Liba
Tchenakal, V.
Terzian, Yervant
Theodossiou, Efstratios
Tignalli, Horacio
Tobin, William
Townes, Charles
Treder, H.
Trimble, Virginia
Tripathy, Sushanta
Tsvetkov, Milcho
Urama, Johnson
Usher, Peter
Vahia, Mayank
Vakoch, Douglas
Valdes Parra, Jose
van Gent, Robert
van Woerden, Hugo
Vass, Gheorghe
Vats, Hari
Vaughan, Alan
Vavilova, Iryna
Venkatakrishnan, P.
Verdet, Jean-Pierre
Verdun, Andreas
Viollier, Raoul
Voigt, Hans
Volyanska, Margaryta
Wainscoat, Richard
WANG, Dechang
WANG, Guangchao
WANG, Rongbin
Warner, Brian
Watson, Frederick
Weiss, Werner
Wendt, Harry
Whitaker, Ewen
White, Graeme
Whiteoak, John
Whitrow, Gerald
Wielen, Roland
Wilkins, George
Williams, Thomas
Wilson, Curtis
Wolfschmidt, Gudrun
Woudt, Patrick
Wright, Alan
Wünsch, Johann
XI, Zezong
XIONG, Jianning
YAMAOKA, Hitoshi
YANG, Hong-Jin
Yau, Kevin
Yeomans, Donald
YUMIN, Wang
Zanini, Valeria
Zeilik, Michael
ZHANG, Peiyu
ZHANG, Shouzhong
ZHOU, Yonghong
Zsoldos, Endre

Division G - Commission 42 Close Binary Stars

President: Mercedes T. Richards
Vice-President: Theodor Pribulla

Organizing Committee Members:

Horst Drechsel
Carla Maceroni
Andrej Prša
John Southworth
Shay Zucker
Tomaž Zwitter
Colin David Scarfe
Ulisse Munari
Joanna Mikołajewska
David H. Bradstreet

Members:

AK, Tansel
Akashi, Muhammad
Al-Naimiy, Hamid
Andersen, Johannes
Andronov, Ivan
Antipova, Lyudmila
Antokhin, Igor
Antonopoulou, Evgenia
Anupama, G.
Aquilano, Roberto
Arbutina, Bojan
Arefiev, Vadim
Aungwerojwit, Amornrat
Awadalla, Nabil
BABA, Hajime
Babkovskaia, Natalia
Bailyn, Charles
Balman, Solen
Baptista, Raymundo
Baran, Andrzej
Barkin, Yuri
Barone, Fabrizio
Bartolini, Corrado
Bath, Geoffrey
Batten, Alan
Bear, Ealeal
Bell, Steven
Benacquista, Matthew
Bianchi, Luciana
Blair, William
Blundell, Katherine
Boffin, Henri
Bolton, Charles
Bonazzola, Silvano
Bopp, Bernard
Borisov, Nikolay
Bortoletto, Alexandre
Boyd, David
Boyle, Stephen
Bozic, Hrvoje
Bradstreet, David
Brandi, Elisande
Breinhorst, Robert
Broglia, Pietro
Brown, David
Brownlee, Robert
Bruch, Albert
Bruhweiler, Frederick
Budaj, Jan
Budding, Edwin
Bunner, Alan
Burderi, Luciano
Burikham, Piyabut
Busa', Innocenza
Busso, Maurizio
Callanan, Paul
Canalle, Joao
Catalano, Santo
Cester, Bruno
Chambliss, Carlson
Chapman, Robert
Chaty, Sylvain
Chaubey, Uma
CHEN, An-Le
CHEN, Xuefei
Cherepashchuk, Anatolij
Chochol, Drahomir
CHOI, Chul-Sung
CHOI, Kyu Hong
CHOU, Yi
Ciardi, David
Cillie, G.
Claria, Juan
Cornelisse, Remon
Corradi, Romano
Cowley, Anne
Crause, Lisa
Cropper, Mark
CUI, Wenyuan
Cutispoto, Giuseppe
Dadaev, Aleksandr
DAI, Zhibin
Dall, Thomas
D'Amico, Nicolo
D'Antona, Francesca
Day-Jones, Avril
de Greve, Jean-Pierre
de Groot, Mart
Delgado, Antonio
de Loore, Camiel
Del Santo, Melania
de Mink, Selma
Demircan, Osman
Derekas, Aliz
Diaz, Marcos
Djurasevic, Gojko
Dobrotka, Andrej
Dobrzycka, Danuta
Dorfi, Ernst
Dougherty, Sean
Drechsel, Horst
Dubus, Guillaume

Dupree, Andrea
Durisen, Richard
Duschl, Wolfgang
Eaton, Joel
Edalati Sharbaf, Mohammad
Ederoclite, Alessandro
Eggleton, Peter
Eldridge, John
Elias II, Nicholas
Engelbrecht, Chris
Etzel, Paul
Eyres, Stewart
Fabiani, Sergio
Fabrika, Sergei
Falize, Emeric
Farrell, Sean
Faulkner, John
Fekel, Francis
Ferluga, Steno
Fernández Lajús, Eduardo
Ferrario, Lilia
Ferrer, Osvaldo
Firmani, Claudio
Flannery, Brian
Fors, Octavi
Fracastoro, Mario
Frank, Juhan
Fredrick, Laurence
Gänsicke, Boris
Gallagher III, John
Gamen, Roberto
Garcia, Lia
García de María, Juan
Garcia-Lorenzo, Maria
Garmany, Katy
Gasiprong, Nipon
Geldzahler, Barry
Geyer, Edward
Giannone, Pietro
Gibson, David
Gies, Douglas
Giovannelli, Franco
Goldman, Itzhak
Gomboc, Andreja
GONG, Biping
González, Gabriela
González Martínez-Pais, Ignacio
Gosset, Eric
Graffagnino, Vito
Groot, Paul
Grygar, Jiri
GU, Wei-Min
Guinan, Edward
Gulliver, Austin
Gün, Gulnur
Gunn, Alastair
GUO, Jianheng
Gursky, Herbert
Haas, Martin
Hadrava, Petr
Hakala, Pasi
Hall, Douglas
Hallinan, Gregg
Hammerschlag-Hensberge, Godelieve
HANAWA, Tomoyuki
Hantzios, Panayiotis
Harmanec, Petr
Hassall, Barbara
Haswell, Carole
HAYASAKI, Kimitake
Hazlehurst, John
HE, Jinhua
Hegedues, Tibor
Hellier, Coel
Helt, Bodil
Hempelmann, Alexander
Hensler, Gerhard
Hilditch, Ronald
Hill, Graham
Hills, Jack
Hillwig, Todd
HIROSE, Shigenobu
Hoard, Donald
Holmgren, David
Holt, Stephen
Honeycutt, R.
HORIUCHI, Ritoku
Hric, Ladislav
Hrivnak, Bruce
Hube, Douglas
Hutchings, John
Ibanoglu, Cafer
Ikhsanov, Nazar
Imamura, James
Imbert, Maurice
Ioannou, Zacharias
Ivanova, Natalia
Izzard, Robert
Jabbar, Sabeh
Jasniewicz, Gerard
Jeffers, Sandra
JEONG, Jang-Hae
Jin, Liping
JIN, Zhenyu
Jones, David
Jonker, Peter
Joss, Paul
Kadouri, Talib
Kafka, Styliani (Stella)
Kaitchuck, Ronald
Kalomeni, Belinda
Kałużny, Janusz
KANG, Young Woon
Karami, Kayoomars
Karetnikov, Valentin
KATO, Taichi
KAWABATA, Shusaku
KENJI, Nakamura
Kenny, Harold
Kenyon, Scott
Khalesseh, Bahram
KIM, Chun-Hwey
KIM, Ho-il
KIM, Woong-Tae
KIM, Young-Soo
King, Andrew
Kjurkchieva, Diana
Kley, Wilhelm
Kolb, Ulrich
Kolesnikov, Sergey
Komonjinda, Siramas
Konacki, Maciej
KONDO, Yoji
Kopal, ZdenÃ«k
Koubsky, Pavel
Kraft, Robert
Kraicheva, Zdravka
Krautter, Joachim
Kreiner, Jerzy
Kreykenbohm, Ingo
Kriwattanawong, Wichean
Kruchinenko, Vitaliy
Kruszewski, Andrzej
Krzeminski, Wojciech
Kudashkina, Larisa
Kumsiashvily, Mzia
Kwee, K.

Lacy, Claud
Lamb Jr, Donald
Landolt, Arlo
Lanning, Howard
Lapasset, Emilio
Larsson, Stefan
Larsson-Leander, Gunnar
Lavrov, Mikhail
Lee, William
LEE, Chung-Uk
LEE, Jae Woo
LEE, Woo baik
LEE, Yong Sam
Leedjarv, Laurits
Leung, Kam
LI, Ji
LI, Lifang
LI, Zhi
LI, Zhongyuan
LIM, Jeremy
LIN, Yi-qing
Linnell, Albert
Linsky, Jeffrey
LIU, Jifeng
LIU, Qingyao
LIU, Qingzhong
Livio, Mario
Lloyd, Huw
Lucy, Leon
Luque-Escamilla, Pedro
MacDonald, James
Maceroni, Carla
Malasan, Hakim
Malkov, Oleg
Mammano, Augusto
Mandel, Ilya
Manimanis, Vassilios
Mardirossian, Fabio
Marilli, Ettore
Markoff, Sera
Markworth, Norman
Marsh, Thomas
Martayan, Christophe
Mason, Paul
Mathieu, Robert
Mauder, Horst
Mayer, Pavel
Mazeh, Tsevi
McCluskey Jr, George
Meibom, Soren
Meintjes, Petrus
Melia, Fulvio
Meliani, Mara
Mennickent, Ronald
Mereghetti, Sandro
Merrill, John
Meyer-Hofmeister, Emmi
Mezzetti, Marino
Migaszewski, Cezary
Mikołajewska, Joanna
Mikulášek, Zdeněk
Milano, Leopoldo
Millour, Florentin
Milone, Eugene
MINESHIGE, Shin
Miszalski, Brent
MIYAJI, Shigeki
Mochnacki, Stefan
Monard, Libert
Montgomery, Michele
Morales Rueda, Luisa
Morgan, Thomas
Morrell, Nidia
Mouchet, Martine
Mürset, Urs
Mumford, George
Munari, Ulisse
Murray, James
Mutel, Robert
NAKAMURA, Yasuhisa
NAKAO, Yasushi
NARIAI, Kyoji
Naylor, Tim
Neff, James
Nelemans, Gijs
Nelson, Burt
Neustroev, Vitaly
Newsom, Gerald
NHA, Il Seong
Niarchos, Panagiotis
Nitschelm, Christian
Nordström, Birgitta
Norton, Andrew
Ogłoza, Waldemar
OH, Kyu-Dong
OKAZAKI, Akira
Oláh, Katalin
Oliveira, Alexandre
Olson, Edward
OSAKI, Toru
OSAKI, Yoji
Özeren, Ferhat
Ozkan, Mustafa
Padalia, T.
Pandey, Uma
Panei, Jorge
Parimucha, Stefan
PARK, Hong-Seo
Parthasarathy, Mudumba
Patkos, Laszlo
Pavlenko, Elena
Pavlovski, Kresimir
Pearson, Kevin
Peters, Geraldine
Piccioni, Adalberto
Piirola, Vilppu
Plavec, Mirek
Pojmanski, Grzegorz
Polidan, Ronald
Pollacco, Don
Pooley, David
Popov, Sergey
Postnov, Konstantin
Potter, Stephen
Pretorius, Magaretha
Pribulla, Theodor
Pringle, James
Prokhorov, Mikhail
Prša, Andrej
Pustylnik, Izold
Pustõnski, Vladislav-Veniamin
QIAN, Shengbang
QIAO, Guojun
Rafert, James
Rahunen, Timo
Ramsey, Lawrence
Ransom, Scott
Rao, Pasagada
Rasio, Frederic
Refsdal, Sjur
Reglero Velasco, Victor
Rey, Soo-Chang
Richards, Mercedes
Ricker, Paul
Riles, Keith
Ringwald, Frederick
Ritter, Hans
Robb, Russell
Robertson, John

Robinson, Edward
Rodrigues, Claudia
Rovithis, Peter
Rovithis-Livaniou, Helen
Roxburgh, Ian
Rucinski, Slavek
Ruffert, Maximilian
Russo, Guido
Sadik, Aziz
SAIJO, Keiichi
Samec, Ronald
Sanwal, N.
Sanyal, Ashit
Sarty, Gordon
Savonije, Gerrit
Scaltriti, Franco
Scarfe, Colin
Schartel, Norbert
Schiller, Stephen
Schmid, Hans
Schmidtke, Paul
Schmidtobreick, Linda
Schober, Hans
Schoeffel, Eberhard
Seggewiss, Wilhelm
Selam, Selim
Semeniuk, Irena
Shafter, Allen
Shahbaz, Tariq
Shakura, Nikolaj
Shaviv, Giora
Shaw, Simon
SHEN, Liangzhao
Shimansky, Vladislav
SHU, Frank
Sima, Zdislav
Simmons, John
Sinvhal, Shambhu
Sion, Edward
Sistero, Roberto
Skopal, Augustin
Slovak, Mark
Smak, Józef
Smith, Robert
Sobieski, Stanley
Soderhjelm, Staffan
Solheim, Jan
SONG, Liming
Sonti, Sreedhar
Southworth, John
Sowell, James
Sparks, Warren
Srivastava, J.
Srivastava, Ram
Stachowski, Grzegorz
Stagg, Christopher
Stanishev, Vallery
Starrfield, Sumner
Steiman-Cameron, Thomas
Steiner, Joao
Stencel, Robert
Sterken, Christiaan
Stoyanov, Kiril
Stringfellow, Guy
Sudar, Davor
SUGIMOTO, Daiichiro
Szkody, Paula
Taam, Ronald
TAKAHASHI, Rohta
TAN, Huisong
Taş, Günay
Tauris, Thomas
Teays, Terry
TERADA, Yukikatsu
Terrell, Dirk
Tessema, Solomon
Torres, Guillermo
Tout, Christopher
Tremko, Jozef
Trimble, Virginia
Tsesevich, V.
Turolla, Roberto
Tutukov, Aleksandr
Ulla Miguel, Ana
Unda-Sanzana, Eduardo
Ureche, Vasile
Vaccaro, Todd
van den Berg, Maureen
van den Heuvel, Edward
van Hamme, Walter
van't Veer, Frans
Vaz, Luiz Paulo
Vennes, Stephane
Vetesnik, Miroslav
Vierdayanti, Kiki
Vilhu, Osmi
Volo'shina, Irina
Vrielmann, Sonja
Wachter, Stefanie
Wade, Richard
Walder, Rolf
Walker, William
WANG, Bo
WANG, Xunhao
Ward, Martin
Warner, Brian
Webbink, Ronald
Weiler, Edward
Wesson, Roger
Wheatley, Peter
Wheeler, J.
White II, James
Williamon, Richard
Williams, Glen
Williams, Robert
Wilson, Robert
Wittenmyer, Robert
Wood, Janet
Worters, Hannah
XUE, Li
YAMAOKA, Hitoshi
YAMASAKI, Atsuma
YOON, Tae-Seog
YU, Cong
Yüce, Kutluay
Zakirov, Mamnum
Zamanov, Radoslav
Zavala, Robert
Zeilik, Michael
Zejda, Miloslav
Zhai, DiSheng
ZHANG, Bo
ZHANG, Er-Ho
ZHANG, Jintong
ZHANG, Shu
Zharikov, Sergey
Zhilkin, Andrey
ZHOU, Daoqi
ZHOU, Hongnan
ZHU, Liying
Ziółkowski, Janusz
Zoła, Stanisław
Zucker, Shay
Zuiderwijk, Edwardus
Zwitter, Tomaž

Division D - Commission 44 Space & High Energy Astrophysics

President: Christine Jones
Vice-President: Noah Brosch

Organizing Committee Members:

João Braga
Hernan Quintana
Diana Mary Worrall
Hideyo KUNIEDA
Mattheus de Graauw
Haruyuki OKUDA
Marco Salvati
Thierry Montmerle
Kulinder Pal Singh
Matthew G. Baring
Martin Adrian Barstow
Eugene M. Churazov
Jean Eilek
Jayant Murthy
Isabella Pagano

Members:

Abramowicz, Marek
Acharya, Bannanje
Acton, Loren
Aghaee, Alireza
Agrawal, P.
Aguiar, Odylio
Aharonian, Felix
Ahluwalia, Harjit
Ahmad, Imad
Alexander, Joseph
Allington-Smith, Jeremy
Almleaky, Yasseen
Amati, Lorenzo
Andersen, Bo Nyborg
Antonelli, Lucio Angelo
Antoniou, Vallia
Apparao, K.
ARAFUNE, Jiro
Arefiev, Vadim
Arnaud, Monique
Arnould, Marcel
Arons, Jonathan
ASANO, Katsuaki
Aschenbach, Bernd
Asseo, Estelle
Asvarov, Abdul
Audard, Marc
Audley, Michael
Audouze, Jean
Augereau, Jean-Charles
AWAKI, Hisamitsu
Axelsson, Magnus
Ayres, Thomas
Baan, Willem
Badiali, Massimo
Bailyn, Charles
Baki, Paul
Balikhin, Michael
Baliunas, Sallie
Balman, Solen
Bamba, Aya
Baring, Matthew
Barkhouse, Wayne
Barrantes, Marco
Barret, Didier
Barstow, Martin
Baskill, Darren
Baym, Gordon
Bazzano, Angela
Becker, Robert
Becker, Werner
Beckmann, Volker
Begelman, Mitchell
Behar, Ehud
Beiersdorfer, Peter
Beklen, Elif
Belloni, Tomaso
Bender, Peter
Benedict, George
Benford, Gregory
Bennett, Charles
Bennett, Kevin
Benvenuto, Omar
Bergeron, Jacqueline
Bernardini, Federico
Berta, Stefano
Beskin, Gregory
Beskin, Vasily
Bhattacharjee, Pijushpani
Bhattacharya, Dipankar
Bhattacharyya, Sudip
Bianchi, Luciana
Bianchi, Stefano
Bicknell, Geoffrey
Biermann, Peter
Bigdeli, Mohsen
Bignami, Giovanni
Bingham, Robert
Birkmann, Stephan
Blamont, Jacques-Emile
Blandford, Roger
Bleeker, Johan
Bless, Robert
Blinnikov, Sergey
Bloemen, Hans
Blondin, John
Bludman, Sidney
Bocchino, Fabrizio
Bodaghee, Arash
BoÃ«r, Michel
Boggess, Albert
Boggess, Nancy
Bohlin, Ralph
Boksenberg, Alec
Bonazzola, Silvano
Bonnet, Roger

Bonnet-Bidaud, Jean-Marc
Bonometto, Silvio
Borka Jovanović, Vesna
Borozdin, Konstantin
Bosch-Ramon, Valenti
Bougeret, Jean-Louis
Bowyer, C.
Boyarchuk, Alexander
Bradley, Arthur
Braga, João
Braithwaite, Jonathan
Branchesi, Marica
Brandt, John
Brandt, Soeren
Brandt, William
Brecher, Kenneth
Brenneman, Laura
Breslin, Ann
Breton, Rene
Brinkman, Bert
Bromberg, Omer
Brosch, Noah
Brown, Alexander
Bruhweiler, Frederick
Bruner, Marilyn
Brunetti, Gianfranco
Bumba, Vaclav
Bunner, Alan
Buote, David
Burderi, Luciano
Burenin, Rodion
Burger, Marijke
Burikham, Piyabut
Burke, Bernard
Burrows, Adam
Burrows, David
Bursa, Michal
Butler, Christopher
Butler, H.
Butterworth, Paul
Caccianiga, Alessandro
CAI, Michael
CAI, Mingsheng
Camenzind, Max
Campana, Riccardo
Campbell, Murray
Cannon, Kipp
CAO, Li
Cappi, Massimo
Caputi, Karina
Caraveo, Patrizia
Cardini, Daniela
Carlson, Per
Carpenter, Kenneth
Carroll, P.
Casandjian, Jean-Marc
Cash Jr, Webster
Cassano, Rossella
Cassé, Michel
Castro-Tirado, Alberto
Catura, Richard
Cavaliere, Alfonso
Celotti, Anna Lisa
Cenko, Stephen
Cesarsky, Catherine
Chakrabarti, Sandip
Chakraborty, Deo
CHANG, Heon-Young
CHANG, Hsiang-Kuang
Channok, Chanruangrit
Chapman, Robert
Chapman, Sandra
Charles, Philip
Chartas, George
Chechetkin, Valerij
Chelliah Subramonian, Stalin
Chelouche, Doron
CHEN, Lin-wen
Chenevez, JérÃ´me
CHENG, Kwongsang
CHENGMO, Zhang
Chernyakova, Maria
Cheung, Cynthia
Chian, Abraham
Chiappetti, Lucio
CHIKAWA, Michiyuki
Chitre, Shashikumar
Chochol, Drahomir
CHOE, Gwangson
CHOI, Chul-Sung
CHOU, Yi
Chupp, Edward
Churazov, Eugene
Ciotti, Luca
Clark, David
Clark, George
Clark, Thomas
Clay, Roger
Code, Arthur
Cohen, David
Collin, Suzy
Comastri, Andrea
Condon, James
Contopoulos, Ioannis
Corbel, Stéphane
Corbet, Robin
Corbett, Ian
Corcoran, Michael
Cordova, France
Cornelisse, Remon
Costantini, Elisa
Courtes, Georges
Courvoisier, Thierry
Cowie, Lennox
Cowsik, Ramanath
Crannell, Carol
Crocker, Roland
Cropper, Mark
Croston, Judith
Croton, Darren
Cruise, Adrian
Cuadra, Jorge
Cui, Wei
Culhane, John
Cunniffe, John
Curir, Anna
Cusumano, Giancarlo
da Costa, António Armando
da Costa, Jose
Dadhich, Naresh
DAI, Zigao
Dalla Bontà, Elena
D'Amico, Flavio
D Ammando, Filippo
Darriulat, Pierre
da Silveira, Enio
Dautcourt, G.
Davidson, William
Davis, Michael
Davis, Robert
Dawson, Bruce
De Becker, MichaÃ«l
de Felice, Fernando
de Graauw, Mattheus
de Jager, Cornelis
Della Ceca, Roberto
Del Santo, Melania
Del Zanna, Luca

de Martino, Domitilla
Dempsey, Robert
den Herder, Jan-Willem
DeNisco, Kenneth
Dennerl, Konrad
Dennis, Brian
Dermer, Charles
de Ugarte Postigo, Antonio
Dewitt, Bryce
Diaz Trigo, Maria
Di Cocco, Guido
Digel, Seth
Disney, Michael
Dokuchaev, Vyacheslav
Dolan, Joseph
Domingo, Vicente
Dominis Prester, Dijana
Donea, Alina
Dong, Xiao-Bo
DOTANI, Tadayasu
DOU, Jiangpei
Dovciak, Michal
Downes, Turlough
Drake, Frank
Drury, Luke
Dubus, Guillaume
Duorah, Hira
Dupree, Andrea
Durouchoux, Philippe
Duthie, Joseph
Easther, Richard
Edelson, Rick
Edwards, Paul
Edwards, Philip
Ehle, Matthias
Eichler, David
Eilek, Jean
El Raey, Mohamed
Elvis, Martin
Elyiv, Andrii
Emanuele, Alessandro
Enßlin, Torsten
ESAMDIN, Ali
ESIMBEK, Jarken
Ettori, Stefano
Eungwanichayapant, Anant
Evans, Daniel
Evans, W.
Fabian, Andrew
Fabiani, Sergio
Fabricant, Daniel
Falize, Emeric
Faraggiana, Rosanna
Farrell, Sean
Fatkhullin, Timur
Faure, Cécile
Fazio, Giovanni
Feldman, Paul
Fender, Robert
Fendt, Christian
Fenton, Keith
Ferrari, Attilio
Fichtel, Carl
Field, George
Fisher, Philip
Fishman, Gerald
Fitton, Brian
Florido, Estrella
Foing, Bernard
Fomin, Valery
Fonseca Gonzalez, Maria
Forman, William
Foschini, Luigi
Franceschini, Alberto
Frandsen, Soeren
Frank, Juhan
Fransson, Claes
Fraschetti, Federico
Fredga, Kerstin
Fruscione, Antonella
FU, Cheng Qi
FUJIMOTO, Shin-ichiro
FUJITA, Mitsutaka
Furniss, Ian
Fyfe, Duncan
Gabriel, Alan
Gaensler, Bryan
Gaisser, Thomas
Galeotti, Piero
Galloway, Duncan
Gal-Yam, Avishay
Gammie, Charles
Gangadhara, R.T.
GAO, Yu
Garmire, Gordon
Gaskell, C.
Gastaldello, Fabio
Gathier, Roel
Gehrels, Neil
Gendre, Bruce
Georgantopoulos, Ioannis
Geppert, Ulrich
Gezari, Daniel
Ghia, Piera Luisa
Ghirlanda, Giancarlo
Ghisellini, Gabriele
Giacconi, Riccardo
Gilra, Daya
Gioia, Isabella
Giroletti, Marcello
Gitti, Myriam
Glaser, Harold
Goicoechea, Luis
Goldsmith, Donald
Goldwurm, Andrea
Gomboc, Andreja
Gomez, Haley
Gómez de Castro, Ana
Gondhalekar, Prabhakar
GONG, Biping
Gonzales'a, Walter
González, Gabriela
Gordon, Chris
Gotthelf, Eric
Götz, Diego
Graffagnino, Vito
Grebenev, Sergei
Greenhill, John
Gregorio, Anna
Grenier, Isabelle
Grewing, Michael
Greyber, Howard
Griffiths, Richard
Grindlay, Jonathan
Grosso, Nicolas
GU, Minfeng
Guessoum, Nidhal
Gull, Theodore
Gumjudpai, Burin
Gün, Gulnur
Gunn, James
Gursky, Herbert
Gurvits, Leonid
Gutiérrez, Carlos
Guziy, Sergiy
Haddock, Fred
Hakkila, Jon
Halevin, Alexandros
Hall, Andrew
Hallam, Kenneth

Hameury, Jean-Marie
HAN, Zhengzhong
HANG, Hengrong
Hanna, David
Hannah, Iain
Hannikainen, Diana
Hardcastle, Martin
Harms, Richard
Harris, Daniel
Harvey, Christopher
Harvey, Paul
Harwit, Martin
Hasan, Hashima
HATSUKADE, Isamu
Haubold, Hans
Haugbølle, Troels
Hauser, Michael
Hawkes, Robert
Hawking, Stephen
Hawkins, Isabel
HAYAMA, Kazuhiro
Haymes, Robert
Hearn, Anthony
Heckathorn, Harry
Heger, Alexander
Hein, Righini
Heinz, Sebastian
Heise, John
Helfand, David
Helmken, Henry
Helou, George
Hempel, Marc
Heng, Kevin
Henoux, Jean-Claude
Henriksen, Richard
Henry, Richard
Hensberge, Herman
Heske, Astrid
Heyl, Jeremy
Hicks, Amalia
Hill, Adam
Hjalmarsdotter, Linnea
Ho, Wynn
Hoffman, Jeffrey
Holberg, Jay
Holloway, Nigel
Holt, Stephen
Holz, Daniel
Hora, Joseph
Hörandel, Jörg
Horns, Dieter
Hornschemeier, Ann
Hornstrup, Allan
Houziaux, Leo
Howarth, Ian
Hoyng, Peter
HSU, Rue-Ron
HU, Wenrui
Huang, Jiasheng
HUANG, YongFeng
Huber, Martin
Hulth, Per
Hurley, Kevin
Hutchings, John
HWANG, Chorng-Yuan
Ibrahim, Alaa
ICHIMARU, Setsuo
Ikhsanov, Nazar
Illarionov, Andrei
Imamura, James
Imhoff, Catherine
INOUE, Hajime
INOUE, Makoto
in't Zand, Johannes
IOKA, Kunihito
Ipser, James
ISHIDA, Manabu
Israel, Werner
ITO, Kensai
ITOH, Masayuki
Iyengar, K.
Jackson, John
Jaffe, Walter
Jakobsson, Pall
Jamar, Claude
Janka, Hans
Jaranowski, Piotr
Jenkins, Edward
JI, Li
Johns, Bethany
Jokipii, Jack
Jones, Christine
Jones, Thomas
Jonker, Peter
Jordan, Carole
Jordan, Stuart
Joss, Paul
Juliusson, Einar
Kafatos, Menas
Kahabka, Peter
Kalemci, Emrah
KAMENO, Seiji
Kaminker, Alexander
KANEDA, Hidehiro
Kaper, Lex
Kapoor, Ramesh
Karakas, Amanda
Karami, Kayoomars
Karpov, Sergey
Kaspi, Victoria
Kasturirangan, K.
Katarzyński, Krzysztof
KATO, Yoshiaki
KATO, Tsunehiko
Katsova, Maria
Katz, Jonathan
KAWAI, Nobuyuki
Kellermann, Kenneth
Kellogg, Edwin
Kelly, Brandon
Kembhavi, Ajit
KENJI, Nakamura
Kessler, Martin
Khumlemlert, Thiranee
KII, Tsuneo
Killeen, Neil
KIM, Minsun
KIM, Yonggi
Kimble, Randy
KINUGASA, Kenzo
Kirk, John
KIYOSHI, Hayashida
Klinkhamer, Frans
Klose, Sylvio
Knapp, Johannes
KO, Chung-Ming
Kobayashi, Shiho
Koch-Miramond, Lydie
Kohmura, Takayoshi
KOIDE, Shinji
KOJIMA, Yasufumi
Kokubun, Motohide
Kolb, Edward
KONDO, Yoji
KONDO, Masaaki
Kong, Albert
KOSHIBA, Masatoshi
KOSUGI, George
Koupelis, Theo
Kovář, Jiří

KOYAMA, Katsuji
Kozłowski, Maciej
Kozma, Cecilia
Kraft, Ralph
Kreisel, E.
Kretschmar, Peter
Kreykenbohm, Ingo
Kristiansson, Krister
Krittinatham, Watcharawuth
Kryvdyk, Volodymyr
Kudryavtseva, Nadezhda (Nadia)
Kuiper, Lucien
Kulsrud, Russell
KUMAGAI, Shiomi
Kuncic, Zdenka
Kundt, Wolfgang
KUNIEDA, Hideyo
Kunz, Martin
Kurt, Vladimir
KUSUNOSE, Masaaki
La Franca, Fabio
Lagache, Guilaine
Lal, Dharam
Lamb, Frederick
Lamb, Susan
Lamb Jr, Donald
Lamers, Henny
Lampton, Michael
La Mura, Giovanni
Lapi, Andrea
Lapington, Jon
Lasher, Gordon
Lattimer, James
Lea, Susan
Leckrone, David
Lee, William
LEE, Sang-Sung
LEE, Wo-Lung
Leighly, Karen
Lemaire, Philippe
Lena, Pierre
Levenson, Nancy
Levin, Yuri
Levine, Robyn
Lewin, Walter
LI, Li-Xin
LI, Min
LI, Tipei
LI, Xiangdong
LI, Yuanjie
LI, Zhongyuan
LI, Zongwei
Liang, Edison
LIN, Xuan-bin
Linsky, Jeffrey
Linsley, John
Liu, Bifang
LIU, Guoqing
LIU, Jifeng
LIU, Siming
Loaring, Nicola
Lochner, James
Long, Knox
Longair, Malcolm
Lopes de Oliveira, Raimundo
Loubser, Ilani
Lovelace, Richard
LU, Fangjun
LU, Jufu
LU, Ye
LU, Youjun
Lüst, Reimar
Luigi, P.
Luminet, Jean-Pierre
Luo, Qinghuan
Luque-Escamilla, Pedro
Lutovinov, Alexander
Lynden-Bell, Donald
Lyubarsky, Yury
MA, YuQian
Maccacaro, Tommaso
Maccarone, Thomas
Macchetto, Ferdinando
MACHIDA, Mami
Maciesiak, Krzysztof
Maggio, Antonio
Maguire, Kate
Mainieri, Vincenzo
Majumdar, Subhabrata
Makarov, Valeri
Malaise, Daniel
Malesani, Daniele
Malitson, Harriet
Malkan, Matthew
Manara, Alessandro
Mandolesi, Nazzareno
Mangano, Vanessa
Maran, Stephen
Marar, T.
Maričić, Darije
Marino, Antonietta
Markoff, Sera
Marov, Mikhail
Marranghello, Guilherme
Martin, Inácio
Martínez Bravo, Oscar
Martinis, Mladen
MASAI, Kuniaki
Masnou, Jean-Louis
Mason, Glenn
Mason, Keith
Massaro, Francesco
Mather, John
MATSUMOTO, Hironori
MATSUMOTO, Ryoji
MATSUOKA, Masaru
MATSUSHITA, Kyoko
Matt, Giorgio
Matz, Steven
Mazurek, Thaddeus
McBreen, Brian
McBride, Vanessa
McCluskey Jr, George
McCray, Richard
McWhirter, R.
Medina, Jose
Medina Tanco, Gustavo
Meier, David
Meiksin, A.
Melatos, Andrew
Melia, Fulvio
Melikidze, Giorgi
Melnick, Gary
Melnyk, Olga
Melrose, Donald
Méndez, Mariano
Mereghetti, Sandro
Merlo, David
Mestel, Leon
Meszaros, Peter
Meyer, Friedrich
Meyer, Jean-Paul
Micela, Giuseppina
Michel, F.
Miller, Eric
Miller, Guy
Miller, John

Miller, Michael
Mineo, Teresa
Miroshnichenko, Alla
Miyaji, Takamitsu
MIYAJI, Shigeki
MIYAMOTO, Sigenori
Miyata, Emi
MIZUMOTO, Yoshihiko
MIZUNO, Yosuke
MIZUTANI, Kohei
Moderski, Rafał
Modisette, Jerry
Molla, Mercedes
Monet, David
Monfils, Andre
Montmerle, Thierry
Moodley, Kavilan
MOON, Shin-Haeng
Moos, Henry
Morgan, Thomas
MORI, Koji
MORI, Masaki
MOROKUMA, Tomoki
Morsony, Brian
Morton, Donald
Mota, David
Motch, Christian
MOTIZUKI, Yuko
Mourão, Ana Maria
Mucciarelli, Paola
Mukhopadhyay, Banibrata
Mulchaey, John
MURAKAMI, Hiroshi
MURAKAMI, Toshio
MURAYAMA, Hitoshi
Murdock, Thomas
Murtagh, Fionn
Murthy, Jayant
NAGATAKI, Shigehiro
Naidenov, Victor
NAITO, Tsuguya
Nakar, Ehud
NAKAYAMA, Kunji
Neff, Susan
Ness, Norman
Neuhaeuser, Ralph
Neupert, Werner
Neustroev, Vitaly
Nichols, Joy
Nicollier, Claude
Nielsen, Krister
Nikołajuk, Marek
NISHIMURA, Osamu
NITTA, Shin-ya
Nityananda, Rajaram
NOMOTO, Ken'ichi
Norci, Laura
Nordh, Lennart
Norman, Colin
Novick, Robert
Noyes, Robert
Nulsen, Paul
Nymark, Tanja
O'Brien, Paul
O'Connell, Robert
ODA, Naoki
Oertel, Goetz
OGAWARA, Yoshiaki
Okeke, Pius
OKUDA, Haruyuki
OKUDA, Toru
Olthof, Henk
O'Mongain, Eon
Onken, Christopher
OOHARA, Ken-ichi
Oozeer, Nadeem
Orellana, Mariana
Orford, Keith
Orio, Marina
Orlandini, Mauro
Orlando, Salvatore
Osborne, Julian
Oskinova, Lidia
Osten, Rachel
Ostriker, Jeremiah
Ostrowski, Michał
O'Sullivan, Denis
Ott, Juergen
Owen, Tobias
Owers, Matthew
OZAKI, Masanobu
Özel, Mehmet
Pacholczyk, Andrzej
Paciesas, William
Pagano, Isabella
Page, Clive
Page, Mathew
PAK, Soojong
Paltani, Stéphane
Palumbo, Giorgio
Pandey, Uma
Panessa, Francesca
Papadakis, Iossif
Papitto, Alessandro
Paragi, Zsolt
Pareschi, Giovanni
PARK, Myeong Gu
Parker, Eugene
Parkinson, William
Patten, Brian
Paul, Biswajit
Pauliny, Toth
Pavlov, George
Peacock, Anthony
Pearce, Mark
Pearson, Kevin
Pellegrini, Silvia
Pellizza, Leonardo
PENG, Qingyu
PENG, Qiuhe
Perez, Mario
Pérez-González, Pablo
Perola, Giuseppe
Perry, Peter
Peters, Geraldine
Peterson, Bradley
Peterson, Bruce
Peterson, Laurence
Pethick, Christopher
Petkaki, Panagiota
Petro, Larry
Petrosian, Vahe
Peytreman, Eric
Phillips, Kenneth
Pian, Elena
Pinkau, K.
Pinto, Philip
Pipher, Judith
Piran, Tsvi
Piro, Luigi
Pitkin, Matthew
Polidan, Ronald
Polletta, Maria del Carmen
Pooley, David
Popov, Sergey
Porquet, Delphine
Pottschmidt, Katja
Pounds, Kenneth
Poutanen, Juri
Pozanenko, Alexei

Prasanna, A.
Preuss, Eugen
Produit, Nicolas
Prokof'ev, Vladimir
Protheroe, Raymond
Prouza, Michael
Prusti, Timo
Pshirkov, Maxim
Pustil'nik, Lev
QIU, Yulei
QU, Jinlu
QU, Qinyue
Quintana, Hernan
Quirós, Israel
Raiteri, Claudia
Ramadurai, Souriraja
Ramírez, Jose
Ranalli, Piero
Rao, Arikkala
Rao, Ramachandra
Rasmussen, Ib
Rasmussen, Jesper
Raubenheimer, Barend
Ray, Paul
Raychaudhury, Somak
Razdan, Hiralal
Rea, Nanda
Reale, Fabio
Rees, Martin
Reeves, Edmond
Reeves, Hubert
Reichert, Gail
Reig, Pablo
Reimer, Olaf
Reiprich, Thomas
Reisenegger, Andreas
Reitze, David
Rengarajan, Thinniam
Revnivtsev, Mikhail
Rhoads, James
Ricker, Paul
Riegler, Guenter
Riles, Keith
Risaliti, Guido
Robba, Natale
Roberts, Timothy
Roman, Nancy
Romano, Patrizia
Roming, Peter
Rosendhal, Jeffrey
Rosner, Robert
Rossi, Elena
Rovero, Adrián
Rowell, Gavin
Różańska, Agata
Rubino-Martin, Jose Alberto
Rubio-Herrera, Eduardo
Ruder, Hanns
Ruffini, Remo
Ruffolo, David
Russell, Alexander
Ruszkowski, Mateusz
Rutledge, Robert
Sabau-Graziati, Lola
Safi-Harb, Samar
Sagdeev, Roald
Sahlén, Martin
Saiz, Alejandro
Sakano, Masaaki
Sakelliou, Irini
Salpeter, Edwin
Salvati, Marco
Samimi, Jalal
Sanchez, Norma
Sanders, Gary
Sanders III, Wilton
Santos, Nilton
Santos-Lleó, Maria
Sari, Re'em
Sartori, Leo
Saslaw, William
SATO, Katsuhiko
Savage, Blair
Savedoff, Malcolm
Sazonov, Sergey
Sbarufatti, Boris
Scargle, Jeffrey
Schaefer, Gerhard
Schartel, Norbert
Schatten, Kenneth
Schilizzi, Richard
Schmitt, Juergen
Schnopper, Herbert
Schoeneich, W.
Schreier, Ethan
Schulz, Norbert
Schure, Klara
Schwartz, Daniel
Schwartz, Steven
Schwehm, Gerhard
Sciortino, Salvatore
Scott, John
Seielstad, George
Selvelli, Pierluigi
Semerák, Oldrich
SEON, Kwang il
Sequeiros, Juan
Setti, Giancarlo
Severgnini, Paola
Seward, Frederick
Shaham, Jacob
Shahbaz, Tariq
Shakhov, Boris
Shakura, Nikolaj
Shalchi, Andreas
Sharma, Prateek
Shaver, Peter
Shaviv, Giora
Shaw, Simon
SHEN, Zhiqiang
SHIBAI, Hiroshi
Shibanov, Yuri
SHIBAZAKI, Noriaki
Shields, Gregory
SHIGEYAMA, Toshikazu
SHIMURA, Toshiya
SHIN, Watanabe
SHIRASAKI, Yuji
Shivanandan, Kandiah
Shoemaker, David
Shukre, C.
Shustov, Boris
Signore, Monique
Sikora, Marek
Silvestro, Giovanni
Simic, Sasa
Simon, Paul
Simon, Vojtech
Sims, Mark
Simunac, Kristin
Singh, Kulinder Pal
Skilling, John
Skinner, Stephen
SkjÃ¦raasen, Olaf
Slan, Petr
Smale, Alan
Smida, Radomír
Smith, Bradford
Smith, Francis

Smith, Linda
Smith, Nigel
Smith, Peter
Snow, Theodore
Sofia, Sabatino
Sokolov, Vladimir
Somasundaram, Seetha
SONG, Qian
Sonneborn, George
Sood, Ravi
Soria, Roberto
Spada, Gianfranco
Spallicci di Filottrano, Alessandro
Speer, R.
Sreekumar, Parameswaran
Srinivasan, Ganesan
Srivastava, Dhruwa
Staubert, Rüdiger
Stecher, Theodore
Stecker, Floyd
Steigman, Gary
Steinberg, Jean-Louis
Steiner, Joao
Stencel, Robert
Stephens, S.
Stergioulas, Nikolaos
Stern, Robert
Stevens, Ian
Stier, Mark
Still, Martin
Stockman Jr, Hervey
Stoehr, Felix
Stone, R.
Straumann, Norbert
Stringfellow, Guy
Strohmayer, Tod
Strong, Ian
Struminsky, Alexei
Stuchlík, Zdenek
Sturrock, Peter
SU, Cheng-yue
Šubr, Ladislav
Suleimanov, Valery
SUMIYOSHI, Kosuke
SUN, Wei-Hsin
Sunyaev, Rashid
SUZUKI, Hideyuki
Swank, Jean
Tagliaferri, Gianpiero
TAKAHARA, Fumio
TAKAHASHI, Masaaki
TAKAHASHI, Rohta
TAKAHASHI, Tadayuki
TAKAKURA, Tatsuo
TAKEI, Yoh
Tammi, Joni
Tanaka, Yasuo
TASHIRO, Makoto
TATEHIRO, Mihara
Tavecchio, Fabrizio
Tempel, Elmo
Templeton, Matthew
Teng, Stacy
TERADA, Yukikatsu
TERASHIMA, Yuichi
Tessema, Solomon
Thöne, Christina
Thomas, Roger
Thorne, Kip
Thronson Jr, Harley
TIAN, Wenwu
TOMIMATSU, Akira
TOMINAGA, Nozomu
Török, Gabriel
Torres, Carlos Alberto
Torres, Diego
Tousey, Richard
Tovmassian, Hrant
Traub, Wesley
Tresse, Laurence
Trimble, Virginia
Truemper, Joachim
Truran, James
Trussoni, Edoardo
TSAI, An-Li
Tsujimoto, Masahiro
TSUNEMI, Hiroshi
TSURU, Takeshi
Tsuruta, Sachiko
Tsygan, Anatolij
Tsygankov, Sergey
Türler, Marc
Tylka, Allan
UEDA, Yoshihiro
Ulyanov, Oleg
Underwood, James
Upson, Walter
URATA, Yuji
Uslenghi, Michela
Usov, Vladimir
Uttley, Philip
Vahia, Mayank
Väliviita, Jussi-Pekka
Valnicek, Boris
Valtonen, Mauri
van Beek, Frank
van den Berg, Maureen
van den Heuvel, Edward
van der Hucht, Karel
van der Walt, Diederick
van Duinen, R.
van Putten, Maurice
van Riper, Kenneth
van Speybroeck, Leon
Vaughan, Simon
Venter, Christo
Venters, Tonia
Vercellone, Stefano
Vestergaard, Marianne
Vial, Jean-Claude
Vidal, Nissim
Vidal-Madjar, Alfred
Vierdayanti, Kiki
Vignali, Cristian
Vikhlinin, Alexey
Vilhu, Osmi
Villata, Massimo
Vink, Jacco
Viollier, Raoul
Viotti, Roberto
Völk, Heinrich
Volonteri, Marta
Vrtilek, Saeqa
Wagner, Alexander
Wagner, Robert
Walker, Helen
Walker, Simon
Wanas, Mamdouh
Wandel, Amri
Wang, Yi-ming
WANG, Chen
WANG, Ding-Xiong
WANG, Feilu
WANG, Hong-Guang
WANG, jiancheng
WANG, Sen
WANG, Shiang-Yu
WANG, Shouguan
WANG, Shui

WANG, Shujuan
WANG, Zhenru
Warner, John
Watanabe, Ken
WATARAI, Kenya
Watts, Anna
Waxman, Eli
Weaver, Kimberly
Weaver, Thomas
Webster, Adrian
Wehrle, Ann
WEI, Daming
Weiler, Edward
Weiler, Kurt
Weinberg, Jerry
Weisheit, Jon
Weisskopf, Martin
Wells, Donald
WEN, Linqing
Wesselius, Paul
Wheatley, Peter
Wheeler, J.
Wheeler, John
Whitcomb, Stanley
White, Nicholas
Wijers, Ralph
Wijnands, Rudy
Will, Clifford
Willis, Allan
Willner, Steven
Wilms, Jörn
Wilson, Andrew
Wilson, Gillian
Wilson, James
Wilson, R.
Winkler, Christoph
Winter, Lisa
Wise, Michael
Wolfendale FRS, Sir Arnold
Wolstencroft, Ramon
Wolter, Anna
Woltjer, Lodewijk
WOO, Jong-Hak
Worrall, Diana
Wray, James
Wu, Chi
WU, Jiun-Huei
WU, Shaoping
WU, Xue-Feng
WU, Xuejun
Wunner, Guenter
XIANGYU, Wang
XU, Dawei
XU, Renxin
Yadav, Jagdish
Yamada, Shoichi
YAMAMOTO, Yoshiaki
Yamasaki, Tatsuya
YAMASAKI, Noriko
YAMASHITA, Kojun
YAMAUCHI, Makoto
YAMAUCHI, Shigeo
YANG, Lantian
YANG, Pibo
YI, Insu
Yock, Philip
YOICHIRO, Suzuki
YOKOYAMA, Takaaki
YONETOKU, Daisuke
YOSHIDA, Atsumasa
YOU, Junhan
YU, Cong
YU, Qingjuan
YU, Wenfei
Yuan, Weimin
YUAN, Feng
YUAN, Ye-fei
Zacchei, Andrea
Zamorani, Giovanni
Zane, Silvia
Zannoni, Mario
Zdziarski, Andrzej
Zezas, Andreas
Zhang, Jie
Zhang, Yuying
ZHANG, Chengmin
ZHANG, Heqi
ZHANG, Jialu
ZHANG, JiangShui
ZHANG, Jingyi
ZHANG, Li
ZHANG, Shu
ZHANG, Shuang Nan
ZHANG, William
ZHANG, Yanxia
ZHANG, You-Hong
ZHANG, Zhen-Jiu
ZHANG, Zhibin
ZHENG, Wei
ZHENG, Xiaoping
ZHOU, Jianfeng
ZHOU, Xia
Zombeck, Martin
ZOU, Huicheng
Zwintz, Konstanze

Division G - Commission 45 Stellar Classification

President: Birgitta Nordström
Vice-President: Caroline Soubiran

Organizing Committee Members:

Sandy K. Leggett
Harinder P. Singh
Werner W. Weiss
Carlos Allende Prieto
Richard O. Gray
Adam J. Burgasser
Ranjan Gupta
Margaret Murray Hanson

Members:

AK, Serap
Allende Prieto, Carlos
Ardeberg, Arne
Arellano Ferro, Armando
Babu, G.S.D.
Baglin, Annie
Barbier-Brossat, Madeleine
Bartaya, R.
Bartkevicius, Antanas
Bear, Ealeal
Buchhave, Lars
Burgasser, Adam
Buser, Roland
Cester, Bruno
Cherepashchuk, Anatolij
Christy, James
Claria, Juan
Coluzzi, Regina
Corral, Luis
Cowley, Anne
Crawford, David
Creech-Eakman, Michelle
Dal Ri Barbosa, Cassio
Divan, Lucienne
Drilling, John
Duflot, Marcelle
Eglitis, Ilgmars
Egret, Daniel
Eyer, Laurent
Faraggiana, Rosanna
Feast, Michael
Feltzing, Sofia
Fitzpatrick, Edward
FUKUDA, Ichiro
Gamen, Roberto
García García, Miriam
Garmany, Katy
Garrison, Robert
Gerbaldi, Michele
Geyer, Edward
Giorgi, Edgard
Giridhar, Sunetra
Gizis, John
Glagolevskij, Yurij
Golay, Marcel
Goswami, Aruna
Gray, Richard
Grenon, Michel
Grosso, Monica
Guetter, Harry
Gupta, Ranjan
Hallam, Kenneth
Hanson, Margaret
Hauck, Bernard
Hayes, Donald
Heck, Andre
Holmberg, Johan
Houk, Nancy
HUANG, Lin
Humphreys, Roberta
Irwin, Michael
Jaschek, Carlos
Jaschek, Mercedes
Johnson, H.
KATO, Ken-ichi
Kurtanidze, Omar
Kurtz, Donald
Kurtz, Michael
Labhardt, Lukas
Lasala Jr., Gerald
Lattanzio, John
Laugalys, Vygandas
LEE, Sang-Gak
Leggett, Sandy
Lepine, Sebastien
Levato, Orlando
LI, Jinzeng
Lobel, Alex
Lodén, Kerstin
Low, Frank
Lu, Phillip
LUO, Ali
Luri, Xavier
Lutz, Julie
MacConnell, Darrell
MAEHARA, Hideo
Maíz Apellániz, Jesús
Malagnini, Maria
Malaroda, Stella
McClure, Robert
McConnell, David
Mendoza, V.
Montes, David
Morgan, W.
Morossi, Carlo
Morrell, Nidia
Nicolet, Bernard
Nordström, Birgitta
North, Pierre
Notni, Peter
Oja, Tarmo
Olsen, Erik
Osborn, Wayne
Oswalt, Terry
Pakhomov, Yury
Paletou, Frédéric
Parsons, Sidney
Pasinetti, Laura
Philip, A.G.
Pizzichini, Graziella
Preston, George
Prša, Andrej
Pulone, Luigi
Rautela, B.
Redman, R.
Reid, Warren
Rizzi, Luca
Roman, Nancy
Rountree, Janet
Sanwal, N.
Schild, Rudolph
Schmidt-Kaler, Theodor
Sharpless, Stewart
Shore, Steven
Shvelidze, Teimuraz
Singh, Harinder
Sinnerstad, Ulf
Sion, Edward
Smith, J.
Sonti, Sreedhar
Soubiran, Caroline
Steinlin, Uli
Straižys, Vytautas
Strobel, Andrzej
TAKAHASHI, Hidenori
Úbeda, Leonardo
Ueta, Toshiya
Upgren, Arthur
von Hippel, Theodore
Walborn, Nolan
Walker, Gordon

Warren Jr, Wayne
Weaver, William
Weiss, Werner
Westerlund, Bengt
Williams, John
Wing, Robert
Wright, Nicholas
WU, Hsin-Heng
Wyckoff, Susan
YAMASHITA, Yasumasa
YUMIKO, Oasa
Zdanavičius, Kazimeras
Zdanavičius, Justas
ZHANG, Yanxia

Division C - Commission 46 Astronomy Education & Development

President: Jean-Pierre de Greve
Vice-President: Beatriz Elena García

Organizing Committee Members:

Magda G. Stavinschi (Romania)
Roger Ferlet (France)
John B. Hearnshaw (New Zealand)
Silvia Torres-Peimbert (Mexico)
Rosa M. Ros (Spain)
Jay M. Pasachoff (United States)
Barrie W. Jones (United Kingdom)
George Kildare Miley (Netherlands)
Laurence A. Marschall (United States)
Edward F. Guinan (United States)
Rajesh Kochhar (India)

Members:

Acker, Agnes
Aghaee, Alireza
Aguilar, Maria
Airapetian, Vladimir
Albanese, Lara
Alexandrov, Yuri
Al-Naimiy, Hamid
Alonso-Herrero, Almudena
Alsabti, Abdul Athem
Alvarez, Rodrigo
Alvarez Pomares, Oscar
Alves, Virgínia
Anandaram, Mandayam
Andrews, Frank
Ansari, S.M.
Aquilano, Roberto
Arbutina, Bojan
Arcidiacono, Carmelo
Arellano Ferro, Armando
Arion, Douglas
Asanok, Kitiyanee
Aslan, Zeki
Atanackovic, Olga
Aubier, Monique
Babu, G.S.D.
Badescu, Octavian
BAEK, Chang Hyun
Bailey, Katherine
Barclay, Charles
BARET, Bruny
Barlow, Nadine
Barrantes, Marco
Barthel, Peter
Baskill, Darren
Batten, Alan
Benavidez, Alberto
Ben Lakhdar, Zohra
Benson, Priscilla
Berger, Jean-Philippe
Bernabeu, Guillermo
Bhavsar, Suketu
Birlan, Mirel
Bittar, Jamal
Black, Adam
Bobrowsky, Matthew
Bojurova, Eva
Booth, Roy
Borchkhadze, Tengiz
Botez, Elvira
Bottinelli, Lucette
Braes, Lucien
Brammer, Gabriel
Bretones, Paulo
Brieva, Eduardo
Brosch, Noah
Budding, Edwin
Cabanac, Remi
CAI, Michael
Calvet, Nuria
Cannon, Wayne
Capaccioli, Massimo
Caretta, Cesar
Carter, Brian
Cassan, Arnaud
Catala, Poch
Celebre, Cynthia

Chakrabarti, Supriya
Chamcham, Khalil
CHEN, Alfred
CHEN, An-Le
CHEN, Dongni
CHEN, Lin-wen
CHEN, Xinyang
Chernyakova, Maria
Chitre, Dattakumar
Christensen, Lars
Christlieb, Norbert
Chung, Eduardo
Ciroi, Stefano
Clarke, David
Coffey, Deirdre
Cohen, David
Colafrancesco, Sergio
Conti, Alberto
Cora, Alberto
Corbally, Christopher
Cottrell, Peter
Couper, Heather
Courtois, Helene
Couto da Silva, Telma
Covone, Giovanni
Cracco, Valentina
Craig, Nahide
Crawford, David
Cui, Wei
CUI, Zhenhua
Cunningham, Maria
Dall'Ora, Massimo
Daniel, Jean-Yves
Danner, Rolf
DARHMAOUI, Hassane
Darriulat, Pierre
de Greve, Jean-Pierre
de Grijs, Richard
Delsanti, Audrey
Del Santo, Melania
Demircan, Osman
DeNisco, Kenneth
de Swardt, Bonita
Devaney, Martin
Diego, Francisco
Dole, Herve
Donahue, Megan
Ducati, Jorge
Dukes Jr., Robert
Dupuy, David
Duval, Marie-France
Dworetsky, Michael
Eastwood, Kathleen
El Eid, Mounib
Esguerra, Jose Perico
Eze, Romanus
Fairall, Anthony
Falceta-Goncalves, Diego
Feitzinger, Johannes
Ferlet, Roger
Fernández, Julio
Fernandez-Figueroa, M.
Ferraz-Mello, Sylvio
Ferrusca, Daniel
Fienberg, Richard
Fierro, Julieta
Figueiredo, Newton
Fleck, Robert
Florsch, Alphonse
Floyd, David
Forbes, Douglas
Forero Villao, Vicente
Frew, David
FU, Hsieh-Hai
Gabriel, Carlos
Gallino, Roberto
Gangui, Alejandro
Ganguly, Rajib
Gasiprong, Nipon
Gavrilov, Mikhail
Gay, Pamela
Gerbaldi, Michele
Germany, Lisa
Ghobros, Roshdy
Gill, Peter
Gimenez, Álvaro
Gingerich, Owen
Girard, Julien
Gomez, Edward
Gouguenheim, Lucienne
Govender, Kevindran
Gray, Richard
Gregorio, Anna
Gregorio-Hetem, Jane
Grundstrom, Erika
Guglielmino, Salvatore
Guinan, Edward
Gurm, Hardev
Hafizi, Mimoza
Haque, Shirin
Haubold, Hans
Havlen, Robert
Hawkins, Isabel
Haywood, J.
Hearnshaw, John
Hemenway, Mary
Heudier, Jean-Louis
Heydari-Malayeri, Mohammad
Hicks, Amalia
Hidayat, Bambang
Hobbs, George
Hockey, Thomas
Hönig, Sebastian
Hollow, Robert
Horn, Martin
Hotan, Aidan
Houziaux, Leo
HSU, Rue-Ron
Huan, Nguyen
Huertas-Company, Marc
Huettemeister, Susanne
Hughes, Stephen
Ibrahim, Alaa
Ilyas, Mohammad
Impey, Christopher
Inglis, Michael
Isaak, Kate
ISHIZAKA, Chiharu
Iwaniszewska, Cecylia
Izzard, Robert
Jafelice, Luiz
Jarrett, Alan
Johnston-Hollitt, Melanie
Jubier, Xavier
Kablak, Nataliya
Kalemci, Emrah
Kammeyer, Peter
Karetnikov, Valentin
Karttunen, Hannu
Kay, Laura
Keeney, Brian
Keller, Hans-Ulrich
Khan, J
Khodachenko, Maxim
Kiasatpour, Ahmad
Kikwaya Eluo, Jean-Baptiste
Kim, Yoo Jea
Kitchin, Christopher

Klinglesmith III, Daniel
Knight, Matthew
Koechlin, Laurent
Kolenberg, Katrien
Kolka, Indrek
Kolomiyets, Svitlana
Komonjinda, Siramas
KONG, Xu
Kononovich, Edvard
Kotulla, Ralf
Kouwenhoven, M.B.N. (Thijs)
KOZAI, Yoshihide
Kramer, Busaba
Kreiner, Jerzy
Krishna, Gopal
Krupp, Edwin
KUAN, Yi-Jehng
Kuiper, Rolf
Kunth, Daniel
Lago, Maria
Lai, Sebastiana
Lanciano, Nicoletta
Lanfranchi, Gustavo
LAS VERGNAS, Olivier
LEE, Kang Hwan
LEE, Yong Bok
Letarte, Bruno
Leung, Chun
Leung, Kam
Levato, Orlando
LI, Min
LI, Zongwei
LIN, Chuang-Jia
LIN, Weipeng
Linden-Vørnle, Michael
Little-Marenin, Irene
LIU, Xiaoqun
Lomb, Nicholas
Lopes de Oliveira, Raimundo
Loubser, Ilani
Lowenthal, James
Luck, John
MA, Er
MA, Xingyuan
Maciel, Walter
Maddison, Ronald
Madjarska, Maria
Madsen, Claus
Mahoney, Terence
Malasan, Hakim
Mallamaci, Claudio
Mamadazimov, Mamadmuso
Mancìni, Dario
Marco, Olivier
Marschall, Laurence
Marsh, Julian
Marshalov, Dmitriy
Martinet, Louis
Martinez, Peter
Martínez Bravo, Oscar
Massey, Robert
Maza, José
Mazumdar, Anwesh
McKinnon, David
McNally, Derek
Meidav, Meir
Melbourne, Jason
Merlo, David
Metaxa, Margarita
Meyer, Michael
MIAO, Yongkuan
Miley, George
MIZUNO, Takao
Montgomery, Michele
Moreels, Guy
Morrell, Nidia
Mothé-Diniz, Thais
Murphy, John
Najid, Nour-Eddine
Nammahachak, Suwit
Nandi, Dibyendu
Narlikar, Jayant
Navone, Hugo
Nayar, S.R.Prabhakaran
Nguyễn, Lan
Nguyễn, Khanh
Nguyễn, Phuong
Nguyen-Quang, Rieu
NHA, Il Seong
Nicolson, Iain
Ninković, Slobodan
Noels, Arlette
Norton, Andrew
Nymark, Tanja
Oberst, Thomas
Ödman, Carolina
O'Donoghue, Aileen
Oja, Heikki
Okeke, Pius
Olive, Don
Olsen, Hans
Oluseyi, Hakeem
Onuora, Lesley
Oozeer, Nadeem
Orchiston, Wayne
Ortiz Gil, Amelia
Osborn, Wayne
Osório, José
Oswalt, Terry
Othman, Mazlan
OWAKI, Naoaki
Özeren, Ferhat
Page, Thornton
Pandey, Uma
Pantoja, Carmen
Parenti, Susanna
Parisot, Jean-Paul
Pasachoff, Jay
Pat-El, Igal
Penston, Margaret
Percy, John
Pérez Torres, Miguel
Perozzi, Ettore
Perrin, Marshall
PHAM, Diep
PHAN, Dong
Picazzio, Enos
Pokorny, Zdenek
Pompea, Stephen
Popov, Sergey
Porras Juárez, Bertha
Pović, Mirjana
Price, Charles
Proverbio, Edoardo
Pustil'nik, Lev
Quamar, Jawaid
Querci, Francois
Quirós, Israel
Raboud, Didier
Radeva, Veselka
Ramadurai, Souriraja
Rassat, Anais
Ravindranath, Swara
Reboul, Henri
Reid, Michael
Rekola, Rami
Reyes, Reinabelle

Rigutti, Mario
Rijsdijk, Case
Ringuelet, Adela
Roberts, Morton
Roca Cortés, Teodoro
Rojas, Gustavo
Ros, Rosa
Rosa González, Daniel
Rosenzweig-Levy, Patrica
Różańska, Agata
Rubio-Herrera, Eduardo
Russo, Pedro
Sabra, Bassem
Sadat, Rachida
Saenz, Eduardo
Saffari, Reza
Safko, John
Samodurov, Vladimir
Sampson, Russell
Sánchez-Blázquez, Patricia
Sandqvist, Aage
Sandrelli, Stefano
Saraiva, Maria de Fatima
Sattarov, Isroil
Saucedo Morales, Julio
Sawicki, Marcin
Saxena, P.
Sbirkova-Natcheva, Temenuzhka
Schleicher, David
Schlosser, Wolfhard
Schmitter, Edward
Schroeder, Daniel
Scorza, Cecilia
Seaton, Daniel
Seeds, Michael
Sese, Rogel Mari
Shelton, Ian
SHEN, Chun-Shan
Shipman, Harry
Sigismondi, Costantino
SIHER, El Arbi
Simons, Douglas
Slater, Timothy
Smail, Ian
Smith, Francis
Sobreira, Paulo
Solheim, Jan
Soriano, Bernardo
Stachowski, Grzegorz
Stavinschi, Magda
Štefl, Vladimir
Stenholm, Björn
Stoev, Alexey
Stonkutė, Rima
Straižys, Vytautas
Strubbe, Linda
Sukartadiredja, Darsa
Svestka, Jiri
Swarup, Govind
Szostak, Roland
Taborda, Jose
Tessema, Solomon
Tignalli, Horacio
Tolbert, Charles
Torres, Jesus Rodrigo
Torres-Peimbert, Silvia
Touma, Jihad
trân, Ha
Trinidad, Miguel
TSUBOTA, Yukimasa
Tugay, Anatoliy
Úbeda, Leonardo
UEDA, Haruhiko
Ueta, Toshiya
Ugolnikov, Oleg
Ulla Miguel, Ana
Unda-Sanzana, Eduardo
Urama, Johnson
Urban, Sean
Valdes Parra, Jose
van den Heuvel, Edward
van Groningen, Ernst
van Santvoort, Jacques
Vauclair, Sylvie
Verma, Aprajita
Vierdayanti, Kiki
Vilinga, Jaime
Vilks, Ilgonis
Villar Martin, Montserrat
Viñuales Gavín, Ederlinda
Voelzke, Marcos
Vujnovic, Vladis
Wadadekar, Yogesh
Walker, Constance
Walsh, Wilfred
WANG, Shouguan
WANG, Shunde
Ward, Richard
West, Michael
West, Richard
Whelan, Emma
White II, James
Whitelock, Patricia
Williamon, Richard
Willmore, A.
Winter, Lisa
Wittenmyer, Robert
Wolfschmidt, Gudrun
WOO, Jong
XIE, Xianchun
XIONG, Jianning
Yair, Yoav
YE, Shuhua
YIM, Hong Suh
YUMIKO, Oasa
Zadnik, Marjan
Zakirov, Mamnum
Zealey, William
Zeilik, Michael
ZHANG, Yang
ZHANG, Yong
ZHANG, You-Hong
ZHAO, Jun Liang
ZHENG, Xiaonian

Division J - Commission 47 Cosmology

President: Brian P. Schmidt

Organizing Committee Members:

Andrew J. Bunker
Ofer Lahav
Douglas Scott
Yipeng JING
Anton M. Koekemoer
Olivier Le Fèvre
Benedetta Ciardi
David C. Koo

Members:

Abbas, Ummi
Abu Kassim, Hasan
Adami, Christophe
Adams, Jenni
Aghaee, Alireza
Aharonian, Felix
Ajhar, Edward
Akashi, Muhammad
Alard, Christophe
Alcaniz, Jailson
Alimi, Jean-Michel
Allan, Peter
Allington-Smith, Jeremy
Almaini, Omar
Amendola, Luca
Ammons, Stephen
Andersen, Michael
Andreani, Paola
Aretxaga, Itziar
Argüeso, Francisco
Atrio Barandela, Fernando
Audouze, Jean
Auluck, Faqir
Aussel, Herve
Avelino, Pedro
AZUMA, Takahiro
Babul, Arif
Baddiley, Christopher
Bagla, Jasjeet
Bahcall, Neta
Bajtlik, Stanisław
Baker, Andrew
Baki, Paul
Balbi, Amedeo
Balland, Christophe
Bamford, Steven
Banday, Anthony
Banerji, Sriranjan
Banhatti, Dilip
Bannikova, Elena
Barberis, Bruno
Barbuy, Beatriz
Bardeen, James
Bardelli, Sandro
Barger, Amy
Barkana, Rennan
Barkhouse, Wayne
Barr, Jordi
Barrow, John
Bartelmann, Matthias
Barthel, Peter
Barton, Elizabeth
Baryshev, Andrey
Basa, Stephane
Bassett, Bruce
Basu, Kaustuv
Battye, Richard
Bechtold, Jill
Beckman, John
Beesham, Aroonkumar
Behar, Ehud
Belinski, Vladimir
Bennett, Charles
Bennett, David
Bergeron, Jacqueline
Bergvall, Nils
Berman, Marcelo
Berta, Stefano
Bertola, Francesco
Bertschinger, Edmund
Betancor Rijo, Juan
Bharadwaj, Somnath
Bhavsar, Suketu
Bianchi, Simone
Bicknell, Geoffrey
Bignami, Giovanni
Binetruy, Pierre
Birkinshaw, Mark
Biviano, Andrea
BjÃ¦lde, Ole
Blakeslee, John
Blanchard, Alain
Bleyer, Ulrich
Bludman, Sidney
Blundell, Katherine
Böhringer, Hans
Boksenberg, Alec
Bolzonella, Micol
Bond, John
Bongiovanni, Angel
Bonnor, W.
Boquien, Médéric
Borgani, Stefano
Boschin, Walter
Bouchet, François
Bouwens, Rychard
Bowen, David
Boyle, Brian
Branchesi, Marica
Brecher, Kenneth
Bridle, Sarah
Brinchmann, Jarle
Brough, Sarah
Brown, Michael
Brunner, Robert
Bryant, Julia
Bunker, Andrew
Buote, David
Burikham, Piyabut
Burns, Jack
Cabanac, Remi

CAI, Mingsheng
Calvani, Massimo
Campusano, Luis
Canavezes, Alexandre
CAO, Li
Cappi, Alberto
Caputi, Karina
Carr, Bernard
Carretti, Ettore
Cassano, Rossella
Castagnino, Mario
Cattaneo, Andrea
Cavaliere, Alfonso
Cesarsky, Diego
CHAE, Kyu Hyun
CHANG, Heon-Young
CHANG, Kyongae
Charlot, Stephane
Chelliah Subramonian, Stalin
CHEN, DaMing
CHEN, Dongni
CHEN, Fuhua
CHEN, Hsiao-Wen
CHEN, Jiansheng
CHEN, Lin-wen
CHEN, Pisin
CHEN, Xuelei
CHENG, Fuzhen
Chiang, Hsin
CHIBA, Takeshi
Chincarini, Guido
Chitre, Dattakumar
Chodorowski, Michał
Choudhury, Tirthankar
Christensen, Lise
CHU, Yaoquan
Ciardi, Benedetta
Ciliegi, Paolo
Claeskens, Jean-François
Claria, Juan
Clarke, Tracy
Clarkson, Chris
Clocchiatti, Alejandro
Clowe, Douglas
Clowes, Roger
Cocke, William
Cohen, Ross
Colafrancesco, Sergio
Cole, Shaun
Coles, Peter
Colless, Matthew
Colombi, Stephane
Condon, James
Conti, Alberto
Cooray, Asantha
Copin, Yannick
Cora, Sofia
Corsini, Enrico
Courbin, Frederic
Courteau, Stéphane
Courtois, Helene
Covone, Giovanni
Crane, Patrick
Crane, Philippe
Crawford, Steven
Cristiani, Stefano
Croom, Scott
Crosta, Mariateresa
Croton, Darren
Csabai, Istvan
Cui, Wei
Curran, Stephen
Cypriano, Eduardo
da Costa, Luiz
Da Costa, Gary
Dadhich, Naresh
Dahle, Håkon
Daigne, Frederic
DAISUKE, Iono
Dalla Bontà, Elena
Danese, Luigi
Dannerbauer, Helmut
Da Rocha, Cristiano
Das, P.
Dautcourt, G.
Davidson, William
Davies, Paul
Davies, Roger
Davis, Marc
Davis, Michael
De Bernardis, Paolo
Dekel, Avishai
de Lapparent, Valérie
de Lima, José
Dell'Antonio, Ian
Del Popolo, Antonino
De Lucia, Gabriella
Demarco, Ricardo
Demiański, Marek
de Petris, Marco
de Ruiter, Hans
Désert, François-Xavier
de Silva, Lindamulage
de Ugarte Postigo, Antonio
Deustua, Susana
de Vaucouleurs, Gérard
de Zotti, Gianfranco
Dhurandhar, Sanjeev
Diaferio, Antonaldo
Diaz, Ruben
Díaz-Santos, Tanio
Dietrich, Jörg
Dionysiou, Demetrios
Djorgovski, Stanislav
Dobbs, Matt
Dobrzycki, Adam
D' Odorico, Valentina
Dole, Herve
Domínguez, Mariano
Dong, Xiao-Bo
Doroshkevich, Andrei
Dressler, Alan
Drinkwater, Michael
Dultzin-Hacyan, Deborah
Dunlop, James
Dunsby, Peter
Dyer, Charles
Eales, Stephen
Easther, Richard
Edsjö, Joakim
Efstathiou, George
Ehlers, Jürgen
Einasto, Jaan
Elgarøy, Øystein
Elizalde, Emilio
Ellis, George
Ellis, Richard
Ellison, Sara
Elvis, Martin
Elyiv, Andrii
Enginol, Turan
Ettori, Stefano
Eungwanichayapant, Anant
Faber, Sandra
Fairall, Anthony
Falk Jr, Sydney
Fall, S.
FAN, Zuhui
Fassnacht, Christopher

Fatkhullin, Timur
Faure, Cécile
Fedeli, Cosimo
Fedorova, Elena
FENG, Long Long
Ferreira, Pedro
Field, George
Figueiredo, Newton
Filippenko, Alexei
Firmani, Claudio
Florides, Petros
Focardi, Paola
Fong, Richard
Ford, Holland
Forman, William
FOUCAUD, Sebastien
Fouqué, Pascal
Fox, Andrew
Franceschini, Alberto
Frenk, Carlos
Friaca, Amancio
Frutos-Alfaro, Francisco
FUJIMOTO, Mitsuaki
FUKUGITA, Masataka
FUKUI, Takao
Furlanetto, Steven
Füzfa, Andre
Fynbo, Johan
Gallazzi, Anna
Gangui, Alejandro
Ganguly, Rajib
Garilli, Bianca
Garrido, César
Garrison, Robert
Gastaldello, Fabio
Gavignaud, Isabelle
Geller, Margaret
GENG, Lihong
Germany, Lisa
Ghirlanda, Giancarlo
Giallongo, Emanuele
Gilbank, David
Gioia, Isabella
Gitti, Myriam
Glazebrook, Karl
Glover, Simon
Goicoechea, Luis
Goldsmith, Donald
GONG, Biping
González Sánchez, Alejandro
Gonzalez-Solares, Eduardo
Goobar, Ariel
Gordon, Chris
Goret, Philippe
Gosset, Eric
Gottlöber, Stefan
GOUDA, Naoteru
Govinder, Keshlan
Goyal, Ashok
Grainge, Keith
Granato, Gian Luigi
Gray, Meghan
Gray, Richard
Green, Anne
Gregorio, Anna
Gregory, Stephen
Greve, Thomas
Greyber, Howard
Griest, Kim
Grishchuk, Leonid
Gudmundsson, Einar
Gumjudpai, Burin
Gunn, James
Gutiérrez, Carlos
Guzzo, Luigi
Haas, Martin
Haehnelt, Martin
Hagen, Hans-Juergen
Hall, Patrick
Hamilton, Andrew
Hanna, David
Hannestad, Steen
Hansen, Frode
Hardy, Eduardo
Harms, Richard
Hau, George
Haugbølle, Troels
Hawking, Stephen
Hayakawa, Satio
Hayes, Matthew
HAZUMI, Masashi
HE, XiangTao
Heavens, Alan
Heinamaki, Pekka
Hellaby, Charles
Heller, Michał
Hendry, Martin
Henriksen, Mark
Hernández, Xavier
Hernández-Monteagudo, Carlos
Hervik, Sigbjorn
Hewett, Paul
Hewitt, Adelaide
Heyrovsk, David
Hicks, Amalia
HIDEKI, Asada
Hirv, Anti
Hnatyk, Bohdan
Hoekstra, Hendrik
Holz, Daniel
Hu, Esther
HU, Hongbo
Huang, Jiasheng
Hudson, Michael
Huertas-Company, Marc
Hughes, David
Hütsi, Gert
Huynh, Minh
HWANG, Chorng-Yuan
HWANG, Jai-chan
Ibata, Rodrigo
Icke, Vincent
IKEUCHI, Satoru
Iliev, Ilian
IM, Myungshin
Impey, Christopher
INADA, Naohisa
Iovino, Angela
ISHIHARA, Hideki
Ivanov, Pavel
IWATA, Ikuru
Iyer, Balasubramanian
Jahnke, Knud
Jakobsson, Pall
Jannuzi, Buell
Jaroszyński, Michał
Jauncey, David
Jaunsen, Andreas
Jedamzik, Karsten
Jelić, Vibor
Jensen, Brian
Jetzer, Philippe
Jha, Saurabh
JIANG, Shuding
JING, Yipeng
Johnston-Hollitt, Melanie
Jones, Bernard

Jones, Christine
Jones, Heath
Jones, Laurence
Jordán, Andrés
Joshi, Mohan
Jovanović, Predrag
Junkkarinen, Vesa
Kaisin, Serafim
KAJINO, Toshitaka
Kaminker, Alexander
Kanekar, Nissim
KANG, Hyesung
KANG, Xi
Kapoor, Ramesh
Karachentsev, Igor
Karami, Kayoomars
Karouzos, Marios
Kasper, U.
KATO, Shoji
Kauffmann, Guinevere
Kaul, Chaman
Kausch, Wolfgang
Kaviraj, Sugata
KAWABATA, Kiyoshi
KAWASAKI, Masahiro
KAYO, ISSHA
Kellermann, Kenneth
Kembhavi, Ajit
Keshet, Uri
Khare, Pushpa
Khmil, Sergiy
KIM, Jik
King, Lindsay
Kirilova, Daniela
KIYOTOMO, Ichiki
Kneib, Jean-Paul
KOBAYASHI, Masakazu
KODAMA, Hideo
Koekemoer, Anton
Koivisto, Tomi
Kokkotas, Konstantinos
Kolb, Edward
Kompaneets, Dmitrij
Koo, David
Koopmans, Leon
Kormendy, John
Kovalev, Yuri
Kovetz, Attay
KOZAI, Yoshihide
Kozlovsky, Ben
Kozłowski, Szymon
Krasiński, Andrzej
Kriss, Gerard
Kristiansen, Jostein
Kudrya, Yury
Kunth, Daniel
Kunz, Martin
KUSAKABE, Motohiko
La Barbera, Francesco
Lacey, Cedric
Lachieze-Rey, Marc
La Franca, Fabio
Lagache, Guilaine
Lahav, Ofer
Lake, George
Lake, Kayll
Lanfranchi, Gustavo
Lapi, Andrea
Larionov, Mikhail
Lasota-Hirszowicz, Jean-Pierre
Lausberg, Andre
Layzer, David
Lĕo, Joao Rodrigo
LEE, Wo-Lung
Le Fèvre, Olivier
Lehnert, Matthew
Lequeux, James
Leubner, Manfred
Levin, Yuri
Levine, Robyn
Lewis, Geraint
LI, Li-Xin
LI, Xiaoqing
LIANG, Yanchun
Liddle, Andrew
Liebscher, Dierck-E
Lilje, Per
Lilly, Simon
LIN, Weipeng
LIN, Yen-Ting
LIOU, Guo Chin
LIU, Yongzhen
Łokas, Ewa
Lombardi, Marco
Longair, Malcolm
Longo, Giuseppe
Lonsdale, Carol
Lopes, Paulo
Lopez, Sebastian
Lopez-Corredoira, Martin
Loveday, Jon
Lowenthal, James
LU, Youjun
Lubin, Lori
Lukash, Vladimir
Luminet, Jean-Pierre
Lynden-Bell, Donald
MA, Jun
Maartens, Roy
Maccagni, Dario
MacCallum, Malcolm
Mackey, Alasdair
Maddox, Stephen
MAEDA, Kei-ichi
Maguire, Kate
Maharaj, Sunil
Maia, Marcio
Maier, Christian
Mainieri, Vincenzo
Majumdar, Subhabrata
Malesani, Daniele
Mamon, Gary
Mandolesi, Nazzareno
Mangalam, Arun
Mann, Robert
Manrique, Alberto
Mansouri, Reza
Mao, Shude
Maoz, Dan
Marano, Bruno
Mardirossian, Fabio
Marek, John
Maris, Michele
Marleau, Francine
Marranghello, Guilherme
Martinez-Gonzalez, Enrique
Martinis, Mladen
Martins, Carlos
Massardi, Marcella
Masters, Karen
Mather, John
Mathez, Guy
MATSUMOTO, Toshio
MATSUMURA, Tomotake
Matzner, Richard
Mavrides, Stamatia
McCracken, Henry
McKean, John

McVittie, George
Meiksin, A.
Mellier, Yannick
Melnyk, Olga
Melott, Adrian
Meneghetti, Massimo
Menéndez-Delmestre, Karin
Merighi, Roberto
Meszaros, Attila
Meszaros, Peter
Meyer, David
Meyer, Martin
Meylan, Georges
Meza, Andres
Mezzetti, Marino
Miralda-Escudé, Jordi
Miralles, Joan-Marc
Miranda, Oswaldo
Misner, Charles
Miyazaki, Satoshi
Miyoshi, Shigeru
Mo, Houjun
Mohr, Joseph
Molla, Mercedes
Monaco, Pierluigi
Moodley, Kavilan
Moore, Ben
Moreau, Olivier
MOROKUMA, Tomoki
Mörtsell, Edvard
Moscardini, Lauro
Mota, David
Motta, Veronica
Mourão, Ana Maria
Muanwong, Orrarujee
Mücket, Jan
Müller, Andreas
Mukhopadhyay, Banibrata
Muller, Richard
MURAKAMI, Izumi
MURAYAMA, Hitoshi
Murphy, John
Murphy, Michael
NAKAMICHI, Akika
NAMBU, Yasusada
Narasimha, Delampady
NARDETTO, Nicolas
Nariai, Hideichi
Narlikar, Jayant
Naselsky, Pavel
Nasr-Esfahani, Bahram
Neves de Araujo, Jose
Nguyễn, Lan
Nicoll, Jeffrey
NISHIDA, Minoru
Noerdlinger, Peter
NOH, Hyerim
Noonan, Thomas
Norman, Colin
Norman, Dara
Noterdaeme, Pasquier
Nottale, Laurent
Novello, Mario
Novikov, Igor
Novosyadlyj, Bohdan
Novotn, Jan
Nozari, Kourosh
Nuza, Sebastian
O'Connell, Robert
Ocvirk, Pierre
Oemler Jr, Augustus
Ogando, Ricardo
Oliver, Sebastian
Olowin, Ronald
Omnes, Roland
Onuora, Lesley
Oscoz, Alejandro
Ostorero, Luisa
OTA, Naomi
Ott, Heinz-Albert
Oukbir, Jamila
OYA, Shin
Ozsváth, Istvan
Pachner, Jaroslav
Page, Don
PAN, Rong-Shi
Paragi, Zsolt
Parnovsky, Sergei
Partridge, Robert
Peacock, John
Pecker, Jean-Claude
Pedersen, Kristian
Pedrosa, Susana
Peebles, P.
Peel, Michael
Pello, Roser
Pen, Ue-Li
Pentericci, Laura
Penzias, Arno
Pérez-Garrido, Antonio
Peroux, Céline
Perryman, Michael
Persides, Sotirios
Persson, Carina
Peterson, Bradley
Peterson, Bruce
Petitjean, Patrick
Petrosian, Vahe
PHAM, Diep
Pimbblet, Kevin
Pitkin, Matthew
Plionis, Manolis
Podolsk, Jiri
Polletta, Maria del Carmen
Pompei, Emanuela
Popescu, Nedelia
Portinari, Laura
Power, Chris
Poznanski, Dovi
Prandoni, Isabella
Premadi, Premana
Press, William
Puetzfeld, Dirk
Puget, Jean-Loup
Puy, Denis
QIN, Bo
QU, Qinyue
Quirós, Israel
Rahvar, Sohrab
Ramella, Massimo
Ranalli, Piero
Rassat, Anais
Rauch, Michael
Ravindranath, Swara
Read, Justin
Rebolo, Rafael
Reboul, Henri
Rees, Martin
Reeves, Hubert
Refsdal, Sjur
Reiprich, Thomas
Reisenegger, Andreas
Reitze, David
Revnivtsev, Mikhail
Rey, Soo-Chang
Reyes, Reinabelle
Rhodes, Jason
Riazi, Nematollah
Riazuelo, Alain

Ribeiro, André Luis
Ribeiro, Marcelo
Richard, Johan
Richter, Philipp
Ricker, Paul
Ricotti, Massimo
Ridgway, Susan
Rindler, Wolfgang
Rivolo, Arthur
Roberts, David
Robinson, I.
Rocca-Volmerange, Brigitte
Roeder, Robert
Romano-Diaz, Emilio
Romeo, Alessandro
Romer, A.
Rosa González, Daniel
Rosquist, Kjell
Rothberg, Barry
Röttgering, Huub
Rowan-Robinson, Michael
Roxburgh, Ian
Rubin, Vera
Rubino-Martin, Jose Alberto
Rudnick, Lawrence
Ruffini, Remo
Ruszkowski, Mateusz
RÅ¯žička, Adam
Saar, Enn
Sadat, Rachida
Saffari, Reza
Sahlén, Martin
Sahni, Varun
Saiz, Alejandro
Salvador-Sole, Eduardo
Salzer, John
Santos-Lleó, Maria
Sanz, Jose
Sapar, Arved
Saracco, Paolo
SASAKI, Misao
SASAKI, Shin
SATO, Humitaka
SATO, Katsuhiko
SATO, Shinji
SATO, Jun'ichi
Savage, Ann
Saviane, Ivo
Sawicki, Marcin
Sazhin, Mikhail
Scaramella, Roberto
Schartel, Norbert
Schaye, Joop
Schechter, Paul
Schindler, Sabine
Schmalzing, Jens
Schmidt, Brian
Schmidt, Maarten
Schneider, Donald
Schneider, Jean
Schneider, Peter
Schneider, Raffaella
Schramm, Thomas
Schuch, Nelson
Schucking, Engelbert
Schuecker, Peter
Schumacher, Gerard
Scodeggio, Marco
Scott, Douglas
Seielstad, George
Semerák, Oldrich
Sergijenko, Olga
Serjeant, Stephen
SETO, Naoki
Setti, Giancarlo
Severgnini, Paola
Seymour, Nicholas
Shandarin, Sergei
Shanks, Thomas
SHAO, Zhengyi
Sharp, Nigel
Shaver, Peter
Shaviv, Giora
Shaya, Edward
SHIBATA, Masaru
Shimon, Meir
SHIRAHATA, Mai
SHIRASAKI, Yuji
Shivanandan, Kandiah
Siebenmorgen, Ralf
Signore, Monique
Silk, Joseph
Silva, Laura
Simon, Rene
Siringo, Giorgio
Sironi, Giorgio
Sistero, Roberto
Slosar, Anze
Smail, Ian
Smette, Alain
Smith, Nigel
Smith, Rodney
Smoot III, George
Sokołowski, Lech
Sollerman, Jesper
SONG, Doo-Jong
SONG, Yong-Seon
Souradeep, Tarun
Souriau, Jean-Marie
Spinoglio, Luigi
Spyrou, Nicolaos
Squires, Gordon
Srianand, Raghunathan
Stacy, Athena
Stadel, Joachim
Stecker, Floyd
Steigman, Gary
Stewart, John
Stoehr, Felix
Stolyarov, Vladislav
Storrie-Lombardi, Lisa
Stott, John
Straumann, Norbert
Strauss, Michael
Stritzinger, Maximilian
Struble, Mitchell
Strukov, Igor
Stuchlík, Zdenek
Stuik, Remko
Subrahmanya, C.
SUGINOHARA, Tatsushi
SUGIYAMA, Naoshi
Suhhonenko, Ivan
Sullivan, Mark
Sunyaev, Rashid
Surdej, Jean
SUSA, Hajime
Sutherland, William
SUTO, Yasushi
Szalay, Alex
Szydłowski, Marek
TAGOSHI, Hideyuki
TAKADA, Masahiro
TAKAHARA, Fumio
Tammann, Gustav
TANABE, Kenji
TANAKA, Masayuki
Tarter, Jill
TARUYA, Atsushi

TATEKAWA, Takayuki
Taylor, Angela
Tempel, Elmo
Temporin, Sonia
Tepper Garcia, Thorsten
Teyssier, Romain
Thanjavur, Karunananth
Thöne, Christina
Thorne, Kip
Thuan, Trinh
Tifft, William
Tipler, Frank
Toffolatti, Luigi
Toft, Sune
TOMIMATSU, Akira
TOMITA, Kenji
Tonry, John
Tormen, Giuseppe
TOTANI, Tomonori
Tozzi, Paolo
Treder, H.
Tremaine, Scott
Trenti, Michele
Tresse, Laurence
Treu, Tommaso
Trevese, Dario
Trimble, Virginia
Trotta, Roberto
Trujillo Cabrera, Ignacio
Tsamparlis, Michael
Tugay, Anatoliy
Tully, Richard
Turner, Edwin
Turner, Michael
Turnshek, David
Tyson, John
Tytler, David
Tyul'bashev, Sergei
UEDA, Haruhiko
Ugolnikov, Oleg
UMEMURA, Masayuki
Uson, Juan
Vaccari, Mattia
Vagnetti, Fausto
Väisänen, Petri
Valdivielso, Luisa
Väliviita, Jussi-Pekka
Valls-Gabaud, David
Valluri, Monica
Valtchanov, Ivan
van der Laan, Harry
van Eymeren, Janine
van Haarlem, Michiel
van Kampen, Eelco
Vedel, Henrik
Vega, Olga
Venters, Tonia
Verdoes Kleijn, Gijsbert
Vergani, Daniela
Verma, Aprajita
Vestergaard, Marianne
Vettolani, Giampaolo
Viana, Pedro
Viel, Matteo
Vishniac, Ethan
Vishveshwara, C.
Voit, Gerard
Volonteri, Marta
von Borzeszkowski, H.
Waddington, Ian
Wagner, Robert
Wagoner, Robert
Wainwright, John
Wambsganß, Joachim
Wanas, Mamdouh
WANG, Huiyuan
WANG, Yu
Watson, Darach
Webb, Tracy
Webster, Adrian
Webster, Rachel
Weilbacher, Peter
Weinberg, Steven
Wendt, Martin
Wesson, Paul
Wheeler, John
White, Simon
Whiting, Alan
Whitrow, Gerald
Widrow, Lawrence
Will, Clifford
Willis, Jon
Wilson, Andrew
Wilson, Gillian
Windhorst, Rogier
Wold, Margrethe
Woltjer, Lodewijk
WOO, Jong-Hak
Woszczyna, Andrzej
Wright, Edward
WU, Jianghua
WU, Jiun-Huei
WU, Wentao
WU, Xiangping
WU, Xue-Feng
Wucknitz, Olaf
Wulandari, Hesti
Wyithe, Stuart
XIANG, Shouping
XINLIAN, Luo
XU, Chongming
XU, Dawei
YAMAMOTO, Hiroaki
YANG, Lantian
YASUDA, Naoki
Yeşilyurt, Ibrahim
YI, Sukyoung
YOICHIRO, Suzuki
YOKOYAMA, Jun'ichi
YONETOKU, Daisuke
YOSHIDA, Hiroshi
YOSHII, Yuzuru
YOSHIKAWA, Kohji
YOSHIMURA, Motohiko
YOSHIOKA, Satoshi
YU, Qingjuan
Yushchenko, Alexander
Zacchei, Andrea
Zamorani, Giovanni
Zanichelli, Alessandra
Zannoni, Mario
Zaroubi, Saleem
Zeldovich, Ya.
Zhang, Yuying
ZHANG, Jialu
ZHANG, Tong-Jie
ZHANG, Zhen-Jiu
ZHAO, Donghai
ZHOU, Hongyan
ZHOU, Youyuan
ZHU, Shichang
ZHU, Xingfeng
ZHU, Zong-Hong
Zhuk, Alexander
Zibetti, Stefano
Zięba, Stanisław
Zirm, Andrew
ZOU, Zhenlong
Zucca, Elena
Zuiderwijk, Edwardus

Division E - Commission 49 Interplanetary Plasma & Heliosphere

President: Ingrid Mann
Vice-President: P. K. Manoharan

Organizing Committee Members:

Igor V. Chashei
Olga Malandraki
Natchimuthuk Gopalswamy
Sarah Gibson
Yoichiro Hanaoka
Carine Briand
David Lario

Members:

Ahluwalia, Harjit
Andretta, Vincenzo
Balikhin, Michael
Banerjee, Dipankar
Barnes, Aaron
Barrantes, Marco
Barrow, Colin
Barta, Miroslav
Benz, Arnold
Bertaux, Jean-Loup
Blandford, Roger
Blum, Peter
Bochsler, Peter
Bonnet, Roger
Bougeret, Jean-Louis
Brandt, John
Briand, Carine
Browning, Philippa
Bruno, Roberto
Burlaga, Leonard
Buti, Bimla
Cairns, Iver
Cecconi, Baptiste
Chakrabarti, Supriya
Channok, Chanruangrit
Chapman, Sandra
Chashei, Igor
Chassefiere, Eric
CHEN, Biao
Chitre, Shashikumar
CHOE, Gwangson
CHOU, Chih-Kang
Couturier, Pierre
Cramer, Neil
Cuperman, Sami
Daglis, Ioannis
Dalla, Silvia
Damé, Luc
Dasso, Sergio
de Jager, Cornelis
De Keyser, Johan
Del Zanna, Luca
de Toma, Giuliana
Dinulescu, Simona
Dolginov, Arkady
Dorotovic, Ivan
Dryer, Murray
Duldig, Marcus
Durney, Bernard
Ergma, Ene
Eshleman, Von
Eviatar, Aharon
Fahr, Hans
Fernandes, Francisco
Feynman, Joan
Fichtner, Horst
Field, George
Foullon, Claire
Fraenz, Markus
Fraschetti, Federico
Frutos-Alfaro, Francisco
Galvin, Antoinette
Gangadhara, R.T.
Gedalin, Michael
Gibson, Sarah
Gleisner, Hans
Goldman, Martin
Gopalswamy, Natchimuthuk
Gosling, John
Grzędzielski, Stanisław
Guglielmino, Salvatore
Habbal, Shadia
Hanaoka, Yoichiro
Harvey, Christopher
HAYASHI, Keiji
Heras, Ana
Hewish, Antony
Heynderickx, Daniel
Hollweg, Joseph
Holzer, Thomas
Huber, Martin
Humble, John
HWANG, Junga
INAGAKI, Shogo
Ivanov, Evgenij
Jacobs, Carla
Jokipii, Jack
Joselyn, Jo
KAKINUMA, Takakiyo
Kasiviswanathan, Sankarasubramanian
Keller, Horst
Khan, J
KO, Chung-Ming
KOJIMA, Masayoshi
Kozak, Lyudmyla
Kretzschmar, Matthieu
Krittinatham, Watcharawuth
Lafon, Jean-Pierre

Lai, Sebastiana
Lallement, Rosine
Landi, Simone
Lapenta, Giovanni
Lario, David
Levy, Eugene
Li, Bo
LIU, Siming
Livadiotis, George
Lotova, Natalja
Lüst, Reimar
Lundstedt, Henrik
MA, Guanyi
MacQueen, Robert
Malandraki, Olga
Malara, Francesco
Mangeney, André
Mann, Ingrid
Manoharan, P.
Marsch, Eckart
Mason, Glenn
Matsuura, Oscar
Mavromichalaki, Helen
Meister, Claudia
Melrose, Donald
Mendis, Devamitta
Mestel, Leon
Michel, F.
Moncuquet, Michel
Morabito, David
Moussas, Xenophon
MUNETOSHI, Tokumaru
NAKAGAWA, Yoshinari
Nandi, Dibyendu
Nickeler, Dieter
Nozawa, Satoshi
OH, Suyeon
Pandey, Birendra
Parenti, Susanna
Paresce, Francesco
Parhi, Shyamsundar
Parker, Eugene
Perkins, Francis
Pflug, Klaus
Podesta, John
Pustil'nik, Lev
Quémerais, Eric
Raadu, Michael
Readhead, Anthony
Reinard, Alysha
Reshetnyk, Volodymyr
Rickett, Barnaby
Riddle, Anthony
Riley, Pete
Ripken, Hartmut
Robinson, Peter
Rosa, Reinaldo
Rosner, Robert
Roth, Ilan
Roth, Michel
Roxburgh, Ian
Ruffolo, David
Russell, Alexander
Russell, Christopher
Sagdeev, Roald
Saiz, Alejandro
Sarris, Emmanuel
Sastri, Hanumath
Sawyer, Constance
Scherb, Frank
Schindler, Karl
Schreiber, Roman
Schwartz, Steven
Schwenn, Rainer
Seaton, Daniel
Setti, Giancarlo
Shalchi, Andreas
Shea, Margaret
SHIBATA, Kazunari
Simunac, Kristin
Smith, Dean
Sonett, Charles
Srivastava, Nandita
Stone, R.
Struminsky, Alexei
Sturrock, Peter
Suess, Steven
Tritakis, Basil
Tyul'bashev, Sergei
Usoskin, Ilya
Vainshtein, Leonid
Vandas, Marek
Verheest, Frank
Viall, Nicholeen
Vidotto, Aline
Vinod, S.
Voitsekhovska, Anna
Vucetich, Héctor
Walker, Simon
Wang, Yi-ming
WANG, Shunde
WATANABE, Takashi
WATARI, Shinichi
Webb, David
Weller, Charles
Wild, John
Willes, Andrew
Wood, Brian
Wu, Chin-Chun
Wu, Shi
YANG, Jing
YANG, Lei
Yeh, Tyan
Yeşilyurt, Ibrahim
YI, Yu
Zhang, Jie
Zhang, Tielong

Division B - Commission 50 Protection of Existing & Potential Observatory Sites

President: Richard F. Green
Vice-President: Constance Elaine Walker

Organizing Committee Members:

Beatriz Elena Garcia
Ramotholo R. Sefako
Ferdinando Patat
Elizabeth M. Alvarez del Castillo
Masatoshi OHISHI
Anastasios Tzioumis
Margarita Metaxa

Members:

Alvarez del Castillo, Elizabeth
Ardeberg, Arne
Arsenijevic, Jelisaveta
Baan, Willem
Baddiley, Christopher
Baskill, Darren
Bazzano, Angela
Benkhaldoun, Zouhair
Bensammar, Slimane
Bertola, Francesco
Blanco, Carlo
Boonstra, Albert
Brown, Robert
Cabanac, Remi
Carramiñana, Alberto
Carrasco, Bertha
Cayrel, Roger
Cinzano, Pierantonio
Clegg, Andrew
Colas, François
Costero, Rafael
Coyne, S.J, George
Crawford, David
Davis, Donald
de Greiff, J.
Dukes Jr., Robert
Edwards, Paul
FAN, Yufeng
Galan, Maximino
Garcia, Beatriz
Garcia-Lorenzo, Maria
Gergely, Tomas
Gibson, David
Goebel, Ernst
Green, Richard
Hänel, Andreas
Heck, Andre
Helmer, Leif
Hempel, Marc
Hidayat, Bambang
Ilin, Gennadii
Ilyasov, Sabit
Ioannou, Zacharias
Jefferies, John
JIANG, Shi-Yang
Kadiri, Samir
Keeney, Brian
Kołomański, Sylwester
Kontizas, Evangelos
Kontizas, Mary
Kovalevsky, Jean
KOZAI, Yoshihide
Kramer, Busaba
Leibowitz, Elia
Lewis, Brian
Lomb, Nicholas
Malin, David
Mancìni, Dario
Markkanen, Tapio
Masciadri, Elena
Mattig, W.
McNally, Derek
Mendoza-Torress, Jose-Eduardo
Menzies, John
Metaxa, Margarita
Mitton, Jacqueline
Murdin, Paul
Nelson, Burt
Osório, José
Owen, Frazer
Özel, Mehmet
Oezisik, Tuncay
Pankonin, Vernon
Patat, Ferdinando
Percy, John
Posch, Thomas
Pound, Marc
Sánchez, Francisco
Sanders, Gary
Schilizzi, Richard
Sefako, Ramotholo
Sese, Rogel Mari
Shetrone, Matthew
Siebenmorgen, Ralf
Smith, Francis
Smith, Malcolm
Smith, Robert
Stencel, Robert
Storey, Michelle
Sullivan, III, Woodruff
Suntzeff, Nicholas
Torres, Carlos Alberto
Tremko, Jozef
Tzioumis, Anastasios
Upgren, Arthur
van den Bergh, Sidney
van Driel, Wim
Vernin, Jean
Vetesnik, Miroslav
Wainscoat, Richard

Walker, Constance
Walker, Merle
WANG, Xunhao
Whiteoak, John
Woolf, Neville
WU, Mingchan
YANO, Hajime
ZHANG, Haiyan
ZHENG, Xiaonian

Division F - Commission 51 Bio-Astronomy

President: Pascale Ehrenfreund
Vice-President: Sun Kwok

Organizing Committee Members:

Nils G. Holm
Ray Jayawardhana
Charles H. Lineweaver
Muriel Gargaud
Nader Haghighipour
Masatoshi OHISHI
Douglas Galante
Anny-Chantal Levasseur-Regourd

Members:

ALIBERT, Yann
Allard, France
Almár, Ivan
Al-Naimiy, Hamid
Alonso Sobrino, Roi
Alsabti, Abdul Athem
ANDO, Hiroyasu
Apai, Daniel
Baki, Paul
Balázs, Bela
Balbi, Amedeo
Ball, John
Bania, Thomas
Basu, Baidyanath
Basu, Kaustuv
Beaudet, Gilles
Beckman, John
Beckwith, Steven
Beebe, Reta
Benest, Daniel
Bennett, David
Berendzen, Richard
Berger, Jean-Philippe
Billingham, John
Biraud, François
Bless, Robert
Bond, Ian
Boss, Alan
Bowyer, C.
Boyce, Peter
Brandeker, Alexis
Bretones, Paulo
Broderick, John
Brown, Ronald
Buccino, Andrea
Burke, Bernard
CAI, Kai
Calvin, William
Campbell, Bruce
Campusano, Luis
Cardoso Santos, Nuno
Carlson, John
Cassan, Arnaud
Chaisson, Eric
Chung, Eduardo
Cirkovic, Milan
Connes, Pierre
Cosmovici, Cristiano
Cottin, Hervé
Coudé du Foresto, Vincent
Couper, Heather
Coustenis, Athena
Cuesta Crespo, Luis
Cunningham, Maria
Currie, Douglas
Daigne, Gerard
da Silveira, Enio
Davis, Michael
De Becker, MichaÃ«l
Deeg, Hans
de Jager, Cornelis
de Jonge, J.
de Loore, Camiel
Delsemme, Armand
Dent, William
Despois, Didier
de Vincenzi, Donald
Dick, Steven
Domiciano de Souza, Armando
Dorschner, Johann
Doubinskij, Boris
Downs, George
Drake, Frank
Dreher, John
Dutil, Yvan
Dyson, Freeman
Eccles, Michael
Ehrenfreund, Pascale
Ehrenreich, David
Ellis, George
Epstein, Eugene
Evans, Neal
Fazio, Giovanni
Feldman, Paul
Field, George
Firneis, Maria
Fisher, Philip
Fraser, Helen
Fredrick, Laurence

Freire Ferrero, Rubens
FUJIMOTO, Masa-Katsu
FUJIMOTO, Mitsuaki
Gargaud, Muriel
Gatewood, George
Ghigo, Francis
Gillon, MichaÃ«l
Giovannelli, Franco
Golden, Aaron
Goldsmith, Donald
Gott, J.
Goudis, Christos
Gregory, Philip
Gulkis, Samuel
Gunn, James
Gurm, Hardev
Haddock, Fred
Haghighipour, Nader
Haisch, Bernard
Hale, Alan
Hart, Michael
Heck, Andre
Heidmann, Jean
Hershey, John
Heudier, Jean-Louis
Hinners, Noel
HIRABAYASHI, Hisashi
Hoang, Binh
Högbom, Jan
Hollis, Jan
Holm, Nils
Horowitz, Paul
Howard, Andrew
Hsieh, Henry
Hunter, James
Hysom, Edmund
Irvine, William
Israel, Frank
Jastrow, Robert
Jayawardhana, Ray
Jeffers, Stanley
Jones, Eric
Kafatos, Menas
KAIFU, Norio
Kane, Stephen
Kardashev, Nicolay
Kaufmann, Pierre
Kawada, Mitsunobu
Keay, Colin
Keheyan, Yeghis
Keller, Hans-Ulrich
Kellermann, Kenneth
Kilston, Steven
Klahr, Hubert
Knowles, Stephen
Kocer, Dursun
Koeberl, Christian
Korpela, Eric
Ksanfomality, Leonid
KUAN, Yi-Jehng
Kuiper, Thomas
Kuzmin, Arkadij
Kwok, Sun
Lafon, Jean-Pierre
Lammer, Helmut
Lamontagne, Robert
Laques, Pierre
Laufer, Diana
Lazio, Joseph
LEE, Sang-Gak
Leger, Alain
Levasseur-Regourd, Anny-Chantal
Lilley, Edward
Lineweaver, Charles
Lippincott Zimmerman, Sarah
LIU, Sheng-Yuan
Loden, Lars
Lodieu, Nicolas
Lyon, Ian
Margrave Jr, Thomas
Marov, Mikhail
Martin, Anthony
Martin, Maria
Martinez-Frias, Jesus
Marzo, Giuseppe
Matsakis, Demetrios
MATSUDA, Takuya
Matthews, Clifford
Mayor, Michel
McAlister, Harold
McDonough, Thomas
Meech, Karen
Melott, Adrian
Mendoza, V.
Merín Martín, Bruno
Milet, Bernard
MINN, Young-Ki
Minniti, Dante
Mirabel, Igor
Mokhele, Khotso
Moore, Marla
Morris, Mark
Morrison, David
Mousis, Olivier
Muller, Richard
Naef, Dominique
NAKAGAWA, Yoshinari
NARITA, Norio
Nelson, Robert
Neuhaeuser, Ralph
Niarchos, Panagiotis
Norris, Raymond
Nuth, Joseph
OHISHI, Masatoshi
Ollongren, A.
Ostriker, Jeremiah
Owen, Tobias
Pallé Bagó, Enric
Parijskij, Yurij
Pascucci, Ilaria
Pasinetti, Laura
Perek, Luboš
Pilling, Sergio
Pollacco, Don
Ponsonby, John
Prochazka, Franz
QIU, Puzhang
QIU, Yaohui
Quanz, Sascha
Quintana, Hernan
Quintana, José
Quirrenbach, Andreas
Rajamohan, R.
Rawlings, Mark
Reay, Newrick
Rees, Martin
Reyes-Ruiz, Mauricio
Rodriguez, Luis
Rowan-Robinson, Michael
Russell, Jane
SAKURAI, Kunitomo
Sanchez Bejar, Victor
Sancisi, Renzo
Sarre, Peter
Scargle, Jeffrey
Schild, Rudolph
Schneider, Jean
Schober, Hans

Schuch, Nelson
Seielstad, George
Semenov, Dmitry
SHEN, Chun-Shan
SHIMIZU, Mikio
Shostak, G.
Sims, Mark
Singh, Harinder
Sivaram, C.
Slysh, Vyacheslav
Snyder, Lewis
SOFUE, Yoshiaki
SONG, In-Ok
Sozzetti, Alessandro
Sparks, William
Stalio, Roberto
Stein, John
Sterzik, Michael
Stone, Remington
Straižys, Vytautas
Sturrock, Peter
Sullivan, III, Woodruff
TAKABA, Hiroshi
TAKADA-HIDAI, Masahide
Tarter, Jill
Tavakol, Reza
Tedesco, Edward
Tejfel, Victor
Terzian, Yervant
Thaddeus, Patrick
TIAN, Feng
Tolbert, Charles
Tovmassian, Hrant
Townes, Charles
Trimble, Virginia
Troitsky, V.
Turner, Edwin
Turner, Kenneth
Udry, Stephane
Vakoch, Douglas
Valbousquet, Armand
Vallée, Jacques
van Flandern, Tom
Varshalovich, Dmitrij
Vauclair, Gérard
Vázquez, Manuel
Vázquez, Roberto
Venugopal, V.
Verschuur, Gerrit
Vogt, Nikolaus
von Braun, Kaspar
von Hippel, Theodore
Vukotic, Branislav
Wallis, Max
Walsh, Andrew
Walsh, Wilfred
Wandel, Amri
Watson, Frederick
Welch, William
Wellington, Kelvin
Wesson, Paul
Wielebinski, Richard
Williams, Iwan
Willson, Robert
Wilson, Thomas
Wolstencroft, Ramon
Womack, Maria
Wright, Alan
Wright, Ian
XU, Weibiao
YE, Shuhua
Zadnik, Marjan
Zapatero-Osorio, Maria Rosa
Zuckerman, Benjamin

Division A - Commission 52 Relativity in Fundamental Astronomy

President: Michael H. Soffel
Vice-President: Sergei M Kopeikin

Organizing Committee Members:

Daniel Hestroffer
William M. Folkner
Neil Ashby
Jin-he TAO
Gérard Petit

Members:

Abbas, Ummi
Aharonian, Felix
Antonelli, Lucio Angelo
Aquilano, Roberto
ARAKIDA, Hideyoshi
Ashby, Neil
Bazzano, Angela
Boucher, Claude
Bromberg, Omer
Brumberg, Victor
Burikham, Piyabut
Bursa, Michal
Calabretta, Mark
Cannon, Kipp
Capitaine, Nicole
CHAE, Kyu Hyun
Clarkson, Chris
Crosta, Mariateresa

de Felice, Fernando
Domínguez, Mariano
Efroimsky, Michael
Evans, Daniel
Fienga, Agnès
Folkner, William
Foschini, Luigi
Frutos-Alfaro, Francisco
FUKUSHIMA, Toshio
Gangadhara, R.T.
Garrido, César
Giorgini, Jon
González, Gabriela
Gray, Norman
Guinot, Bernard
Gumjudpai, Burin
Hackman, Christine
Hernández-Monteagudo, Carlos
Hestroffer, Daniel
Hilton, James
Ho, Wynn
Hobbs, David
Hobbs, George
Hohenkerk, Catherine
Holz, Daniel
Horn, Martin
HSU, Rue-Ron
HU, Hongbo
HUANG, Cheng
HUANG, Tianyi
Iorio, Lorenzo
Ivanov, Pavel
Kaplan, George
Kausch, Wolfgang
Khumlemlert, Thiranee
Klioner, Sergei
Koivisto, Tomi
Kopeikin, Sergei
Kovalevsky, Jean
Lammers, Uwe
Le Poncin-Lafitte, Christophe
LI, Li-Xin
Luzum, Brian
Maciesiak, Krzysztof
Manchester, Richard
Mandel, Ilya
Marranghello, Guilherme
McCarthy, Dennis
Melnyk, Olga
Mignard, François
Minazzoli, Olivier
MIZUNO, Yosuke
Morabito, David
Mota, David
Müller, Andreas
Mukhopadhyay, Banibrata
NAGATAKI, Shigehiro
Orellana, Mariana
Osório, José
Panessa, Francesca
Petit, Gérard
Pireaux, Sophie
Pitjeva, Elena
Pitkin, Matthew
Podolsk, Jiri
Potapov, Vladimir
Ray, James
Reisenegger, Andreas
Reitze, David
Riles, Keith
Rosińska, Dorota
Rushton, Anthony
Saffari, Reza
Sanders, Gary
Schartel, Norbert
Seidelmann, P.
Shoemaker, David
Sigismondi, Costantino
Smirnova, Tatiana
Soffel, Michael
Soria, Roberto
Standish, E.
Stergioulas, Nikolaos
TAKAHASHI, Rohta
TAO, Jin-he
van Leeuwen, Joeri
Vityazev, Veniamin
Wallace, Patrick
Watts, Anna
WEN, Linqing
WU, Jiun-Huei
Wucknitz, Olaf
XIE, Yi
ZHANG, Zhibin

Division F - Commission 53 Extrasolar Planets (WGESP)

President: Alain Lecavelier des Etangs
Vice-President: Dante Minniti

Organizing Committee Members:

Gang ZHAO
Ray Jayawardhana
Heike Rauer
Didier Queloz
Eiichiro KOKUBO
Rosemary A. Mardling
Andrew Collier Cameron
Peter Bodenheimer

Members:

Absil, Olivier
Adams, Fred
Aigrain, Suzanne
ALIBERT, Yann
Allan, Alasdair
Allard, France
Alonso Sobrino, Roi
Ammons, Stephen
ANDO, Hiroyasu
Apai, Daniel
Ardila, David
Artymowicz, Pawel
Augereau, Jean-Charles
Baines, Ellyn
Baki, Paul
Baran, Andrzej
Barge, Pierre
Barrado Navascués, David
Baryshev, Andrey
Bear, Ealeal
Beaulieu, Jean-Philippe
Benest, Daniel
Bennett, David
Berger, Jean-Philippe
Biazzo, Katia
Bodaghee, Arash
Bodenheimer, Peter
Bodewits, Dennis
Bond, Ian
Bonfils, Xavier
Bordé, Pascal
Boss, Alan
Boyajian, Tabetha
Brandeker, Alexis
Brandner, Wolfgang
Briot, Danielle
Buccino, Andrea
Buchhave, Lars
Burrows, Adam
Cameron, Andrew
Canto Martins, Bruno
Cardoso Santos, Nuno
casassus, simon
Casoli, Fabienne
Cassan, Arnaud
Chabrier, Gilles
Chaisson, Eric
Chakrabarti, Supriya
Chakraborty, Deo
Chauvin, Gael
CHEN, Xinyang
Ciardi, David
Cieza, Lucas
Claudi, Riccardo
Connolly, Harold
Cosmovici, Cristiano
Cottini, Valeria
Coudé du Foresto, Vincent
Coustenis, Athena
Creech-Eakman, Michelle
Cuesta Crespo, Luis
Cuntz, Manfred
DAI, Zhibin
Dall, Thomas
Danchi, William
Day-Jones, Avril
Deeg, Hans
DELEUIL, Magali
Delplancke, Francoise
Demory, Brice-Olivier
Dent, William
de Val-Borro, Miguel
Dominis Prester, Dijana
DOU, Jiangpei
Downes Wallace, Juan
Doyle, Laurance
Doyon, Rene
Dreher, John
Ehrenreich, David
Elias II, Nicholas
Emel'yanenko, Vacheslav
Esenoğlu, Hasan
FAN, Yufeng
Ferlet, Roger
Fernández Lajús, Eduardo
Ferraz-Mello, Sylvio
Fletcher, Leigh
Ford, Eric
Forveille, Thierry
Fouqué, Pascal
Fraser, Helen
Fridlund, Malcolm
Fromang, Sebastien
FUKAGAWA, Misato
Gábor, Pavel
Gänsicke, Boris
Gandolfi, Davide
Gawroński, Marcin
Geller, Aaron
Giani, Elisabetta
Gillon, MichaÃ«l
Girard, Julien
Golden, Aaron
Goldsmith, Donald
Gomez, Edward
GoÅºdziewski, Krzysztof
Gregory, Philip
Grillmair, Carl
Guenther, Eike
Gulkis, Samuel
Haghighipour, Nader
Hale, Alan
Hallinan, Gregg
HAN, Inwoo
Hatzes, Artie
Helled, Ravit
Heng, Kevin
HERSANT, Franck
HIROSE, Shigenobu
Homeier, Derek
HOU, Xiyun
Hovenier, J.
Howard, Andrew
HU, Zhong wen
Hubbard, William
Ianna, Philip
Ireland, Michael
Iro, Nicolas
Irvine, William
Isaak, Kate
ITOH, Yoichi
Ivanov, Pavel
Janes, Kenneth
Jayawardhana, Ray
Jeffers, Sandra
Jehin, Emmanuel
JI, Jianghui
Jin, Liping
Johansen, Anders
Jura, Michael
Kabath, Petr

Kafka, Styliani (Stella)
Kaltenegger, Lisa
Kane, Stephen
Karoff, Christoffer
Kenworthy, Matthew
Kern, Pierre
Khodachenko, Maxim
Kibblewhite, Edward
Kim, Joo Hyeon
Kim, Yoo Jea
KIM, Seung-Lee
Kitiashvili, Irina
Klahr, Hubert
Kley, Wilhelm
Koeberl, Christian
KOKUBO, Eiichiro
Konacki, Maciej
Kostama, Veli-Petri
Kovács, József
Kozak, Lyudmyla
Kozłowski, Szymon
Ksanfomality, Leonid
Kürster, Martin
Lacour, Sylvestre
Lagage, Pierre-Olivier
Lammer, Helmut
Lamontagne, Robert
Lanza, Antonino
Latham, David
Lawson, Peter
Lazio, Joseph
Lecavelier des Etangs, Alain
LEE, Chung-Uk
LEE, Jae Woo
LEE, Myung Gyoon
Leger, Alain
Li, Jian-Yang
Lin, Douglas
Lineweaver, Charles
Lintott, Chris
Lissauer, Jack
LIU, Michael
LIU, Yujuan
Livio, Mario
Lodieu, Nicolas
Longmore, Andrew
Lovis, Christophe
Malbet, Fabien
Mamajek, Eric
Mancìni, Dario
Marchi, Simone
Mardling, Rosemary
Marley, Mark
Marois, Christian
Martinez Fiorenzano, Aldo
Marzari, Francesco
Masciadri, Elena
Mawet, Dimitri
Mayor, Michel
Mazeh, Tsevi
Meech, Karen
Meibom, Soren
Mendillo, Michael
Menzies, John
Merín Martín, Bruno
Meyer, Michael
Migaszewski, Cezary
Mignard, François
Millan-Gabet, Rafael
Minniti, Dante
Moerchen, Margaret
Montes, David
Mookerjea, Bhaswati
Moór, Attila
Mordasini, Christoph
Mousis, Olivier
MURAKAMI, Naoshi
Naef, Dominique
Namouni, Fathi
NARITA, Norio
Neuhaeuser, Ralph
Niedzielski, Andrzej
Novaković, Bojan
Oberst, Thomas
Owen, Tobias
Pál, András
Pallé Bagó, Enric
Papaloizou, John
PARK, Byeong-Gon
Patten, Brian
Peale, Stanton
Penny, Alan
Penny, Matthew
Pepe, Francesco
Perets, Hagai
Perez, Mario
Pérez-Garrido, Antonio
Perrin, Marshall
Persson, Carina
Pilat-Lohinger, Elke
Pilling, Sergio
Plavchan, Peter
Podolak, Morris
Pollacco, Don
Prša, Andrej
Puxley, Phil
Puzia, Thomas
Quanz, Sascha
Queloz, Didier
Quirrenbach, Andreas
Rank-Lueftinger, Theresa
Rasio, Frederic
Rauer, Heike
Reddy, Bacham
Redman, Stephen
Reffert, Sabine
Reiners, Ansgar
Reyes-Ruiz, Mauricio
Richardson, Derek
Ridgway, Stephen
Roberts Jr, Lewis
Rojo, Patricio
Saffe, Carlos
Sahu, Kailash
Sanchez Bejar, Victor
Sari, Re'em
Sarre, Peter
SATO, Bunei
Schneider, Jean
Seiradakis, John
Selam, Selim
Sengupta, Sujan
Shankland, Paul
Silvotti, Roberto
Snik, Frans
Sozzetti, Alessandro
Stam, Daphne
Stapelfeldt, Karl
Sterzik, Michael
Stewart-Mukhopadhyay, Sarah
Stringfellow, Guy
Süli, Áron
Szabó, Gyula
TAKADA-HIDAI, Masahide
Tarter, Jill
TIAN, Feng
Traub, Wesley

Trimble, Virginia
Tsvetanov, Zlatan
Turner, Edwin
Udry, Stephane
Unda-Sanzana, Eduardo
Valdivielso, Luisa
van Belle, Gerard
Vavrek, Roland
Vidal-Madjar, Alfred
Vidotto, Aline
Villaver, Eva
von Braun, Kaspar
von Hippel, Theodore
Wallace, James
WANG, Shiang-Yu
WANG, Xiao-bin
Whelan, Emma
Williams, Iwan
Winn, Joshua
Wittenmyer, Robert
Wolszczan, Alexander
Womack, Maria
WU, Zhen-Yu
XIAO, Dong
Yeşilyurt, Ibrahim
YU, Cong
YU, Qingjuan
Yüce, Kutluay
YUMIKO, Oasa
YUTAKA, Shiratori
Zapatero-Osorio, Maria Rosa
Zarka, Philippe
ZHANG, You-Hong
ZHAO, Gang
Zinnecker, Hans
Zucker, Shay
Zuckerman, Benjamin

Division B - Commission 54 Optical & Infrared Interferometry

President: Gerard T. van Belle
Vice-President: Denis Mourard

Organizing Committee Members:

Michael James Ireland
Theo A. ten Brummelaar
John Thomas Armstrong
Markus Wittkowski
Bruno Lopez
Karine Perraut
Éric M. Thiébaut
John Stephen Young

Members:

Absil, Olivier
Acke, Bram
Akeson, Rachel
Allen, Ronald
Ambrocio-Cruz, Silvia
Arcidiacono, Carmelo
Armstrong, John
Aufdenberg, Jason
Augereau, Jean-Charles
Babkovskaia, Natalia
Baddiley, Christopher
Baines, Ellyn
Bakker, Eric
Ballester, Pascal
Bamford, Steven
Barrientos, Luis
Beklen, Elif
Bendjoya, Philippe
Benisty, Myriam
Bensammar, Slimane
Benson, James
Berger, Jean-Philippe
Blazit, Alain
Boboltz, David
Boden, Andrew
Bonneau, Daniel
Boonstra, Albert
Bordé, Pascal
Boyajian, Tabetha
Brandner, Wolfgang
Bryant, Julia
Buscher, David
Carciofi, Alex
Carpenter, Kenneth
Cenko, Stephen
Chelli, Alain
CHEN, Xinyang
CHEN, Zhiyuan
Ciardi, David
Ciliegi, Paolo
Coffey, Deirdre
Coudé du Foresto, Vincent
Crause, Lisa
Crawford, Steven
Creech-Eakman, Michelle
Cruzalèbes, Pierre
Cuby, Jean-Gabriel
Cuillandre, Jean-Charles
Damé, Luc
Danchi, William
de Lange, Gert
Delplancke, Francoise
Demory, Brice-Olivier
den Herder, Jan-Willem

de Petris, Marco
Dieleman, Pieter
Dionatos, Odysseas
Domiciano de Souza, Armando
Dougados, Catherine
Dravins, Dainis
Dutrey, Anne
Duvert, Gilles
Eisner, Josh
Elias II, Nicholas
Faure, Cécile
Florido, Estrella
Forveille, Thierry
Foy, Renaud
Fridlund, Malcolm
Gabanyi, Krisztina
Gábor, Pavel
Gai, Mario
Garcia, Paulo
Genzel, Reinhard
Giani, Elisabetta
Girard, Julien
Glindemann, Andreas
GONG, Xuefei
Groh, Jose
GU, Xuedong
Haniff, Christopher
HAO, Lei
Hatziminaoglou, Evanthia
Henning, Thomas
Hönig, Sebastian
Hora, Joseph
HU, Zhong wen
Hummel, Christian
Hutter, Donald
Ireland, Michael
IWATA, Ikuru
Jordán, Andrés
Jorgensen, Anders
Kalomeni, Belinda
Kawada, Mitsunobu
Kenworthy, Matthew
Kern, Pierre
Kervella, Pierre
Kibblewhite, Edward
KIM, Ji Hoon
KIM, Young-Soo
Köhler, Rainer
Konacki, Maciej
Kóspál, Ágnes
Kraus, Stefan
Labeyrie, Antoine
Lacour, Sylvestre
Lawson, Peter
Lebohec, Stephan
Le Bouquin, Jean-Baptiste
Lehnert, Matthew
Leinert, Christoph
Leisawitz, David
Lena, Pierre
Ligori, Sebastiano
Lopez, Bruno
Maillard, Jean-Pierre
Mason, Brian
Mawet, Dimitri
Millan-Gabet, Rafael
Millour, Florentin
Monnier, John
Mosoni, Laszlo
Mourard, Denis
MURAKAMI, Naoshi
MURAYAMA, Hitoshi
NARDETTO, Nicolas
Nguyễn, Khanh
Oberst, Thomas
Ohnaka, Keiichi
Paresce, Francesco
Pauls, Thomas
Paumard, Thibaut
Percheron, Isabelle
Pérez Torres, Miguel
Perraut, Karine
Perrin, Guy
Pinte, Christophe
Pott, Jörg-Uwe
Queloz, Didier
Quirrenbach, Andreas
Rabbia, Yves
Rajagopal, Jayadev
Rastegaev, Denis
Richichi, Andrea
Ridgway, Stephen
Rivinius, Thomas
Robertson, James
Rouan, Daniel
Rousset, Gérard
Saha, Swapan
Schinckel, Antony
Schmitt, Henrique
Schneider, Jean
Schödel, Rainer
Schöller, Markus
Schuller, Peter
SHANG, Hsien
Simon, Michal
Stée, Philippe
Surdej, Jean
Tallon, Michel
Tallon-Bosc, Isabelle
ten Brummelaar, Theo
Thevenin, Frederic
Thiébaut, Éric
Traub, Wesley
TRIPPE, Sascha
Tristram, Konrad
Tuthill, Peter
Tycner, Christopher
Unwin, Stephen
Urquhart, James
Vakili, Farrokh
van Belle, Gerard
van Kampen, Eelco
Vaňko, Martin
Vega, Olga
Verhoelst, Tijl
von Braun, Kaspar
Wallace, James
WANG, Guomin
WANG, Jingyu
WANG, Sen
WANG, Shen
Warner, Peter
Weigelt, Gerd
Wittkowski, Markus
Woillez, Julien
XIAO, Dong
YANG, Dehua
Young, John
YUAN, Xiangyan
Zavala, Robert
ZHANG, Yong
ZHANG, You-Hong
ZHOU, Jianfeng
Zinnecker, Hans

Division C - Commission 55 Communicating Astronomy with the Public

President: Lars Lindberg Christensen
Vice-President: Pedro Russo

Organizing Committee Members:

Zi ZHU
Ian E. Robson
Carolina Johanna Ödman
Kimberley Kowal Arcand
Sze-leung Cheung
Kazuhiro SEKIGUCHI
Jin ZHU

Members:

Accomazzi, Alberto
Ajhar, Edward
Allan, Alasdair
Alvarez Pomares, Oscar
Alves, Virgínia
Apai, Daniel
Argo, Megan
Arion, Douglas
Augusto, Pedro
Axelsson, Magnus
Babul, Arif
BAEK, Chang Hyun
Bailey, Katherine
Balbi, Amedeo
Bamford, Steven
Barlow, Nadine
Bartlett, Jennifer
Bauer, Amanda
Beck, Sara
Becker, Werner
Beckmann, Volker
Berrilli, Francesco
Botti, Thierry
Bretones, Paulo
Briand, Carine
Cassan, Arnaud
Cattaneo, Andrea
CHEN, Dongni
Chernyakova, Maria
Christensen, Lars
Coffey, Deirdre
Conti, Alberto
Copin, Yannick
Cora, Alberto
Couto da Silva, Telma
Crabtree, Dennis
Croft, Steve
Crosta, Mariateresa
Cuesta Crespo, Luis
Cui, Wei
DAISUKE, Iono
Damineli Neto, Augusto
DARHMAOUI, Hassane
de Grijs, Richard
de Lange, Gert
Delsanti, Audrey
Demarco, Ricardo
DeNisco, Kenneth
de Swardt, Bonita
Domínguez, Mariano
Donahue, Megan
Doran, Rosa
Easther, Richard
Ehle, Matthias
Ekström García Nombela, Sylvia
Engelbrecht, Chris
English, Jayanne
Esguerra, Jose Perico
Espenak, Fred
Feain, Ilana
Fienberg, Richard
Figueiredo, Newton
Forero Villao, Vicente
Gangui, Alejandro
GAO, Jian
Garcia-Lorenzo, Maria
Gavrilov, Mikhail
Gay, Pamela
George, Martin
Gilbank, David
Gills, Martins
Girard, Julien
Goldes, Guillermo
Gomez, Edward
Govender, Kevindran
Green, Anne
Gumjudpai, Burin
Günthardt, Guillermo
Hannah, Iain
Hau, George
Haverkorn, Marijke
Hempel, Marc
Heydari-Malayeri, Mohammad
HIRAMATSU, Masaaki
Hjalmarsdotter, Linnea
Hollow, Robert
Horn, Martin
Ibrahim, Alaa
Iliev, Ilian
ITOH, Yoichi
Jafelice, Luiz
Jahnke, Knud
JEON, Young Beom
Jiménez-Vicente, Jorge
Johns, Bethany
Johnston, Helen
Jones, David
Kalemci, Emrah
Kalirai, Jason
Karouzos, Marios

Keeney, Brian
Kembhavi, Ajit
Kołomański, Sylwester
Komonjinda, Siramas
Koschny, Detlef
Kostama, Veli-Petri
Kotulla, Ralf
Kristiansen, Jostein
Kuiper, Rolf
LAS VERGNAS, Olivier
Laufer, Diana
LEE, Jun
Letarte, Bruno
Levenson, Nancy
Lintott, Chris
Loaring, Nicola
Longo, Giuseppe
Lopes de Oliveira, Raimundo
López-Sánchez, Angel
Lowenthal, James
Madsen, Claus
Majumdar, Subhabrata
Marchi, Simone
Mariĉić, Darije
Martínez-Delgado, David
Mason, Helen
Masters, Karen
Medina, Etelvina
Melbourne, Jason
Merín Martín, Bruno
Miller, Eric
Minier, Vincent
Morris, Patrick
Mourão, Ana Maria
Müller, Andreas
Mújica, Raul
MURAYAMA, Hitoshi
NARITA, Norio
Nesvadba, Nicole
Nguyễn, Khanh
Niemczura, Ewa
Nota, Antonella
Nymark, Tanja
Ocvirk, Pierre
Ödman, Carolina
O'Donoghue, Aileen
Olive, Don
Olivier, Enrico
Önel, Hakan
Oozeer, Nadeem
Ortiz Gil, Amelia
Özeren, Ferhat
Oezisik, Tuncay
PALACIOS, Ana
Pantoja, Carmen
Pat-El, Igal
Pavani, Daniela
Peel, Michael
Perrin, Marshall
PHAN, Dong
Politi, Romolo
Price, Charles
Price, Daniel
Rassat, Anais
Reid, Michael
Rekola, Rami
Reyes, Reinabelle
Reynolds, Cormac
Roberts, Douglas
Robson, Ian
Rodríguez Hidalgo, Inés
Rojas, Gustavo
Román-Zúñiga, Carlos
Różańska, Agata
Russo, Pedro
Saffari, Reza
Sampson, Russell
Sandrelli, Stefano
Santos-Lleó, Maria
Sawicki, Marcin
SEKIGUCHI, Kazuhiro
Sese, Rogel Mari
Sharp, Nigel
Shelton, Ian
SHIMOIKURA, Tomomi
Simons, Douglas
Snik, Frans
Srivastava, Nandita
Stam, Daphne
Stavinschi, Magda
Strubbe, Linda
SUMI, Takahiro
SUNG, Hyun-Il
TANAKA, Ichi Makoto
Tautvaisiene, Gražina
Templeton, Matthew
Teng, Stacy
Tignalli, Horacio
Tomasella, Lina
trân, Ha
Tugay, Anatoliy
Unda-Sanzana, Eduardo
Vakoch, Douglas
Valdivielso, Luisa
van Eymeren, Janine
Verdoes Kleijn, Gijsbert
Vergani, Daniela
Vierdayanti, Kiki
Villar Martin, Montserrat
Voit, Gerard
Wadadekar, Yogesh
Wagner, Robert
Walker, Constance
Walsh, Robert
WANG, Chen
West, Michael
Whitelock, Patricia
Wilson, Gillian
Winter, Lisa
WU, Jiun-Huei
WU, Zhen-Yu
YAN, Yihua
Yanamandra-Fisher, Padma
ZHU, Jin
ZHU, Zi